Fr. 20.98

LINEAR CONTROL
SYSTEM ANALYSIS
AND DESIGN

McGRAW-HILL ELECTRICAL AND ELECTRONIC ENGINEERING SERIES

FREDERICK EMMONS TERMAN, Consulting Editor
W. W. HARMAN, J. G. TRUXAL, AND R. A. ROHRER, Associate Consulting Editors

**McGRAW-HILL
BOOK COMPANY**

New York
St. Louis
San Francisco
Auckland
Düsseldorf
Johannesburg
Kuala Lumpur
London
Mexico
Montreal
New Delhi
Panama
Paris
São Paulo
Singapore
Sydney
Tokyo
Toronto

**JOHN J. D'AZZO
CONSTANTINE H. HOUPIS**
*Professors of Electrical Engineering
Department of Electrical Engineering
School of Engineering
Air Force Institute of Technology*

Linear Control System Analysis and Design

CONVENTIONAL AND MODERN

This book was set in Times New Roman.
The editors were Kenneth J. Bowman and M. E. Margolies;
the cover was designed by Nicholas Krenitsky;
the production supervisor was Dennis J. Conroy.
New drawings were done by J & R Services, Inc.
Kingsport Press, Inc., was printer and binder.

Library of Congress Cataloging in Publication Data

D'Azzo, John Joachim.
    Linear control system analysis and design: conventional and modern

    (McGraw-Hill electrical and electronic engineering series)
    1. Automatic control. 2. Control theory. 3. Electric engineering—Mathematics.
I. Houpis, Constantine H., joint author. II. Title.
TJ213.D33   629.8'32   74–19012
ISBN 0–07–016179–8

**LINEAR CONTROL
SYSTEM ANALYSIS
AND DESIGN**
CONVENTIONAL AND MODERN

34567890 FGRFGR 798

CONTENTS

PREFACE

This textbook is intended to solve the pressing problem of providing a clear, understandable, and motivated account of the subject which spans both conventional and modern control theory. The authors have tried to exert meticulous care with explanations, diagrams, calculations, tables, and symbols. They have tried to ensure that the student is made aware that rigor is necessary for more advanced control work. Also stressed is the importance of clearly understanding the concepts which provide the rigorous foundations of modern control theory. The text provides a strong, comprehensive, and illuminating account of those elements of conventional control theory which have relevance in the current design and analysis of control systems. The presentation of a variety of different techniques contributes to the development of the student's working understanding of what A. T. Fuller has called "the enigmatic control system." To provide a coherent development of the subject, an attempt is made to eschew formal proofs and lemmas with an organization that draws the perceptive student steadily and surely onto the demanding theory of optimal control. It is the opinion of the authors that a student who has reached this point is fully equipped to undertake with confidence the challenges presented by the more advanced control theories.

The establishment of appropriate differential equations to describe the performance of physical systems, networks, and devices is set forth in Chapter 2, which also introduces some elementary matrix algebra, the block diagram, and the transfer function. The essential concept of modern control theory, the state space is dealt with also. The approach used is the simultaneous derivation of the state-vector differential equation with the single-input single-output differential equation for a chosen physical system. The relationship of the transfer function to the state equation of the system is deferred until Chapter 6. The derivation of an adequate description of a physical system by using Lagrange equations is also given.

The first half of Chapter 3 deals with the classical method of solving differential equations and with the nature of the resulting response in a fashion almost identical to our earlier book.* However, once the state-variable equation has been introduced, careful account is given of its solution. The central importance of the state transition matrix is brought out, and the state transition equation is derived. The idea of an eigenvalue is next explained and this theory is used with the Cayley-Hamilton and Sylvester theorems to evaluate the state transition matrix. A thorough explanation of the use of state-variable diagrams brings the chapter to a close.

The early part of Chapter 4 presents a comprehensive account of Laplace transform methods and pole-zero maps. Some further aspects of matrix algebra are dealt with before dealing with the solution of the state equation by the use of Laplace transforms. Finally the evaluation of transfer matrices is clearly explained.

Chapter 5 begins with system representation by the conventional block-diagram approach. It is followed by a straightforward account of the determination of the state transition equation by the use of signal flow graphs. By deriving parallel state diagrams from system transfer functions, the advantages of having the state equation in uncoupled form are established. This is followed by the methods of diagonalizing the system coefficient matrix. A feature of this chapter is the clear treatment of how to transform an **A** matrix which has complex eigenvalues into a suitable alternative form.

In Chapter 6 the system characteristics are treated in identical fashion to the earlier book, but those very important concepts of modern control theory—controllability and observability—are treated in a simple, straightforward, and correct manner. Although the treatment is brief, it provides sufficient coverage of these topics for the requirements of the remainder of the book. The chapter closes with a short presentation of the sensitivity concepts of Bode.

Chapters 7 to 11 are shortened versions of the same chapters in the earlier book but present substantially the same material. In Chapter 7 the details of the root-locus method of analysis are presented. Then the frequency-response method of analysis is given in Chapters 8 and 9, using both the log and the polar plots. These chapters include the following topics: Nyquist stability criterion; correlation between the s plane, frequency domain, and time domain; and gain setting to achieve a desired output response peak value. Chapters 10 and 11 describe the possible improvements in system performance, along with examples of the technique for applying cascade

---

* J. J. D'Azzo and C. H. Houpis, "Feedback Control System Analysis and Synthesis," 2d ed., McGraw-Hill Book Company, 1966.

and feedback compensators. Both the root-locus and frequency-response methods of designing compensators are covered.

The techniques of achieving desired system characteristics by pole placement and using complete state-variable feedback are developed thoroughly and carefully in Chapter 12. This provides a useful foundation for the work of Chapter 14.

Additional matrix algebra is presented in Chapter 13 with particular emphasis on quadratic forms. This material is used in the presentation of a short account of some of the important aspects of stability considered from the Liapunov point of view. An account of trajectories in the state space and some associated phase plane techniques is also given. A feature of this approach is that the arguments are extended to nonlinear systems. The treatment of such systems by linearization is presented. The use of a Liapunov function is presented for the determination of system stability and instability. It serves as an introduction to its use in establishing a performance index, as shown in Chapter 14.

Chapter 14 starts with a careful and comprehensive treatment of the use of performance indexes for the parameter-optimization methods for single-input single-output systems. The chapter goes on to deal with the nature of the problem of optimal control. The solution to the infinite-time linear quadratic problem is dealt with in extenso, but the results are derived using the Liapunov function approach rather than the calculus of variations of the method proposed by Pontryagin et al. The heuristic advantages of this approach for beginning students are very great. A treatment of the same problem from the standpoints of the Bode diagram and root square locus underscores the essential unity of the subject and the mutuality of the modern and conventional control theories.

The last chapter presents some methods of optimal linear system design. The relationship of these methods to the conventional methods is stressed and evaluated. Although the account is limited, for heuristic convenience, to single-input systems, the subsequent correlation provides the student with the opportunity to develop an invaluable insight into the nature of the linear quadratic optimal control problem.

The authors have tried to provide students of control engineering with a clear, unambiguous, and relevant account of appropriate and contemporary control theory. It is suitable as an introductory and bridging text for undergraduate and graduate students.

The text is arranged so that it can be used for self-study by the engineer in practice. Included are as many examples of feedback control systems in various areas of practice (electrical, aeronautical, mechanical, etc.) as space permits while maintaining a strong basic feedback control text that can be used for study in any of the various branches of engineering. To make the text meaningful and valuable to all engineers, the authors have attempted to unify the treatment of physical control systems through use of mathematical and block-diagram models common to all. A large portion of the text has been class-tested, thus enhancing its value for classroom and self-study use.

The authors express their thanks to the students who have used this book and to the faculty who have reviewed it for their helpful comments and corrections. Appreciation is expressed to Dr. R. E. Fontana, Head of the Electrical Engineering Department, Air Force Institute of Technology, for the encouragement he has given, and to Dr. T. J. Higgins, Professor of Electrical Engineering, University of Wisconsin, for his thorough review of the manuscript.

Especial appreciation is expressed to Dr. Donald McLean, a Visiting Professor at the Air Force Institute of Technology during the time this manuscript was being prepared in final form. He provided an in-depth review of the complete book and used portions of it in his courses. His perception and insight have contributed extensively to the clarity and rigor of the presentation. The association with him has been an enlightening and refreshing experience that will not be forgotten. He is currently a Lecturer at the University of Technology, Loughborough, England.

JOHN J. D'AZZO
CONSTANTINE H. HOUPIS

**LINEAR CONTROL
SYSTEM ANALYSIS
AND DESIGN**

INTRODUCTION

1-1 INTRODUCTION

The art of automatic control systems permeates life in all advanced societies today. Such systems act as a catalyst in promoting progress and development. The automatic toaster, the thermostat, the washer and dryer, the space vehicles, and the control systems that have speeded up the production and quality of manufactured goods—all have influenced our way of life. Control systems are an integral component of any industrial society and are necessary for the production of goods required by an increasing world population. Technological developments have made it possible to travel to the moon and outer space. The successful operation of space vehicles depends on the proper functioning of the large number of control systems used in such ventures.

1-2 INTRODUCTION TO CONTROL SYSTEMS

Assume that a toaster is set for the desired darkness of the toasted bread. The setting of the "darkness," or timer, knob represents the input quantity, and the degree of darkness of the toast produced is the output quantity. If the degree of darkness is not satisfactory, because of the condition of the bread or for some similar reason, this condition can in no way automatically alter the length of time heat is

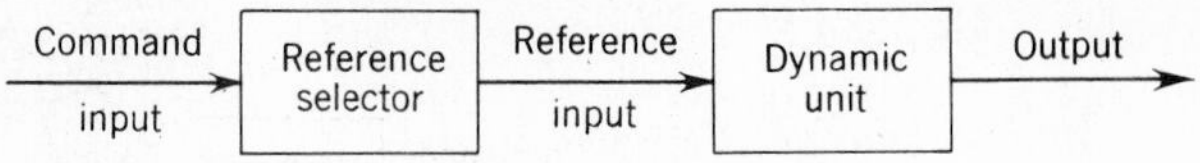

FIGURE 1-1
Functional block diagram of an open-loop control system.

applied. Thus it can be said that the output quantity has no influence on the input quantity. The heater portion of the toaster, excluding the timer unit, represents the dynamic part of the overall system.

Another example is a dc shunt motor. For a given value of field current, a certain value of voltage is applied to the armature to produce the desired value of motor speed. In this case the motor is the dynamic part of the system, the applied armature voltage is the input quantity, and the speed is the output quantity. A variation of the speed from the desired value, due to a change of mechanical load on the shaft, can in no way cause a change in the value of the applied armature voltage to maintain the desired speed. In this example it can also be said that the output quantity has no influence on the input quantity.

Systems in which the output quantity has no effect upon the input quantity are called *open-loop control systems*. The two examples just cited can be represented symbolically by a functional block diagram, as shown in Fig. 1-1. The desired darkness of the toast or the desired speed of the motor is the command input; the selection of the value of time on the toaster timer or the value of voltage applied to the motor armature is represented by the reference-selector block; and the output of this block is identified as the reference input. The reference input is applied to the dynamic unit that performs the desired control function, and the output of this block is the desired output.

A human being can be added to the systems above for the purpose of sensing the actual value of the output with respect to the reference input. If the output does not have the desired value, a person can alter the reference-selector position to achieve this value. Addition of the person provides a means through which the output is fed back and by which the output is compared with the input. Any necessary change is then made in order to cause the output to equal the desired value. The *feedback* action has controlled the input to the dynamic unit. Systems in which the output has an effect upon the input quantity are called *closed-loop control systems*.

To improve the performance of the closed-loop system with respect to maintaining the output quantity as close to the desired quantity as possible, the person is replaced by a mechanical, electrical, or other form of comparison unit. The functional block diagram of a closed-loop control system is illustrated in Fig. 1-2. The comparison between the reference input and the feedback signals results in an actuating signal that is the difference between these two quantities. The actuating signal acts to maintain the output at the desired value.

This system may now be properly called a *closed-loop control system*. The designation *closed-loop* implies the action resulting from the comparison between the output and input quantities in order to maintain the output at the desired value. Thus, the output is controlled in order to maintain the desired value.

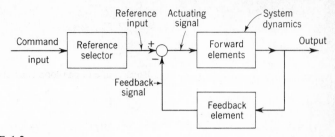

FIGURE 1-2
Functional block diagram of a closed-loop control system.

Desired performance can often be achieved for many systems that are structured as indicated in Fig. 1-2. The design methods for these systems are covered in Chaps. 6 to 11. Some systems require a precision in their performance that cannot be achieved by the structure of Fig. 1-2. Also, systems exist for which there are multiple inputs and/or multiple outputs. The design methods for such systems are based on a representation of the system in terms of *state variables*. For example, position, velocity, and acceleration may represent the state variables of a position-control system. The definition of state variables and their use in representing systems are contained in Chaps. 2, 3, and 5. The use of state-variable methods to achieve improved or sometimes optimal performance is presented in Chaps 12 to 15.

1-3   DEFINITIONS

From the preceding discussion the following definitions are evolved, based in part on the standards of the IEEE.[1]†

> *System*   A combination of components that act together to perform a function not possible with any of the individual parts. The word *systems* as used herein is interpreted to include physical, biological, organizational, and other entities, and combinations thereof, which can be represented through a common mathematical symbolism. The formal name *systems engineering* may also be assigned to this definition of the word *systems*. Thus, the study of feedback control systems is essentially a study of an important aspect of systems engineering.
>
> *Command input*   The motivating input signal to the system, which is independent of the output of the system and exercises complete control over it (if the system is completely controllable).
>
> *Reference selector* (*reference input element*)   The unit that establishes the value of the reference input. The reference selector is calibrated in terms of the desired value of the system output.

† Superscript numbers refer to items in the References at the end of the chapter.

Reference input   The reference signal produced by the reference selector, i.e., the command expressed in a form directly usable by the system. It is the actual signal input to the control system.

Forward element (*system dynamics*)   The unit that reacts to an actuating signal to produce a desired output. This unit does the work of controlling the output and thus may be a power amplifier.

Output (*controlled variable*)   The quantity that must be maintained at a prescribed value, i.e., following the command input.

Open-loop control system   A system in which the output has no effect upon the input signal.

Feedback element   The unit that provides the means for feeding back the output quantity, or a function of the output, in order to compare it with the reference input.

Actuating signal   The signal that is the difference between the reference input and the feedback signal. It actuates the control unit in order to maintain the output at the desired value.

Closed-loop control system   A system in which the output has an effect upon the input quantity in such a manner as to maintain the desired output value.

Note that the fundamental difference between the open- and closed-loop systems is the *feedback action*, which may be continuous or discontinuous. Continuous control implies that the output is continuously fed back and compared with the reference input. In one form of discontinuous control the input and output quantities are periodically sampled and compared; i.e., the control action is discontinuous in time. This type is commonly called a *discrete-data* or *sampled-data* feedback control system. In another form of discontinuous control system, the actuating signal must reach a prescribed value before the system dynamics reacts to it; i.e., the control action is discontinuous in amplitude rather than in time. This type of discontinuous control system is commonly called an *on-off* or *relay* feedback control system. Both forms may be present in a system. In this text only continuous control systems are considered since they lend themselves readily to a basic understanding of feedback control systems.

With the above introductory material, it seems proper to state a definition[1] of a feedback control system: "A control system that operates to achieve prescribed relationships between selected system variables by comparing functions of these variables and using the comparison to effect control." In other books and papers on this subject the following terms may also be used.

Servomechanism (*often abbreviated as servo*)   This term is often used to refer to a mechanical system in which the steady-state error is zero for a constant input signal. Sometimes by generalization it is used to refer to any feedback control system.

Regulator   This term is used to refer to systems in which there is a constant steady-state output for a constant input signal. The name is derived from the early speed and voltage controls, called speed and voltage regulators.

The reader is cautioned to ascertain the meaning of a particular author. Throughout this text an attempt is made to conform to the IEEE definitions.[1]

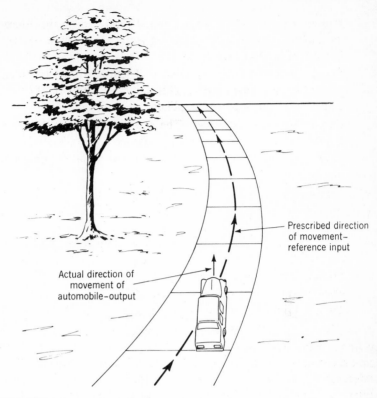

Prescribed direction
of movement-
reference input

Actual direction of
movement of
automobile-output

FIGURE 1-3
A pictorial demonstration of an automobile as a feedback control system.

1-4 HISTORICAL BACKGROUND[2]

The action of steering an automobile to maintain a prescribed direction of movement satisfies the definition of a feedback control system. In Fig. 1-3, the prescribed direction is the reference input. The eyes perform the function of comparing the actual direction of movement with the prescribed direction, the desired output. The eyes transmit a signal to the brain, which interprets this signal and transmits a signal to the arms to turn the steering wheel, adjusting the actual direction of movement to bring it in line with the desired direction. Thus, steering an automobile constitutes a feedback control system.

One of the earliest open-loop control systems was Hero's device for opening the doors of a temple. The command input to the system (see Fig. 1-4) was lighting a fire upon the altar. The expanding hot air under the fire drove the water from the container into the bucket. As the bucket became heavier, it descended and turned the door spindles by means of a rope, causing the counterweight to rise. The door could be closed by dousing the fire. As the air in the container cooled and the pressure was thereby reduced, the water from the bucket siphoned back into the storage container. Thus the bucket became lighter and the counterweight, being heavier, moved down, thereby closing the door. This occurs as long as the bucket is higher than the

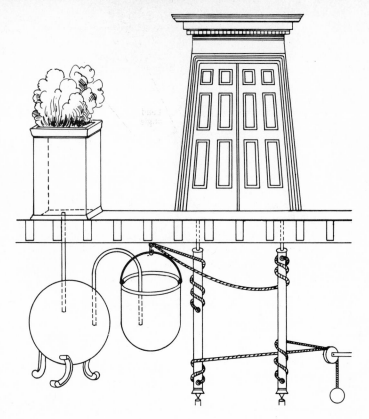

FIGURE 1-4
Hero's device for opening temple doors.

container. The device was probably actuated when the ruler and his entourage started to ascend the temple steps. The system for opening the door was not visible or known to the masses. Thus it created an air of mystery and demonstrated the power of the Olympian gods.

James Watt's flyball governor for controlling speed, developed in 1788, can be considered the first widely used feedback control system not involving a human being. Maxwell, in 1868, made an analytic study of the stability of the flyball governor. This was followed by a more detailed solution of the stability of a third-order flyball governor in 1876 by the Russian engineer Wischnegradsky.[3] Minorsky made one of the early deliberate applications of nonlinear elements in closed-loop systems in his study of automatic ship steering about 1922.[4]

A significant date in the history of automatic feedback control systems is 1934, when Hazen's paper Theory of Servomechanisms was published in the *Journal of the Franklin Institute,* marking the beginning of the very intense modern interest in this new field. It was in this paper that the word *servomechanism* originated, from the words *servant* (or slave) and *mechanism.* The word *servomechanism* thus implies a slave mechanism. It is interesting to note that in the same year Black's important

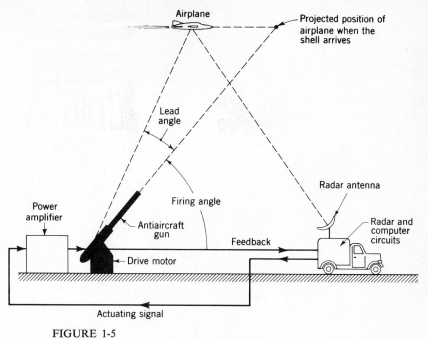

FIGURE 1-5
Antiaircraft radar tracking control system.

paper on feedback amplifiers appeared.[5] During the following 6 years, further basic work was accomplished. Owing to World War II security restrictions, the developments in the period 1940 to 1945 were obscured, delaying rapid progress in this field. During this time three important laboratories organized at the Massachusetts Institute of Technology—the Servomechanisms, Radiation, and Instrumentation Laboratories—contributed much to the advancement of the control field. The research done by many companies during this period also helped to strengthen the foundation of this new science. After the lifting of wartime security restrictions in 1945, rapid progress was made in the control field. Since then many books and thousands of articles and technical papers have been written, and the application of control systems in the industrial and military fields has been extensive. This rapid growth of feedback control systems was accelerated by the equally rapid development and widespread use of computers.

An early example of a military application of a feedback control system is the antiaircraft radar tracking control system shown in Fig. 1-5. The radar antenna detects the position and velocity of the target airplane, and the computer circuit takes this information and determines the correct firing angle for the gun. This angle includes the necessary lead angle so that the shell reaches the projected position at the same time as the airplane. The output signal of the computer, a function of the firing angle, is fed into an amplifier which provides power for the drive motor. The motor then aims the gun at the necessary firing angle. A feedback signal proportional to the gun position ensures correct alignment with the position determined by the

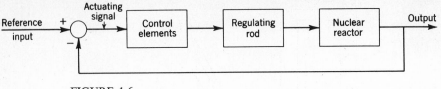

FIGURE 1-6
A nuclear-reactor power-level control system.

computer. Since the gun must be positioned both horizontally and vertically, this system has two drive motors, which are parts of two separate feedback loops.

The advent of the nuclear reactor was a milestone in the advancement of science and technology. For proper operation the power level of the reactor must be maintained at a desired value or must vary in a prescribed manner. This must be accomplished automatically with minimum supervision by people. Figure 1-6 is a simplified block diagram of a feedback control system for controlling the power output level of a reactor. If the power output level differs from the reference input value, the actuating signal produces a signal at the output of the control elements. This, in turn, moves the regulating rod in the proper direction to achieve the desired power level of the nuclear reactor. The position of the regulating rod determines the rate of nuclear fission and therefore the total power generated. This output nuclear power can be converted into steam power, for example, which is then used for generating electric energy.

The control theory developed through the late 1950s may be categorized as *conventional* control theory. Conventional control theory can still be applied to many control design problems, especially to systems with a single input signal and single output signal. The control theory that has been developed since the late fifties for the design of more complicated systems and for multiple-input multiple-output systems is called *modern* control theory. The advances that have been made in space travel were possible only because of the advent of modern control theory. Areas such as trajectory optimization and minimum-time and/or minimum-fuel problems, which are very important in space travel, can be readily handled only by modern control theory.

An example for which only the modern theory can be utilized is the transport of an *unfueled* space missile from the hangar, where it is assembled, to its launch pad. This example is illustrated in Fig. 1-7, where the missile can be represented by a chain whose individual links are coupled by means of pivots and which are free to rotate. The pivot points on the chain represent the nodes of the bending motion of the missile as it is being transported. The design objective of the transport control system is to minimize the bending of the missile while it is being transported. Thus, the displacements at the pivots with respect to the transport must be restricted to remain within design limits in order to prevent structural damage. The signals from the sensors, located at each node, are fed to the controller. The controller, designed by means of modern control theory, produces signals that actuate the drive that controls the position x and its velocity in such a manner as to satisfy the design objective (performance index). This system has many outputs which are represented by the displacements of the pivots of the chain-link representation.

The advent of the steam engine and the subsequent industrial revolution provided man with larger quantities of controlled power than he had had at his

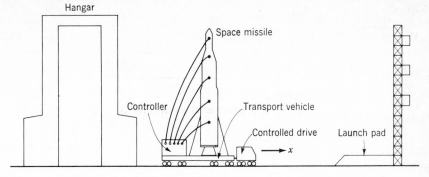

FIGURE 1-7
Transport control system of a space missile from hangar to launch pad.

command before. This created the need for controlling large amounts of power by means of low-power input signals and paved the way for the development of feedback control system analysis. Man's desire to span the universe has given an impetus to the development of modern control theory. These same principles applied to many industrial processes have revolutionized manufacturing methods, in terms of quality, quantity, and technical sophistication of the goods produced. The result has been to completely change the way of life in the industrialized nations.

As is frequently true of a science, the large number of individuals with many different backgrounds who have worked in this field has resulted in a large number of terms and definitions. An effort has been made by several major engineering societies to eliminate this confusion and to establish a set of definitions that can serve as a standard.[1]

The development of the control concept in the engineering field has been extended to the realm of human engineering. The basic concept of feedback control has been utilized extensively in the field of business management. The field of medicine is also one to which the principles of control systems and systems engineering are being applied extensively. Thus, standards of optimum performance are established in all areas of endeavor: the actual performance is compared with the standard, and any difference between the two is used to bring them into closer agreement.

1-5 MATHEMATICAL BACKGROUND

The early studies of control systems were based upon the solution of differential equations by classical means. Other than for simple systems, the analysis in this approach is tedious and does not readily indicate what changes should be made to improve system performance. Use of the Laplace transform simplifies this analysis somewhat. Nyquist's paper[6] published in 1932 dealt with the application of steady-state frequency-response calculations to feedback amplifier design. This work was extended by Black[5] and Bode.[7] Hall[8] and Harris[9] applied frequency-response analysis in the study of feedback control systems, which furthered the development of control theory as a whole.

Another advance occurred in 1948, when Evans[10] presented his root-locus theory. This theory affords a graphical display of the stability properties of a system and permits the graphical evaluation of the frequency response. Laplace transform theory and network theory are joined in the root-locus calculation. In the conventional control theory portion of this text the reader will come to appreciate the simplicity and value of this phase of analysis.

Laplace transform theory and linear-algebra theory are utilized in the application of modern control theory to system analysis and design. The nth-order differential equation describing the system can be converted into a set of n first-order differential equations expressed in terms of the state variables. These equations can be expressed in matrix notation for simpler mathematical manipulation. The matrix equations lend themselves very well to computer computation. It is this characteristic that has enabled modern control theory to solve many problems, such as nonlinear and optimization problems, that could not be solved by conventional control theory.

Throughout the various phases of linear analysis presented in this text, mathematical models are used. Once a physical system has been described by a set of mathematical equations, they are manipulated to achieve an appropriate mathematical model. When this has been done, the subsequent method of analysis is independent of the nature of the physical system; i.e., it does not matter whether the system is electrical, mechanical, etc. This technique helps the designer to spot similarities from previous experience.

As the reader covers the various phases of conventional control-theory analysis presented here, he should keep in mind that no single aspect is intended to be used to the exclusion of the others. Depending upon the known factors and the simplicity (or complexity) of a control-system problem, a designer may use either one aspect exclusively or a combination. With experience in the design of feedback control systems comes the ability to utilize the advantages of each method to a greater extent.

The modern control theory presented in this text is intended as an introduction to the area of system performance optimization. In addition, one widely used method of system optimization, by solution of the algebraic Riccati equation, is presented in detail. This is extended to incorporate the manner of achieving some specified values of the conventional control theory figures of merit along with an optimal performance.

As mentioned earlier, the use of computers greatly aids the designer in making a synthesis of his control problem. For a complicated design problem an engineer must write his own digital-computer program especially geared to helping him achieve a satisfactory system performance.

1-6 GENERAL NATURE OF THE ENGINEERING CONTROL PROBLEM

In general, a control problem can be divided into the following steps:

1 A set of performance specifications is established.

2 As a result of the performance specifications a control problem exists.

3 A set of differential equations that describe the physical system is formulated.

4 Using the conventional control-theory approach:

    a The performance of the basic (or original) system is determined by application of one of the applicable methods of analysis (or a combination of them).

    b If the performance of the original system does not meet the required specifications, equipment must be added to improve the response.

5 Using the modern control-theory approach, the designer specifies an optimal performance index for the system. The design yields the necessary structure to minimize the specified performance index, thus producing an optimal system.

Design of the system to obtain the desired performance is the control problem. The necessary basic equipment is then assembled into a system to perform the desired control function. To a varying extent, most systems are nonlinear. In many cases the nonlinearity is small enough to be neglected, or the limits of operation are small enough to allow a linear analysis to be made. In this textbook linear systems or those that can be approximated as linear systems are considered. Because of the relative simplicity and straightforwardness of this approach, the reader can obtain a thorough understanding of linear systems. After mastering the terminology, definitions, and methods of analysis for linear control systems, the engineer will find it easier to undertake a study of nonlinear systems. A method of linearizing a nonlinear system is included in Chap. 13.

A basic system has the minimum amount of equipment necessary to accomplish the control function. The differential equations that describe the physical system are derived, and an analysis of the basic system is made. If the analysis indicates that the desired performance has not been achieved with this basic system, additional equipment must be inserted into the system. Generally this analysis also indicates the characteristics for the additional equipment that are necessary to achieve the desired performance. After the system is synthesized to achieve the desired performance, based upon a linear analysis, final adjustments can be made on the actual system to take into account the nonlinearities that were neglected. It should be noted that a computer is generally used in the design depending upon the complexity of the system.

1-7 OUTLINE OF TEXT

The first few chapters deal with the mathematics underlying the analysis of control systems. In conjunction with this presentation, various basic physical units are discussed. Once the technique of writing the system equations (and, in turn, the Laplace transforms) that describe the performance of a dynamic system has been mastered, the ideas of block and simulation diagrams and transfer functions are developed. When physical systems are described in terms of block diagrams and transfer functions, they exhibit basic servo characteristics. These characteristics are described and discussed. The concept of state is introduced, and the system equations are developed in the standard matrix format. The necessary linear algebra required to manipulate the matrix equations is included. The very important concepts of modern control theory, controllability and observability, are treated in Chap. 6. Then follows a presentation of the various methods of analysis that can be utilized in the study of feedback control systems. These methods use the root-locus and steady-state frequency-response analysis. How a basic system can be improved by the use of compensators if it does not meet the desired specifications is then presented.

The remaining portion of the text deals with single-input single-output system design utilizing modern control theory. Topics such as pole placement via state-variable feedback, phase-plane analysis, linearization of nonlinear systems, Liapunov stability concepts, the use of performance indexes for parameter optimization, the solution of the infinite-time linear quadratic problem (LQP), etc., are presented. The last chapter presents some methods of optimal linear-system design. These methods indicate how conventional control figures of merit can be achieved while satisfying the LQP. In the appendix is a table of Laplace transform pairs, a section dealing with the use of the Spirule, and some nomograms for the LQP that are useful to a feedback control system engineer.

It seems appropriate to close this chapter by pointing out that a feedback control engineer is essentially a "system engineer," i.e., a person whose primary concern is with the design and synthesis of the overall system. To an extent depending on his own background and experience, he relies on and works closely with engineers in the various recognized branches of engineering (electrical, mechanical, aeronautical, etc.) to furnish him with the transfer functions and/or system equations of various portions of a control system.

In closing this introductory chapter it is important to stress that both conventional and optimal control theory may be utilized for designing practical control systems. The two theories produce system designs that have different performance and require different implementation in terms of cost and equipment. The following design policy includes factors that are worthy of consideration:

1 Use proven design methods.
2 Select the system design which has the minimum complexity.
3 Use minimum specifications or requirements that yield a satisfactory system response. Compare the cost with the performance and select the fully justified system implementation.
4 Perform a complete and adequate testing of the system.

REFERENCES

1 "IEEE Standard Dictionary of Electrical and Electronics Terms," Wiley-Interscience, New York, 1972.
2 Mayr, Otto: "Origins of Feedback Control," M.I.T. Press, Cambridge, Mass., 1971.
3 Trinks, W.: "Governors and the Governing of Prime Movers," Van Nostrand, Princeton, N.J., 1919.
4 Minorsky, N.: Directional Stability and Automatically Steered Bodies, *J. Am. Soc. Nav. Eng.*, vol. 34, p. 280, 1922.
5 Black, H. S.: Stabilized Feedback Amplifiers, *Bell Syst. Tech. J.*, 1934.

6 Nyquist, H.: Regeneration Theory, *Bell Syst. Tech. J.*, 1932.

7 Bode, H. W.: "Network Analysis and Feedback Amplifier Design," Van Nostrand, Princeton, N.J., 1945.

8 Hall, A. C.: Application of Circuit Theory to the Design of Servomechanisms, *J. Franklin Inst.*, 1946.

9 Harris, H.: The Frequency Response of Automatic Control Systems, *Trans. AIEE*, vol. 65, pp. 539–546, 1946.

10 Evans, W. R.: Graphical Analysis of Control Systems, *Trans. AIEE*, vol. 67, pt. II, pp. 547–551, 1948.

2
WRITING SYSTEM EQUATIONS

2-1 INTRODUCTION

Before one can analyze a dynamic system one must be able to determine its performance. Both this ability and the precision of the results depend on how well the characteristics of each component can be expressed mathematically. Numerous exact techniques are available for solving linear equations with constant coefficients. However, when the equations are time-varying or nonlinear, it is more difficult to solve them analytically. In general, except for some low-order equations, the solution of a time-varying or nonlinear equation often requires a numerical, a graphic, or computer-effected procedure. The systems considered in this text are those which are described completely by a set of linear *constant-coefficient* differential equations. Such systems are said to be *time-invariant*;[1] i.e., the relationship between the system input and system output is independent of time. Thus, the system structure does not change with time. Accordingly, the output is independent of the time at which the input is applied.

This chapter presents methods for writing the differential and state equations for a variety of electrical, mechanical, thermal, and hydraulic systems.[2] This is the first step that must be mastered by the would-be control systems engineer. The basic physical laws are given for each system, and the associated parameters are defined. Examples are included to show the application of the basic laws to physical equipment. The result is a differential equation or a set of differential equations that describes the system. The equations derived are limited to linear systems or to those systems which can be represented by linear equations over their useful operating range. The important concepts of system state and of state variables are also introduced. The system equations, expressed in terms of state variables, are called *state equations*. The analytical tools from linear algebra are introduced as they are needed in the development of the state equations and their solutions. The solutions of both the differential and the state equations are covered in Chap. 3.

The analysis of system behavior and the development of the system equations are enhanced by using the block-diagram representation of the system. Control systems generally have many components. Complete drawings of all the detailed parts are frequently too congested to show the specific functions that are performed. To simplify the picture of the complete system, it is common to use a block diagram, in which each function in the system is represented by a block. Each block is labeled with the name of the component, and the blocks are appropriately interconnected by line segments. This type of diagram removes excess detail from the picture and shows the functional operation of the system.

A block diagram[3] represents the flow of information and the functions performed by each component in the system. The primary concern is the dynamic behavior of a complete control system. The use of a block diagram provides a simple means by which the functional relationship of the various components can be shown and reveals the operation of the system more readily than observation of the physical system itself. The simple functional block diagram shows clearly that apparently different physical systems can be analyzed by the same techniques. Since a block diagram is involved not with the physical characteristics of the system but only with the functional relationship between various points in the system, it can reveal the similarity between apparently unrelated physical systems.

A further step taken to increase the information supplied by the block diagram is to label the input quantity into each block and the output quantity from each block. Arrows are used to show the direction of the flow of information. The block represents the function or dynamic characteristics of the component.

The complete block diagram shows how the functional components are connected and the mathematical equations that determine the response of each component. Examples of block diagrams are shown throughout this chapter.

In general, time-varying quantities are indicated by small letters. These are sometimes indicated by the form $x(t)$, but more often this is written just as x. There are some exceptions, because of established convention in the use of certain symbols.

To simplify the writing of differential equations, the D operator notation is used.[4] The symbols D and $1/D$ are defined by

$$Dy \equiv \frac{dy(t)}{dt} \qquad D^2y \equiv \frac{d^2y(t)}{dt^2} \qquad (2\text{-}1)$$

$$D^{-1}y \equiv \frac{1}{D}y \equiv \int_0^t y(t)\,dt + \int_{-\infty}^0 y(t)\,dt = \int_0^t y(t)\,dt + Y_0 \qquad (2\text{-}2)$$

where Y_0 represents the value of the integral at time $t = 0$, that is, the initial value of the integral.

2-2   ELECTRIC CIRCUITS AND COMPONENTS[5]

The equations for an electric circuit obey Kirchhoff's laws, which state:

1   The algebraic sum of the potential differences around a closed circuit equals zero. This may be restated as follows: in traversing any closed loop the sum of the voltage rises equals the sum of the voltage drops.
2   The algebraic sum of the currents at a junction, or node, equals zero. In other words, the sum of the currents entering the junction equals the sum of the currents leaving the junction.

Both these laws are used in examples in this chapter.

The voltage sources are generators. The usual direct-current (dc) voltage source is a battery or dc generator. The voltage drops appear across the three basic electrical elements: resistors, inductors, and capacitors.

The voltage drop across a resistor is given by Ohm's law, which states that the voltage drop across a resistor is equal to the product of the current through the resistor and its resistance. Resistors absorb energy from the system. Symbolically, this voltage is written as

$$v_R = Ri \qquad (2\text{-}3)$$

The voltage drop across an inductor is given by Faraday's law, which is written

$$v_L = L\frac{di}{dt} \equiv L\,Di \qquad (2\text{-}4)$$

This equation states that the voltage drop across an inductor is equal to the product of the inductance and the time rate of increase of current. A positive-valued derivative implies an increasing current and thus a positive voltage drop; and a negative-valued derivative implies a decreasing current and thus a negative voltage drop.

The positively directed voltage drop across a capacitor is defined in magnitude as the ratio of the magnitude of the positive electric charge on its positive plate to the value of its capacitance. Its direction is from the positive plate to the negative plate. The charge on a capacitor plate is equal to the time integral from the initial instant to the arbitrary time t of the current entering the plate, plus the initial value of the charge. The capacitor voltage is written in the form

$$v_C = \frac{q}{C} = \frac{1}{C}\int_0^t i\,dt + \frac{Q_0}{C} = \frac{i}{CD} \qquad (2\text{-}5)$$

The mks units for these electrical quantities in the practical system are given in Table 2-1.

Series Resistor-Inductor Circuit

The voltage source e in Fig. 2-1 is a function of time. Setting the voltage rise equal to the sum of the voltage drops produces

$$v_R + v_L = e$$

$$Ri + L\frac{di}{dt} = Ri + L\,Di = e \qquad (2\text{-}6)$$

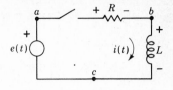

FIGURE 2-1
Series resistor-inductor circuit.

The voltage v_L across the inductor can be obtained in the following manner. The voltage across the inductor is

$$v_L = L\,Di$$

The current through the inductor is therefore

$$i = \frac{1}{LD}\,v_L$$

Substituting these values into the original equation gives

$$\frac{R}{LD}\,v_L + v_L = e \qquad (2\text{-}7)$$

The node method is also convenient for writing the system equations directly in terms of the voltages. The junctions of any two elements are called *nodes*. This circuit has three nodes, labeled a, b, and c (see Fig. 2-1). One node is used as a reference point; in this circuit it will be node c. The voltages at the other nodes are all considered with respect to the reference node. Thus v_{ac} is the voltage drop from node a to node c, and v_{bc} is the voltage drop from node b to the reference node c. For simplicity, these voltages are written just as v_a and v_b.

The source voltage $v_a = e$ is known; therefore there is only one unknown voltage, v_b, and only one node equation is necessary. Kirchhoff's second law, that the algebraic sum of the currents at a node must equal zero, will be applied to node b.

The current from node b to node a through the resistor R is $(v_b - v_a)/R$. The current from node b to node c through the inductor L is $(1/LD)v_b$. The sum of these currents must equal zero:

$$\frac{v_b - v_a}{R} + \frac{1}{LD}\,v_b = 0 \qquad (2\text{-}8)$$

Table 2-1 **ELECTRICAL SYMBOLS AND UNITS**

Symbol	Quantity	Units
e or v	Voltage	Volts
i	Current	Amperes
L	Inductance	Henrys
C	Capacitance	Farads
R	Resistance	Ohms

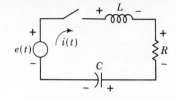

FIGURE 2-2
Series RLC circuit.

Rearranging terms gives

$$\left(\frac{1}{R} + \frac{1}{LD}\right) v_b - \frac{1}{R} v_a = 0 \qquad (2\text{-}9)$$

Except for the use of different symbols, this is the same as Eq. (2-7). Note that the node method required writing only one equation.

Series Resistor-Inductor-Capacitor Circuit

For the series RLC circuit shown in Fig. 2-2, the applied voltage is equal to the sum of the voltage drops when the switch is closed:

$$v_L + v_R + v_C = e$$

$$L \, Di + Ri + \frac{1}{CD} i = e \qquad (2\text{-}10)$$

The circuit equation can be written in terms of the voltage drop across any circuit element. For example, in terms of the voltage across the resistor, $v_R = Ri$, Eqs. (2-10) become

$$v_R + \frac{L}{R} Dv_R + \frac{1}{RCD} v_R = e \qquad (2\text{-}11)$$

Multiloop Electric Circuits

Multiloop electric circuits (see Fig. 2-3) may be solved by either loop or nodal equations. The following example illustrates both methods. The problem is to solve for the output voltage v_0.

Loop method   A loop current is drawn in each closed loop; then Kirchhoff's voltage equation is written for each loop:

$$\left(R_1 + \frac{1}{CD}\right) i_1 - R_1 i_2 - \frac{1}{CD} i_3 = e \qquad (2\text{-}12)$$

$$-R_1 i_1 + (R_1 + R_2 + LD)i_2 - R_2 i_3 = 0 \qquad (2\text{-}13)$$

$$-\frac{1}{CD} i_1 - R_2 i_2 + \left(R_2 + R_3 + \frac{1}{CD}\right) i_3 = 0 \qquad (2\text{-}14)$$

The output voltage is

$$v_0 = R_3 i_3 \qquad (2\text{-}15)$$

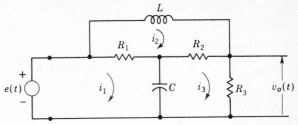

FIGURE 2-3
Multiloop network.

These four equations must be solved simultaneously to obtain $v_0(t)$ in terms of the input voltage $e(t)$ and the circuit parameters.

Node method   The junctions, or nodes, are labeled by letters in Fig. 2-4. Kirchhoff's current equations are written for each node in terms of the node voltages, where node d is taken as reference. The voltage v_{bd} is the voltage of node b with reference to node d. For simplicity, the voltage v_{bd} is written just as v_b. There are two unknown voltages v_b and v_o, and therefore two equations are required:

For node b: $$i_1 + i_2 + i_3 = 0 \qquad (2\text{-}16)$$

For node c: $$-i_3 + i_4 + i_5 = 0 \qquad (2\text{-}17)$$

In terms of the node voltages, these equations are

$$\frac{v_b - v_a}{R_1} + C\,Dv_b + \frac{v_b - v_o}{R_2} = 0 \qquad (2\text{-}18)$$

$$\frac{v_o - v_b}{R_2} + \frac{v_o}{R_3} + \frac{1}{LD}(v_o - e) = 0 \qquad (2\text{-}19)$$

Rearranging the terms to systematize the form of the equations gives

$$\left(\frac{1}{R_1} + CD + \frac{1}{R_2}\right) v_b - \frac{1}{R_2} v_o = \frac{1}{R_1} e \qquad (2\text{-}20)$$

$$-\frac{1}{R_2} v_b + \left(\frac{1}{R_2} + \frac{1}{R_3} + \frac{1}{LD}\right) v_o = \frac{1}{LD} e \qquad (2\text{-}21)$$

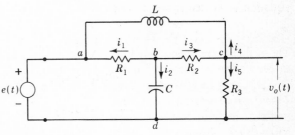

FIGURE 2-4
Multinode network.

For this example, only two nodal equations are needed to solve for the potential at node c. An additional equation must be used if the current in R_3 is required. With the loop method, three equations must be solved simultaneously to obtain the current in any branch; an additional equation must be used if the voltage across R_3 is required. The method that requires the solution of the fewest equations should be used. This varies with the circuit.

The rules for writing the node equations can be summarized as follows:

1 The number of equations required is equal to the number of unknown node voltages.

2 An equation is written for each node.

3 Each equation includes:

    a The node voltage multiplied by the sum of all the admittances that are connected to this node. This term is positive.

    b The node voltage at the other end of each branch multiplied by the admittance connected between the two nodes. This term is negative.

The reader should learn to apply these rules so that Eqs. (2-20) and (2-21) can be written directly for the circuit of Fig. 2-4.

2-3   BASIC LINEAR ALGEBRA[6]

Before proceeding with a discussion of other physical systems it is appropriate to introduce some basic matrix concepts needed in the development of the state method of describing and analyzing physical systems.

Matrix   A *matrix* is a rectangular array of elements. The elements may be real or complex numbers or variables of time or frequency. A matrix with α rows and β columns is called an $\alpha \times \beta$ matrix or is said to be of *order* $\alpha \times \beta$. Sometimes the matrix is said to have size or dimension $\alpha \times \beta$. If $\alpha = \beta$, the matrix is called a *square matrix*. Boldface capital letters are used to denote rectangular matrices, and boldface small letters are used to denote column matrices. A general expression for an $\alpha \times \beta$ matrix is

$$\mathbf{M} = \begin{bmatrix} m_{11} & m_{12} & \cdots & m_{1\beta} \\ m_{21} & m_{22} & \cdots & m_{2\beta} \\ \cdots\cdots\cdots\cdots\cdots\cdots \\ m_{\alpha 1} & m_{\alpha 2} & \cdots & m_{\alpha\beta} \end{bmatrix} = [m_{ij}] \qquad (2\text{-}22)$$

The elements are denoted by lowercase letters. A double-subscript notation is utilized to denote the location of the element in the matrix; thus, the element m_{ij} is located in the ith row and the jth column. A scalar is a matrix of order 1.

Transpose   The transpose of a matrix $\mathbf{M}$ is denoted by $\mathbf{M}^T$. The matrix $\mathbf{M}^T$ is obtained by interchanging the rows and columns of the matrix $\mathbf{M}$. In general, with $\mathbf{M} = [m_{ij}]$, the transpose matrix is $\mathbf{M}^T = [m_{ji}]$.

Vector   A vector is a matrix which has either one row or one column. An $\alpha \times 1$ matrix is called a *column vector*, written,

$$\mathbf{x} = \begin{bmatrix} x_1 \\ x_2 \\ \vdots \\ x_\alpha \end{bmatrix} \qquad (2\text{-}23)$$

and a $1 \times \beta$ matrix is called a *row vector*, written

$$\mathbf{x}^T = \begin{bmatrix} x_1 & x_2 & \cdots & x_\beta \end{bmatrix} \qquad (2\text{-}24)$$

Thus, the transpose of a column vector $\mathbf{x}$ yields the row vector $\mathbf{x}^T$.

Addition and subtraction of matrices   The sum or difference of two matrices $\mathbf{M}$ and $\mathbf{N}$, both of order $\alpha \times \beta$, is a matrix $\mathbf{W}$ of order $\alpha \times \beta$. The element w_{ij} of $\mathbf{W} = \mathbf{M} \pm \mathbf{N}$ is

$$w_{ij} = m_{ij} \pm n_{ij} \qquad (2\text{-}25)$$

These operations are commutative and associative, that is:

Commutative: $\qquad\qquad\qquad \mathbf{M} \pm \mathbf{N} = \pm\mathbf{N} + \mathbf{M}$

Associative: $\qquad\qquad\qquad (\mathbf{M} + \mathbf{N}) + \mathbf{Q} = \mathbf{M} + (\mathbf{N} + \mathbf{Q})$

EXAMPLE 1   *Addition of matrices*

$$\begin{bmatrix} 3 & 1 & 2 \\ 1 & 0 & -4 \\ 0 & 5 & 7 \end{bmatrix} + \begin{bmatrix} 1 & -3 & 1 \\ 2 & 1 & 5 \\ -4 & -1 & 0 \end{bmatrix} = \begin{bmatrix} 4 & -2 & 3 \\ 3 & 1 & 1 \\ -4 & 4 & 7 \end{bmatrix}$$

Multiplication of matrices   The multiplication of two matrices $\mathbf{MN}$ can be performed *if and only if* (iff) they *conform*. If the orders of $\mathbf{M}$ and $\mathbf{N}$ are $\alpha \times \beta$ and $\gamma \times \delta$, respectively, they are conformable iff $\beta = \gamma$, that is, the number of columns of $\mathbf{M}$ must be equal to the number of rows of $\mathbf{N}$. Under this condition $\mathbf{MN} = \mathbf{W}$, where the elements of $\mathbf{W}$ are defined by

$$w_{ij} = \sum_{k=1}^{\beta} m_{ik} n_{kj} \qquad (2\text{-}26)$$

That is, each element of the ith row of $\mathbf{M}$ is multiplied by the corresponding element of the jth column of $\mathbf{N}$, and these products are summed to yield the ijth element of $\mathbf{W}$. The dimension or order of the resulting matrix $\mathbf{W}$ is $\alpha \times \delta$.

Matrix multiplication operations are summarized as follows:

1   An $\alpha \times \beta$ $\mathbf{M}$ matrix times a $\beta \times \gamma$ $\mathbf{N}$ matrix yields an $\alpha \times \gamma$ $\mathbf{W}$ matrix; that is, $\mathbf{MN} = \mathbf{W}$.

2   An $\alpha \times \beta$ $\mathbf{M}$ matrix times a $\beta \times \alpha$ $\mathbf{N}$ matrix yields an $\alpha \times \alpha$ $\mathbf{W}$ square matrix; that is, $\mathbf{MN} = \mathbf{W}$. Note that although $\mathbf{NM} = \mathbf{Y}$ is also a square matrix, it is of order $\beta \times \beta$.

3 When **M** is of order $\alpha \times \beta$ and **N** is of order $\beta \times \alpha$, each of the products **MN** and **NM** is conformable. However, in general,

$$\mathbf{MN} \neq \mathbf{NM}$$

The product of **M** and **N** is said to be *noncommutable*.

4 The product **MN** can be referred to as **N** *premultiplied* by **M** or as **M** *postmultiplied* by **N**. In other words, premultiplication or postmultiplication is used to indicate whether one matrix is multiplied by another from the left or from the right.

5 A $1 \times \alpha$ row vector times an $\alpha \times \beta$ matrix yields a $1 \times \beta$ row vector; that is, $\mathbf{x}^T\mathbf{M} = \mathbf{y}^T$.

6 A $\beta \times \alpha$ matrix times an $\alpha \times 1$ column vector yields a $\beta \times 1$ column vector; that is, $\mathbf{Mx} = \mathbf{y}$.

7 A $1 \times \alpha$ row vector times an $\alpha \times 1$ column vector yields a 1×1 matrix, that is, $\mathbf{x}^T\mathbf{y} = \mathbf{w}$. The 1×1 **w** matrix is a scalar quantity and possesses all the properties of a scalar.

8 The k-fold multiplication of a square matrix **M** by itself is indicated by $\mathbf{M}^k$.

These operations are illustrated by the following examples.

EXAMPLE 2 *Matrix multiplication*

$$\begin{bmatrix} 2 & 1 & 3 \\ 1 & 4 & 1 \end{bmatrix} \begin{bmatrix} 2 & 1 & 0 \\ 0 & 3 & 1 \\ 1 & 2 & 1 \end{bmatrix} = \begin{bmatrix} 7 & 11 & 4 \\ 3 & 15 & 5 \end{bmatrix}$$

$$\begin{bmatrix} 2 & 1 & 3 \\ 1 & 4 & 1 \end{bmatrix} \begin{bmatrix} 2 & 1 \\ 0 & 3 \\ 1 & 2 \end{bmatrix} = \begin{bmatrix} 7 & 11 \\ 3 & 15 \end{bmatrix}$$

$$\begin{bmatrix} 2 & 1 \\ 0 & 3 \\ 1 & 2 \end{bmatrix} \begin{bmatrix} 2 & 1 & 3 \\ 1 & 4 & 1 \end{bmatrix} = \begin{bmatrix} 5 & 6 & 7 \\ 3 & 12 & 3 \\ 4 & 9 & 5 \end{bmatrix}$$

$$\begin{bmatrix} 2 & 1 \\ 1 & 1 \end{bmatrix} \begin{bmatrix} 1 & 0 \\ 2 & 3 \end{bmatrix} = \begin{bmatrix} 4 & 3 \\ 3 & 3 \end{bmatrix}$$

$$\begin{bmatrix} 1 & 0 \\ 2 & 3 \end{bmatrix} \begin{bmatrix} 2 & 1 \\ 1 & 1 \end{bmatrix} = \begin{bmatrix} 2 & 1 \\ 7 & 5 \end{bmatrix}$$

$$\begin{bmatrix} 1 & 3 \end{bmatrix} \begin{bmatrix} 2 & 1 \\ 0 & 3 \end{bmatrix} = \begin{bmatrix} 2 & 10 \end{bmatrix}$$

$$\begin{bmatrix} 2 & 1 \\ 0 & 3 \end{bmatrix} \begin{bmatrix} 1 \\ 3 \end{bmatrix} = \begin{bmatrix} 5 \\ 9 \end{bmatrix}$$

$$\begin{bmatrix} 1 & 3 \end{bmatrix} \begin{bmatrix} 2 \\ 1 \end{bmatrix} = \begin{bmatrix} 5 \end{bmatrix} = 5$$

Multiplication of a matrix by a scalar The multiplication of matrix **M** by a scalar k is effected by multiplying each element m_{ij} by k, that is,

$$k\mathbf{M} = \mathbf{M}k = [km_{ij}] \qquad (2\text{-}27)$$

EXAMPLE 3

$$2 \begin{bmatrix} 1 & 0 \\ 5 & -7 \end{bmatrix} = \begin{bmatrix} 1 & 0 \\ 5 & -7 \end{bmatrix} 2 = \begin{bmatrix} 2 & 0 \\ 10 & -14 \end{bmatrix}$$

Unit or identity matrix   A unit matrix, denoted **I**, is a diagonal matrix in which each element on the principal diagonal is unity. Sometimes the notation $\mathbf{I}_n$ is used to indicate an identity matrix of order n. An example of a unit matrix is

$$\mathbf{I} = \begin{bmatrix} 1 & 0 & 0 & 0 \\ 0 & 1 & 0 & 0 \\ 0 & 0 & 1 & 0 \\ 0 & 0 & 0 & 1 \end{bmatrix}$$

The premultiplication or postmultiplication of a matrix **M** by the unit matrix **I** leaves the matrix unchanged.

$$\mathbf{MI} = \mathbf{IM} = \mathbf{M}$$

Differentiation of a matrix   The differentiation of a matrix with respect to a scalar is effected by differentiating each element of the matrix with respect to the indicated variable:

$$\frac{d}{dt}[\mathbf{M}(t)] = \dot{\mathbf{M}}(t) = [\dot{m}_{ij}] = \begin{bmatrix} \dot{m}_{11} & \dot{m}_{12} & \cdots & \dot{m}_{1\beta} \\ \dot{m}_{21} & \dot{m}_{22} & \cdots & \dot{m}_{2\beta} \\ \cdots & \cdots & \cdots & \cdots \\ \dot{m}_{\alpha 1} & \dot{m}_{\alpha 2} & \cdots & \dot{m}_{\alpha\beta} \end{bmatrix}$$

The derivative of a product of matrices follows rules similar to those for the derivative of a scalar product, with preservation of order. Thus, typically,

$$\frac{d}{dt}[\mathbf{M}(t)\mathbf{N}(t)] = \mathbf{M}(t)\dot{\mathbf{N}}(t) + \dot{\mathbf{M}}(t)\mathbf{N}(t)$$

Integration of a matrix   The integration of a matrix is effected by integrating each element of the matrix with respect to the indicated variable:

$$\int \mathbf{M}(t)\, dt = \left[\int m_{ij}\, dt\right]$$

2-4   STATE CONCEPTS

With basic matrix concepts, the system state concept and the method of writing and solving the state equations can be presented. The state of a system (henceforth referred to only as state) is defined by Kalman[7] as follows.

State   The *state* of a system is a mathematical structure containing a set of n variables $x_1(t), x_2(t), \ldots, x_i(t), \ldots, x_n(t)$, called the *state variables*, such that the initial values

$x_i(t_0)$ of this set and the system inputs $u_j(t)$ are sufficient to uniquely describe the system's future response for $t \geq t_0$. There is a minimum set of state variables which is required to represent the system accurately. The r inputs, $u_1(t)$, $u_2(t)$, . . . , $u_j(t)$, . . . , $u_r(t)$, are deterministic, i.e., they have specific values for all values of time $t \geq t_0$.

Generally the initial starting time t_0 is taken to be zero. The state variables need not be physically observable and measurable quantities; they may be purely mathematical quantities. As a consequence of the definition of state, the following additional definitions are generated.

State vector   The set of state variables $x_i(t)$ represents the elements or components of the n-dimensional state vector $\mathbf{x}(t)$; that is,

$$\mathbf{x}(t) \equiv \begin{bmatrix} x_1(t) \\ x_2(t) \\ \vdots \\ x_n(t) \end{bmatrix} = \begin{bmatrix} x_1 \\ x_2 \\ \vdots \\ x_n \end{bmatrix} \equiv \mathbf{x} \qquad (2\text{-}28)$$

When all the inputs $u_j(t)$ to a given system are specified for $t > t_0$, the resulting state vector uniquely determines the system behavior for any $t > t_0$.

State space   State space is defined as the n-dimensional space in which the components of the state vector represent its coordinate axes.

State trajectory   State trajectory is defined as the path produced in the state space by the state vector $\mathbf{x}(t)$ as it changes with the passage of time. State space and state trajectory in the two-dimensional case are referred to as the *phase plane* and *phase trajectory*, respectively. Examples of state trajectories are presented in Chap. 13.

The first step in applying these definitions to a physical system is the selection of the system variables that are to represent the state of the system. It should be clearly understood that there is no unique way of making this selection. The three common techniques for expressing the system state are the *physical, phase,* and *canonical state-variable* methods. The physical state-variable method is introduced in this chapter. The other two methods are introduced in later chapters. The remainder of this section deals with the application of the state definitions to the physical systems of the previous section.

The selection of the state variables for the physical-variable method is based upon the energy-storage elements of the system. Table 2-2 lists some common energy-storage elements that exist in physical systems and the corresponding energy equations. The physical variable in the energy equation for each energy-storage element *can* be selected as a state variable of the system. Only independent physical variables are chosen to be state variables. *Independent state variables* are those state variables which cannot be expressed in terms of the remaining assigned state variables. In some systems it may be necessary to identify more state variables than just the energy-storage variables. This is illustrated in some of the following examples where velocity is a state variable. When position, the integral of this state variable, is of interest, it must also be assigned as a state variable.

EXAMPLE 1   *Series resistor-inductor circuit* (Fig. 2-1)   This circuit contains only one energy-storage element, the inductor; thus only one state variable is required. From Table 2-2, the state variable is $x_1 = i$. The equation desired is one which

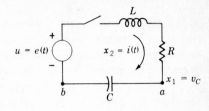

FIGURE 2-5
Series *RLC* circuit.

contains the first derivative of the state variable. The loop equation, Eq. (2-6), can be rewritten, letting $u = e$, as

$$Rx_1 + L\dot{x}_1 = u$$

$$\dot{x}_1 = -\frac{R}{L}x_1 + \frac{1}{L}u \qquad (2\text{-}29)$$

The letter u is the standard notation for the input forcing function and is called the *control variable*. Equation (2-29) is called the *state equation* of the system. There is only one state equation because this is a first-order system, $n = 1$.

EXAMPLE 2 *Series resistor-inductor-capacitor circuit* (Fig. 2-2) This circuit contains two energy-storage elements, the inductor and capacitor. From Table 2-2, the two assigned state variables are immediately identified as $x_1 = v_C$ (the voltage across the capacitor) and $x_2 = i$ (the current in the inductor). Thus two state equations are required.

To obtain an equation containing the derivative of the current in the inductor, a loop or branch equation is written. To obtain an equation containing the derivative

Table 2-2 ENERGY-STORAGE ELEMENTS

Element	Energy	Physical variable
Capacitor C	$\dfrac{Cv^2}{2}$	Voltage v
Inductor L	$\dfrac{Li^2}{2}$	Current i
Mass M	$\dfrac{Mv^2}{2}$	Translational velocity v
Moment of inertia J	$\dfrac{J\omega^2}{2}$	Rotational velocity ω
Spring K	$\dfrac{Kx^2}{2}$	Displacement x
Fluid compressibility $\dfrac{V}{K_B}$	$\dfrac{VP_L{}^2}{2K_B}$	Pressure P_L
Fluid capacitor $C = \rho A$	$\dfrac{\rho A h^2}{2}$	Height h
Thermal capacitor C	$\dfrac{C\theta^2}{2}$	Temperature θ

of the capacitor voltage, a node equation is written. The number of loop equations that must be written is equal to the number of state variables representing currents in inductors. The number of equations involving node voltages that must be written is equal to the number of state variables representing voltages across capacitors. These are usually, but not always, node equations. These loop and node equations are written in terms of the inductor *branch* currents and capacitor *branch* voltages. From these equations, it is necessary to determine which of the assigned physical variables are independent.

Figure 2-2 is redrawn in Fig. 2-5 with node b as the reference node. The node equation for node a and the loop equation are, respectively,

$$C\dot{x}_1 = x_2 \qquad (2\text{-}30)$$

$$L\dot{x}_2 + Rx_2 + x_1 = u \qquad (2\text{-}31)$$

Rearranging terms yields

$$\dot{x}_1 = \frac{1}{C} x_2 \qquad (2\text{-}32)$$

$$\dot{x}_2 = -\frac{1}{L} x_1 - \frac{R}{L} x_2 + \frac{1}{L} u \qquad (2\text{-}33)$$

Equations (2-32) and (2-33) represent the state equations of the system containing two independent state variables. Note that they are *first-order* linear differential equations and are $n = 2$ in number. They are the minimum number of state equations required to represent the system's future performance.

The following definition is based upon these two examples.
State equation The state equations of a system are a set of n first-order differential equations, where n is the number of independent states.

The state equations represented by Eqs. (2-32) and (2-33) are expressed in matrix notation as

$$\begin{bmatrix} \dot{x}_1 \\ \dot{x}_2 \end{bmatrix} = \begin{bmatrix} 0 & \dfrac{1}{C} \\ -\dfrac{1}{L} & -\dfrac{R}{L} \end{bmatrix} \begin{bmatrix} x_1 \\ x_2 \end{bmatrix} + \begin{bmatrix} 0 \\ \dfrac{1}{L} \end{bmatrix} u \qquad (2\text{-}34)$$

It can be expressed in a more compact form as

$$\dot{\mathbf{x}} = \mathbf{A}\mathbf{x} + \mathbf{b}u \qquad (2\text{-}35)$$

where

$$\dot{\mathbf{x}} = \begin{bmatrix} \dot{x}_1 \\ \dot{x}_2 \end{bmatrix} \qquad \text{an } n \times 1 \text{ } column \text{ } vector$$

$$\mathbf{A} = \begin{bmatrix} a_{11} & a_{12} \\ a_{21} & a_{22} \end{bmatrix} = \begin{bmatrix} 0 & \dfrac{1}{C} \\ -\dfrac{1}{L} & -\dfrac{R}{L} \end{bmatrix} \qquad \text{an } n \times n \text{ } plant \text{ } coefficient \text{ } matrix$$

$$\mathbf{x} = \begin{bmatrix} x_1 \\ x_2 \end{bmatrix} \qquad \text{an } n \times 1 \text{ } \textit{state vector}$$

$$\mathbf{b} = \begin{bmatrix} b_1 \\ b_2 \end{bmatrix} = \begin{bmatrix} 0 \\ \dfrac{1}{L} \end{bmatrix} \qquad \text{an } n \times 1 \text{ } \textit{control matrix}$$

and u, in this case, is a one-dimensional control vector. In Eq. (2-34) matrices $\mathbf{A}$ and $\mathbf{x}$ are conformable.

If the output quantity $y(t)$ for the circuit of Fig. 2-2 is the voltage across the capacitor v_c, then

$$y(t) = v_c = x_1$$

Thus the matrix *system output* equation is

$$y(t) = \mathbf{c}^T \mathbf{x} = \begin{bmatrix} 1 & 0 \end{bmatrix} \begin{bmatrix} x_1 \\ x_2 \end{bmatrix} \qquad (2\text{-}36)$$

where

$$\mathbf{c}^T = \begin{bmatrix} 1 & 0 \end{bmatrix}$$

is a 1×2 *row vector* and y, in this example, is a one-dimensional *output vector*.

Equations (2-35) and (2-36) are for a single-input single-output system. For a multiple-input multiple-output system, with r inputs and m outputs, these equations become

$$\dot{\mathbf{x}} = \mathbf{Ax} + \mathbf{Bu} \qquad (2\text{-}37)$$
$$\mathbf{y} = \mathbf{Cx} \qquad (2\text{-}38)$$

where $\mathbf{B}$ is an $n \times r$ *control matrix*
$\quad \mathbf{C}$ is an $m \times n$ *output matrix*
$\quad \mathbf{u}$ is an r-dimensional *control vector*
$\quad \mathbf{y}$ is an m-dimensional *output vector*

EXAMPLE 3   Obtain the state equation for the circuit of Fig. 2-6, where i_2 is considered to be the output of this system. The assigned state variables are $x_1 = i_1$, $x_2 = i_2$, and $x_3 = v_c$. Thus two loop and one node equations are written:

$$R_1 x_1 + L_1 \dot{x}_1 + x_3 = u \qquad (2\text{-}39)$$
$$-x_3 + L_2 \dot{x}_2 + R_2 x_2 = 0 \qquad (2\text{-}40)$$
$$-x_1 + x_2 + C \dot{x}_3 = 0 \qquad (2\text{-}41)$$

The three state variables are independent, and the system state and output equations are

$$\dot{\mathbf{x}} = \begin{bmatrix} -\dfrac{R_1}{L_1} & 0 & -\dfrac{1}{L_1} \\[2ex] 0 & -\dfrac{R_2}{L_2} & \dfrac{1}{L_2} \\[2ex] \dfrac{1}{C} & -\dfrac{1}{C} & 0 \end{bmatrix} \mathbf{x} + \begin{bmatrix} \dfrac{1}{L_1} \\[2ex] 0 \\[2ex] 0 \end{bmatrix} u \qquad (2\text{-}42)$$

$$y = \begin{bmatrix} 0 & 1 & 0 \end{bmatrix} \mathbf{x} \qquad (2\text{-}43)$$

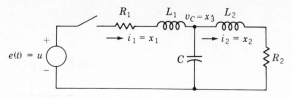

FIGURE 2-6
An electric circuit.

EXAMPLE 4   Obtain the state equations for the circuit of Fig. 2-7. The output is the voltage v_1. The input or control variable is a current source $i(t)$. The assigned state variables are i_1, i_2, i_3, v_1, and v_2. Three loop equations and two node equations are written:

$$v_1 = L_1 \, Di_1 \qquad (2\text{-}44)$$

$$v_2 = L_2 \, Di_2 + v_1 \qquad (2\text{-}45)$$

$$v_2 = L_3 \, Di_3 \qquad (2\text{-}46)$$

$$i_2 = C_1 \, Dv_1 + i_1 \qquad (2\text{-}47)$$

$$i = i_3 + C_2 \, Dv_2 + i_2 \qquad (2\text{-}48)$$

Substituting from Eqs. (2-44) and (2-46) into Eq. (2-45) or writing the loop equation through L_1, L_2, and L_3 and then integrating (multiplying by $1/D$) gives

$$L_3 i_3 = L_2 i_2 + L_1 i_1 \qquad (2\text{-}49)$$

This equation reveals that one inductor current is dependent upon the other two inductor currents. Thus, this circuit has only *four* independent physical state variables, two inductor currents and two capacitor voltages. The four independent state variables are designated as $x_1 = v_1$, $x_2 = v_2$, $x_3 = i_1$, and $x_4 = i_2$, and the control variable is $u = i$. The three state equations are obtainable from Eqs. (2-44), (2-45), and (2-47). The fourth equation is obtained by eliminating the dependent current i_3 from Eqs. (2-48) and (2-49). The result is

$$\dot{\mathbf{x}} = \begin{bmatrix} 0 & 0 & -\dfrac{1}{C_1} & \dfrac{1}{C_1} \\[2ex] 0 & 0 & -\dfrac{L_1}{L_3 C_2} & -\dfrac{L_2 + L_3}{L_3 C_2} \\[2ex] \dfrac{1}{L_1} & 0 & 0 & 0 \\[2ex] -\dfrac{1}{L_2} & \dfrac{1}{L_2} & 0 & 0 \end{bmatrix} \mathbf{x} + \begin{bmatrix} 0 \\[2ex] \dfrac{1}{C_2} \\[2ex] 0 \\[2ex] 0 \end{bmatrix} u \qquad (2\text{-}50)$$

$$y = \begin{bmatrix} 1 & 0 & 0 & 0 \end{bmatrix} \mathbf{x} \qquad (2\text{-}51)$$

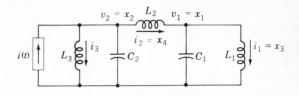

FIGURE 2-7
Electric circuit.

The dependence of i_3 on i_1 and i_2 as shown by Eq. (2-49) may not be readily observed. In that case the matrix state equation for this example would be written with five state variables.

The examples considered in this section are fairly straightforward. In general it is necessary to write more than just the number of state equations because other system variables appear in them. These equations are solved simultaneously to eliminate all internal variables in the circuit except for the state variables. This is necessary in some of the problems for this chapter. For more complex circuits it is possible to introduce more generalized and systematized methods using linear graphs for obtaining the state equations. They are not included in this book.

2-5 TRANSFER FUNCTION AND BLOCK DIAGRAM

A quantity that plays an important role in control theory is the system *transfer function*, defined as follows.

Transfer function If the differential equation is linear, the ratio of the output variable to the input variable, where the variables are expressed as functions of the D operator, is called the transfer function.

In the *RLC* circuit of Fig. 2-2, consider the system output as $v_c = y$. Substituting $i = C\,Dv_c$ into Eq. (2-10) yields

$$(LCD^2 + RCD + 1)v_c(D) = e(D) \qquad (2\text{-}52)$$

The system transfer function is

$$G(D) = \frac{y(t)}{u(t)} = \frac{v_c(t)}{e(t)} = \frac{1}{LCD^2 + RCD + 1} \qquad (2\text{-}53)$$

The notation $G(D)$ is used to denote a transfer function when it is expressed in terms of the D operator. It may also be written simply as G.

The *block-diagram* representation of this system (Fig. 2-8) represents the mathematical operation $G(D)u(t) = y(t)$; that is, the transfer function times the input is equal to the output of the block. The resulting equation is the differential equation of the system.

FIGURE 2-8
Block-diagram representation of Fig. 2-2.

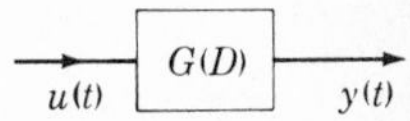

2-6   MECHANICAL TRANSLATION SYSTEMS

Mechanical systems obey the basic law that the sum of the forces equals zero. This is known as Newton's law and may be restated as follows: the sum of the applied forces must be equal to the sum of the reactive forces. The following analysis includes only linear functions. Static friction, coulomb friction, and other nonlinear friction terms are not included. The three qualities characterizing elements in a mechanical translation† system are mass, elastance, and damping. Basic elements entailing these qualities are represented as network elements,[8] and a mechanical network is drawn for each mechanical system to facilitate writing the differential equations.

The mass M is the inertial element. A force applied to a mass produces an acceleration of the mass. The reaction force f_M is equal to the product of mass and acceleration and is opposite in direction to the applied force. In terms of displacement x, velocity v, and acceleration a, the force equation is

$$f_M = Ma = M\,Dv = M\,D^2x \qquad (2\text{-}54)$$

The network representation of mass is shown in Fig. 2-9a. One terminal, a, has the motion of the mass; and the other terminal, b, is considered to have the motion of the reference. The reaction force f_M is a function of time and acts "through" M.

The elastance, or stiffness, K provides a restoring force as represented by a spring. Thus, if stretched, the spring tries to contract; if compressed, it tries to expand to its normal length. The reaction force f_K on each end of the spring is the same and is equal to the product of the stiffness K and the amount of deformation of the spring.

The network representation of a spring is shown in Fig. 2-9b. The displacement of each end of the spring is measured from the original or equilibrium position. End c has a position x_c, and end d has a position x_d, measured from the respective equilibrium positions. The force equation, in accordance with Hooke's law, is

$$f_K = K(x_c - x_d) \qquad (2\text{-}55)$$

If the end d is stationary, then $x_d = 0$ and the above equation reduces to

$$f_K = Kx_c \qquad (2\text{-}56)$$

The damping, or viscous friction, B characterizes the element that absorbs energy. The damping force is proportional to the difference in velocity of two bodies, and the assumption is made that the viscous friction is linear. This assumption simpli-

† Translation means motion in a straight line.

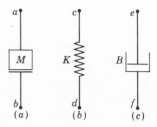

FIGURE 2-9
Network elements of mechanical translation systems.

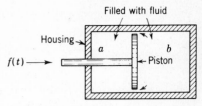

FIGURE 2-10
Dashpot construction.

fies the solution of the dynamic equation. The network representation of damping action is a dashpot, as shown in Fig. 2-9c. It should be realized that damping may either be intentional or occur unintentionally and is present because of physical construction.

The reaction damping force f_B is equal to the product of damping B and the relative velocity of the two ends of the dashpot. The direction of this force, given by Eq. (2-57), depends on the relative magnitudes and directions of the velocities Dx_e and Dx_f:

$$f_B = B(v_e - v_f) = B(Dx_e - Dx_f) \qquad (2\text{-}57)$$

Damping may be added to a system by use of a dashpot. The basic operation of a dashpot, in which the housing is filled with a fluid, is shown in Fig. 2-10. If a force f is applied to the shaft, the piston presses against the fluid, increasing the pressure on side b and decreasing the pressure on side a. As a result, the fluid flows around the piston from side b to side a. If necessary, a small hole can be drilled through the piston to provde a positive path for the flow of fluid. The force required to move the piston inside the housing is given by Eq. (2-57), where the damping B depends on the dimensions and the fluid used.

Before writing the differential equations of a complete system, the first step is to draw the mechanical network. This is done by connecting the terminals of elements that have the same displacement. Then the force equation is written for each node or position by equating the sum of the forces at each position to zero. The equations are similar to the node equations in an electric circuit, with force analogous to current, velocity analogous to voltage, and the mechanical elements with their appropriate operators analogous to admittance. Several examples are shown in the following pages. The reference positions in all cases should be taken from the static equilibrium positions. The force of gravity therefore does not appear in the system equations. The English and metric systems of units are shown in Table 2-3.

Simple Mechanical Translation System

The system shown in Fig. 2-11 is initially at rest. The end of the spring and the mass have positions denoted as the reference positions, and any displacement from these reference positions is labeled x_a or x_b, respectively. A force f applied at the end of the spring must be balanced by a compression of the spring. The same force is also transmitted through the spring and acts at point x_b.

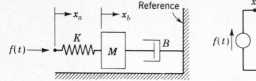

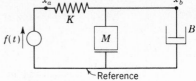

FIGURE 2-11
Simple mass-spring-damper mechanical system.

FIGURE 2-12
Mechanical network corresponding to Fig. 2-11.

To draw the mechanical network, the points x_a and x_b and the reference are located. The network elements are then connected between these points. For example, one end of the spring has the position x_a, and the other end has the position x_b. Therefore the spring is connected between these points. The complete mechanical network is drawn in Fig. 2-12.

The displacements x_a and x_b are nodes of the circuit. At each node the sum of the forces must add to zero. Accordingly, the equations may be written

$$f = f_K = K(x_a - x_b) \qquad (2\text{-}58)$$

$$f_K = f_M + f_B = M\,D^2 x_b + B\,D x_b \qquad (2\text{-}59)$$

These two equations can be solved for the two displacements x_a and x_b and their respective velocities $v_a = Dx_a$ and $v_b = Dx_b$.

It is possible to obtain one equation relating x_a to f, x_b to x_a, or x_b to f by combining Eqs. (2-58) and (2-59):

$$K(MD^2 + BD)x_a = (MD^2 + BD + K)f \qquad (2\text{-}60)$$

$$(MD^2 + BD + K)x_b = Kx_a \qquad (2\text{-}61)$$

$$(MD^2 + BD)x_b = f \qquad (2\text{-}62)$$

Table 2-3   MECHANICAL TRANSLATION SYMBOLS AND UNITS

Symbol	Quantity	English units	Metric units
f	Force	Pounds	Newtons
x	Distance	Feet	Meters
v	Velocity	Feet/second	Meters/second
a	Acceleration	Feet/second^2	Meters/second^2
M†	Mass	Slugs $= \dfrac{\text{pound-second}^2}{\text{foot}}$	Kilograms
K	Stiffness coefficient	Pounds/foot	Newtons/meter
B	Damping coefficient	Pounds/(foot/second)	Newtons/(meter/second)

† Mass M in the English system above has the dimensions of slugs. Sometimes it is given in units of pounds. If so, then in order to use the consistent set of units above, the mass must be expressed in slugs by using the conversion factor 1 slug $= 32.2$ lb.

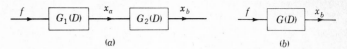

FIGURE 2-13
(a) Detailed and (b) overall block-diagram representation of Fig. 2-11.

The solution of Eq. (2-61) shows the motion x_b resulting from a given motion x_a. Also, the solutions of Eqs. (2-60) and (2-62) show the motions x_a and x_b, respectively, resulting from a given force f. From each of these three equations the following transfer functions are obtained:

$$G_1 = \frac{x_a}{f} = \frac{MD^2 + BD + K}{K(MD^2 + BD)} \qquad (2\text{-}63)$$

$$G_2 = \frac{x_b}{x_a} = \frac{K}{(MD^2 + BD + K)} \qquad (2\text{-}64)$$

$$G = \frac{x_b}{f} = \frac{1}{MD^2 + BD} \qquad (2\text{-}65)$$

Note that the last equation is equal to the product of the first two, i.e.,

$$G = G_1 G_2 = \frac{x_a}{f}\frac{x_b}{x_a} = \frac{x_b}{f} \qquad (2\text{-}66)$$

The block diagram of Fig. 2-11 representing the mathematical operation of Eq. (2-66) is shown in Fig. 2-13. Figure 2-13a is a detailed representation which indicates all variables in the system. The two blocks G_1 and G_2 are said to be in *cascade*. Figure 2-13b, called the *overall block-diagram* representation, shows only the input f and the output x_b, where x_b is considered the output variable of the system of Fig. 2-11.

The multiplication of transfer functions, as in Eq. (2-66), is valid as long as there is no coupling or loading between the two blocks in Fig. 2-13a. The signal x_a is unaffected by the presence of the block having the transfer function G_2; thus the multiplication is valid. When electric circuits are coupled, the transfer functions may not be independent unless they are isolated by an electronic amplifier with a very high input impedance.

EXAMPLE 1   Determine the state equations for Eq. (2-59).

Equation (2-59) involves only one energy-storage element, the mass M, whose energy variable is v_b. The output quantity in this system is the position $y = x_b$. Since this quantity is not one of the physical or energy-related state variables, it is necessary to increase the number of state variables to 2. Equation (2-59) is of second order, which confirms that two state variables are required. Note that the spring constant does not appear in this equation since the force f is transmitted through the spring to the mass. With $x_1 = x_b$, $x_2 = v_b = \dot{x}_1$, $u = f$, and $y = x_1$, the state and output equations are

$$\dot{\mathbf{x}} = \begin{bmatrix} 0 & 1 \\ 0 & -\dfrac{B}{M} \end{bmatrix} \mathbf{x} + \begin{bmatrix} 0 \\ \dfrac{1}{M} \end{bmatrix} u \qquad y = \begin{bmatrix} 1 & 0 \end{bmatrix} \mathbf{x}$$

EXAMPLE 2   Determine the state equations for Eq. (2-61).

Equation (2-61) involves two energy-storage elements, K and M, whose energy variables are x_b and v_b, respectively. Note that the spring constant K does appear in this equation since x_a is the input u. Therefore a state variable must be associated with this energy element. Thus the assigned state variables are x_b and $v_b = Dx_b$, which are independent state variables of this system. Let $x_1 = x_b = y$ and $x_2 = v_b = \dot{x}_1$. Equation (2-61) converted into state-variable form yields the two equations

$$\dot{x}_1 = x_2$$

$$\dot{x}_2 = -\frac{K}{M} x_1 - \frac{B}{M} x_2 + \frac{K}{M} u$$

or

$$\dot{\mathbf{x}} = \mathbf{A}\mathbf{x} + \mathbf{b}u$$

where

$$\mathbf{A} = \begin{bmatrix} 0 & 1 \\ -\dfrac{K}{M} & -\dfrac{B}{M} \end{bmatrix} \qquad \mathbf{b} = \begin{bmatrix} 0 \\ \dfrac{K}{M} \end{bmatrix}$$

$$y = x_1 = \begin{bmatrix} 1 & 0 \end{bmatrix} \begin{bmatrix} x_1 \\ x_2 \end{bmatrix} = \mathbf{c}^T \mathbf{x}$$

Multiple-Element Mechanical Translation System

A force $f(t)$ is applied to the mass M_1 of Fig. 2-14. The sliding friction between the masses M_1 and M_2 and the surface is indicated by the viscous-friction coefficients B_1 and B_2. The system equations can be written in terms of the two displacements x_a and x_b. The mechanical network is drawn by connecting the terminals of the elements that have the same displacement (see Fig. 2-15). Since the forces at each node must add to zero, the equations are written according to the rules for node equations:

For node a: $\qquad (M_1 D^2 + B_1 D + B_3 D + K_1)x_a - (B_3 D)x_b = f$ (2-67)

For node b: $\qquad -(B_3 D)x_a + (M_2 D^2 + B_2 D + B_3 D + K_2)x_b = 0$ (2-68)

A definite pattern to these equations can be detected. Observe that K_1, M_1, B_1, and B_3 are connected to node a and that Eq. (2-67), for node a, contains all four of these terms as coefficients of x_a. Notice that element B_3 is also connected to node b and that the term $-B_3$ appears as a coefficient of x_b. When this pattern is used Eq. (2-68) can be written directly. Thus, since K_2, M_2, B_2, and B_3 are connected to

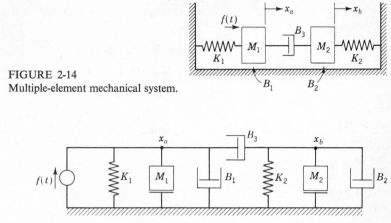

FIGURE 2-14
Multiple-element mechanical system.

FIGURE 2-15
Mechanical network for the mechanical system of Fig. 2-14.

node b, they appear as coefficients of x_b. B_3 is also connected to node a, and $-B_3$ appears as the coefficient of x_a.

The node equations for a mechanical system follow directly from the mechanical network of Fig. 2-15. They are similar in form to the node equations for an electric circuit and follow the same rules.

The block diagram representing the system in Fig. 2-14 is also given by Fig. 2-13, where again x_b is considered to be the system output. The transfer functions G_1 and G_2 are obtained by solving the equations for this system.

EXAMPLE 3 Obtain the state equations for the system of Fig. 2-14, where x_b is the system output. There are four energy-storage elements, thus the four assigned state variables are x_a, x_b, Dx_a, and Dx_b. Analyzing Eqs. (2-67) and (2-68) shows that this is a fourth-order system. Let

$$x_1 = x_b \quad \text{for spring } K_2$$

$$x_2 = \dot{x}_1 = v_b \quad \text{for mass } M_2$$

$$x_3 = x_a \quad \text{for spring } K_1$$

$$x_4 = \dot{x}_3 = v_a \quad \text{for mass } M_1$$

$$u = f$$

$$y = x_b = x_1$$

The four state variables are independent, i.e., no one state variable can be expressed in terms of the remaining state variables. The system equations are

$$\dot{\mathbf{x}} = \mathbf{A}\mathbf{x} + \mathbf{b}u \qquad y = \mathbf{c}^T\mathbf{x}$$

where

$$\mathbf{A} = \begin{bmatrix} 0 & 1 & 0 & 0 \\ -\dfrac{K_2}{M_2} & -\dfrac{B_2 + B_3}{M_2} & 0 & \dfrac{B_3}{M_2} \\ 0 & 0 & 0 & 1 \\ 0 & \dfrac{B_3}{M_1} & -\dfrac{K_1}{M_1} & -\dfrac{B_1 + B_3}{M_1} \end{bmatrix}$$

$$\mathbf{b} = \begin{bmatrix} 0 \\ 0 \\ 0 \\ \dfrac{1}{M_1} \end{bmatrix} \qquad \mathbf{c}^T = \begin{bmatrix} 1 & 0 & 0 & 0 \end{bmatrix}$$

2-7   ANALOGOUS CIRCUITS

Analogous circuits represent systems for which the differential equations have the same form. The corresponding variables and parameters in two circuits represented by equations of the same form are called *analogs*. An electric circuit can be drawn that looks like the mechanical circuit and is represented by node equations that have the same mathematical form as the mechanical equations. The analogs are listed in Table 2-4.

In this table the force f and the current i are analogs and are classified as "through" variables. There is a physical similarity since a measuring instrument must be placed in series in both cases; i.e., an ammeter and a force indicator must be placed in series with the system. Also, the velocity "across" a mechanical element is analogous to voltage across an electrical element. Again, the physical similarity is present since a measuring instrument must be placed across the system in both cases. A voltmeter must be placed across a circuit to measure voltage; it must have a point of reference. A velocity indicator must also have a point of reference. Nodes in the mechanical network are analogous to nodes in the electric network.

Table 2-4   ELECTRICAL AND MECHANICAL ANALOGS

Mechanical translation element		Electrical element	
Symbol	Quantity	Symbol	Quantity
f	Force	i	Current
$v = Dx$	Velocity	e or v	Voltage
M	Mass	C	Capacitance
K	Stiffness coefficient	$\dfrac{1}{L}$	Reciprocal inductance
B	Damping coefficient	$G = \dfrac{1}{R}$	Conductance

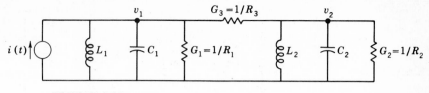

FIGURE 2-16
Electrical analog of Figs. 2-14 and 2-15.

By using the analogs of Table 2-4, the mechanical network of Fig. 2-14 is drawn in Fig. 2-16. Note that the circuit diagram of the electrical analog in Fig. 2-16 and the mechanical network in Fig. 2-15 are similar. The node equations for Fig. 2-16 are written by inspection as

$$\left(C_1 D + G_1 + G_3 + \frac{1}{L_1 D} \right) v_1 - G_3 v_2 = i \qquad (2\text{-}69)$$

$$-G_3 v_1 + \left(C_2 D + G_2 + G_3 + \frac{1}{L_2 D} \right) v_2 = 0 \qquad (2\text{-}70)$$

Equations (2-67) and (2-68) can be written in terms of the velocities instead of the displacements for a better comparison with the electric-circuit equations:

$$\left(M_1 D + B_1 + B_3 + \frac{K_1}{D} \right) D x_a - B_3 D x_b = f \qquad (2\text{-}71)$$

$$-B_3 D x_a + \left(M_2 D + B_2 + B_3 + \frac{K_2}{D} \right) D x_b = 0 \qquad (2\text{-}72)$$

A comparison of Eqs. (2-69) and (2-70) with Eqs. (2-71) and (2-72) shows that they have the same mathematical form. The solutions of the dependent variables in either set of equations must therefore also have the same mathematical form.

The advantage of the electrical analog is that it can be set up very easily in the laboratory. Also, a change in any parameter is accomplished more readily in the electric circuit. The electric circuit can readily be adjusted to produce a desired response. Then the parameters in the mechanical system must be changed by a corresponding amount to achieve the same desired response.

2-8 MECHANICAL ROTATIONAL SYSTEMS

The equations characterizing rotational systems are similar to those for translation systems. Writing torque equations parallels the writing of force equations, with the displacement, velocity, and acceleration terms now angular quantities. The applied torque is equal to the sum of the reaction torques. The three elements in a rotational system are inertia, the spring, and the dashpot. The mechanical-network representation of these elements is shown in Fig. 2-17.

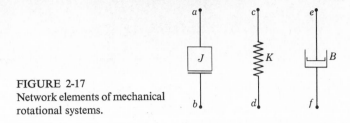

FIGURE 2-17
Network elements of mechanical
rotational systems.

The torque applied to a body having a moment of inertia J produces an angular acceleration. The reaction torque T_J is opposite to the direction of the applied torque and is equal to the product of moment of inertia and acceleration. In terms of angular displacement θ, angular velocity ω, or angular acceleration α, the torque equation is

$$T_J = J\alpha = J\,D\omega = J\,D^2\theta \qquad (2\text{-}73)$$

When a torque is applied to a spring, the spring is twisted by an angle θ. The applied torque is transmitted through the spring and appears at the other end. The reaction spring torque T_K that is produced is equal to the product of the *stiffness*, or *elastance*, K of the spring and the angle of twist. By denoting the positions of the two ends of the spring, measured from the neutral position, as θ_c and θ_d, the reaction torque is

$$T_K = K(\theta_c - \theta_d) \qquad (2\text{-}74)$$

Damping occurs whenever a body moves through a fluid, which may be a liquid or a gas such as air. To produce motion of the body a torque must be applied to overcome the reaction damping torque. The damping is represented as a dashpot with a *viscous-friction coefficient* B. The damping torque T_B is equal to the product of damping B and the relative angular velocity of the ends of the dashpot. The reaction torque of a damper is

$$T_B = B(\omega_e - \omega_f) = B(D\theta_e - D\theta_f) \qquad (2\text{-}75)$$

By drawing the mechanical network for the system first, the work of writing the differential equations can be simplified. This is done by first designating nodes that correspond to each angular displacement. Then each element is connected between the nodes that correspond to the two motions at each end of that element. The inertia elements are connected from the reference node to the node representing its position. The spring and dashpot elements are connected to the two nodes that represent the position of each end of the element. Then the torque equation is written for each node by equating the sum of the torques at each node to zero. These equations are similar to those for mechanical translation and are analogous to those for electric circuits. Several examples are shown in the following pages. The units for these elements are given in Table 2-5.

An electrical analog can be obtained for a mechanical rotational system in the manner described in Sec. 2-7. The changes in Table 2-4 are due only to the fact that rotational quantities are involved. Thus torque is the analog of current, and the moment of inertia is the analog of capacitance.

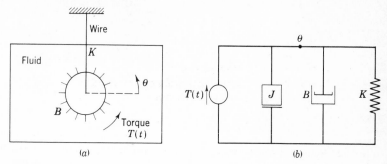

FIGURE 2-18
(a) Simple rotational system; (b) mechanical network.

Simple Mechanical Rotational System

The system shown in Fig. 2-18 has a mass with a moment of inertia J immersed in a fluid. A torque T is applied to the mass. The wire produces a reactive torque proportional to a stiffness K and to the angle of twist. The fins moving through the fluid have a damping B, which requires a torque proportional to the rate at which they are moving. The mechanical network is drawn in Fig. 2-18b. There is one node having a displacement θ; therefore, only one equation is necessary:

$$J\,D^2\theta + B\,D\theta + K\theta = T(t) \qquad (2\text{-}76)$$

Multiple-Element Mechanical Rotational System

The system represented by Fig. 2-19a has two disks which have damping between them and also between each of them and the frame. The mechanical network is drawn in Fig. 2-19b by connecting the terminals of those elements which have the same angular displacement. The torques at each node must add to zero.

Table 2-5 MECHANICAL ROTATIONAL SYMBOLS AND UNITS

Symbol	Quantity	English units	Metric units
T	Torque	Pound-feet	Newton-meters
θ	Angle	Radians	Radians
ω	Angular velocity	Radians/second	Radians/second
α	Angular acceleration	Radians/second²	Radians/second²
J†	Moment of inertia	Slug-feet² (or pound-feet-second²)	Kilogram-meter²
K	Stiffness coefficient	Pound-feet/radian	Newton-meters/radian
B	Damping coefficient	Pound-feet/(radian/second)	Newton-meters/(radian/second)

† The moment of inertia J has the dimensions of mass-distance² which have the units slug-feet² in the English system. Sometimes the units are given as pound-feet². To use the consistent set of units above, the moment of inertia must be expressed in slug-feet² by using the conversion factor 1 slug = 32.2 lb.

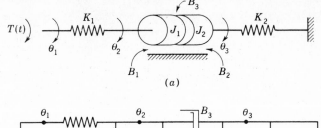

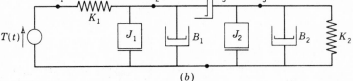

FIGURE 2-19
(*a*) Rotational system; (*b*) corresponding mechanical network.

The equations are written directly in systematized form:

For node 1:

$$K_1\theta_1 - K_1\theta_2 = T(t) \qquad (2\text{-}77)$$

For node 2:

$$-K_1\theta_1 + [J_1D^2 + (B_1 + B_3)D + K_1]\theta_2 - (B_3D)\theta_3 = 0 \qquad (2\text{-}78)$$

For node 3:

$$-(B_3D)\theta_2 + [J_2D^2 + (B_2 + B_3)D + K_2]\theta_3 = 0 \qquad (2\text{-}79)$$

These three equations can be solved simultaneously for θ_1, θ_2, and θ_3 as a function of the applied torque. If θ_3 is the system output, these equations can be solved for the following four transfer functions:

$$G_1 = \frac{\theta_1}{T} \qquad G_2 = \frac{\theta_2}{\theta_1} \qquad G_3 = \frac{\theta_3}{\theta_2}$$

$$G = G_1G_2G_3 = \frac{\theta_1}{T}\frac{\theta_2}{\theta_1}\frac{\theta_3}{\theta_2} = \frac{\theta_3}{T} \qquad (2\text{-}80)$$

The detailed and overall block-diagram representations of Fig. 2-19 are shown in Fig. 2-20. The overall transfer function given by Eqs. (2-80) is the product of the three transfer functions, which are said to be *in cascade*. This is general and applies to any number of elements in cascade when there is no loading between the blocks.

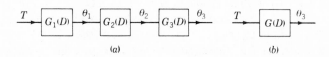

FIGURE 2-20
The detailed and overall block-diagram representations of Fig. 2-19.

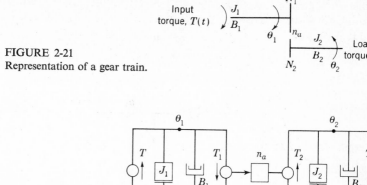

FIGURE 2-21
Representation of a gear train.

FIGURE 2-22
Mechanical network for the gear train of
Fig. 2-21.

EXAMPLE   Determine the state variables for the system of Fig. 2-19, where θ_3 is the system output y and torque T is the system input u. The spring K_1 is effectively not in the system because the torque is transmitted through the spring. Since there are three energy-storage elements J_1, J_2, and K_2, the assigned state variables are $D\theta_2$, $D\theta_3$, and θ_3. Analyzing Eqs. (2-77) to (2-79) reveals that these variables are independent state variables. Therefore, let $x_1 = \theta_3$, $x_2 = D\theta_3$, and $x_3 = D\theta_2$. If the system input is $u = \theta_1$, then the spring K_1 is included, resulting in four state variables (see Prob. 2-8).

Effective Moment of Inertia and Damping of a Gear Train

When the load is coupled to a drive motor through a gear train, the moment of inertia and damping relative to the motor are important. Since the shaft length between gears is very short, the stiffness may be considered infinite. A simple representation of a gear train is shown in Fig. 2-21. The following definitions are used:

N = number of teeth on each gear
$\omega = D\theta$ = velocity of each gear
n_a = ratio of (speed of driving shaft)/(speed of driven shaft)
θ = angular position

The mechanical network for the gear train is shown in Fig. 2-22. At each gear pair two torques are produced. For example, a restraining torque T_1 is produced on gear 1 by the rest of the gear train. There is also produced a driving torque T_2 on gear 2. T_1 is the load on gear 1 produced by the rest of the gear train. T_2 is the torque transmitted to gear 2 to drive the rest of the gear train. These torques are inversely proportional to the speeds of the respective gears. The block labeled n_a between T_1 and T_2 is used to show the relationship between them; that is, $T_2 = n_a T_1$. A transformer may be considered as the electrical analog of the gear train, with angular velocity analogous to voltage and torque analogous to current. The equations describing the system are

$$J_1 D^2\theta_1 + B_1 D\theta_1 + T_1 = T$$
$$J_2 D^2\theta_2 + B_2 D\theta_2 + T_L = T_2$$
$$n_a = \frac{\omega_1}{\omega_2} = \frac{\theta_1}{\theta_2} = \frac{N_2}{N_1} \qquad \theta_2 = \frac{\theta_1}{n_a} \qquad \text{(2-81)}$$
$$T_2 = n_a T_1$$

The equations can be combined to produce

$$J_1 D^2\theta_1 + B_1 D\theta_1 + \frac{1}{n_a}(J_2 D^2\theta_2 + B_2 D\theta_2 + T_L) = T \qquad \text{(2-82)}$$

This equation can be expressed in terms of the input position θ_1 only:

$$\left(J_1 + \frac{J_2}{n_a{}^2}\right) D^2\theta_1 + \left(B_1 + \frac{B_2}{n_a{}^2}\right) D\theta_1 + \frac{T_L}{n_a} = T \qquad \text{(2-83)}$$

Equation (2-83) represents the system performance as a function of a single dependent variable. An equivalent system is one having an equivalent moment of inertia and damping equal to

$$J_{eq} = J_1 + \frac{J_2}{n_a{}^2} \qquad \text{(2-84)}$$

$$B_{eq} = B_1 + \frac{B_2}{n_a{}^2} \qquad \text{(2-85)}$$

If the solution for θ_2 is wanted, the equation can be altered by the substitution of $\theta_1 = n_a\theta_2$. This system can be generalized for any number of gear stages. When the gear-reduction ratio is large, the load moment of inertia may contribute a negligible value to the equivalent moment of inertia.

2-9   THERMAL SYSTEMS[9]

A limited number of thermal systems can be represented by linear differential equations. The basic requirement is that the temperature of a body be considered uniform. When the body is small, this approximation is valid. Also, when the region consists of a body of air or liquid, the temperature can be considered uniform if there is perfect mixing of the fluid.

The necessary condition of equilibrium requires that the heat added to the system equal the heat stored plus the heat carried away. This can also be expressed in terms of rate of heat flow.

The symbols shown in Table 2-6 are used for thermal systems.

A thermal-system network is drawn for each system in which the thermal capacitance and thermal resistance are represented by network elements. From the thermal network the differential equations can be written.

Table 2-6 THERMAL SYMBOLS AND UNITS

Symbol	Quantity	English units	Metric units
q	Rate of heat flow	Btu/minute	Joules/second
M	Mass	Pounds	Kilograms
S	Specific heat	Btu/(pound)(°F)	Joules/(kilogram)(°C)
C	Thermal capacitance $C = MS$	Btu/°F	Joules/°C
K	Thermal conductance	Btu/(minute)(°F)	Joules/(second)(°C)
R	Thermal resistance	Degrees/(Btu/minute)	Degrees/(joule/second)
θ	Temperature	°F	°C
h	Heat energy	Btu	Joules

The additional heat stored in a body whose temperature is raised from θ_1 to θ_2 is given by

$$h = \frac{q}{D} = C(\theta_2 - \theta_1) \qquad (2\text{-}86)$$

In terms of rate of heat flow, this equation can be written as

$$q = C\,D(\theta_2 - \theta_1)$$

The thermal capacitance determines the amount of heat stored in a body. It is analogous to the electric capacitance of a capacitor in an electric circuit, which determines the amount of charge stored. The network representation of thermal capacitance is shown in Fig. 2-23a.

Rate of heat flow through a body in terms of the two boundary temperatures θ_3 and θ_4 is

$$q = \frac{\theta_3 - \theta_4}{R} \qquad (2\text{-}87)$$

The thermal resistance determines the rate of heat flow through the body. This is analogous to the resistance of a resistor in an electric circuit, which determines the current flow. The network representation of thermal resistance is shown in Fig. 2-23b.

In the thermal network the temperature is analogous to potential.

Simple Mercury Thermometer

Consider a thin glass-walled thermometer filled with mercury which has stabilized at a temperature θ_1. It is plunged into a bath of temperature θ_0 at $t = 0$. In its

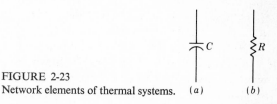

FIGURE 2-23
Network elements of thermal systems.   (a)   (b)

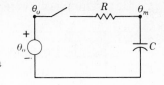

FIGURE 2-24
Simple network representation of a
thermometer.

simplest form, the thermometer can be considered to have a capacitance C which stores heat and a resistance R which limits the heat flow. The temperature of the mercury is θ_m. The flow of heat into the thermometer is

$$q = \frac{\theta_0 - \theta_m}{R}$$

The heat entering the thermometer is stored in the thermal capacitance and is given by

$$h = \frac{q}{D} = C(\theta_m - \theta_1)$$

These equations can be combined to form

$$\frac{\theta_0 - \theta_m}{RD} = C(\theta_m - \theta_1) \qquad (2\text{-}88)$$

Differentiating Eq. (2-88) and rearranging terms gives

$$RC\,D\theta_m + \theta_m = \theta_0 \qquad (2\text{-}89)$$

The thermal network is drawn in Fig. 2-24. The node equation for this circuit, with the temperature considered as a voltage, gives Eq. (2-89) directly. From this equation the transfer function is $G = \theta_m/\theta_0$. Since there is one energy-storage element, the independent state variable is $x_1 = \theta_m$ and the input is $u = \theta_0$. Thus the state equation is

$$\dot{x}_1 = -\frac{1}{RC}x_1 + \frac{1}{RC}u$$

More Exact Analysis of a Mercury Thermometer

A more exact analysis of the thermometer takes into account both the resistance R_g of the glass and the resistance R_m of the mercury. Also the glass has a capacitance

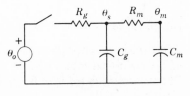

FIGURE 2-25
More exact network representation of a
thermometer.

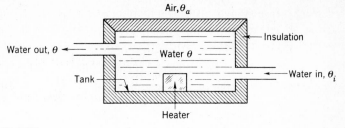

FIGURE 2-26
Electric water heater.

C_g, and the mercury has a capacitance C_m. The thermal network can be represented by Fig. 2-25, where θ_s represents the temperature at the inner surface between the glass case and the mercury. The temperature of the mercury is θ_m. The equations can be written from the thermal network:

For node s:
$$\left(\frac{1}{R_g} + \frac{1}{R_m} + C_g D\right)\theta_s - \frac{1}{R_m}\theta_m = \frac{\theta_0}{R_g} \qquad (2\text{-}90)$$

For node m:
$$-\frac{1}{R_m}\theta_s + \left(\frac{1}{R_m} + C_m D\right)\theta_m = 0 \qquad (2\text{-}91)$$

The two equations can be combined to eliminate θ_s:

$$\left[C_g C_m D^2 + \left(\frac{C_g}{R_m} + \frac{C_m}{R_g} + \frac{C_m}{R_m}\right)D + \frac{1}{R_g R_m}\right]\theta_m = \frac{1}{R_m R_g}\theta_0 \qquad (2\text{-}92)$$

Equations (2-90) to (2-92) yield the transfer functions $G_1 = \theta_s/\theta_0$, $G_2 = \theta_m/\theta_s$, and $G = \theta_m/\theta_0$, which can be represented, with the appropriate change in variable notation, by the block diagrams of Fig. 2-13.

The two energy-storage elements in Fig. 2-25 yield the assigned state variables θ_s and θ_m, which are independent. With $y = x_1 = \theta_m$, $x_2 = \theta_s$, and $u = \theta_0$, the state equation is

$$\dot{\mathbf{x}} = \begin{bmatrix} -\dfrac{1}{R_m C_m} & \dfrac{1}{R_m C_m} \\[2ex] \dfrac{1}{R_m C_g} & -\dfrac{1}{R_g C_g} - \dfrac{1}{R_m C_g} \end{bmatrix} \mathbf{x} + \begin{bmatrix} 0 \\[2ex] \dfrac{1}{R_g C_g} \end{bmatrix} u$$

Simple Heat-Transfer System

The electric water heater used in many homes to supply hot water is a good example of a simple heat-transfer problem. A sketch of such a heater is shown in Fig. 2-26. The tank is insulated to reduce heat loss to the surrounding air. The electrical heating element is turned on and off by a thermostatic switch to maintain a reference temperature. A demand from any faucet in the house causes hot water to leave and cold water to enter the tank. The necessary simplifying assumptions are as follows:

1 There is no heat storage in the insulation. This is valid since the specific heat of the insulation is small and the water-temperature variation is small.

2 All the water in the tank is at a uniform temperature. This requires perfect mixing of the water.

Definitions of the system parameters and variations are as follows:

q = rate of heat flow of heating element
q_t = rate of heat flow into water in tank
q_o = rate of heat flow carried out by hot water leaving tank
q_i = rate of heat flow carried in by cold water entering tank
q_e = rate of heat flow through tank insulation
θ = temperature of water in tank
θ_i = temperature of water entering tank
θ_a = temperature of air surrounding tank
C = thermal capacitance of water in tank
R = thermal resistance of insulation
n = water flow from tank
S = specific heat of water

There are three temperatures, θ, θ_i, and θ_a; therefore the thermal network must have three nodes. The network can be drawn first or after the equations are written. The equilibrium equation for rate of heat flow is

$$q_t + q_o - q_i + q_e = q \qquad (2\text{-}93)$$

where
$$q_t = C\,D\theta \qquad q_o = nS\theta \qquad q_i = nS\theta_i \qquad q_e = \frac{\theta - \theta_a}{R}$$

Combining these equations gives

$$C\,D\theta + nS(\theta - \theta_i) + \frac{\theta - \theta_a}{R} = q \qquad (2\text{-}94)$$

The thermal network is shown in Fig. 2-27*a*. The heat loss, due to the difference $\theta - \theta_i$, can be represented as occurring through the resistor of value $1/nS$. In this thermal network the rate of heat flow from the heater is analogous to the current from a constant-current source in an electric circuit.

There are four variables in this system, θ, θ_i, θ_a, and n. Three of them must be specified in order to solve the problem. For the special case in which n is a constant and $\theta_a = \theta_i$, Eq. (2-94) can be simplified. In terms of θ, which now is the temperature above the reference θ_a, the equation is

$$C\,D\theta + \left(nS + \frac{1}{R}\right)\theta = q \qquad (2\text{-}95)$$

The thermal network is drawn in Fig. 2-27*b*.

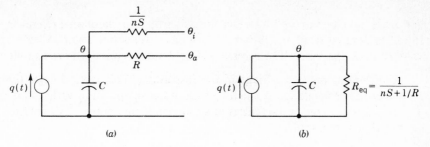

FIGURE 2-27
(a) Thermal network of an electric water heater; (b) simplified network.

2-10 HYDRAULIC LINEAR ACTUATOR

The valve-controlled hydraulic actuator is used in many applications as a power amplifier. Very little power is required to position the valve, but a large power output is controlled. The hydraulic unit is relatively small, which makes its use very attractive. Figure 2-28 shows a simple hydraulic actuator in which motion of the valve regulates the flow of oil to either side of the main cylinder. An input motion x of a few thousandths of an inch results in a large change of oil flow. The resulting difference in pressure on the main piston causes motion of the output shaft. The oil flowing in is supplied by a source which maintains a constant high pressure P_h, and the oil on the opposite side of the piston flows into the drain at low pressure P_s. The load-induced pressure P_L is the difference between the pressures on each side of the main piston:

$$P_L = P_1 - P_2 \qquad (2\text{-}96)$$

The flow of fluid through an inlet orifice is given by[10]

$$q = ca \sqrt{2g \frac{\Delta p}{\omega}} \qquad (2\text{-}97)$$

where c = orifice coefficient
a = orifice area
ω = specific weight of fluid
Δp = pressure drop across orifice
g = gravitational acceleration constant
q = rate of flow of fluid

Simplified Analysis

As a first-order approximation, it can be assumed that the orifice coefficient and the pressure drop across the orifice are constant and independent of valve position. Also, the orifice area can be expressed in terms of the valve displacement x. Equation (2-97), which gives the rate of flow of hydraulic fluid through the valve, can be rewritten as

$$q = C_x x \qquad (2\text{-}98)$$

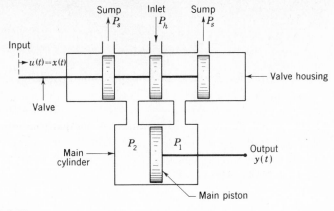

FIGURE 2-28
Hydraulic actuator.

where x is the displacement of the valve. The displacement of the main piston is proportional to the volume of fluid that enters the main cylinder. By neglecting the compressibility of the fluid and the leakage around the valve and main piston, the equation of motion of the main piston is

$$q = C_b\, Dy \qquad (2\text{-}99)$$

Combining the two equations gives

$$Dy = \frac{C_x}{C_b} x = C_1 x \qquad (2\text{-}100)$$

This analysis is essentially correct when the load reaction is small.

More Complete Analysis

When the load reaction is not negligible, a more complete analysis should take into account the pressure drop across the orifice, the leakage of oil around the piston, and the compressibility of the oil.

The pressure drop Δp across the orifice is a function of the source pressure P_h and the load pressure P_L. Since P_h is assumed constant, the flow equation is a function of valve displacement x and load pressure P_L:

$$q = f(x, P_L) \qquad (2\text{-}101)$$

The differential dq, expressed in terms of partial derivatives, is

$$dq = \frac{\partial q}{\partial x}\, dx + \frac{\partial q}{\partial P_L}\, dP_L \qquad (2\text{-}102)$$

If q, x, and P_L are measured from zero values as reference points, and if the partial derivatives are constant at the values they have at zero, the integration of Eq. (2-102) gives

$$q = \left(\frac{\partial q}{\partial x}\right)_0 x + \left(\frac{\partial q}{\partial P_L}\right)_0 P_L \qquad (2\text{-}103)$$

By defining

$$C_x \equiv \left(\frac{\partial q}{\partial x}\right)_0 \quad \text{and} \quad C_p \equiv \left(\frac{-\partial q}{\partial P_L}\right)_0$$

the flow equation for fluid entering the main cylinder can be written as

$$q = C_x x - C_p P_L \quad (2\text{-}104)$$

Both C_x and C_p have positive values. A comparison with Eq. (2-98) shows that the load pressure reduces the flow into the main cylinder. The flow of fluid into the cylinder must satisfy the continuity conditions of equilibrium. This flow is equal to the sum of the components:

$$q = q_o + q_l + q_c \quad (2\text{-}105)$$

where q_o = incompressible component (causes motion of piston)
q_l = leakage component
q_c = compressible component

The component q_o, which produces a motion y of the main piston, is

$$q_o = C_b \, Dy \quad (2\text{-}106)$$

The compressible component is derived in terms of the bulk modulus of elasticity, which is defined as the ratio of incremental stress to incremental strain. Thus

$$K_B = \frac{\Delta P_L}{\Delta V / V}$$

Solving for ΔV and dividing both sides of the equation by Δt gives

$$\frac{\Delta V}{\Delta t} = \frac{V}{K_B} \frac{\Delta P_L}{\Delta t}$$

Taking the limit as Δt approaches zero and letting $q_c = dV/dt$ gives

$$q_c = \frac{V}{K_B} DP_L \quad (2\text{-}107)$$

where V is the effective volume of fluid under compression and K_B is the bulk modulus of the hydraulic oil. The volume V at the middle position of the piston stroke is often used in order to linearize the differential equation.

The leakage component is

$$q_l = LP_L \quad (2\text{-}108)$$

where L is the leakage coefficient of the whole system.

Combining these equations gives

$$q = C_x x - C_p P_L = C_b \, Dy + \frac{V}{K_B} DP_L + LP_L \quad (2\text{-}109)$$

and rearranging terms gives

$$C_b \, Dy + \frac{V}{K_B} DP_L + (L + C_p)P_L = C_x x \quad (2\text{-}110)$$

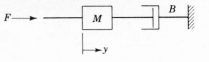

FIGURE 2-29
Load on a hydraulic piston.

The force developed by the main piston is

$$F = n_F A P_L = C P_L \qquad (2\text{-}111)$$

where n_F is the force conversion efficiency of the unit and A is the area of the main actuator piston.

An example of a specific type of load consisting of a mass and a dashpot is shown in Fig. 2-29. The equation for this system is obtained by equating the force produced by the piston, which is given by Eq. (2-111) to the reactive load forces:

$$F = M\,D^2 y + B\,Dy = C P_L \qquad (2\text{-}112)$$

Substituting the value of P_L from Eq. (2-112) into Eq. (2-110) gives the equation relating the input motion x to the response y:

$$\frac{MV}{CK_B} D^3 y + \left[\frac{BV}{CK_B} + \frac{M}{C}(L + C_p) \right] D^2 y + \left[C_b + \frac{B}{C}(L + C_p) \right] Dy = C_x x \qquad (2\text{-}113)$$

The analysis above is based on perturbations about the reference set of values $x = 0$, $q = 0$, $P_L = 0$. For the entire range of motion x of the valve, the quantities $\partial q / \partial x$ and $-\partial q / \partial P_L$ can be determined experimentally. Although they are not constant at values equal to the values C_x and C_p at the zero reference point, average values can be assumed in order to simulate the system by linear equations. For conservative design the volume V is determined for the main piston at the midpoint.

To write the state equation for the hydraulic actuator and load of Figs. 2-28 and 2-29 the energy-related variables must be determined. The mass M yields one energy-storage variable, the output velocity Dy. The compressible component q_c represents an energy-storage element in a hydraulic system. The compression of a fluid produces stored energy, just as in the compression of a spring. The equation for hydraulic energy is

$$E(t) = \int_0^t P(\tau) q(\tau)\, d\tau \qquad (2\text{-}114)$$

where $P(t)$ is the pressure and $q(t)$ is the rate of flow of fluid. The energy storage in a compressed fluid is obtained in terms of the bulk modulus of elasticity K_B. Combining Eq. (2-107) with Eq. (2-114) for a constant volume yields

$$E_c(P_L) = \int_0^{P_L} \frac{V}{K_B} P_L\, dP_L = \frac{V}{2K_B} P_L^2 \qquad (2\text{-}115)$$

The stored energy in a compressed fluid is proportional to the pressure P_L squared; thus P_L may be used as a physical state variable.

Since the output quantity in this system is the position y, it is necessary to increase the state variables to three. Further evidence of the need for three state variables is

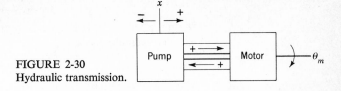

FIGURE 2-30
Hydraulic transmission.

the fact that Eq. (2-113) is a third-order equation. Therefore, in this example, let $x_1 = y$, $x_2 = Dy = \dot{x}_1$, $x_3 = P_L$, and $u = x$. Then, from Eqs. (2-110) and (2-112), the state and output equations are

$$\dot{\mathbf{x}} = \begin{bmatrix} 0 & 1 & 0 \\ 0 & -\dfrac{B}{M} & \dfrac{C}{M} \\ 0 & -\dfrac{C_b K_B}{V} & -\dfrac{K_B(L + C_L)}{V} \end{bmatrix} \mathbf{x} + \begin{bmatrix} 0 \\ 0 \\ \dfrac{K_B C_x}{V} \end{bmatrix} u \qquad (2\text{-}116)$$

$$y = \begin{bmatrix} 1 & 0 & 0 \end{bmatrix}\mathbf{x} = x_1 \qquad (2\text{-}117)$$

The effect of augmenting the state variables by adding the piston displacement $x_1 = y$ is to produce a singular system; that is, $|\mathbf{A}| = 0$. This property does not appear if a spring is added to the load, as shown in Prob. 2-10. In that case $x_1 = y$ is an independent state variable.

2-11 POSITIVE-DISPLACEMENT ROTATIONAL HYDRAULIC TRANSMISSION[11]

When a large torque is required in a control device, it is possible to use a hydraulic transmission. The transmission contains a variable-displacement pump driven at constant speed. It pumps a quantity of oil that is proportional to a control stroke and independent of back pressure. The direction of fluid flow is determined by the direction of displacement of the control stroke. The hydraulic motor has an angular velocity proportional to the volumetric flow rate and in the direction of the oil flow from the pump.

The assumption is made that over a limited range of operation the hydraulic transmission is linear. A schematic picture of the system is shown in Fig. 2-30.

The following symbols are used:

q_p = total volumetric flow rate from pump
q_m = volumetric flow rate through motor
q_l = volumetric leakage flow rate of both pump and motor
q_c = compressibility flow rate
x = control stroke (x varies from 0 to ± 1)
ω_p = angular velocity of pump shaft (constant)
ω_m = angular velocity of motor shaft (variable)

FIGURE 2-31
Inertia load on a hydraulic transmission.

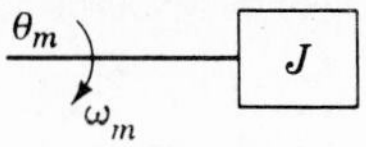

$\theta_m = $ angular position of motor shaft
$d_p = $ volumetric displacement (at $x = 1$) per unit angular displacement
$d_m = $ volumetric motor displacement per unit angular displacement
$L = $ leakage coefficient of complete system, $(\text{ft}^3/\text{s})/(\text{lb}/\text{ft}^2)$
$V = $ total volume of liquid under compression, ft^3
$K_B = $ bulk modulus of oil, lb/ft^2
$P_L = $ load-induced pressure drop across motor, lb/ft^2
$C = $ motor torque constant, ft^3
$n_T = $ torque conversion efficiency of motor

The basic equation is based on the fact that the fluid flow rate from the pump equals the sum of the flow rates in the system. This is given by

$$q_p = q_m + q_l + q_c \qquad (2\text{-}118)$$

where

$$q_p = x\, d_p\omega_p \qquad q_m = d_m\omega_m \qquad q_l = LP_L \qquad q_c = DV = \frac{V}{K_B} DP_L$$

Combining these flow rates into the original equation produces

$$d_m\omega_m + LP_L + \frac{V}{K_B} DP_L = d_p\omega_p x \qquad (2\text{-}119)$$

The torque produced at the motor shaft is

$$T = n_T\, d_m P_L = CP_L \qquad (2\text{-}120)$$

Since the torque required depends on the load, two cases are considered.

Case 1   Inertia Load

Figure 2-31 shows an inertia load on a hydraulic transmission. Equating the generated torque to the load reaction torque gives

$$T = J\, D^2\theta_m = CP_L \qquad (2\text{-}121)$$

When P_L from Eq. (2-121) is inserted in Eq. (2-119), the result is

$$\frac{VJ}{K_BC} D^3\theta_m + \frac{LJ}{C} D^2\theta_m + d_m\, D\theta_m = d_p\omega_p x \qquad (2\text{-}122)$$

This equation can be solved for the motor position θ_m in terms of the stroke position x.

Based upon the analogs for the linear hydraulic actuator, the three state variables for the rotational hydraulic system are $x_1 = \theta_m$, $x_2 = \omega_m = D\theta_m = \dot{x}_1$, and $x_3 = P_L$. The state equation can be obtained in the usual manner.

FIGURE 2-32
Spring and inertia load on a hydraulic
transmission.

Case 2 Inertia Load Coupled Through a Spring

Figure 2-32 shows a spring and inertia load on a hydraulic transmission. Equating the torque generated by the hydraulic motor to the load torque and then solving for P_L in terms of θ_m gives

$$T = K(\theta_m - \theta_L) = J\,D^2\theta_L = CP_L \qquad (2\text{-}123)$$

$$P_L = \frac{KJ\,D^2\theta_m}{(JD^2 + K)C} \qquad (2\text{-}124)$$

Using the value of P_L from Eq. (2-124) in the system equation (2-119) relates the motor output position θ_m to the stroke input x by

$$\left[\left(\frac{J}{K} + \frac{VJ}{K_B Cd_m}\right)D^3 + \frac{LJ}{Cd_m}D^2 + D\right]\theta_m = \frac{d_p\omega_p}{d_m}\left(\frac{J}{K}D^2 + 1\right)x \qquad (2\text{-}125)$$

The solution of these differential equations is considered in the next chapter. It will be found that the leakage L is essential for stable response. Without leakage the system would have a sustained steady-state oscillation. Therefore, rather than rely on accidental leakage, the designer provides for a positive and finite leakage of hydraulic fluid around the pistons of the motor. This leakage can also serve to lubricate the piston joints.

The transfer function for this system, obtained from Eq. (2-125), is

$$G = \frac{(d_p\omega_p/d_m)(JD^2/K + 1)}{(J/K + VJ/K_B Cd_m)D^3 + (LJ/Cd_m)D^2 + D} \qquad (2\text{-}126)$$

Note that this transfer function contains the operator D in the numerator. In such systems the physical variables may not always be used as the state variables. This problem is discussed in more detail in Chaps. 3 and 4, in which the phase and canonical state variables are introduced.

2-12 LIQUID-LEVEL SYSTEM[12]

Figure 2-33 represents a two-tank liquid-level control system. Definitions of the system parameters are

$$q_i, q_1, q_2 = \text{rates of flow of fluid}$$
$$R_1, R_2 = \text{flow resistance}$$
$$h_1, h_2 = \text{heights of fluid level}$$
$$A_1, A_2 = \text{cross-sectional tank areas}$$

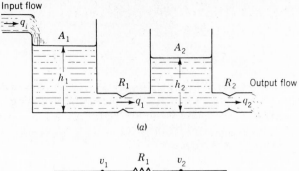

(a)

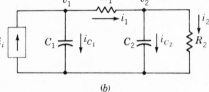

(b)

FIGURE 2-33
Liquid-level system and its equivalent electrical analog.

The following basic linear relationships hold for this system:

$$q = \frac{h}{R} = \text{rate of flow through orifice} \qquad (2\text{-}127)$$

$$q_n = \text{(tank input rate of flow)} - \text{(tank output rate of flow)}$$
$$= \text{net tank rate of flow} = A \, Dh \qquad (2\text{-}128)$$

Applying Eq. (2-128) to tanks 1 and 2 yields, respectively,

$$q_{n_1} = A_1 \, Dh_1 = q_i - q_1 = q_i - \frac{h_1 - h_2}{R_1} \qquad (2\text{-}129)$$

$$q_{n_2} = A_2 \, Dh_2 = q_1 - q_2 = \frac{h_1 - h_2}{R_1} - \frac{h_2}{R_2} \qquad (2\text{-}130)$$

These equations can be solved simultaneously to obtain the transfer functions h_1/q_i and h_2/q_i.

The energy stored in each tank represents potential energy, which is equal to $\rho A h^2/2$, where ρ is the fluid density coefficient. Since there are two tanks, the system has two energy-storage elements, whose energy-storage variables are h_1 and h_2. Letting $x_1 = h_1$, $x_2 = h_2$, and $u = q_i$ in Eqs. (2-129) and (2-130) reveals that x_1 and x_2 are independent state variables. Thus the state equation is

$$\dot{\mathbf{x}} = \begin{bmatrix} -\dfrac{1}{R_1 A_1} & \dfrac{1}{R_1 A_1} \\[3ex] \dfrac{1}{R_1 A_2} & -\dfrac{1}{R_1 A_2} - \dfrac{1}{R_2 A_2} \end{bmatrix} \mathbf{x} + \begin{bmatrix} \dfrac{1}{A_1} \\[2ex] 0 \end{bmatrix} u \qquad (2\text{-}131)$$

The levels of the two tanks are the outputs of the system. Letting $y_1 = x_1 = h_1$ and $y_2 = x_2 = h_2$ yields

$$\mathbf{y} = \begin{bmatrix} 1 & 0 \\ 0 & 1 \end{bmatrix} \mathbf{x} \qquad (2\text{-}132)$$

The potential energy of a tank can be represented as a capacitor whose stored potential energy is $Cv_c{}^2/2$; thus the electrical analog of h is v_c. As a consequence, an analysis of Eqs. (2-129) and (2-130) yields the analogs between the hydraulic and electrical quantities listed in Table 2-7. The analogous electrical equations are

$$C_1 \, Dv_1 = i_i - \frac{v_1 - v_2}{R_1} \qquad (2\text{-}133)$$

$$C_2 \, Dv_2 = \frac{v_1 - v_2}{R_1} - \frac{v_2}{R_2} \qquad (2\text{-}134)$$

These two equations yield the equivalent electric circuit of Fig. 2-33b.

2-13   ROTATING POWER AMPLIFIERS[13,14]

A dc generator can be used as a power amplifier in which the power required to excite the field circuit is lower than the power output rating of the armature circuit. The voltage e_g induced in the armature circuit is directly proportional to the product of the flux ϕ set up by the field and the speed of rotation ω of the armature. This is expressed by

$$e_g = K_1 \phi \omega \qquad (2\text{-}135)$$

The flux is a function of field current and the type of iron used in the field. A typical magnetization curve showing flux as a function of field current is given in Fig. 2-34. Up to saturation the relation is approximately linear, and the flux is directly proportional to field current:

$$\phi = K_2 i_f \qquad (2\text{-}136)$$

Combining these equations gives

$$e_g = K_1 K_2 \omega i_f \qquad (2\text{-}137)$$

When used as a power amplifier the armature is driven at constant speed, and this equation becomes

$$e_g = K_g i_f \qquad (2\text{-}138)$$

Table 2-7   HYDRAULIC AND ELECTRICAL ANALOGS

Hydraulic element		Electrical element	
Symbol	Quantity	Symbol	Quantity
q_i	Input flow rate	i_t	Current source
h	Height	v_c	Capacitor voltage
A	Tank area	C	Capacitance
R	Flow resistance	R	Resistance

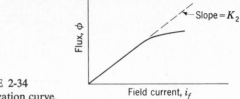

FIGURE 2-34
Magnetization curve.

A generator is represented schematically in Fig. 2-35, in which L_f and R_f and L_g and R_g are the inductance and resistance of the field and armature circuits, respectively. The equations for the generator are

$$e_f = (L_fD + R_f)i_f$$
$$e_g = K_g i_f \qquad (2\text{-}139)$$
$$e_t = e_g - (L_gD + R_g)i_a$$

The armature current depends on the load connected to the generator terminals. Combining the first two equations gives

$$(L_fD + R_f)e_g = K_g e_f \qquad (2\text{-}140)$$

Equation (2-140) relates the generated voltage e_g to the input field voltage e_f.

2-14   DC SERVOMOTOR

A current-carrying conductor located in a magnetic field experiences a force proportional to the magnitude of the flux, the current, the length of the conductor, and the sine of the angle between the conductor and the direction of the flux. When the conductor is a fixed distance from an axis about which it can rotate, a torque is produced that is proportional to the product of the force and the radius. In a motor the resultant torque is the sum of the torques produced by each conductor. For any given motor the only two adjustable quantities are the flux and armature current. The developed torque can be expressed as

$$T(t) = K_3 \phi i_m \qquad (2\text{-}141)$$

In this case, to avoid confusion with t for time, the capital letter T is used to indicate torque. It may denote either a constant or a function that varies with time. There are two modes of operation of a servomotor. In one mode the field current is held

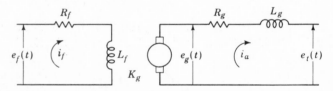

FIGURE 2-35
Schematic diagram of a generator.

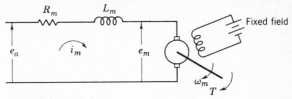

FIGURE 2-36
Circuit diagram of a dc motor.

constant, and an adjustable voltage is applied to the armature. In the second mode the armature current is held constant, and an adjustable voltage is applied to the field. These methods of operation are considered separately.

Armature Control

A constant field current is obtained by separately exciting the field from a fixed dc source. The flux is produced by the field current and is therefore constant. Thus the torque is proportional only to the armature current and is given as

$$T(t) = K_T i_m \qquad (2\text{-}142)$$

When the motor armature is rotating, there is induced a voltage e_m which is proportional to the product of flux and speed. Because the polarity of this voltage opposes the applied voltage e_a, it is commonly called the *back emf*. Since the flux is held constant, the induced voltage e_m is directly proportional to the speed ω_m:

$$e_m = K_1 \phi \omega_m = K_b \omega_m = K_b \, D\theta_m \qquad (2\text{-}143)$$

Control of the motor speed is obtained by adjusting the voltage applied to the armature. Its polarity determines the direction of the armature current and therefore the direction of the torque generated. This, in turn, determines the direction of rotation of the motor. A circuit diagram of the armature-controlled dc motor is shown in Fig. 2-36. The armature inductance and resistance are labeled L_m and R_m. The voltage equation of the armature circuit is

$$L_m \, Di_m + R_m i_m + e_m = e_a \qquad (2\text{-}144)$$

The current in the armature produces the required torque according to Eq. (2-142). The required torque depends on the load connected to the motor shaft. If the load consists only of a moment of inertia and damper (friction), as shown in Fig. 2-37, the torque equation can be written:

$$J \, D\omega_m + B\omega_m = T(t) \qquad (2\text{-}145)$$

The required armature current i_m can be obtained by equating the generated torque of Eq. (2-142) to the required load torque of Eq. (2-145). Inserting this current and the

FIGURE 2-37
Inertia and friction as a motor load.

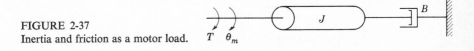

back emf from Eq. (2-143) into Eq. (2-144) produces the system equation in terms of the velocity ω_m:

$$\frac{L_m J}{K_T} D^2 \omega_m + \frac{L_m B + R_m J}{K_T} D\omega_m + \left(\frac{R_m B}{K_T} + K_b\right)\omega_m = e_a \qquad (2\text{-}146)$$

This equation can also be written in terms of motor position θ_m:

$$\frac{L_m J}{K_T} D^3 \theta_m + \frac{L_m B + R_m J}{K_T} D^2 \theta_m + \frac{R_m B + K_b K_T}{K_T} D\theta_m = e_a \qquad (2\text{-}147)$$

There are two energy-storage elements, J and L_m, for the system represented by Figs. 2-36 and 2-37. Designating the independent state variables as $x_1 = \omega_m$ and $x_2 = i_m$ and the input as $u = e_a$ yields the state equation

$$\dot{\mathbf{x}} = \begin{bmatrix} -\dfrac{B}{J} & \dfrac{K_T}{J} \\[2ex] -\dfrac{K_b}{L_m} & -\dfrac{R_m}{L_m} \end{bmatrix} \mathbf{x} + \begin{bmatrix} 0 \\[2ex] \dfrac{1}{L_m} \end{bmatrix} u \qquad (2\text{-}148)$$

If motor position θ_m is the output, another differential equation is required; that is, $\dot{\theta}_m = \omega$. If the solution of θ_m is required, then the system is of third order. A third state variable, $x_3 = \theta_m$, must therefore be added.

The transfer functions ω_m/e_a and θ_m/e_a can be obtained from Eqs. (2-146) and (2-147). The armature inductance is small and can usually be neglected, $L_m \approx 0$. Equation (2-147) is thus reduced to a second-order equation. The corresponding transfer function has the form

$$G = \frac{\theta_m}{e_a} = \frac{K_m}{D(T_m D + 1)} \qquad (2\text{-}149)$$

Field Control

If the armature current i_m is constant, the torque $T(t)$ is proportional only to the flux ϕ. In the unsaturated region (see Fig. 2-34) the flux is directly proportional to the field current [see Eq. (2-136)]. Therefore the torque equation is written as

$$T(t) = K_3 \phi i_m = K_3 K_2 i_m i_f = K_f i_f \qquad (2\text{-}150)$$

Control of the motor speed is obtained by adjusting the voltage applied to the field. Its magnitude and polarity determine the magnitude of the torque and the direction of rotation. A circuit diagram of the field-controlled dc motor is shown in Fig. 2-38. The field winding inductance and resistance are labeled L_f and R_f. The voltage equation of the field circuit is

$$L_f D i_f + R_f i_f = e_f \qquad (2\text{-}151)$$

If the load consists of moment of inertia and viscous friction, as shown in Fig. 2-37, the torque required to drive the load is given by Eq. (2-145). By combining Eqs. (2-145), (2-150), and (2-151), the system equation in terms of the motor velocity is

$$(L_f D + R_f)(JD + B)\omega_m = K_f e_f \qquad (2\text{-}152)$$

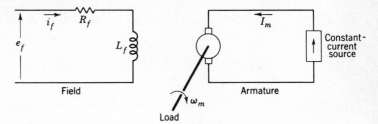

FIGURE 2-38
Circuit diagram of a field-controlled dc motor.

An advantage of field control over armature control is that the power required by the field is much smaller than that required by the armature. Since this power is usually supplied by an amplifier, smaller power capacity is required in the amplifier.

Although a constant armature current i_m is specified in the derivation above, this is not easily achieved. If the armature is connected to a fixed dc voltage source E_a, the armature current depends on the back emf, as shown by

$$i_m = \frac{E_a - e_m}{R_m} = \frac{E_a - K_2\phi\omega_m}{R_m}$$

For the range of speed that the motor undergoes, the armature circuit may be designed so that $E_a \gg e_m$. Thus

$$i_m \approx \frac{E_a}{R_m} \qquad (2\text{-}153)$$

A practical circuit for field control of a dc motor with a split field is shown in Fig. 2-39. The field is energized by a balanced amplifier which has the input e. When this input is zero, the currents i_1 and i_2 are equal. Since they flow in opposite directions, the net flux is zero, thus the torque is zero and the motor is stationary. If e is not zero, one of the currents increases and the other decreases in proportion to the magnitude of e. The resulting flux is proportional to the magnitude of e, and its direction depends on the polarity of e. The size and direction of the generated torque and the resulting speed therefore respond to the magnitude and polarity of the input e.

In this example, using the independent state variables $x_1 = \omega_m$ and $x_2 = i_f$, the state equation can be readily obtained.

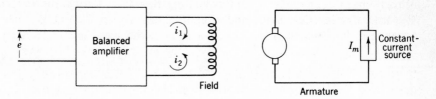

FIGURE 2-39
Circuit diagram of a split-field-controlled dc motor.

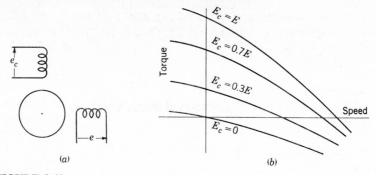

FIGURE 2-40
(*a*) Schematic diagram of a two-phase induction motor; (*b*) servomotor characteristics.

2-15 AC SERVOMOTOR[15]

An ac servomotor is basically a two-phase induction motor that has two stator field coils placed 90 electrical degrees apart, as shown in Fig. 2-40*a*. In a two-phase motor the ac voltages e and e_c are equal in magnitude and separated by a phase angle of 90°. A two-phase induction motor runs at a speed slightly below the synchronous speed and is essentially a constant-speed motor. The synchronous speed is determined by the number of poles produced by the stator windings and the frequency of the voltage applied to the stator windings.

When the unit is used as a servomotor, the speed must be proportional to an input voltage. The two-phase motor can be used as a servomotor by applying an ac voltage e of fixed amplitude to one of the motor windings. When the other voltage e_c is varied, the torque and speed are a function of this voltage. Figure 2-40*b* shows a set of torque-speed curves for various control voltages.

It is important to note that the curve for zero control-field voltage goes through the origin and that the slope is negative. This means that when the control-field voltage becomes zero, the motor develops a decelerating torque, causing it to stop. The curves show a large torque at zero speed. This is a requirement for a servomotor in order to provide rapid acceleration. It is accomplished in an induction motor by building the rotor with a high resistance.

The torque-speed curves are not straight lines. Therefore, a linear differential equation cannot be used to represent the exact motor characteristics. Sufficient accuracy may be obtained by approximating the characteristics by straight lines. The following analysis is based on this approximation.

The torque generated is a function of both the speed ω and the control-field voltage e_c. In terms of partial derivatives, the torque equation is approximated by effecting a double Taylor series expansion of $T(e_c,\omega)$ about the origin and keeping only the linear terms:

$$\left.\frac{\partial T}{\partial e_c}\right|_{\text{origin}} e_c + \left.\frac{\partial T}{\partial \omega}\right|_{\text{origin}} \omega = T(e_c,\omega) \qquad (2\text{-}154)$$

If the torque-speed motor curves are approximated by parallel straight lines, the partial-derivative coefficients of Eq. (2-154) are constants which can be evaluated from the graph. Let

$$\frac{\partial T}{\partial e_c} = K_c \quad \text{and} \quad \frac{\partial T}{\partial \omega} = K_\omega \quad (2\text{-}155)$$

For a load consisting of a moment of inertia and damping, the load torque required is

$$T_L = J\,D\omega + B\omega \quad (2\text{-}156)$$

Since the generated and load torques must be equal, Eqs. (2-154) and (2-156) are equated:

$$K_c e_c + K_\omega \omega = J\,D\omega + B\omega$$

Rearranging terms gives

$$J\,D\omega + (B - K_\omega)\omega = K_c e_c \quad (2\text{-}157)$$

In terms of position θ, this equation can be written as

$$J\,D^2\theta + (B - K_\omega)\,D\theta = K_c e_c \quad (2\text{-}158)$$

In order for the system to be stable (see Chap. 3) the coefficient $B - K_\omega$ must be positive. Observation of the motor characteristics shows that $K_\omega = \partial T/\partial \omega$ is negative; therefore the stability requirement is satisfied.

Analyzing Eq. (2-157) reveals that this system has only one energy-storage element J. Thus the state equation, where $x_1 = \omega$ and $u = e_c$, is

$$\dot{x}_1 = -\frac{B - K_\omega}{J}\,x_1 + \frac{K_c}{J}\,u \quad (2\text{-}159)$$

2-16 LAGRANGE'S EQUATION

Previous sections show the application of Kirchhoff's laws for writing the differential equations of electric networks and the application of Newton's laws for writing the equations of motion of mechanical systems. These laws can be applied, depending on the complexity of the system, with relative ease. In many instances there are systems that contain both electric and mechanical components. The use of Lagrange's equation provides a systematic unified approach for handling a broad class of physical systems, no matter how complex their structure.[16]

Lagrange's equation is given by

$$\frac{d}{dt}\left(\frac{\partial T}{\partial \dot{q}_n}\right) - \frac{\partial T}{\partial q_n} + \frac{\partial D}{\partial \dot{q}_n} + \frac{\partial V}{\partial q_n} = Q_n \quad n = 1, 2, 3, \ldots \quad (2\text{-}160)$$

where T = total kinetic energy of system
$\ D$ = dissipation function of system
$\ V$ = total potential energy of system
$\ Q_n$ = generalized applied force at the coordinate n
$\ q_n$ = generalized coordinate
$\ \dot{q}_n = dq_n/dt$ (generalized velocity)

and $n = 1, 2, 3, \ldots$ denote the number of independent coordinates or degrees of freedom which exist in the system. The total kinetic energy T includes all energy terms, regardless of whether they are electrical or mechanical. The dissipation function D represents one-half the rate at which energy is dissipated as heat; dissipation is produced by friction in mechanical systems and by resistance in electric circuits. The total potential energy stored in the system is designated by V. The forcing functions applied to a system are designated by Q_n; they take the form of externally applied forces or torques in mechanical systems and appear as voltage or current sources in electric circuits. The application of Lagrange's equation is illustrated by the following example.

EXAMPLE *Electromechanical system with capacitive coupling* Figure 2-41 shows a system in which mechanical motion is converted into electric energy. This represents the action which takes place in a capacitor microphone. Plate a of the capacitor is fastened rigidly to the frame. Sound waves impinge upon and exert a force on plate b of mass M, which is suspended from the frame by a spring K and which has damping B. The output voltage which appears across the resistor R is intended to reproduce electrically the sound-wave patterns which strike the plate b.

At equilibrium, with no external force exerted on plate b, there is a charge q_0 on the capacitor. This produces a force of attraction between the plates so that the spring is stretched by an amount x_1 and the space between the plates is x_0. When sound waves exert a force on plate b there will be a resulting motion x which is measured from the equilibrium position. The distance between the plates will then be $x_0 - x$, and the charge on the plates will be $q_0 + q$.

The capacitance is approximated by

$$C = \frac{\varepsilon A}{x_0 - x} \quad \text{and} \quad C_0 = \frac{\varepsilon A}{x_0}$$

where ε is the dielectric constant for air and A is the area of the plate.

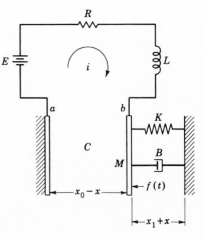

FIGURE 2-41
Electromechanical system with
capacitive coupling.

The energy expressions for this system are

$$T = \tfrac{1}{2}L\dot{q}^2 + \tfrac{1}{2}M\dot{x}^2$$

$$D = \tfrac{1}{2}R\dot{q}^2 + \tfrac{1}{2}B\dot{x}^2$$

$$V = \frac{1}{2C}(q_0 + q)^2 + \tfrac{1}{2}K(x_1 + x)^2$$

$$= \frac{1}{2\varepsilon A}(x_0 - x)(q_0 + q)^2 + \tfrac{1}{2}K(x_1 + x)^2$$

The method is simple and direct. It is merely necessary to include all the energy terms, whether electrical or mechanical. The electromechanical coupling in this example appears in the potential energy. Here, the presence of charge on the plates of the capacitor exerts a force on the mechanical system. Also, motion of the mechanical system produces an equivalent emf in the electric circuit.

The two degrees of freedom are the displacement x of plate b and the charge q on the capacitor. Applying Lagrange's equation twice gives

$$M\ddot{x} + B\dot{x} - \frac{1}{2\varepsilon A}(q_0 + q)^2 + K(x_1 + x) = f(t) \qquad (2\text{-}161)$$

$$L\ddot{q} + R\dot{q} + \frac{1}{\varepsilon A}(x_0 - x)(q_0 + q) = E \qquad (2\text{-}162)$$

These equations are nonlinear. However, a good linear approximation can be obtained when it is realized that x and q are very small quantities and therefore the x^2, q^2, and xq terms can be neglected. This gives

$$(q_0 + q)^2 \approx q_0{}^2 + 2q_0 q$$

$$(x_0 - x)(q_0 + q) \approx x_0 q_0 - q_0 x + x_0 q$$

With these approximations the system equations become

$$M\ddot{x} + Kx_1 + Kx - \frac{q_0{}^2}{2\varepsilon A} - \frac{2q_0 q}{2\varepsilon A} + B\dot{x} = f(t) \qquad (2\text{-}163)$$

$$L\ddot{q} + \frac{x_0 q_0}{\varepsilon A} - \frac{q_0 x}{\varepsilon A} + \frac{x_0 q}{\varepsilon A} + R\dot{q} = E \qquad (2\text{-}164)$$

From the first equation, by setting $f(t) = 0$ and taking steady-state conditions, the result is

$$Kx_1 - \frac{q_0{}^2}{2\varepsilon A} = 0$$

This simply equates the force on the spring and the force due to the charges at the equilibrium condition. Similarly, in the second equation at equilibrium

$$\frac{x_0 q_0}{\varepsilon A} = \frac{q_0}{C_0} = E$$

Therefore the two system equations can be written in linearized form as

$$M\ddot{x} + B\dot{x} + Kx - \frac{q_0}{\varepsilon A}q = f(t) \qquad (2\text{-}165)$$

$$L\ddot{q} + R\dot{q} + \frac{q}{C_0} - \frac{q_0}{\varepsilon A}x = 0 \qquad (2\text{-}166)$$

These equations show that $q_0/\varepsilon A$ is the coupling factor between the electrical and mechanical portions of the system.

Another form of electromechanical coupling exists when current in a coil produces a force which is exerted on a mechanical system and, simultaneously, motion of a mass induces an emf in an electric circuit. In that case the kinetic energy includes a term

$$T = l\beta N_c xi = Uxi \qquad (2\text{-}167)$$

where l = length of coil
$\quad \beta$ = flux density produced by a permanent magnet which links the coil
$\quad N_c$ = number of turns in coil
$\quad x$ = displacement of coil
$\quad i$ = current through coil

The energy for this system, which is shown in Table 2-8, may also be considered potential energy. This influences the sign on the corresponding term in the differential equation.

The main advantage of Lagrange's equation is the use of a single systematic procedure, eliminating the need to consider separately Kirchhoff's laws for the

Table 2-8 IDENTIFICATION OF ENERGY FUNCTIONS FOR
ELECTROMECHANICAL ELEMENTS

Definition	Element and symbol	Kinetic energy T
x = relative motion i = current in coil $U = l\beta n$ (electromechanical coupling constant) β = flux density produced by permanent magnet l = length of coil N_c = no. of turns in coil	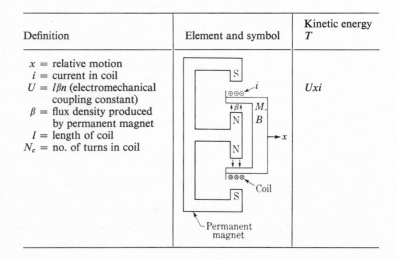	Uxi

electrical aspects and Newton's law for the mechanical aspects of the system in formulating the statements of equilibrium. Once this procedure is mastered, the differential equations which describe the system are readily obtained.

2-17 SUMMARY

The examples in this chapter cover many of the basics of control systems. In order to write the differential and state equations, the basic laws governing performance are first stated for electric, mechanical, thermic, and hydraulic systems. These basic laws are then applied to specific devices, and their differential equations of performance are obtained. The basic matrix, state, transfer-function, and block-diagram concepts are introduced. Lagrange's equation has been introduced to provide a systematized method for writing the differential equations of electrical, mechanical, and electromechanical systems. This chapter constitutes a reference for the reader who wants to review the fundamental concepts involved in writing differential equations of performance.

REFERENCES

1 DeRusso, P. M., et al.: "State Variables for Engineers," Wiley, New York, 1965.

2 Blackburn, J. F. (ed.): "Components Handbook," McGraw-Hill, New York, 1948.

3 Stout, T. M.: A Block Diagram Approach to Network Analysis, *Trans. AIEE*, vol. 71, pp. 255–260, 1952.

4 Wylie, C. R., Jr.: "Advanced Engineering Mathematics," 4th ed., McGraw-Hill, New York, 1975.

5 Kinariwala, B., et al.: "Linear Circuits and Computation," Wiley, New York, 1973.

6 Gantmacher, F. R.: "Applications of the Theory of Matrices," Wiley-Interscience, New York, 1959.

7 Kalman, R. E.: Mathematical Description of Linear Dynamical Systems, *J. Soc. Ind. Appl. Math.*, ser. A, *Control*, vol. 1, no. 2, 1963.

8 Gardner, M. F., and J. L. Barnes: "Transients in Linear Systems," chap. 2, Wiley, New York, 1942.

9 Hornfeck, A. J.: Response Characteristics of Thermometer Elements, *Trans. ASME*, vol. 71, pp. 121–132, 1949.

10 "Flow Meters: Their Theory and Application," American Society of Mechanical Engineers, New York, 1937.

11 Newton, G. C., Jr.: Hydraulic Variable Speed Transmissions as Servomotors, *J. Franklin Inst.*, vol. 243, no. 6, pp. 439–469, June 1947.

12 Takahashi, Y., et al.: "Control and Dynamic Systems," Addison-Wesley, Reading, Mass., 1970.

13 Saunders, R. M.: The Dynamo Electric Amplifier: Class A Operation, *Trans. AIEE*, vol. 68, pp. 1368–1373, 1949.

14 Litman, B.: An Analysis of Rotating Amplifiers, *Trans. AIEE*, vol. 68, pt. II, pp. 1111–1117, 1949.

15 Hopkin, A. M.: Transient Response of Small Two-Phase Servomotors, *Trans. AIEE*, vol. 70, pp. 881–886, 1951.

16 Ogar, G. W., and J. J. D'Azzo: A Unified Procedure for Deriving the Differential Equations of Electrical and Mechanical Systems, *IRE Trans. Educ.*, vol. E-5, no. 1, pp. 18–26, March 1962.

3

SOLUTION OF DIFFERENTIAL EQUATIONS

3-1 INTRODUCTION

The general solution of a linear differential equation[1,2] is the sum of two components, the particular integral and the complementary function. Those terms of the general solution which disappear as t approaches infinity constitute the transient or natural component of the solution. If the ultimate component is periodic (with finite period) or constant (a periodic function of infinite period), it can be termed the steady-state component. Often the particular integral is the steady-state component of the solution of the differential equation; and the complementary function, which is the solution of the corresponding homogeneous equation, is the transient component of the solution. Often the steady-state component of the response has the same form as the driving function. In this book the particular integral is called the steady-state solution even when it is not periodic. The form of the transient component of the response depends only on the roots of the characteristic equation and sometimes on the initial conditions. The instantaneous value of the transient component depends on the initial conditions, the roots of the characteristic equation, and the instantaneous value of the steady-state component.

This chapter covers methods of determining the steady-state and the transient components of the solution. These components are first determined separately and then added to form the complete solution. Analysis of the transient solution should develop in the student a feel for the solution to be expected.

The method of solution is next applied to the matrix state equation. A general format is obtained for the complementary solution in terms of the state transition matrix (STM). Then the complete solution is obtained as a function of the STM and the input forcing function. Simulation diagrams are presented as a means of representing a system for which an overall differential equation is given. These block diagrams are used to formulate several different sets of state variables.

3-2   INPUT TO CONTROL SYSTEMS

For some control systems the input has a specific form which may be represented either by an analytical expression or as a specific curve. An example of the latter is the pattern used in a machining operation where the cutting tool is required to follow the path indicated by the pattern outline.

For other control systems the input may be random in shape. In this case it cannot be expressed analytically and is not repetitive. An example is the camera platform used in a photographic airplane. The airplane flies at a fixed altitude and speed, and the camera takes a series of pictures of the terrain below it, which are then fitted together to form one large picture of the area. This requires that the camera platform remain level regardless of the motion of the airplane. Since the attitude of the airplane varies with wind gusts and depends on the stability of the airplane itself, the input to the camera platform is obviously a random function.

It is important to have a basis of comparison for various systems. One way of doing this is comparing the response with a standardized input. The input or inputs used as a basis of comparison must be determined from the required response of the system and the actual form of its input. The following standard inputs, with unit amplitude, are often used in checking the response of a system:

1	Sinusoidal function	$r = \cos \omega t$
2	Power-series function	$r = a_0 + a_1 t + a_2 t^2 + \cdots$
3	Step function	$r = u_{-1}(t)$
4	Ramp (step velocity) function	$r = u_{-2}(t) = tu_{-1}(t)$
5	Parabolic (step acceleration) function	$r = u_{-3}(t) = \dfrac{t^2}{2} u_{-1}(t)$
6	Impulse function	$r = u_0(t)$

Functions 3 to 6 are called *singularity functions* (Fig. 3-1). The singularity functions can be obtained from one another by successive differentiation or integration. For example, the derivative of the parabolic function is the ramp function, the derivative of the ramp function is the step function, and the derivative of the step function is the impulse function.

For each of these inputs a complete solution of the differential equation is determined in this chapter. First, generalized methods are developed to determine the steady-state output for each type of input. These methods are applicable to differential equations of any order. Next, the method of evaluating the transient component of response is determined, and it is shown that the form of the transient component of response depends on the characteristic equation. Addition of the steady-state

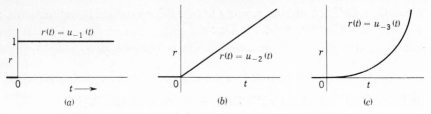

FIGURE 3-1

Singularity functions: (a) step function, $u_{-1}(t)$; (b) ramp function, $u_{-2}(t)$; (c) parabolic function, $u_{-3}(t)$.

component and the transient component gives the complete solution. The coefficients of the transient terms are determined by the instantaneous value of the steady-state component, the roots of the characteristic equation, and the initial conditions. Several examples are used to illustrate these principles.

The solution of the differential equation with a pulse input is postponed until the next chapter, where the Laplace transform is used.

3-3   STEADY-STATE RESPONSE: SINUSOIDAL INPUT

The input quantity r is assumed to be a sinusoidal function of the form

$$r(t) = R \cos (\omega t + \alpha) \qquad (3\text{-}1)$$

The general integrodifferential equation to be solved is of the form

$$A_v D^v c + A_{v-1} D^{v-1} c + \cdots + A_0 D^0 c + A_{-1} D^{-1} c + \cdots$$

$$+ A_{-w} D^{-w} c = r \qquad (3\text{-}2)$$

The steady-state solution can be obtained directly by use of Euler's identity,

$$e^{j\omega t} = \cos \omega t + j \sin \omega t$$

The input can then be written

$$r = R \cos (\omega t + \alpha) = \text{real part of } (Re^{j(\omega t + \alpha)}) = \text{Re } (Re^{j(\omega t + \alpha)})$$

$$= \text{Re } (Re^{j\alpha} e^{j\omega t}) = \text{Re } (\mathbf{R} e^{j\omega t}) \qquad (3\text{-}3)$$

For simplicity, the phrase *real part of* or its symbolic equivalent Re is often omitted, but it must be remembered that the real part is intended. The quantity $\mathbf{R} = Re^{j\alpha}$ is the phasor representation of the input; i.e., it has both a magnitude R and an angle α. The magnitude R represents the maximum value of the input quantity $r(t)$. For simplicity the angle $\alpha = 0°$ usually is chosen for $\mathbf{R}$. The input r from Eq. (3-3) is inserted in Eq. (3-2). Then, in order for the expression to be an equality, the response c must be of the form

$$c_{ss}(t) = C \cos (\omega t + \phi) = \text{Re } (Ce^{j\phi} e^{j\omega t}) = \text{Re } (\mathbf{C} e^{j\omega t}) \qquad (3\text{-}4)$$

where $\mathbf{C} = Ce^{j\phi}$ is a phasor quantity having the magnitude C and the angle ϕ. The nth derivative of c with respect to time is

$$D^n c_{ss}(t) = \mathrm{Re}\left[(j\omega)^n \mathbf{C} e^{j\omega t}\right] \qquad (3\text{-}5)$$

Inserting c and its derivatives from Eqs. (3-4) and (3-5) into Eq. (3-2) gives

$$\mathrm{Re}\left[A_v(j\omega)^v \mathbf{C} e^{j\omega t} + A_{v-1}(j\omega)^{v-1}\mathbf{C} e^{j\omega t} + \cdots + A_{-w}(j\omega)^{-w}\mathbf{C} e^{j\omega t}\right]$$
$$= \mathrm{Re}\,(\mathbf{R} e^{j\omega t}) \qquad (3\text{-}6)$$

Canceling $e^{j\omega t}$ from both sides of the equation and solving for $\mathbf{C}$ gives

$$\mathbf{C} = \frac{\mathbf{R}}{A_v(j\omega)^v + A_{v-1}(j\omega)^{v-1} + \cdots + A_0 + A_{-1}(j\omega)^{-1} + \cdots + A_{-w}(j\omega)^{-w}} \qquad (3\text{-}7)$$

where $\mathbf{C}$ is the phasor representation of the output; i.e., it has a magnitude C and an angle ϕ. Since the values of C and ϕ are functions of the frequency ω, it may be written as $\mathbf{C}(j\omega)$ to show this relationship. Similarly, $\mathbf{R}(j\omega)$ denotes the fact that the input is sinusoidal and may be a function of frequency.

When Eqs. (3-2) and (3-6) are compared, it can be seen that one equation can be determined easily from the other. Substituting $j\omega$ for D, $\mathbf{C}(j\omega)$ for c, and $\mathbf{R}(j\omega)$ for r in Eq. (3-2) results in Eq. (3-6). The reverse is also true and is independent of the order of the equation. It should be realized that this is simply a *rule of thumb* which yields the desired expression.

The time response can be obtained directly from the phasor response. The output is

$$c_{ss}(t) = \mathrm{Re}\,(\mathbf{C} e^{j\omega t}) = |\mathbf{C}|\,\cos\,(\omega t + \phi) \qquad (3\text{-}8)$$

As an example consider the rotational hydraulic transmission described in Sec. 2-11, which has an input x and an output angular position θ_m. Equation (2-122), which relates input to output in terms of system parameters, is repeated below:

$$\frac{VJ}{K_B C}D^3\theta_m + \frac{LJ}{C}D^2\theta_m + d_m\,D\theta_m = d_p\omega_p x \qquad (3\text{-}9)$$

When the input is sinusoidal, $x = X\sin\omega t$, the corresponding phasor equation can be obtained from Eq. (3-9) by replacing $x(t)$ by $\mathbf{X}(j\omega)$, θ_m by $\mathbf{\Theta}_m(j\omega)$, and D by $j\omega$. The ratio of phasor output to phasor input is termed the *frequency transfer function*, often designated by $\mathbf{G}(j\omega)$. This ratio, in terms of the steady-state sinusoidal phasors, is

$$\mathbf{G}(j\omega) = \frac{\mathbf{\Theta}_m(j\omega)}{\mathbf{X}(j\omega)} = \frac{d_p\omega_p/d_m}{j\omega[(VJ/K_B Cd_m)(j\omega)^2 + (LJ/Cd_m)(j\omega) + 1]} \qquad (3\text{-}10)$$

3-4 STEADY-STATE RESPONSE: POLYNOMIAL INPUT

A general development of the steady-state solution of a differential equation with a general polynomial input is first given. Particular cases of the power series are then covered in detail.

The general differential equation is repeated here:

$$A_v D^v c + A_{v-1} D^{v-1} c + \cdots + A_0 c + A_{-1} D^{-1} c + \cdots$$
$$+ A_{-w} D^{-w} c = r \qquad (3\text{-}11)$$

The polynomial input is of the form

$$r(t) = a_0 + a_1 t + \frac{a_2 t^2}{2!} + \cdots + \frac{a_k t^k}{k!} \qquad (3\text{-}12)$$

where the highest-order term in the input is $a_k t^k / k!$. For time $t < 0$ the value of $r(t)$ is zero. The problem is to find the steady-state or particular solution of the dependent variable c. The method used is to assume a polynomial solution of the form

$$c_{ss}(t) = b_0 + b_1 t + \frac{b_2 t^2}{2!} + \cdots + \frac{b_y t^y}{y!} \qquad (3\text{-}13)$$

The assumed solution is then substituted into the differential equation. The coefficients $b_0, b_1, b_2, \ldots$ of the polynomial solution are evaluated by equating the coefficients of like powers of t on both sides of the equation. The highest power of t on the right side of Eq. (3-11) is k; therefore t^k must also appear on the left side of this equation. The highest power of t on the left side of the equation is produced by the lowest-order derivative term and is equal to y minus the order of the lowest derivative. With this information, the value of the highest-order exponent of t to use in the assumed solution of Eq. (3-11) is

$$y = k + X \qquad y \geq 0 \qquad (3\text{-}14)$$

where k is the highest exponent appearing in the input and X is the order of the lowest derivative appearing in the differential equation. When integral terms are present, X is a negative number. X is the lowest-order exponent of the differential operator D appearing in the differential equation. For the general differential equation (3-11), the value of X is equal to $-w$. Equation (3-14) is valid only for positive values of y. For each of the examples the response is a polynomial since the input is of that form. However, the highest power in the response may not be the same as that of the input.

Step-Function Input

A convenient input $r(t)$ to a system is an abrupt change represented by a unit step function, as shown in Fig. 3-1a. This type of input cannot always be put into a system since it may take a definite length of time to make the change in input, but it represents a good mathematical input for checking system response.

The servomotor described in Sec. 2-14 is used as an example. The response of motor velocity ω_m in terms of the voltage e_a applied to the armature, as given by Eq. (2-146), is of the form

$$A_2 D^2 x + A_1 D x + A_0 x = r \qquad (3\text{-}15)$$

When the input is the unit step function $r = u_{-1}(t)$, it is a polynomial in which the highest exponent of t is $k = 0$. The method of solution for a polynomial can therefore be used, which means that the steady-state response is also a polynomial. The lowest-order derivative in Eq. (3-15) is $X = 0$; therefore the steady-state response is of the form

$$x = b_0 \qquad (3\text{-}16)$$

The derivatives are $Dx = 0$ and $D^2x = 0$. Inserting these values into Eq. (3-15) yields

$$x_{ss}(t) = b_0 = \frac{1}{A_0} \qquad (3\text{-}17)$$

Ramp-Function Input (Step Function of Velocity)

The ramp function is a fixed rate of change of a variable as a function of time. This input and its rate of change are shown in Fig. 3-1b. The input is expressed mathematically as

$$r(t) = u_{-2}(t) = tu_{-1}(t) \qquad Dr = u_{-1}(t)$$

A ramp input is a polynomial input where the highest power of t is $k = 1$. When the input $r(t)$ in Eq. (3-15) is the ramp function $u_{-2}(t)$, the highest power of t in the polynomial output is $y = k + X = 1$. The output is therefore

$$x_{ss}(t) = b_0 + b_1 t \qquad (3\text{-}18)$$

The derivatives are $Dx = b_1$ and $D^2x = 0$. Inserting these values into Eq. (3-15) and equating coefficients of t raised to the same power yields

$$t^0: \qquad A_1 b_1 + A_0 b_0 = 0 \qquad b_0 = -\frac{A_1 b_1}{A_0} = -\frac{A_1}{A_0^2}$$

$$t^1: \qquad A_0 b_1 = 1 \qquad b_1 = \frac{1}{A_0}$$

Thus, the steady-state solution is

$$x_{ss}(t) = -\frac{A_1}{A_0^2} + \frac{1}{A_0} t \qquad (3\text{-}19)$$

Parabolic-Function Input (Step Function of Acceleration)

A parabolic function has a constant second derivative, which means that the first derivative is a ramp. The input $r(t)$ is expressed mathematically as

$$r = u_{-3}(t) = \frac{t^2}{2} u_{-1}(t) \qquad Dr = u_{-2}(t) = tu_{-1}(t) \qquad D^2r = u_{-1}(t)$$

A parabolic input is a polynomial input where the highest power of t is $k = 2$. A steady-state solution is found in the conventional manner for a polynomial input. Consider Eq. (3-15) with a parabolic input. The order of the lowest derivative in the system equation is $X = 0$. The value of y is therefore equal to 2, and the steady-state response is of the form

$$x_{ss}(t) = b_0 + b_1 t + \frac{b_2 t^2}{2} \qquad (3\text{-}20)$$

The derivatives of x are $Dx = b_1 + b_2 t$ and $D^2 x = b_2$. Inserting these values into Eq. (3-15) and equating the coefficients of t raised to the same power yields $b_2 = 1/A_0$, $b_1 = -A_1/A_0^2$, and $b_0 = A_1^2/A_0^3 - A_2/A_0^2$.

3-5 TRANSIENT RESPONSE: CLASSICAL METHOD

The classical method of solving for the complementary function or transient response of a differential equation requires, first, the writing of the homogeneous equation. The general differential equation has the form

$$b_v\, D^v c + b_{v-1}\, D^{v-1}c + \cdots + b_0\, D^0 c + b_{-1}\, D^{-1}c + \cdots$$
$$+ b_{-w}\, D^{-w}c = r \qquad (3\text{-}21)$$

where r is the forcing function and c is the response.

The homogeneous equation is formed by letting the right-hand side of the differential equation equal zero:

$$b_v\, D^v c_t + b_{v-1}\, D^{v-1}c_t + \cdots + b_0 c_t + b_{-1}\, D^{-1}c_t + \cdots$$
$$+ b_{-w}\, D^{-w}c_t = 0 \qquad (3\text{-}22)$$

where c_t is the transient component of the general solution.

The general expression for the transient response, which is the solution of the homogeneous equation, is obtained by assuming a solution of the form

$$c_t = A_m e^{mt} \qquad (3\text{-}23)$$

where m is a constant yet to be determined. Substituting this value of c_t into Eq. (3-22) and factoring e^{mt} from all terms gives

$$A_m e^{mt}(b_v m^v + b_{v-1}m^{v-1} + \cdots + b_0 + \cdots + b_{-w}m^{-w}) = 0 \qquad (3\text{-}24)$$

Equation (3-24) must be satisfied for $A_m e^{mt}$ to be a solution. Since e^{mt} cannot be zero for all values of time t, it is necessary that

$$Q(m) = b_v m^v + b_{v-1}m^{v-1} + \cdots + b_0 + \cdots + b_{-w}m^{-w} = 0 \qquad (3\text{-}25)$$

This is purely an algebraic equation and is termed the *characteristic equation*. There are $v + w$ roots, or *eigenvalues*, of the characteristic equation; therefore the complete transient solution contains the same number of terms of the form $A_m e^{mt}$ if all the roots are simple. Thus the transient component, where there are no multiple roots, is

$$c_t = A_1 e^{m_1 t} + A_2 e^{m_2 t} + \cdots + A_k e^{m_k t} + \cdots + A_{v+w}e^{m_{v+w}t} \qquad (3\text{-}26)$$

If there is a root m_q of multiplicity r, the transient includes corresponding terms of the form

$$A_{q1}e^{m_q t} + A_{q2}te^{m_q t} + \cdots + A_{qr}t^{r-1}e^{m_q t} \qquad (3\text{-}27)$$

Instead of using the detailed procedure outlined above, the characteristic equation is usually obtained directly from the homogeneous equation by substituting m for Dc_t, m^2 for $D^2 c_t$, etc.

Since the coefficients of the transient solution must be determined from the initial conditions, there must be $v + w$ known initial conditions. These conditions are values of the variable c and of its derivatives which are known at specific times. The $v + w$ initial conditions are used to set up $v + w$ simultaneous equations of c and its derivatives. The value of c includes both the steady-state and transient components. Since determination of the coefficients includes consideration of the steady-state component, the input affects the value of the coefficient of each exponential term.

Complex Roots

If all values of m_k are real, the transient terms can be evaluated as indicated above. Frequently, some values of m_k are complex. When this happens, they always occur in pairs that are complex conjugates and are of the form

$$m_k = \sigma + j\omega_d \qquad m_{k+1} = \sigma - j\omega_d \qquad (3\text{-}28)$$

The transient terms corresponding to these values of m are

$$A_k e^{(\sigma + j\omega_d)t} + A_{k+1} e^{(\sigma - j\omega_d)t} \qquad (3\text{-}29)$$

These terms are combined to a more useful form by factoring the term $e^{\sigma t}$:

$$e^{\sigma t}(A_k e^{j\omega_d t} + A_{k+1} e^{-j\omega_d t}) \qquad (3\text{-}30)$$

By using the Euler identity $e^{\pm j\omega_d t} = \cos \omega_d t \pm j \sin \omega_d t$ and combining terms, expression (3-30) can be put in the form

$$e^{\sigma t}(B_1 \cos \omega_d t + B_2 \sin \omega_d t) \qquad (3\text{-}31)$$

This can be converted into the form

$$A e^{\sigma t} \sin (\omega_d t + \phi) \qquad (3\text{-}32)$$

where $A = \sqrt{B_1{}^2 + B_2{}^2}$ and $\phi = \tan^{-1}(B_1/B_2)$. This is a very convenient form for plotting the transient response. The student must learn to use this equation directly without deriving it each time. Often the constants in the transient term can be evaluated more readily from the initial conditions by using the form of Eq. (3-31).

This transient term is called an *exponentially damped sinusoid*; it consists of a sine wave of frequency ω_d whose magnitude is $A e^{\sigma t}$; that is, it is decreasing exponentially with time if σ is negative. It has the form shown in Fig. 3-2. The curves $\pm A e^{\sigma t}$ constitute the *envelope*, and σ is called the *damping coefficient*. The plot of the time solution always remains between the two branches of the envelope.

In the analysis above, the complex roots are given in the form

$$m_{k,k+1} = \sigma \pm j\omega_d$$

When σ is negative, the transient decays with time and eventually dies out. This represents a stable system. When σ is positive, the transient increases with time and will destroy the equipment unless otherwise restrained. This represents the undesirable case of an unstable system. Control systems must be designed so that they are always stable.

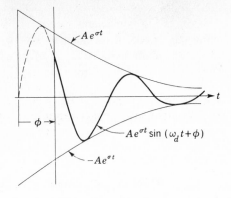

FIGURE 3-2
Sketch of an exponentially
damped sinusoid.

Damping Ratio and Undamped Natural Frequency

When the characteristic equation has a pair of complex-conjugate roots, it has a quadratic factor of the form $b_2 m^2 + b_1 m + b_0$. The roots of this factor are

$$m_{1,2} = -\frac{b_1}{2b_2} \pm j \sqrt{\frac{4b_2 b_0 - b_1{}^2}{4b_2{}^2}} = \sigma \pm j\omega_d \qquad (3\text{-}33)$$

The real part σ is recognized as the exponent of e, and ω_d is the damped natural frequency of the oscillatory portion of the component stemming from this pair of roots, as given by expression (3-32).

The quantity b_1 represents the effective damping constant of the system. If b_1 has the value $2\sqrt{b_2 b_0}$, the two roots m are equal. This represents the critical value of damping constant and is written $b_1' = 2\sqrt{b_2 b_0}$. The damping ratio ζ is defined as the ratio of the actual damping constant to the critical value of damping constant:

$$\zeta = \frac{\text{actual damping constant}}{\text{critical damping constant}} = \frac{b_1}{b_1'} = \frac{b_1}{2\sqrt{b_2 b_0}} \qquad (3\text{-}34)$$

When ζ is positive and less than unity, the roots are complex and the transient is a damped sinusoid of the form of expression (3-32). When ζ is less than unity, the response is said to be *underdamped*. When ζ is greater than unity, the roots are real and the response is *overdamped*; i.e., the transient solution consists of two exponential terms.

The undamped natural frequency ω_n is defined as the frequency of oscillation of the transient if the damping is zero:

$$\omega_n = \sqrt{\frac{b_0}{b_2}} \qquad (3\text{-}35)$$

The case of zero damping constant means that the transient response does not die out; it is a sine wave of constant amplitude.

The quadratic factors are frequently written in terms of the damping ratio and the undamped natural frequency. Underdamped systems are generally analyzed

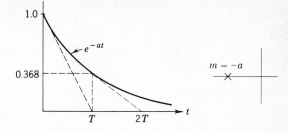

FIGURE 3-3
Plot of the exponential e^{-at} and the root location.

in terms of these two parameters. After factoring b_0, the quadratic factor of the characteristic equation is

$$\frac{b_2}{b_0} m^2 + \frac{b_1}{b_0} m + 1 = \frac{1}{\omega_n{}^2} m^2 + \frac{2\zeta}{\omega_n} m + 1 \qquad (3\text{-}36)$$

When it is multiplied through by $\omega_n{}^2$, the quadratic appears in the form

$$m^2 + 2\zeta\omega_n m + \omega_n{}^2 \qquad (3\text{-}37)$$

The forms given by Eqs. (3-36) and (3-37) are called the *standard forms* of the quadratic factor, and the corresponding roots are

$$m_{1,2} = \sigma \pm j\omega_d = -\zeta\omega_n \pm j\omega_n\sqrt{1 - \zeta^2} \qquad (3\text{-}38)$$

The transient response of Eq. (3-32) for the underdamped case, written in terms of ζ and ω_n, is

$$Ae^{-\zeta\omega_n t} \sin(\omega_n\sqrt{1 - \zeta^2}\, t + \phi) \qquad (3\text{-}39)$$

From this expression the effect on the transient of the terms ζ and ω_n can readily be seen. The larger the product $\zeta\omega_n$, the faster the transient will decay. These terms also affect the damped natural frequency of oscillation of the transient, $\omega_d = \omega_n\sqrt{1 - \zeta^2}$, which varies directly as the undamped natural frequency and decreases with an increase in the damping ratio.

3-6 DEFINITION OF TIME CONSTANT

The transient terms have the exponential form Ae^{mt}. When $m = -a$ is real and negative, the plot of Ae^{-at} has the form shown in Fig. 3-3. The value of time that makes the exponent of e equal to -1 is called the *time constant T*. Thus

$$-aT = -1 \qquad \text{and} \qquad T = \frac{1}{a} \qquad (3\text{-}40)$$

In a duration of time equal to one time constant the exponential e^{-at} decreases from the value 1 to the value 0.368. Geometrically the tangent drawn to the curve Ae^{-at} at $t = 0$ intersects the time axis at the value of time equal to the time constant T.

When $m = \sigma \pm j\omega_d$ is a complex quantity, the transient has the form $Ae^{\sigma t} \sin(\omega_d t + \phi)$. A plot of this function is shown in Fig. 3-2. In the case of the

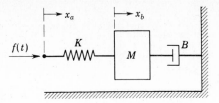

FIGURE 3-4
Simple mechanical system.

damped sinusoid the time constant is defined in terms of the parameter σ that characterizes the envelope $Ae^{\sigma t}$. Thus the time constant T is equal to

$$T = \frac{1}{|\sigma|} \qquad (3\text{-}41)$$

In terms of the damping ratio and undamped natural frequency, the time constant is $T = 1/\zeta\omega_n$. Therefore, the larger the product $\zeta\omega_n$, the greater the instantaneous rate of decay of the transient.

3-7 EXAMPLE: SECOND-ORDER SYSTEM—MECHANICAL

The simple mechanical system of Sec. 2-6 is used as an example and is shown in Fig. 3-4. Equation (2-61) relates the displacement x_b to x_a:

$$M\,D^2x_b + B\,Dx_b + Kx_b = Kx_a \qquad (3\text{-}42)$$

The system is considered to be originally at rest. The function x_a moves 1 unit at time $t = 0$; that is, the input is a unit step function $x_a(t) = u_{-1}(t)$. The problem is to find the motion $x_b(t)$.

The displacement of the mass is given by

$$x_b(t) = x_{b,\text{ss}} + x_{b,t}$$

where $x_{b,\text{ss}}$ is the steady-state solution and $x_{b,t}$ is the transient solution.

The steady-state solution is found first by using the method of Sec. 3-4. In this example it may be easier to consider the following:

1 x_b must reach a fixed steady-state position.
2 When x_b reaches a constant value, the velocity and acceleration become zero.

By putting $D^2x_b = Dx_b = 0$ into Eq. (3-42), the final or steady-state value of x_b is $x_{b,\text{ss}} = x_a$. When the steady-state solution has been found, the transient solution is determined. The characteristic equation is

$$Mm^2 + Bm + K = M\left(m^2 + \frac{B}{M}\,m + \frac{K}{M}\right) = 0$$

Putting this in terms of ζ and ω_n gives

$$m^2 + 2\zeta\omega_n m + \omega_n{}^2 = 0 \qquad (3\text{-}43)$$

for which the roots are $m_{1,2} = -\zeta\omega_n \pm \omega_n\sqrt{\zeta^2 - 1}$. The transient solution depends on whether the damping ratio ζ is (1) greater than unity, (2) equal to unity, or (3) smaller than unity. For ζ greater than unity the roots are real, and the transient response is

$$x_{b,t} = A_1 \exp\left[(-\zeta + \sqrt{\zeta^2 - 1})\omega_n t\right] + A_2 \exp\left[(-\zeta - \sqrt{\zeta^2 - 1})\omega_n t\right] \quad (3\text{-}44)$$

For ζ equal to unity, the roots are real and equal; that is, $m_1 = m_2 = -\zeta\omega_n$. Since there are multiple roots, the transient response is

$$x_{b,t} = A_1 e^{-\zeta\omega_n t} + A_2 t e^{-\zeta\omega_n t} \quad (3\text{-}45)$$

For ζ less than unity, the roots are complex,

$$m_{1,2} = -\zeta\omega_n \pm j\omega_n\sqrt{1 - \zeta^2}$$

and the transient solution, as outlined in Sec. 3-5, is

$$x_{b,t} = A e^{-\zeta\omega_n t} \sin\left(\omega_n\sqrt{1 - \zeta^2}\, t + \phi\right) \quad (3\text{-}46)$$

The complete solution is the sum of the steady-state and transient solutions. For the underdamped case, $\zeta < 1$, the complete solution to Eq. (3-42) is

$$x_b(t) = 1 + A e^{-\zeta\omega_n t} \sin\left(\omega_n\sqrt{1 - \zeta^2}\, t + \phi\right) \quad (3\text{-}47)$$

The two constants A and ϕ must next be determined from the initial conditions. In this example the system was initially at rest; therefore $x_b(0) = 0$. The energy stored in a mass is $W = \frac{1}{2}Mv^2$. From the principle of conservation of energy, *the velocity of a system with mass cannot change instantaneously*; thus $Dx_b(0) = 0$. Two equations are necessary, one for $x_b(t)$ and one for $Dx_b(t)$. Differentiating Eq. (3-47) yields

$$Dx_b(t) = -\zeta\omega_n A e^{-\zeta\omega_n t} \sin\left(\omega_n\sqrt{1 - \zeta^2}\, t + \phi\right)$$
$$+ \omega_n\sqrt{1 - \zeta^2}\, A e^{-\zeta\omega_n t} \cos\left(\omega_n\sqrt{1 - \zeta^2}\, t + \phi\right) \quad (3\text{-}48)$$

Inserting in Eqs. (3-47) and (3-48) the initial conditions

$$x_b(0) = 0 \qquad Dx_b(0) = 0 \qquad t = 0$$

yields

$$0 = 1 + A \sin\phi \qquad 0 = -\zeta\omega_n A \sin\phi + \omega_n\sqrt{1 - \zeta^2}\, A \cos\phi$$

These equations are then solved for A and ϕ:

$$A = \frac{-1}{\sqrt{1 - \zeta^2}} \qquad \phi = \tan^{-1}\frac{\sqrt{1 - \zeta^2}}{\zeta} = \cos^{-1}\zeta$$

Thus the complete solution is

$$x_b(t) = 1 - \frac{e^{-\zeta\omega_n t}}{\sqrt{1 - \zeta^2}} \sin\left(\omega_n\sqrt{1 - \zeta^2}\, t + \cos^{-1}\zeta\right) \quad (3\text{-}49)$$

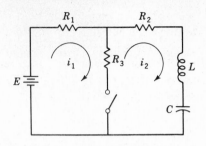

FIGURE 3-5
Electric circuit: $E = 10$ V, $R_1 = 10 \ \Omega$,
$R_2 = 15 \ \Omega$, $R_3 = 10 \ \Omega$, $L = 1$ H,
$C = 0.01$ F.

When the complete solution has been obtained, it should be checked to see that it satisfies the known conditions. For example, putting $t = 0$ into Eq. (3-49) gives $x_b(0) = 0$; therefore the solution checks. In a like manner, the constants can be evaluated for the other two cases of damping.

3-8 EXAMPLE: SECOND-ORDER SYSTEM—ELECTRICAL

The electric circuit of Fig. 3-5 is used to further illustrate the determination of initial conditions. The circuit, as shown, is in the steady state. At time $t = 0$ the switch is closed. The problem is to solve for the current through the inductor for $t \geq 0$.

Two loop equations are written for this circuit:

$$10 = 20i_1 - 10i_2 \qquad (3\text{-}50)$$

$$0 = -10i_1 + \left(D + 25 + \frac{100}{D}\right)i_2 \qquad (3\text{-}51)$$

Eliminating i_1 from these equations yields

$$10 = \left(2D + 40 + \frac{200}{D}\right)i_2 \qquad (3\text{-}52)$$

After differentiating Eq. (3-52), the steady-state solution is found by using the method of Sec. 3-4. Since the input is a step function, the steady-state output is

$$i_{2,ss} = 0 \qquad (3\text{-}53)$$

This can also be deduced from an inspection of the circuit. When a branch contains a capacitor, the steady-state current is always zero for a dc source.

Next the transient solution is determined. The characteristic equation is $m^2 + 20m + 100 = 0$, for which the roots are $m_{1,2} = -10$. Thus the circuit is critically damped, and the current through the inductor can be expressed by

$$i_2(t) = i_{2,t}(t) = A_1 e^{-10t} + A_2 t e^{-10t} \qquad (3\text{-}54)$$

Two equations and two initial conditions are necessary to evaluate the two constants A_1 and A_2. Equation (3-54) and its derivative,

$$Di_2(t) = -10A_1 e^{-10t} + A_2(1 - 10t)e^{-10t} \qquad (3\text{-}55)$$

are utilized in conjunction with the initial conditions $i_2(0^+)$ and $Di_2(0^+)$. In this example the currents just before the switch is closed are $i_1(0^-) = i_2(0^-) = 0$. The energy stored in the magnetic field of a single inductor is $W = \frac{1}{2}Li^2$. Since this energy cannot change instantly, the current through an inductor cannot change instantly either. Therefore $i_2(0^+) = i_2(0^-) = 0$. $Di_2(t)$ is found from the original circuit equation, Eq. (3-51):

$$Di_2(t) = 10i_1(t) - 25i_2(t) - v_c(t) \qquad (3\text{-}56)$$

To determine $Di_2(0^+)$, it is necessary first to determine $i_1(0^+)$, $i_2(0^+)$, and $v_c(0^+)$. Since $i_2(0^+) = 0$, Eq. (3-50) yields $i_1(0^+) = 0.5$ A. Since the energy $W = \frac{1}{2}Cv^2$ stored in a capacitor cannot change instantly, the voltage across the capacitor cannot change instantly either. The steady-state value of capacitor voltage for $t < 0$ is 10 V; thus

$$v_c(0^-) = v_c(0^+) = 10 \text{ V} \qquad (3\text{-}57)$$

Inserting these values into Eq. (3-56) yields $Di_2(0^+) = -5$. Substituting the initial conditions into Eqs. (3-54) and (3-55) results in $A_1 = 0$ and $A_2 = -5$. Therefore the current through the inductor for $t \geq 0$ is

$$i_2(t) = -5te^{-10t} \qquad (3\text{-}58)$$

3-9 SECOND-ORDER TRANSIENTS[2]

The response to a unit step-function input is usually used as a means of evaluating the response of a system. The example of the preceding section is used as an illustrative second-order system. The differential equation is

$$\frac{D^2c}{\omega_n^2} + \frac{2\zeta}{\omega_n} Dc + c = r$$

This is defined as a *simple* second-order equation because there are no derivatives of r on the right side of the equation. The response to a unit step input, subject to zero initial conditions, is derived in Sec. 3-7 and is given by

$$c(t) = 1 - \frac{e^{-\zeta\omega_n t}}{\sqrt{1-\zeta^2}} \sin(\omega_n\sqrt{1-\zeta^2}\,t + \cos^{-1}\zeta) \qquad (3\text{-}59)$$

A family of curves representing this equation is shown in Fig. 3-6, where the abscissa is the dimensionless variable $\omega_n t$. The curves are thus a function only of the damping ratio ζ.

These curves show that the amount of overshoot depends on the damping ratio ζ. For the overdamped and critically damped case, $\zeta \geq 1$, there is no overshoot. For the underdamped case, $\zeta < 1$, the system oscillates around the final value. The oscillations decrease with time, and the system response approaches the final value. The peak overshoot for the underdamped system is the first overshoot.

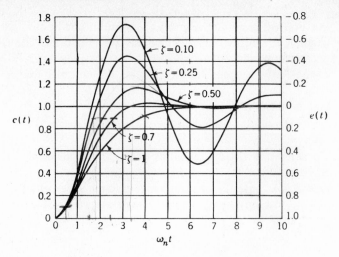

FIGURE 3-6
Simple second-order transients.

The time at which the peak overshoot occurs, t_p, can be found by differentiating $c(t)$ from Eq. (3-59) with respect to time and setting this derivative equal to zero:

$$\frac{dc}{dt} = \frac{\zeta\omega_n e^{-\zeta\omega_n t}}{\sqrt{1 - \zeta^2}} \sin{(\omega_n\sqrt{1 - \zeta^2}\, t + \cos^{-1}\zeta)}$$

$$- \omega_n e^{-\zeta\omega_n t} \cos{(\omega_n\sqrt{1 - \zeta^2}\, t + \cos^{-1}\zeta)} = 0$$

This derivative is zero at $\omega_n\sqrt{1 - \zeta^2}\, t = 0, \pi, 2\pi, \dots$. The peak overshoot occurs at the first value after zero, provided there are zero initial conditions; therefore

$$t_p = \frac{\pi}{\omega_n\sqrt{1 - \zeta^2}} \qquad (3\text{-}60)$$

Inserting this value of time in Eq. (3-59) gives the peak overshoot as

$$M_p = c_p = 1 + \exp\left(- \frac{\zeta\pi}{\sqrt{1 - \zeta^2}}\right) \qquad (3\text{-}61)$$

The per unit overshoot M_o as a function of damping ratio is shown in Fig. 3-7, where

$$M_o = \frac{c_p - c_{ss}}{c_{ss}} \qquad (3\text{-}62)$$

The variation of the frequency of oscillation of the transient with variation of damping ratio is also of interest. In order to represent this variation by one curve, the quantity ω_d/ω_n is plotted against ζ in Fig. 3-8. If the scales of ordinate and abscissa are equal, the curve is an arc of a circle. Note that this curve has been plotted for $\zeta \le 1$. Values of damped natural frequency for $\zeta > 1$ are mathematical only, not physical.

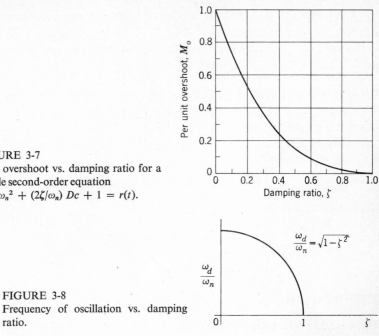

FIGURE 3-7
Peak overshoot vs. damping ratio for a
simple second-order equation
$D^2 c / \omega_n{}^2 + (2\zeta/\omega_n)\, Dc + 1 = r(t)$.

FIGURE 3-8
Frequency of oscillation vs. damping
ratio.

The error in the system is the difference between the input and output; thus
the error equation is

$$e = r - c = \frac{e^{-\zeta\omega_n t}}{\sqrt{1 - \zeta^2}} \sin\left(\omega_n\sqrt{1 - \zeta^2}\, t + \cos^{-1}\zeta\right) \qquad (3\text{-}63)$$

The variation of error with time is sometimes plotted. These curves can be obtained
from Fig. 3-6 by realizing that the curves start at $e(0) = +1$ and have the final value
$e(\infty) = 0$.

Response Characteristics

The transient-response curves for the second-order system show a number of significant characteristics.

The overdamped system is slow-acting and does not oscillate about the final
position. For some applications the absence of oscillations may be necessary. For
example, an elevator cannot be allowed to oscillate at each stop. But for systems
where a fast response is necessary, the slow response of an overdamped system cannot
be tolerated.

The underdamped system reaches the final value faster than the overdamped
system, but the response oscillates about this final value. If this oscillation can be
tolerated, the underdamped system is faster-acting. The amount of permissible overshoot determines the desirable value of damping ratio. A damping ratio $\zeta = 0.4$
has an overshoot of 25.4 percent, and a damping ratio $\zeta = 0.8$ has an overshoot of
1.6 percent.

The settling time is the time required for the oscillations to decrease to a specified absolute percentage of the final value and thereafter remain less than this value. Errors of 2 or 5 percent are common values used to determine settling time. For second-order systems the value of the transient at any time is equal to or less than the exponential $e^{-\zeta\omega_n t}$. The value of this term is given in Table 3-1 for several values of t expressed in a number of time constants T.

The settling time for a 2 percent error criterion is approximately 4 time constants; for a 5 percent error criterion, it is 3 time constants. The percent error criterion used must be determined from the response desired for the system. The time for the envelope of the transient to die out is

$$T_s = \frac{\text{number of time constants}}{\zeta\omega_n} \qquad (3\text{-}64)$$

Since ζ must be determined and adjusted for the permissible overshoot, the undamped natural frequency determines the settling time.

3-10   TIME-RESPONSE SPECIFICATIONS[3]

The desired performance characteristics of a system of any order may be specified in terms of the transient response to a unit step-function input. The performance of a system may be evaluated in terms of the following quantities, as shown in Fig. 3-9.

1 Maximum overshoot c_p is the magnitude of the first overshoot. This may also be expressed in percent of the final value.

2 Time to maximum overshoot t_p is the time required to reach the maximum overshoot.

3 Time to first zero error t_0 is the time required to reach the final value the first time. It is often referred to as *duplicating time*.

4 Settling time t_s is the time required for the output response first to reach and thereafter remain within a prescribed percentage of the final value. This percentage must be specified in the individual case. Common values used for settling time are 2 and 5 percent. As commonly used, the 2 or 5 percent is applied to the envelope which yields T_s. The actual t_s may be smaller than T_s.

5 Frequency of oscillation of the transient, ω_d.

Table 3-1   EXPONENTIAL VALUES

t	$e^{-\zeta\omega_n t}$	Error, %
$1T$	0.368	36.8
$2T$	0.135	13.5
$3T$	0.050	5.0
$4T$	0.018	1.8
$5T$	0.007	0.7

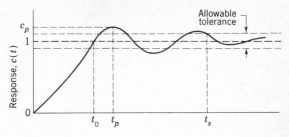

FIGURE 3-9
Typical underdamped response to a step function.

The time response differs for each set of initial conditions. Therefore, to compare the time response of various systems it is necessary to start with standard initial conditions. The most practical standard is to start with the system at rest. Then the response characteristics, such as maximum overshoot and settling time, can be compared significantly.

For some systems these specifications are also applied for a ramp input. In such cases the plot of error with time is used with the definitions. For systems subject to shock inputs the response due to an impulse is used as a criterion of performance.

3-11 STATE-VARIABLE EQUATIONS[4–6]

The differential equations and the corresponding state equations of various physical systems are derived in Chap. 2. The state variables selected in Chap. 2 are restricted to the energy-storage variables. In later sections of this chapter three methods of simulating differential equations are presented. These lead to different formulations of the state variables. In large-scale systems, with many inputs and outputs, the state-variable approach can have distinct advantages over conventional methods, especially when digital computers are used to obtain the solutions. While this text is restricted to linear time-invariant systems, the state-variable approach is applicable to nonlinear and to time-varying systems. In these cases a computer is a practical method for obtaining the solution. A feature of the state-variable method is that it decomposes a complex system into a set of smaller systems which can be normalized to have a minimum interaction and which can be solved individually. Also, it provides a unified approach that is used extensively in modern control theory.

The block diagram of Fig. 3-10 represents a system S which has r inputs, p outputs, and n state variables. The coefficients in the equations representing a linear time-invariant system are constants. The matrix state and output equations are then

$$\dot{\mathbf{x}}(t) = \mathbf{A}\mathbf{x}(t) + \mathbf{B}\mathbf{u}(t) \qquad (3\text{-}65)$$

$$\mathbf{y}(t) = \mathbf{C}\mathbf{x}(t) + \mathbf{D}\mathbf{u}(t) \qquad (3\text{-}66)$$

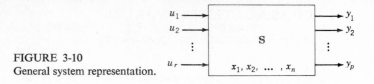

FIGURE 3-10
General system representation.

The variables $\mathbf{x}(t)$, $\mathbf{u}(t)$, and $\mathbf{y}(t)$ are column vectors, and $\mathbf{A}$, $\mathbf{B}$, $\mathbf{C}$, and $\mathbf{D}$ are matrices having constant elements. Equation (3-65) is solved for the state vector $\mathbf{x}(t)$. This result is then used in Eq. (3-66) to determine the output $\mathbf{y}(t)$.

Homogeneous Solution (State Transition Matrix)

The homogeneous state equation, with the input $\mathbf{u}(t) = 0$, is

$$\dot{\mathbf{x}} = \mathbf{A}\mathbf{x} \qquad (3\text{-}67)$$

where $\mathbf{A}$ is a constant $n \times n$ matrix and $\mathbf{x}$ is an $n \times 1$ column vector.

For the scalar first-order equation $\dot{x} = ax$, the solution, in terms of the initial conditions at time $t = 0$, is

$$x(t) = e^{at}x(0) \qquad (3\text{-}68)$$

For any other initial condition, at time $t = t_0$, the solution is

$$x(t) = e^{a(t-t_0)}x(t_0) \qquad (3\text{-}69)$$

Comparing the scalar and the state equations shows the solution of Eq. (3-67) to be analogous to the solution given by Eq. (3-69); it is

$$\mathbf{x}(t) = \exp\left[\mathbf{A}(t - t_0)\right]\mathbf{x}(t_0) \qquad (3\text{-}70)$$

The exponential function of a scalar which appears in Eq. (3-68) can be expressed as the infinite series

$$e^{at} = \exp\left[at\right] = 1 + \frac{at}{1!} + \frac{(at)^2}{2!} + \frac{(at)^3}{3!} + \cdots + \frac{(at)^k}{k!} + \cdots \qquad (3\text{-}71).$$

The analogous exponential function of a square matrix $\mathbf{A}$ which appears in Eq. (3-70), with $t_0 = 0$, is

$$e^{\mathbf{A}t} = \exp\left[\mathbf{A}t\right] = \mathbf{I} + \frac{\mathbf{A}t}{1!} + \frac{(\mathbf{A}t)^2}{2!} + \frac{(\mathbf{A}t)^3}{3!} + \cdots + \frac{(\mathbf{A}t)^k}{k!} + \cdots \qquad (3\text{-}72)$$

Thus $\exp\left[\mathbf{A}t\right]$ is a square matrix of the same order as $\mathbf{A}$. It is more useful when the infinite series of Eq. (3-72) is put in closed form. This is done in Sec. 3-13. It is common to call this the *state transition matrix* or the *fundamental matrix* of the system and to denote it by

$$\mathbf{\Phi}(t - t_0) = \exp\left[\mathbf{A}(t - t_0)\right] \qquad (3\text{-}73)$$

The term *state transition matrix* (STM) is descriptive of the unforced or natural response and is the expression preferred by engineers.

3-12 CHARACTERISTIC VALUES

Consider a system of equations represented by

$$\dot{x} = Ax \qquad (3\text{-}74)$$

The signals $\dot{x}$ and x are column vectors, and A is a square matrix of order n. One case for which a solution of this equation exists is if x and $\dot{x}$ have the same direction in the state space but differ only in magnitude by a scalar proportionality factor λ. The solution must therefore have the form $\dot{x} = \lambda x$. Inserting this into Eq. (3-74) and rearranging terms yields

$$[\lambda I - A]x = 0 \qquad (3\text{-}75)$$

The solution of this equation has a nontrivial solution only if x is not zero. It is therefore required that the determinant of the coefficients of x be zero:

$$Q(\lambda) \equiv |\lambda I - A| = 0 \qquad (3\text{-}76)$$

When A is of order n, the resulting polynomial equation is *characteristic equation*

$$Q(\lambda) = \lambda^n + a_{n-1}\lambda^{n-1} + \cdots + a_1\lambda + a_0 = 0 \qquad (3\text{-}77)$$

The roots λ_i of the characteristic equation are called the *characteristic values* or *eigenvalues* of A. A root may be distinct (simple) or repeated with a multiplicity r. Also, a root may be real or complex. Complex roots must appear in conjugate pairs. The polynomial $Q(\lambda)$ may be written in factored form as

$$Q(\lambda) = (\lambda - \lambda_1)(\lambda - \lambda_2)\cdots(\lambda - \lambda_n) \qquad (3\text{-}78)$$

The product of the *eigenvalues* of a matrix A is equal to its determinant, and the sum of the *eigenvalues* is equal to the sum of the elements on the main diagonal (the trace).

3-13 EVALUATING THE STATE TRANSITION MATRIX

There are several methods for evaluating the STM $\Phi(t) = \exp[At]$ in closed form for a given matrix A. The method illustrated below is based on the Cayley-Hamilton theorem. The Laplace transform method is covered in the next chapter. Consider a general polynomial of the form

$$N(\lambda) = \lambda^m + C_{m-1}\lambda^{m-1} + \cdots + C_1\lambda + C_0 \qquad (3\text{-}79)$$

When the polynomial $N(\lambda)$ is divided by the characteristic polynomial $Q(\lambda)$, the result

$$\frac{N(\lambda)}{Q(\lambda)} = F(\lambda) + \frac{R(\lambda)}{Q(\lambda)}$$

or

$$N(\lambda) = F(\lambda)Q(\lambda) + R(\lambda) \qquad (3\text{-}80)$$

The function $R(\lambda)$ is the remainder. It is a polynomial of order $n - 1$, or 1 less than the order of $Q(\lambda)$. For $\lambda = \lambda_i$ the value $Q(\lambda_i) = 0$; thus

$$N(\lambda_i) = R(\lambda_i) \qquad (3\text{-}81)$$

The matrix polynomial corresponding to Eq. (3-79), using A as the variable, is

$$N(A) = A^m + C_{m-1}A^{m-1} + \cdots + C_1A + C_0I \qquad (3\text{-}82)$$

Since the characteristic equation $Q(\lambda) = 0$ has n roots, there are n equations $Q(\lambda_1) = 0$, $Q(\lambda_2) = 0, \ldots, Q(\lambda_n) = 0$. The analogous matrix equation is

$$Q(A) = A^n + a_{n-1}A^{n-1} + \cdots + a_1A + a_0I = 0$$

where 0 indicates a null matrix of the same order as $Q(A)$. This equation implies the Cayley-Hamilton theorem, which is sometimes expressed as "every square matrix A satisfies its own characteristic equation." The matrix polynomial corresponding to Eqs. (3-80) and (3-81) is therefore

$$N(A) = F(A)Q(A) + R(A) = R(A) \qquad (3\text{-}83)$$

Equations (3-81) and (3-83) are valid when $N(\lambda)$ is a polynomial of any order (or even an infinite series) as long as it is analytic. The exponential function $N(\lambda) = e^{\lambda t}$ is an analytic function which can be represented by an infinite series as shown in Eq. (3-71). Since this function converges in the region of analyticity, it can be expressed in closed form by a polynomial in λ of degree $n - 1$. Thus, for each eigenvalue, from Eq. (3-81)

$$e^{\lambda_i t} = R(\lambda_i) = \alpha_0 + \alpha_1\lambda_i + \cdots + \alpha_k\lambda_i^k + \cdots + \alpha_{n-1}\lambda_i^{n-1} \qquad (3\text{-}84)$$

Inserting the n distinct roots λ_i into Eq. (3-84) yields n equations which can be solved simultaneously for the coefficients α_k. These coefficients may be inserted in the corresponding matrix equation

$$e^{At} = N(A) = R(A) = \alpha_0I + \alpha_1A + \cdots + \alpha_kA^k + \cdots + \alpha_{n-1}A^{n-1} \qquad (3\text{-}85)$$

For the STM this yields

$$\Phi(t) = \exp[At] = \sum_{k=0}^{n-1} \alpha_k A^k \qquad (3\text{-}86)$$

For the case of multiple roots, refer to a text on linear algebra[4] for the evaluation of $\Phi(t)$.

EXAMPLE Find $\exp[At]$ when

$$A = \begin{bmatrix} 0 & 6 \\ -1 & -5 \end{bmatrix}$$

The characteristic equation is

$$Q(\lambda) = |\lambda I - A| = \begin{vmatrix} \lambda & -6 \\ 1 & \lambda + 5 \end{vmatrix} = \lambda^2 + 5\lambda + 6 = 0$$

The roots of this equation are $\lambda_1 = -2$ and $\lambda_2 = -3$. Since A is a second-order matrix, the remainder polynomial given by Eq. (3-84) is

$$N(\lambda_i) = R(\lambda_i) = \alpha_0 + \alpha_1\lambda_i \qquad (3\text{-}87)$$

Substituting $N(\lambda_i) = e^{\lambda_i t}$ and the two roots λ_1 and λ_2 into Eq. (3-87) yields

$$e^{-2t} = \alpha_0 - 2\alpha_1 \quad \text{and} \quad e^{-3t} = \alpha_0 - 3\alpha_1$$

Solving these equations for the coefficients α_0 and α_1 yields

$$\alpha_0 = 3e^{-2t} - 2e^{-3t} \quad \text{and} \quad \alpha_1 = e^{-2t} - e^{-3t}$$

The STM obtained by using Eq. (3-86) is

$$\Phi(t) = \exp[At] = \alpha_0 I + \alpha_1 A = \begin{bmatrix} \alpha_0 & 0 \\ 0 & \alpha_0 \end{bmatrix} + \begin{bmatrix} 0 & 6\alpha_1 \\ -\alpha_1 & -5\alpha_1 \end{bmatrix}$$

$$= \begin{bmatrix} 3e^{-2t} - 2e^{-3t} & 6e^{-2t} - 6e^{-3t} \\ -e^{-2t} + e^{-3t} & -2e^{-2t} + 3e^{-3t} \end{bmatrix} \tag{3-88}$$

Another procedure for evaluating the STM, $\Phi(t)$, is by use of the Sylvester expansion. The format in this method makes it especially useful for use with a digital computer. The method is presented here without proof for the case when A is a square matrix with n distinct eigenvalues. The polynomial $N(A)$ can be written

$$e^{At} = N(A) = \sum_{i=1}^{n} N(\lambda_i) Z_i(\lambda) \tag{3-88a}$$

where

$$Z_i(\lambda) = \frac{\prod_{\substack{j=1 \\ j \neq i}}^{n} (A - \lambda_j I)}{\prod_{\substack{j=1 \\ j \neq i}}^{n} (\lambda_i - \lambda_j)} \tag{3-88b}$$

For the previous example there are two eigenvalues $\lambda_1 = -2$ and $\lambda_2 = -3$. The two matrices Z_1 and Z_2 are evaluated from Eq. (3-88b):

$$Z_1 = \frac{A - \lambda_2 I}{\lambda_1 - \lambda_2} = \frac{\begin{bmatrix} 0+3 & 6 \\ -1 & -5+3 \end{bmatrix}}{-2-(-3)} = \begin{bmatrix} 3 & 6 \\ -1 & -2 \end{bmatrix}$$

$$Z_2 = \frac{A - \lambda_1 I}{\lambda_2 - \lambda_1} = \frac{\begin{bmatrix} 0+2 & 6 \\ -1 & -5+2 \end{bmatrix}}{-3-(-2)} = \begin{bmatrix} -2 & -6 \\ 1 & 3 \end{bmatrix}$$

Using these values in Eq. (3-88a), where $N(\lambda_i) = e^{\lambda_i t}$, yields

$$\Phi(t) = \exp[At] = e^{\lambda_1 t} Z_1 + e^{\lambda_2 t} Z_2$$

Equation (3-88) contains the value of $\Phi(t)$.

3-14 COMPLETE SOLUTION OF THE STATE EQUATION

When an input $u(t)$ is present, the complete solution for $x(t)$ is obtained from Eq. (3-65). The starting point for obtaining the solution is the following equality, obtained by applying the rule for the derivative of the product of two matrices (see Sec. 2-3):

$$\frac{d}{dt}[e^{-At} x(t)] = e^{-At}[\dot{x}(t) - Ax(t)] \tag{3-89}$$

Utilizing Eq. (3-65), $\dot{x} = Ax + Bu$, in the right side of Eq. (3-89) yields

$$\frac{d}{dt}[e^{-At} x(t)] = e^{-At} Bu(t)$$

Integrating this equation gives

$$e^{-\mathbf{A}t}\,\mathbf{x}(t) = \int_0^t e^{-\mathbf{A}\tau}\,\mathbf{B}\mathbf{u}(\tau)\,d\tau + \mathbf{K} \qquad (3\text{-}90)$$

where $\mathbf{K}$ is an arbitrary constant matrix of integration. The lower limit on the integral is the initial time at which $\mathbf{u}(t)$ is applied. Inserting $t = 0$ into Eq. (3-90) yields $K = \mathbf{x}(0)$. Multiplying by $e^{\mathbf{A}t}$ and letting $\beta = t - \tau$ produces

$$\mathbf{x}(t) = e^{\mathbf{A}t}\mathbf{x}(0) + \int_0^t e^{\mathbf{A}(t-\tau)}\,\mathbf{B}\mathbf{u}(\tau)\,d\tau = \mathbf{\Phi}(t)\mathbf{x}(0) + \int_0^t \mathbf{\Phi}(t-\tau)\mathbf{B}\mathbf{u}(\tau)\,d\tau \qquad (3\text{-}91a)$$

$$= \mathbf{\Phi}(t)\mathbf{x}(0) + \int_0^t \mathbf{\Phi}(\beta)\mathbf{B}u(t-\beta)\,d\beta \qquad (3\text{-}91b)$$

The first term on the right is recognized as the complementary solution derived in Sec. 3-11. The second term is the particular solution for an input $\mathbf{u}(t)$. Using the STM and generalizing for initial conditions at time $t = t_0$ gives the solution to the state-variable equation with an input $\mathbf{u}(t)$

$$\mathbf{x}(t) = \mathbf{\Phi}(t - t_0)\mathbf{x}(t_0) + \int_{t_0}^t \mathbf{\Phi}(t - \tau)\mathbf{B}\mathbf{u}(\tau)\,d\tau \qquad t > t_0 \qquad (3\text{-}92)$$

This equation is called the *state transition equation*; i.e., it describes the change of state relative to the initial conditions $\mathbf{x}(t_0)$ and the input $\mathbf{u}(t)$.

EXAMPLE   Solve the following state equation for a unit step-function scalar input, $u(t) = u_{-1}(t)$:

$$\dot{\mathbf{x}} = \begin{bmatrix} 0 & 6 \\ -1 & -5 \end{bmatrix}\mathbf{x} + \begin{bmatrix} 0 \\ 1 \end{bmatrix}u(t)$$

The STM for this equation is given by Eq. (3-88). Thus, the total time solution is obtained by substituting this value of $\mathbf{\Phi}(t)$ into Eq. (3-92), with $t_0 = 0$. Performing the matrix multiplication $\mathbf{\Phi}(t - \tau)\mathbf{B}\mathbf{u}(\tau)$ yields

$$\mathbf{x}(t) = \mathbf{\Phi}(t)\mathbf{x}(0) + \int_0^t \begin{bmatrix} 6e^{-2(t-\tau)} - 6e^{-3(t-\tau)} \\ -2e^{-2(t-\tau)} + 3e^{-3(t-\tau)} \end{bmatrix} d\tau$$

$$= \begin{bmatrix} 3e^{-2t} - 2e^{-3t} & 6e^{-2t} - 6e^{-3t} \\ -e^{-2t} + e^{-3t} & -2e^{-2t} + 3e^{-3t} \end{bmatrix} \begin{bmatrix} x_1(0) \\ x_2(0) \end{bmatrix}$$

$$+ \begin{bmatrix} 1 - 3e^{-2t} + 2e^{-3t} \\ e^{-2t} - e^{-3t} \end{bmatrix} \qquad (3\text{-}93)$$

The integral of a matrix is the integral of each element in the matrix with respect to the indicated scalar variable. This property is used in evaluating the particular integral in Eq. (3-93). Inserting the initial conditions produces the final solution for the state transition equation. The Laplace transform method presented in Chap. 4 is a more direct method for finding this solution.

3-15   SIMULATION DIAGRAMS[4,7]

A block diagram is often used to represent the dynamic equations of a system. The simulation may show physical variables which appear in the system, or it may show

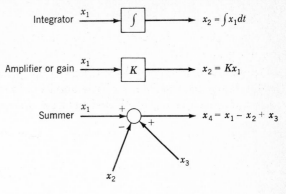

FIGURE 3-11
Elements used in a simulation diagram.

variables that are used purely for mathematical convenience. In either case the over-all response of the system is the same. The simulation diagram is similar to the diagram used to represent the system on an analog computer. The basic elements used are ideal integrators, ideal amplifiers, and ideal summers, shown in Fig. 3-11. Additional elements such as multipliers and dividers may be used for nonlinear systems.

One of the methods used to obtain a simulation diagram includes the following steps.

Step 1 Start with the differential equation.

Step 2 On the left side of the equation put the highest-order derivative of the *dependent* variable. A first-order or higher-order derivative of the input may appear in the equation. In this case the highest-order derivative of the input is also placed on the left side of the equation. All other terms are put on the right side.

Step 3 Start the diagram by assuming that the signal, represented by the terms on the left side of the equation, is available. Then integrate it as many times as needed to obtain all the lower-order derivatives. It may be necessary to add a summer in the simulation diagram to obtain the dependent variable explicitly.

Step 4 Complete the diagram by feeding back the appropriate outputs of the integrators to a summer to generate the original signal of step 2. Include the input function if it is required.

EXAMPLE Draw the simulation diagram for the series *RLC* circuit of Fig. 2-2 in which the output is the voltage across the capacitor.

Step 1 When $y = v_c$ and $u = e$ are used, Eq. (2-10) becomes

$$LC\ddot{y} + RC\dot{y} + y = u \qquad (3\text{-}94)$$

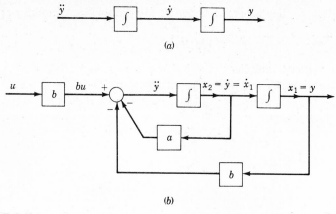

FIGURE 3-12
Simulation diagram for Eq. (3-94).

Step 2 Rearrange terms to the form

$$\ddot{y} = bu - a\dot{y} - by \qquad a = R/L \qquad (3\text{-}95)$$
$$b = 1/LC$$

Step 3 The signal $\ddot{y}$ is integrated twice, as shown in Fig. 3-12*a*.

Step 4 The block or simulation diagram is completed as shown in Fig. 3-12*b* in order to satisfy Eq. (3-95).

The state variables are often selected as the outputs of the integrators in the simulation diagram. In this case they are $x_1 = y$ and $x_2 = \dot{x}_1 = \dot{y}$. The state and output equations are therefore

$$\begin{bmatrix} \dot{x}_1 \\ \dot{x}_2 \end{bmatrix} = \begin{bmatrix} 0 & 1 \\ -\dfrac{1}{LC} & -\dfrac{R}{L} \end{bmatrix} \begin{bmatrix} x_1 \\ x_2 \end{bmatrix} + \begin{bmatrix} 0 \\ \dfrac{1}{LC} \end{bmatrix} u \qquad (3\text{-}96)$$

$$y = \begin{bmatrix} 1 & 0 \end{bmatrix} \begin{bmatrix} x_1 \\ x_2 \end{bmatrix} + \mathbf{0}u \qquad (3\text{-}97)$$

It is common in a physical system for the **D** matrix to be zero, as in this example. These equations are different from Eqs. (2-34) and (2-36), yet they represent the same system. This illustrates that state variables are not unique. When the state variables are the dependent variable and the derivatives of the dependent variable, as in this example, they are called *phase variables*. The phase variables are applicable to differential equations of any order and can be used without drawing the simulation diagram.

Case 1 The general differential equation which contains no derivatives of the input is

$$D^n y + a_{n-1} D^{n-1} y + \cdots + a_1 Dy + a_0 y = u \qquad (3\text{-}98)$$

The phase variables are selected as $x_1 = y$, $x_2 = Dx_1 = \dot{x}_1$, $x_3 = \dot{x}_2, \ldots, x_n = \dot{x}_{n-1}$. When these state variables are used, Eq. (3-98) becomes $\dot{x}_n + a_{n-1}x_n + a_{n-2}x_{n-1} + \cdots + a_1 x_2 + a_0 x_1 = u$. The resulting state and output equations using phase variables are

$$\dot{\mathbf{x}} = \begin{bmatrix} 0 & 1 & & & & & \\ & 0 & 1 & & & \mathbf{0} & \\ & & 0 & 1 & & & \\ & & & 0 & \cdot & & \\ & \mathbf{0} & & & 0 & \cdot & \\ & & & & & 0 & \cdot \\ & & & & & 0 & 1 \\ -a_0 & -a_1 & -a_2 & -a_3 & \cdots & -a_{n-2} & -a_{n-1} \end{bmatrix} \mathbf{x} + \begin{bmatrix} 0 \\ 0 \\ \vdots \\ 0 \\ 1 \end{bmatrix} u \qquad (3\text{-}99)$$

$$y = \begin{bmatrix} 1 & 0 & 0 & \cdots \end{bmatrix} x \qquad (3\text{-}100)$$

This form of the state and output equations can be written directly from the original differential equation (3-98)†. The **A** matrix contains the number 1 in the *superdiagonal* and the negative of the coefficients of the original differential equation in the nth row. In this simple form the matrix **A** is called the *companion matrix*. Also, the **B** matrix takes on the form shown in Eq. (3-99) and is indicated by $\mathbf{B}_0$.

Case 2 When derivatives of $u(t)$ appear in the differential equation, the phase variables may be used as the state variables and Eq. (3-99) still applies; however, the output equation is no longer given by Eq. (3-100). This is shown by considering the differential equation

$$(D^n + a_{n-1}D^{n-1} + \cdots + a_1 D + a_0)y$$
$$= (b_w D^w + b_{w-1}D^{w-1} + \cdots + b_1 D + b_0)u \qquad w \le n \qquad (3\text{-}101)$$

The output y is specified as

$$y = (b_w D^w + b_{w-1}D^{w-1} + \cdots + b_1 D + b_0)x_1 \qquad (3\text{-}102)$$

$$y = b_w \dot{x}_w + b_{w-1}x_w + \cdots + b_1 x_2 + b_0 x_1 \qquad (3\text{-}102a)$$

Substituting Eq. (3-102) into Eq. (3-101) yields

$$(D^n + a_{n-1}D^{n-1} + \cdots + a_1 D + a_0)x_1 = u \qquad (3\text{-}103)$$

The two equations given by Eqs. (3-102) and (3-103) represent the system of Eq. (3-101)†. With phase variables, Eq. (3-103) can be represented in the form $\dot{\mathbf{x}} = \mathbf{Ax} + \mathbf{B}u$, where the matrices **A** and **B** are the same as in Eq. (3-99). The new output equation must satisfy Eq. (3-102a) (see Prob. 3-14). The simulation diagram has the advantage that only two summers are required, regardless of the order of the system, one at the input and one at the output. This representation of a differential equation is considered again in Sec. 5-7. Figure 3-13 shows the simulation diagram which represents

† Replacing u by $K_G u$ in Eqs. (3-98) and (3-101) the state equations can be expressed as in Eq. (12-22). Then in Fig. 3-13 the signal u is replaced by $K_G u$.

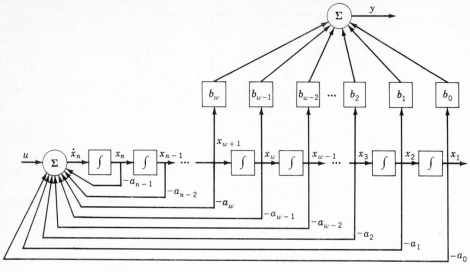

FIGURE 3-13
Simulation diagram representing the system of Eq. (3-101) in terms of Eqs. (3-102a) and (3-103).

the system of Eq. (3-101) in terms of Eqs. (3-102a) and (3-103). It has the desirable feature that differentiation of the input signal $u(t)$ is not required to satisfy the differential equation. Differentiators are avoided in analog and digital-computer simulations since they accentuate any noise present in the signals.

Different sets of state variables may be selected to represent a system. The selection of the state variables determines the **A**, **B**, **C**, and **D** matrices. A general matrix block diagram representing the state and output equations is shown in Fig. 3-14. The matrix **D** is called the *feedforward* matrix.

3-16 GENERAL PROGRAMMING

Another method of converting from a differential equation to state representation is presented in this section. It is a general systematic method which is applicable to

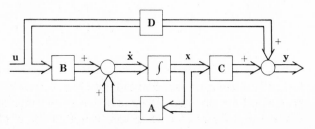

FIGURE 3-14
General matrix block diagram representing the state and output equations.

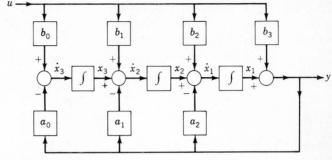

FIGURE 3-15
Simulation diagram for Eq. (3-105).

time-invariant systems with a single input and a single output. It applies even when derivatives of the input $u(t)$ are present. As an example, consider the third-order differential equation

$$D^3 y + a_2 D^2 y + a_1 Dy + a_0 y = b_3 D^3 u + b_2 D^2 u + b_1 Du + b_0 u \qquad (3\text{-}104)$$

Dividing by D^3 and collecting terms gives

$$y = b_3 u + \frac{1}{D}(b_2 u - a_2 y) + \frac{1}{D^2}(b_1 u - a_1 y) + \frac{1}{D^3}(b_0 u - a_0 y)$$

$$= b_3 u + \frac{1}{D}\left\{(b_2 u - a_2 y) + \frac{1}{D}\left[(b_1 u - a_1 y) + \frac{1}{D}(b_0 u - a_0 y)\right]\right\} \qquad (3\text{-}105)$$

The simulation diagram for this equation is shown in Fig. 3-15. The integrator outputs are defined as the state variables. Then, from the diagram, the following equations can be written:

$$y = x_1 + b_3 u$$
$$\dot{x}_1 = x_2 - a_2 y + b_2 u = x_2 - a_2 x_1 + (b_2 - a_2 b_3)u$$
$$\dot{x}_2 = x_3 - a_1 y + b_1 u = x_3 - a_1 x_1 + (b_1 - a_1 b_3)u \qquad (3\text{-}106)$$
$$\dot{x}_3 = \quad\; - a_0 y + b_0 u = \quad\; - a_0 x_1 + (b_0 - a_0 b_3)u$$

The state and output equations are therefore

$$\dot{\mathbf{x}} = \begin{bmatrix} -a_2 & 1 & 0 \\ -a_1 & 0 & 1 \\ -a_0 & 0 & 0 \end{bmatrix} \mathbf{x} + \begin{bmatrix} b_2 - a_2 b_3 \\ b_1 - a_1 b_3 \\ b_0 - a_0 b_3 \end{bmatrix} u \qquad (3\text{-}107)$$

$$\mathbf{y} = \begin{bmatrix} 1 & 0 & 0 \end{bmatrix}\mathbf{x} + [b_3]u \qquad (3\text{-}108)$$

Note that these state variables are not phase variables. The procedure shown can be generalized for any order of equation. An advantage of this method is the simplicity of the **A** and **B** matrices, which can be written directly from the differential equation of Eq. (3-104). The **A** matrix contains the values of 1 on the superdiagonal, and the negative of the coefficients appear in the first column. The elements of the **B** matrix have a simple systematic pattern, but the plant coefficients appear in this matrix and if those coefficients vary, then the elements b_i resulting from this simulation also vary. Therefore, this simulation is useful only for linear systems. The simulation diagram of Fig. 3-15 contains the coefficients of the differential equation as amplifier gains and can easily be implemented on an analog computer.

3-17 STANDARD-FORM PROGRAMMING

Another method that appears frequently in the literature for converting a differential equation to a state-variable representation is presented in this section. A salient feature of this system representation is that derivatives of $u(t)$ do not appear in the resulting state equation and simulation diagram. A general nth-order linear, constant-coefficient differential equation is

$$D^n y + a_{n-1} D^{n-1}y + \cdots + a_1 Dy + a_0 y$$
$$= b_w D^w u + b_{w-1}D^{w-1} u + \cdots + b_1 Du + b_0 u \quad \text{where } w \le n \quad (3\text{-}109)$$

Although derivatives of the input are present in this equation, it is desired to obtain the same $\mathbf{A}$ matrix as in Eq. (3-99). This is achieved by the following procedure, which eliminates the derivatives of u, one at a time. When $w = n$, the term $b_n D^n u$ is combined with $D^n y$ to yield

$$D^n(y - b_n u) = D^n x_1 \quad (3\text{-}110)$$

in which the first state variable is defined by

$$x_1 \equiv y - b_n u \quad (3\text{-}111)$$

The output is therefore

$$y = x_1 + b_n u \quad (3\text{-}112)$$

Inserting this function into Eq. (3-109) and combining terms yields

$$D^n x_1 + a_{n-1} D^{n-1}x_1 + \cdots + a_1 Dx_1 + a_0 x_1$$
$$= (b_{n-1} - a_{n-1}b_n) D^{n-1}u + \cdots + (b_1 - a_1 b_n) Du + (b_0 - a_0 b_n)u \quad (3\text{-}113)$$

Note that defining the first state variable by Eq. (3-111) has resulted in the elimination of the $D^n u$ term as seen in Eq. (3-113). The procedure is continued, combining the terms $D^n x_1$ and $(b_{n-1} - a_{n-1}b_n) D^{n-1}u$ to yield

$$D^{n-1}[Dx_1 - (b_{n-1} - a_{n-1}b_n)u] \equiv D^{n-1}x_2 \quad (3\text{-}114)$$

The second state variable is defined in Eq. (3-114) and yields the state equation

$$Dx_1 = x_2 + (b_{n-1} - a_{n-1}b_n)u \quad (3\text{-}115)$$

Inserting this value into Eq. (3-113) yields

$$D^{n-1}x_2 + a_{n-1} D^{n-2}x_2 + \cdots + a_1 x_2 + a_0 x_1$$
$$= [(b_{n-2} - a_{n-2}b_n) - a_{n-1}(b_{n-1} - a_{n-1}b_n)] D^{n-2}u + \cdots \quad (3\text{-}116)$$

The procedure is continued until all the derivatives of u have been eliminated and n state variables have been defined. The state and output equations are

$$\dot{\mathbf{x}} = \mathbf{A}\mathbf{x} + \mathbf{B}u \qquad y = \mathbf{C}\mathbf{x} + \mathbf{D}u$$

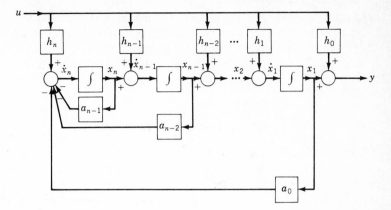

FIGURE 3-16
Simulation diagram for Eq. (3-109) with $w = n$.

where the resulting matrices are

$$\mathbf{A} = \begin{bmatrix} 0 & 1 & & & & & \\ & 0 & 1 & & & \mathbf{0} & \\ & & 0 & \cdot & & & \\ & & & 0 & \cdot & & \\ & \mathbf{0} & & & 0 & \cdot & \\ & & & & & 0 & 1 \\ -a_0 & -a_1 & -a_2 & \cdots & & -a_{n-2} & -a_{n-1} \end{bmatrix} \qquad (3\text{-}117)$$

$$\begin{bmatrix} \mathbf{D} \\ \mathbf{B} \end{bmatrix} = \begin{bmatrix} h_0 \\ h_1 \\ h_2 \\ \cdot \\ h_n \end{bmatrix} = \begin{bmatrix} b_n \\ \hline b_{n-1} - a_{n-1}h_0 \\ b_{n-2} - a_{n-1}h_1 - a_{n-2}h_0 \\ \cdots \cdots \cdots \cdots \cdots \cdots \cdots \cdots \cdots \cdots \\ b_0 - a_{n-1}h_{n-1} - a_{n-2}h_{n-2} - \cdots - a_0 h_0 \end{bmatrix} \qquad (3\text{-}118)$$

$$\mathbf{C} = \begin{bmatrix} 1 & 0 & \cdots & 0 \end{bmatrix} \qquad (3\text{-}119)$$

$$\mathbf{D} = h_0 \qquad (3\text{-}120)$$

The kth state equation is given by

$$\dot{x}_k = x_{k+1} + h_k u \qquad 1 \le k \le n - 1 \qquad (3\text{-}121)$$

The simulation diagram is shown in Fig. 3-16 for $w = n$. This form of the simulation diagram and the corresponding state and output equations are called the *standard form*. Note that the **A** matrix is the same as for the phase-variable method, i.e., a companion matrix.

EXAMPLE   Find the state and output equations in general, standard, and phase-variable forms:

$$\ddot{y} + 2\dot{y} + y = \dot{u} + 2u$$

The coefficients are $a_1 = 2$, $a_0 = 1$, $b_2 = 0$, $b_1 = 1$, $b_0 = 2$.

General form:

$$A = \begin{bmatrix} -2 & 1 \\ -1 & 0 \end{bmatrix} \qquad B = \begin{bmatrix} 1 \\ 2 \end{bmatrix} \qquad C = \begin{bmatrix} 1 & 0 \end{bmatrix} \qquad D = 0$$

Standard form:

$$A = \begin{bmatrix} 0 & 1 \\ -1 & -2 \end{bmatrix} \qquad B = \begin{bmatrix} 1 \\ 0 \end{bmatrix} \qquad C = \begin{bmatrix} 1 & 0 \end{bmatrix} \qquad D = 0$$

Phase-variable form:

$$A = \begin{bmatrix} 0 & 1 \\ -1 & -2 \end{bmatrix} \qquad B = \begin{bmatrix} 0 \\ 1 \end{bmatrix} \qquad C = \begin{bmatrix} 2 & 1 \end{bmatrix} \qquad D = 0$$

It should be recalled (see Sec. 2-11) that the physical variables cannot be used when derivatives of the forcing function appear in the system differential equation. The methods in this chapter are capable of handling this situation.

3-18 SUMMARY

This chapter has established the manner of solving differential equations. The steady-state and transient solutions are determined separately, and then the constants of the transient component of the solution are evaluated to satisfy the initial conditions. The ability to anticipate the form of the response is very important. The solution obtained should be reasonable. The solution can be checked at known times as a check of its reasonableness.

The solution of differential equations has been extended to include solution of the matrix state and output equations. This method is also applicable to multiple-input multiple-output, time-varying, and nonlinear systems. The matrix formulation of these systems lends itself to digital-computer solutions. The solution contains a complementary component which is proportional to the initial values. It also contains a particular solution which is a function of the input forcing function. Obtaining the complete solution by means of the Laplace transform is covered in the next chapter.

Also presented are several techniques for converting an overall system differential equation into state-variable format. Since the state variables are not unique, there are a number of ways of selecting them. The use of physical variables is presented in Chap. 2. Some additional selections for the system state variables have been introduced through the medium of the simulation diagram. This block diagram is similar to an analog-computer diagram and uses integrators, amplifiers, and summers. The outputs of the integrators are the state variables.

REFERENCES

1 Wylie, C. R., Jr.: "Advanced Engineering Mathematics," 4th ed., McGraw-Hill, New York, 1975.
2 Kinariwala, B., et al.: "Linear Circuits and Computation," Wiley, New York, 1973.

3 D'Azzo, J. J., and C. H. Houpis: "Feedback Control System Analysis and Synthesis," 2d ed., McGraw-Hill, New York, 1966.

4 DeRusso, P. M., et al.: "State Variables for Engineers," Wiley, New York, 1965.

5 Schwartz, R. J., and B. Friedland: "Linear Systems," McGraw-Hill, New York, 1965.

6 Ward, J. R., and R. D. Strum: "State Variable Analysis," Prentice-Hall, Englewood Cliffs, N.J., 1970.

7 Wiberg, D. M.: "State Space and Linear Systems," Schaum's Outline Series, McGraw-Hill, New York, 1971.

4

LAPLACE TRANSFORMS

4-1 INTRODUCTION

Solution of differential equations with discontinuous inputs or of higher than second order is laborious by the classical method. Also, inserting the initial conditions to evaluate the constants of integration requires the simultaneous solution of a number of algebraic equations equal to the order of the differential equation. To facilitate and systematize the solution of ordinary constant-coefficient differential equations, the Laplace transform method is used extensively.[1-3] This transform method has found wide acceptance and is used in the technical literature and books on feedback systems. The advantages of this modern transform method are:

1 It includes the boundary or initial conditions.
2 The work involved in the solution is simple algebra.
3 The work is systematized.
4 The use of a table of transforms reduces the labor required.
5 Discontinuous inputs can be treated.
6 The transient and steady-state components of the solution are obtained simultaneously.

The disadvantage of transform methods is that if they are used mechanically, without knowledge of the actual theory involved, they sometimes yield erroneous results. Also, a particular equation can sometimes be solved more simply and with less work by the classical method.

Laplace transforms are applied to the solution of system equations which are in matrix state-variable format. The method for using the state and output equations to obtain the system transfer function is also presented.

4-2   DEFINITION OF THE LAPLACE TRANSFORM

The direct Laplace transformation of a function of time $f(t)$ is given by

$$\mathscr{L}[f(t)] = \int_0^\infty f(t)e^{-st}\, dt = F(s) \qquad (4\text{-}1)$$

where $\mathscr{L}[f(t)]$ is a shorthand notation for the Laplace integral. Evaluation of the integral results in a function $F(s)$ which has s as the parameter. This parameter s is a complex quantity of the form $\sigma + j\omega$. It is to be noted that as the limits of integration are zero and infinity, it is immaterial what value $f(t)$ has for negative or zero time.

There are limitations on the functions $f(t)$ that are Laplace-transformable. Basically, the requirement is that the Laplace integral converge, which means that this integral has a definite functional value. To meet this requirement[3] the function $f(t)$ must be (1) piecewise continuous over every finite interval $0 \le t_1 \le t \le t_2$ and (2) of exponential order. A function is piecewise continuous in a finite interval if that interval can be divided into a finite number of subintervals over each of which the function is continuous and at the ends of each of which $f(t)$ possesses finite right- and left-hand limits. A function $f(t)$ is of exponential order if there exists a constant a such that the product $e^{-at}|f(t)|$ is bounded for all values of t greater than some finite value T. This imposes the restriction that σ, the real part of s, must be greater than a lower bound σ_a for which the product $e^{-\sigma_a t}|f(t)|$ is of exponential order. A linear differential equation with constant coefficients and with a finite number of terms is Laplace-transformable if the driving function is Laplace-transformable.

All cases covered in this book are Laplace-transformable. The basic purpose in using the Laplace transform is to obtain a method of solving differential equations which involves only simple algebraic operations in conjunction with a table or simple analysis.

4-3   DERIVATION OF LAPLACE TRANSFORMS OF SIMPLE FUNCTIONS

A number of examples are presented to show the derivation of the Laplace transform of several time functions. A list of common transform pairs is given in Appendix A.

Step Function $u_{-1}(t)$

The Laplace transform of the unit step function $u_{-1}(t)$ is

$$\mathscr{L}[u_{-1}(t)] = \int_0^\infty u_{-1}(t)e^{-st}\, dt = U_{-1}(s) \qquad (4\text{-}2)$$

Since $u_{-1}(t)$ has the value 1 over the limits of integration,

$$U_{-1}(s) = \int_0^\infty e^{-st}\,dt = -\frac{e^{-st}}{s}\bigg|_0^\infty = \frac{1}{s} \qquad \text{if } \sigma > 0 \qquad (4\text{-}3)$$

The step function is undefined at $t = 0$; but this is immaterial, for it is noted that the integral is defined by a limit process,

$$\int_0^\infty f(t)e^{-st}\,dt = \lim_{\substack{T\to\infty \\ \varepsilon\to 0}} \int_\varepsilon^T f(t)e^{-st}\,dt \qquad (4\text{-}4)$$

and the explicit value at $t = 0$ does not affect the value of the integral. The value of the integral obtained by taking limits is implied in each case but is not written out explicitly.

Decaying Exponential $e^{-\alpha t}$

The exponent α is a positive real number.

$$\mathscr{L}[e^{-\alpha t}] = \int_0^\infty e^{-\alpha t}e^{-st}\,dt = \int_0^\infty e^{-(s+\alpha)t}\,dt$$

$$= -\frac{e^{-(s+\alpha)t}}{s+\alpha}\bigg|_0^\infty = \frac{1}{s+\alpha} \qquad \sigma > -\alpha \qquad (4\text{-}5)$$

Sinusoid $\cos \omega t$

Here ω is a positive real number.

$$\mathscr{L}[\cos \omega t] = \int_0^\infty \cos \omega t\, e^{-st}\,dt \qquad (4\text{-}6)$$

Expressing $\cos \omega t$ in exponential form gives

$$\cos \omega t = \frac{e^{j\omega t} + e^{-j\omega t}}{2}$$

Then $\quad \mathscr{L}[\cos \omega t] = \dfrac{1}{2}\left(\displaystyle\int_0^\infty e^{(j\omega - s)t}\,dt + \int_0^\infty e^{(-j\omega - s)t}\,dt \right)$

$$= \frac{1}{2}\left[\frac{e^{(j\omega - s)t}}{j\omega - s} + \frac{e^{(-j\omega - s)t}}{-j\omega - s} \right]_0^\infty$$

$$= \frac{1}{2}\left(-\frac{1}{j\omega - s} - \frac{1}{-j\omega - s} \right) = \frac{s}{s^2 + \omega^2} \qquad \sigma > 0 \qquad (4\text{-}7)$$

Ramp Function $u_{-2}(t) = tu_{-1}(t)$

$$\mathscr{L}[t] = \int_0^\infty te^{-st}\,dt \qquad \sigma > 0 \qquad (4\text{-}8)$$

This is integrated by parts by using

$$\int_a^b u\, dv = uv\Big|_a^b - \int_a^b v\, du$$

Let $u = t$ and $dv = e^{-st}\, dt$. Then $du = dt$ and $v = -e^{-st}/s$. Thus

$$\int_0^\infty te^{-st}\, dt = -\frac{te^{-st}}{s}\Big|_0^\infty - \int_0^\infty \left(-\frac{e^{-st}}{s}\right) dt$$

$$= 0 - \frac{e^{-st}}{s^2}\Big|_0^\infty = \frac{1}{s^2} \qquad \sigma > 0 \qquad (4\text{-}9)$$

4-4 LAPLACE TRANSFORM THEOREMS

Several theorems that are useful in applying the Laplace transform are presented in this section. In general, they are helpful in evaluating transforms.

Theorem 1 Linearity If a is a constant or is independent of s and t, and if $f(t)$ is transformable, then

$$\mathscr{L}[af(t)] = a\mathscr{L}[f(t)] = aF(s) \qquad (4\text{-}10)$$

Theorem 2 Superposition If $f_1(t)$ and $f_2(t)$ are both Laplace-transformable, the principle of superposition applies:

$$\mathscr{L}[f_1(t) \pm f_2(t)] = \mathscr{L}[f_1(t)] \pm \mathscr{L}[f_2(t)] = F_1(s) \pm F_2(s) \qquad (4\text{-}11)$$

Theorem 3 Translation in time If the Laplace transform of $f(t)$ is $F(s)$ and a is a positive real number, the Laplace transform of the translated function $f(t-a)u_{-1}(t-a)$ is

$$\mathscr{L}[f(t-a)u_{-1}(t-a)] = e^{-as}F(s) \qquad (4\text{-}12)$$

Translation in the positive t direction in the real domain becomes multiplication by the exponential e^{-as} in the s domain.

Theorem 4 Complex differentiation If the Laplace transform of $f(t)$ is $F(s)$, then

$$\mathscr{L}[tf(t)] = -\frac{d}{ds}F(s) \qquad (4\text{-}13)$$

Multiplication by time in the real domain entails differentiation with respect to s in the s domain.

EXAMPLE 1

$$\mathscr{L}[te^{-\alpha t}] = -\frac{d}{ds}\mathscr{L}[e^{-\alpha t}] = -\frac{d}{ds}\left(\frac{1}{s+\alpha}\right) = \frac{1}{(s+\alpha)^2}$$

Theorem 5   Translation in the s domain   If the Laplace transform of $f(t)$ is $F(s)$ and a is either real or complex, then

$$\mathscr{L}[e^{at}f(t)] = F(s - a) \qquad (4\text{-}14)$$

Multiplication by e^{at} in the real domain becomes translation in the s domain.

EXAMPLE 2   Starting with $\mathscr{L}[\sin \omega t] = \omega/(s^2 + \omega^2)$ and applying Theorem 5 gives

$$\mathscr{L}[e^{-\alpha t} \sin \omega t] = \frac{\omega}{(s + \alpha)^2 + \omega^2}$$

Theorem 6   Real differentiation   If the Laplace transform of $f(t)$ is $F(s)$, and if the first derivative of $f(t)$ with respect to time $Df(t)$ is transformable, then

$$\mathscr{L}[Df(t)] = sF(s) - f(0^+) \qquad (4\text{-}15)$$

The term $f(0^+)$ is the value of the right-hand limit of the function $f(t)$ as the origin, $t = 0$, is approached from the right side (thus through positive values of time).   This includes functions, such as the step function, that may be undefined at $t = 0$.   For simplicity, the plus sign following the zero is usually omitted although its presence is implied.

The transform of the second derivative $D^2f(t)$ is

$$\mathscr{L}[D^2f(t)] = s^2F(s) - sf(0) - Df(0) \qquad (4\text{-}16)$$

where $Df(0)$ is the value of the limit of the derivative of $f(t)$ as the origin, $t = 0$, is approached from the right side.

The transform of the nth derivative $D^nf(t)$ is

$$\mathscr{L}[D^nf(t)] = s^nF(s) - s^{n-1}f(0) - s^{n-2}\,Df(0) - \cdots - D^{n-1}f(0) \qquad (4\text{-}17)$$

Note that the transform includes the initial conditions, whereas in the classical method of solution the initial conditions are introduced separately to evaluate the coefficients of the solution of the differential equation. When all initial conditions are zero, the Laplace transform of the nth derivative of $f(t)$ is simply $s^nF(s)$.

Theorem 7   Real integration   If the Laplace transform of $f(t)$ is $F(s)$, its integral

$$D^{-1}f(t) = \int_0^t f(t)\,dt + D^{-1}f(0^+)$$

is transformable and the value of its transform is

$$\mathscr{L}[D^{-1}f(t)] = \frac{F(s)}{s} + \frac{D^{-1}f(0^+)}{s} \qquad (4\text{-}18)$$

The term $D^{-1}f(0^+)$ is the constant of integration and is equal to the value of the integral as the origin is approached from the positive, or right, side. The plus sign is omitted in the remainder of this text.

The transform of the double integral $D^{-2}f(t)$ is

$$\mathscr{L}[D^{-2}f(t)] = \frac{F(s)}{s^2} + \frac{D^{-1}f(0)}{s^2} + \frac{D^{-2}f(0)}{s} \qquad (4\text{-}19)$$

The transform of the nth-order integral $D^{-n}f(t)$ is

$$\mathscr{L}[D^{-n}f(t)] = \frac{F(s)}{s^n} + \frac{D^{-1}f(0)}{s^n} + \cdots + \frac{D^{-n}f(0)}{s} \qquad (4\text{-}20)$$

Theorem 8   Final value   If $f(t)$ and $Df(t)$ are Laplace-transformable, if the Laplace transform of $f(t)$ is $F(s)$, and if the limit $f(t)$ as $t \to \infty$ exists, then

$$\lim_{s \to 0} sF(s) = \lim_{t \to \infty} f(t) \qquad (4\text{-}21)$$

This theorem states that the behavior of $f(t)$ in the neighborhood of $t = \infty$ is related to the behavior of $sF(s)$ in the neighborhood of $s = 0$. If $sF(s)$ has poles [values of s for which $|sF(s)|$ becomes infinite] on the imaginary axis (excluding the origin) or in the right-half s plane, there is no finite final value of $f(t)$ and the theorem cannot be used. If $f(t)$ is sinusoidal, the theorem is invalid, since $\mathscr{L}[\sin \omega t]$ has poles at $s = \pm j\omega$ and $\lim_{t \to \infty} \sin \omega t$ does not exist. However, for poles of $sF(s)$ at the origin, $s = 0$, this theorem gives the final value of $f(\infty) = \infty$. This correctly describes the behavior of $f(t)$ as $t \to \infty$.

Theorem 9   Initial value   If the function $f(t)$ and its first derivative are Laplace-transformable, if the Laplace transform of $f(t)$ is $F(s)$, and if $\lim_{s \to \infty} sF(s)$ exists, then

$$\lim_{s \to \infty} sF(s) = \lim_{t \to 0} f(t) \qquad (4\text{-}22)$$

This theorem states that the behavior of $f(t)$ in the neighborhood of $t = 0$ is related to the behavior of $sF(s)$ in the neighborhood of $|s| = \infty$. There are no limitations on the location of the poles of $sF(s)$.

Theorem 10   Complex integration   If the Laplace transform of $f(t)$ is $F(s)$ and if $f(t)/t$ has a limit as $t \to 0^+$, then

$$\mathscr{L}\left[\frac{f(t)}{t}\right] = \int_{s}^{\infty} F(s)\, ds \qquad (4\text{-}23)$$

This theorem states that division by the variable in the real domain entails integration with respect to s in the s domain.

4-5   APPLICATION OF THE LAPLACE TRANSFORM TO DIFFERENTIAL EQUATIONS

The Laplace transform is now applied to the solution of the differential equation for the simple mechanical system which was solved by the classical method in Sec. 3-7. The differential equation of the system is repeated below:

$$M\,D^2 x_2 + B\,Dx_2 + Kx_2 = Kx_1 \qquad (4\text{-}24)$$

The position $x_1(t)$ undergoes a unit step displacement. This is the input and is called the *driving* function. The unknown quantity for which the equation is to be solved is the output displacement $x_2(t)$, called the *response function*. The Laplace transform of Eq. (4-24) is

$$\mathscr{L}[Kx_1] = \mathscr{L}[M\ D^2x_2 + B\ Dx_2 + Kx_2] \qquad (4\text{-}25)$$

The transform of each term is

$$\mathscr{L}[Kx_1] = KX_1(s)$$
$$\mathscr{L}[Kx_2] = KX_2(s)$$
$$\mathscr{L}[B\ Dx_2] = B[sX_2(s) - x_2(0)]$$
$$\mathscr{L}[M\ D^2x_2] = M[s^2X_2(s) - sx_2(0) - Dx_2(0)]$$

Substituting these terms into Eq. (4-25) and collecting terms gives

$$KX_1(s) = (Ms^2 + Bs + K)X_2(s) - [Msx_2(0) + M\ Dx_2(0) + Bx_2(0)] \qquad (4\text{-}26)$$

Equation (4-26) is the transform equation and shows how the initial conditions—the initial position $x_2(0)$ and the initial velocity $Dx_2(0)$—are incorporated into the equation. The function $X_1(s)$ is called the *driving transform*; the function $X_2(s)$ is called the *response transform*. The coefficient of $X_2(s)$, which is $Ms^2 + Bs + K$, is called the *characteristic function*. The equation formed by setting the characteristic function equal to zero is called the *characteristic equation* of the system. Solving for $X_2(s)$ gives

$$X_2(s) = \frac{K}{Ms^2 + Bs + K}\,X_1(s) + \frac{Msx_2(0) + Bx_2(0) + M\ Dx_2(0)}{Ms^2 + Bs + K} \qquad (4\text{-}27)$$

The coefficient of $X_1(s)$ is defined as the *system transfer function*. The second term on the right side of the equation is called the *initial condition component*. Combining the terms of Eq. (4-27) yields

$$X_2(s) = \frac{KX_1(s) + Msx_2(0) + Bx_2(0) + M\ Dx_2(0)}{Ms^2 + Bs + K} \qquad (4\text{-}28)$$

Finding the function $x_2(t)$ whose transform is given by Eq. (4-28) is symbolized by the inverse transform operator $\mathscr{L}^{-1}$; thus

$$
\begin{aligned}
x_2(t) &= \mathscr{L}^{-1}[X_2(s)] \\
&= \mathscr{L}^{-1}\left[\frac{KX_1(s) + Msx_2(0) + Bx_2(0) + M\ Dx_2(0)}{Ms^2 + Bs + K}\right]
\end{aligned}
\qquad (4\text{-}29)
$$

The function $x_2(t)$ can be found in the table of Appendix A after inserting numerical values into Eq. (4-29). For this example the system is initially at rest. From the principle of conservation of energy the initial conditions are $x_2(0) = 0$ and $Dx_2(0) = 0$. Assume, as in Sec. 3-7, that the damping ratio ζ is less than unity. Since $x_1(t)$ is a step function, $X_1(s) = 1/s$ and

$$x_2(t) = \mathscr{L}^{-1}\left[\frac{K/M}{s(s^2 + Bs/M + K/M)}\right] = \mathscr{L}^{-1}\left[\frac{\omega_n^2}{s(s^2 + 2\zeta\omega_n s + \omega_n^2)}\right] \qquad (4\text{-}30)$$

where $\omega_n = \sqrt{K/M}$ and $\zeta = B/2\sqrt{KM}$. Reference to transform pair 27a in Appendix A provides the solution directly as

$$x_2(t) = 1 - \frac{e^{-\zeta \omega_n t}}{\sqrt{1 - \zeta^2}} \sin\left(\omega_n \sqrt{1 - \zeta^2}\, t + \cos^{-1} \zeta\right)$$

4-6   INVERSE TRANSFORMATION

The application of the Laplace transforms to a differential equation yields an algebraic equation. From the algebraic equation the transform of the response function is readily found. To complete the solution the inverse transform must be found. In some cases the inverse-transform operation

$$f(t) = \mathcal{L}^{-1}[F(s)] \qquad (4\text{-}31)$$

can be performed by direct reference to transform tables. The linearity and translation theorems are useful in extending the tables. When the response transform cannot be found in the tables, the general procedure is to express $F(s)$ as the sum of partial fractions with constant coefficients. The partial fractions have a first-order or quadratic factor in the denominator and are readily found in the table of transforms. The complete inverse transform is the sum of the inverse transforms of each fraction.

The response transform $F(s)$ can be expressed, in general, as the ratio of two polynomials $P(s)$ and $Q(s)$. Consider that these polynomials are of degree w and n, respectively, and are arranged in descending order of the powers of the variable s; thus,

$$F(s) = \frac{P(s)}{Q(s)} = \frac{a_w s^w + a_{w-1} s^{w-1} + \cdots + a_1 s + a_0}{s^n + b_{n-1} s^{n-1} + \cdots + b_1 s + b_0} \qquad (4\text{-}32)$$

The a's and b's are real constants, and the coefficient of the highest power of s in the denominator has been made equal to unity. Only those $F(s)$ which are proper fractions are considered, i.e., those in which n is greater than w.† The first step is to factor $Q(s)$ into first-order and quadratic factors with real coefficients:

$$F(s) = \frac{P(s)}{Q(s)} = \frac{P(s)}{(s - s_1)(s - s_2) \cdots (s - s_k) \cdots (s - s_n)} \qquad (4\text{-}33)$$

The values $s_1, s_2, \ldots, s_n$ in the finite plane that make the denominator equal to zero are called the *zeros* of the denominator. These values of s, which may be either real or complex, also make $|F(s)|$ infinite, and so they are called *poles* of $F(s)$. Therefore, the values $s_1, s_2, \ldots, s_n$ are referred to as zeros of the denominator or poles of the complete function in the finite plane; i.e., there are n poles of $F(s)$. Methods of factoring polynomials exist in the literature. Digital-computer programs are available to perform this operation.

---

† If $n = w$, first divide $P(s)$ by $Q(s)$ to obtain $F(s) = a_w + P_1(s)/Q(s) = a_w + F_1(s)$. Then express $F_1(s)$ as the sum of partial fractions with constant coefficients.

The transform $F(s)$ can be expressed as a series of fractions. If the poles are simple (nonrepeated), the number of fractions is equal to n, the number of poles of $F(s)$. In such case the function $F(s)$ can be expressed as

$$F(s) = \frac{P(s)}{Q(s)} = \frac{A_1}{s - s_1} + \frac{A_2}{s - s_2} + \cdots + \frac{A_k}{s - s_k} + \cdots + \frac{A_n}{s - s_n} \qquad (4\text{-}34)$$

The problem is to evaluate the constants $A_1, A_2, \ldots, A_n$ corresponding to the poles $s_1, s_2, \ldots, s_n$. The coefficients $A_1, A_2, \ldots$ are termed the *residues*† of $F(s)$ at the corresponding poles. Cases of repeated factors and complex factors are treated separately. Several ways of evaluating the constants are taken up in the following section.

4-7   HEAVISIDE PARTIAL-FRACTION EXPANSION THEOREMS

The technique of partial-fraction expansion is set up to take care of all cases systematically. There are four classes of problems, depending on the denominator $Q(s)$. Each of these cases is illustrated separately.

Case 1 $F(s)$ has first-order real poles.

Case 2 $F(s)$ has repeated first-order real poles.

Case 3 $F(s)$ has a pair of complex-conjugate poles (a quadratic factor in the denominator).

Case 4 $F(s)$ has repeated pairs of complex-conjugate poles (a repeated quadratic factor in the denominator).

Case 1   First-Order Real Poles

The position of three real poles of $F(s)$ in the s plane is shown in Fig. 4-1. The poles may be positive, zero, or negative, and they lie on the real axis in the s plane. In this example, s_1 is positive, s_0 is zero, and s_2 is negative. For the poles shown in Fig. 4-1 the transform $F(s)$ and its partial fractions are

$$F(s) = \frac{P(s)}{Q(s)} = \frac{P(s)}{s(s - s_1)(s - s_2)} = \frac{A_0}{s} + \frac{A_1}{s - s_1} + \frac{A_2}{s - s_2} \qquad (4\text{-}35)$$

There are as many fractions as there are factors in the denominator of $F(s)$. Since $s_0 = 0$, the factor $s - s_0$ is written simply as s. The inverse transform of $F(s)$ is

$$f(t) = A_0 + A_1 e^{s_1 t} + A_2 e^{s_2 t} \qquad (4\text{-}36)$$

The pole s_1 is positive; therefore the term $A_1 e^{s_1 t}$ is an increasing exponential and the system is unstable. The pole s_2 is negative, and the term $A_2 e^{s_2 t}$ is a decaying exponen-

---

† More generally the residue is the coefficient of the $(s - s_i)^{-1}$ term in the Laurent expansion of $F(s)$ about $s = s_i$.

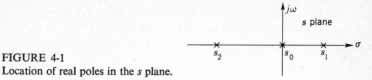

FIGURE 4-1
Location of real poles in the s plane.

tial with a final value of zero. Therefore, for a system to be stable, all real poles that contribute to the complementary solution must be in the left half of the s plane.

To evaluate a typical coefficient A_k, multiply both sides of Eq. (4-34) by the factor $s - s_k$. The result is

$$(s - s_k)F(s) = (s - s_k)\frac{P(s)}{Q(s)} = A_1\frac{s - s_k}{s - s_1} + A_2\frac{s - s_k}{s - s_2} + \cdots + A_k$$

$$+ \cdots + A_n\frac{s - s_k}{s - s_n} \qquad (4\text{-}37)$$

The multiplying factor $s - s_k$ on the left side of the equation and the same factor of $Q(s)$ should be divided out. By letting $s = s_k$, each term on the right side of the equation is zero except A_k. Thus, a general rule for evaluating the constants for single-order real poles is

$$A_k = \left[(s - s_k)\frac{P(s)}{Q(s)}\right]_{s=s_k} = \left[\frac{P(s)}{Q'(s)}\right]_{s=s_k} \qquad (4\text{-}38)$$

where $Q'(s) = dQ(s)/ds = Q(s)/(s - s_k)$. The coefficients A_k are the *residues* of $F(s)$ at the corresponding poles. For the case of

$$F(s) = \frac{s + 2}{s(s + 1)(s + 3)} = \frac{A_0}{s} + \frac{A_1}{s + 1} + \frac{A_2}{s + 3}$$

the constants are

$$A_0 = [sF(s)]_{s=0} = \left[\frac{s + 2}{(s + 1)(s + 3)}\right]_{s=0} = \frac{2}{3}$$

$$A_1 = [(s + 1)F(s)]_{s=-1} = \left[\frac{s + 2}{s(s + 3)}\right]_{s=-1} = -\frac{1}{2}$$

$$A_2 = [(s + 3)F(s)]_{s=-3} = \left[\frac{s + 2}{s(s + 1)}\right]_{s=-3} = -\frac{1}{6}$$

The solution as a function of time is

$$f(t) = \frac{2}{3} - \frac{e^{-t}}{2} - \frac{1}{6}e^{-3t} \qquad (4\text{-}39)$$

Case 2 Multiple-Order Real Poles

The position of real poles of $F(s)$, some of which are repeated, is shown in Fig. 4-2. The notation $]_r$ indicates a pole of order r. All real poles lie on the real axis of the

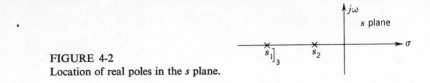

FIGURE 4-2
Location of real poles in the s plane.

s plane. For the poles shown in Fig. 4-2 the transform $F(s)$ and its partial fractions are

$$F(s) = \frac{P(s)}{Q(s)} = \frac{P(s)}{(s - s_1)^3(s - s_2)}$$

$$= \frac{A_{13}}{(s - s_1)^3} + \frac{A_{12}}{(s - s_1)^2} + \frac{A_{11}}{s - s_1} + \frac{A_2}{s - s_2} \qquad (4\text{-}40)$$

The order of $Q(s)$ in this case is 4, and there are four fractions. Note that the multiple pole s_1, which is of order 3, has resulted in three fractions on the right side of Eq. (4-40). To designate the constants in the partial fractions, a single subscript is used for a first-order pole. For multiple-order poles a double-subscript notation is used. The first subscript designates the pole, and the second subscript designates the order of the pole in the partial fraction. The constants associated with first-order denominators in the partial-fraction expansion are called residues; therefore only the constants A_{11} and A_2 are residues of Eq. (4-40).

The inverse transform of $F(s)$ is

$$f(t) = A_{13}\frac{t^2}{2}e^{s_1 t} + A_{12}te^{s_1 t} + A_{11}e^{s_1 t} + A_2 e^{s_2 t} \qquad (4\text{-}41)$$

For the general transform with repeated real roots,

$$F(s) = \frac{P(s)}{Q(s)} = \frac{P(s)}{(s - s_q)^r(s - s_1)\cdots}$$

$$= \frac{A_{qr}}{(s - s_q)^r} + \frac{A_{q(r-1)}}{(s - s_q)^{r-1}} + \cdots + \frac{A_{q(r-k)}}{(s - s_q)^{r-k}} + \cdots$$

$$+ \frac{A_{q1}}{s - s_q} + \frac{A_1}{s - s_1} + \cdots \qquad (4\text{-}42)$$

The constant A_{qr} can be evaluated by simply multiplying both sides of Eq. (4-42) by $(s - s_q)^r$, giving

$$(s - s_q)^r F(s) = \frac{(s - s_q)^r P(s)}{Q(s)}$$

$$= \frac{P(s)}{(s - s_1)\cdots}$$

$$= A_{qr} + A_{q(r-1)}(s - s_q) + \cdots$$

$$+ A_{q1}(s - s_q)^{r-1} + A_1 \frac{(s - s_q)^r}{s - s_1} + \cdots \qquad (4\text{-}43)$$

Notice that the factor $(s - s_q)^r$ is divided out of the left side of the equation.

For $s = s_q$, all terms on the right side of the equation are zero except A_{qr}:

$$A_{qr} = \left[(s - s_q)^r \frac{P(s)}{Q(s)} \right]_{s = s_q} \quad (4\text{-}44)$$

Evaluation of $A_{q(r-1)}$ cannot be performed in a similar manner. Multiplying both sides of Eq. (4-42) by $(s - s_q)^{r-1}$ and letting $s = s_q$ would result in both sides being infinite, which leaves $A_{q(r-1)}$ indeterminate. If the term A_{qr} were eliminated from Eq. (4-43), $A_{q(r-1)}$ could be evaluated. This can be done by differentiating Eq. (4-43) with respect to s:

$$\frac{d}{ds}\left[(s - s_q)^r \frac{P(s)}{Q(s)} \right] = A_{q(r-1)} + 2A_{q(r-2)}(s - s_q) + \cdots \quad (4\text{-}45)$$

Letting $s = s_q$ gives

$$A_{q(r-1)} = \left\{ \frac{d}{ds}\left[(s - s_q)^r \frac{P(s)}{Q(s)} \right] \right\}_{s = s_q} \quad (4\text{-}46)$$

Repeating the differentiation gives the coefficient $A_{q(r-2)}$ as

$$A_{q(r-2)} = \left\{ \frac{1}{2} \frac{d^2}{ds^2}\left[(s - s_q)^r \frac{P(s)}{Q(s)} \right] \right\}_{s = s_q} \quad (4\text{-}47)$$

This process can be repeated until each constant is determined. A general formula for finding these coefficients associated with the repeated real pole of order r is

$$A_{q(r-k)} = \left\{ \frac{1}{k!} \frac{d^k}{ds^k}\left[(s - s_q)^r \frac{P(s)}{Q(s)} \right] \right\}_{s = s_q} \quad (4\text{-}48)$$

For the case of

$$F(s) = \frac{1}{(s + 2)^3 (s + 3)} = \frac{A_{13}}{(s + 2)^3} + \frac{A_{12}}{(s + 2)^2} + \frac{A_{11}}{s + 2} + \frac{A_2}{s + 3} \quad (4\text{-}49)$$

the constants are

$$A_{13} = [(s + 2)^3 F(s)]_{s = -2} = 1$$

$$A_{12} = \left\{ \frac{d}{ds}[(s + 2)^3 F(s)] \right\}_{s = -2} = -1$$

$$A_{11} = \left\{ \frac{d^2}{2ds^2}[(s + 2)^3 F(s)] \right\}_{s = -2} = 1$$

$$A_2 = [(s + 3)F(s)]_{s = -3} = -1$$

and the solution as a function of time is

$$f(t) = \frac{t^2}{2} e^{-2t} - te^{-2t} + e^{-2t} - e^{-3t} \quad (4\text{-}50)$$

Case 3 Complex-Conjugate Poles

The position of complex poles of $F(s)$ in the s plane is shown in Fig. 4-3. Complex poles always are present in complex-conjugate pairs; their real part may be either

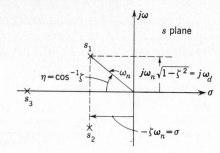

FIGURE 4-3
Location of complex-conjugate poles in
the s plane.

positive or negative. For the poles shown in Fig. 4-3 the transform $F(s)$ and its partial
fractions are

$$F(s) = \frac{P(s)}{Q(s)} = \frac{P(s)}{(s^2 + 2\zeta\omega_n s + \omega_n^2)(s - s_3)}$$

$$= \frac{A_1}{s - s_1} + \frac{A_2}{s - s_2} + \frac{A_3}{s - s_3}$$

$$= \frac{A_1}{s + \zeta\omega_n - j\omega_n\sqrt{1 - \zeta^2}} + \frac{A_2}{s + \zeta\omega_n + j\omega_n\sqrt{1 - \zeta^2}} + \frac{A_3}{s - s_3} \qquad (4\text{-}51)$$

The inverse transform of $F(s)$ is

$$f(t) = A_1 \exp\left[(-\zeta\omega_n + j\omega_n\sqrt{1 - \zeta^2})t\right]$$

$$+ A_2 \exp\left[(-\zeta\omega_n - j\omega_n\sqrt{1 - \zeta^2})t\right] + A_3 e^{s_3 t} \qquad (4\text{-}52)$$

Since the poles s_1 and s_2 are complex conjugates, and since $f(t)$ is a real quantity, the
coefficients A_1 and A_2 must also be complex conjugates. Equation (4-52) can be
written with the first two terms combined to a more useful damped sinusoidal form:

$$f(t) = 2|A_1|e^{-\zeta\omega_n t} \sin(\omega_n\sqrt{1 - \zeta^2}\, t + \phi) + A_3 e^{s_3 t}$$

$$= 2|A_1|e^{\sigma t} \sin(\omega_d t + \phi) + A_3 e^{s_3 t} \qquad (4\text{-}53)$$

where
$$\phi = \text{angle of } A_1 + 90°$$

The values of A_1 and A_3, as found in the manner shown previously, are

$$A_1 = [(s - s_1)F(s)]_{s=s_1} \quad \text{and} \quad A_3 = [(s - s_3)F(s)]_{s=s_3}$$

Since s_1 is complex, the constant A_1 is also complex. *Remember that A_1 is associated
with the complex pole with the positive imaginary part.*

In Fig. 4-3 the complex poles have a negative real part, $\sigma = -\zeta\omega_n$, where the
damping ratio ζ is positive. For this case the corresponding transient response is
known as a damped sinusoid and is shown in Fig. 3-2. Its final value is zero. The angle
η shown in Fig. 4-3 is measured from the negative real axis and is related to the damping ratio by

$$\cos \eta = \zeta$$

If the complex pole has a positive real part, the time response increases exponentially with time and the system is unstable. If the complex roots are in the right half of the s plane, the damping ratio ζ is negative. The angle η for this case is measured from the positive real axis and is given by

$$\cos \eta = |\zeta|$$

For the case of

$$F(s) = \frac{1}{(s^2 + 6s + 25)(s + 2)} = \frac{A_1}{s + 3 - j4} + \frac{A_2}{s + 3 + j4} + \frac{A_3}{s + 2} \quad (4\text{-}54)$$

the constants are

$$A_1 = \left[(s + 3 - j4)\frac{1}{(s^2 + 6s + 25)(s + 2)}\right]_{s=-3+j4}$$

$$= \left[\frac{1}{(s + 3 + j4)(s + 2)}\right]_{s=-3+j4} = 0.0303\underline{/-194°}$$

$$A_3 = \left[(s + 2)\frac{1}{(s^2 + 6s + 25)(s + 2)}\right]_{s=-2}$$

$$= \left(\frac{1}{s^2 + 6s + 25}\right)_{s=-2} = 0.059$$

The solution is

$$f(t) = 0.06e^{-3t} \sin(4t - 104°) + 0.059e^{-2t} \quad (4\text{-}55)$$

The particular function $F(s)$ of Eq. (4-54) appears in Appendix A as transform pair 29. With the notation of Appendix A, the phase angle in the damped sinusoidal term is

$$\phi = \tan^{-1}\frac{b}{c - a} = \tan^{-1}\frac{4}{2 - 3} = \tan^{-1}\frac{4}{-1} = 104° \quad (4\text{-}56)$$

It is important to note that $\tan^{-1}(4/-1) \neq \tan^{-1}(-4/1)$. To get the correct value for the angle ϕ, it is useful to draw a sketch in order to avoid ambiguity and ensure that ϕ is evaluated correctly.

Imaginary poles The position of imaginary poles of $F(s)$ in the s plane is shown in Fig. 4-4. As the real part of the poles is zero, the poles lie on the imaginary axis. This situation is a special case of complex poles, i.e., the damping ratio $\zeta = 0$. For the poles shown in Fig. 4-4 the transform $F(s)$ and its partial fractions are

$$F(s) = \frac{P(s)}{Q(s)} = \frac{P(s)}{(s^2 + \omega_n^2)(s - s_3)} = \frac{A_1}{s - s_1} + \frac{A_2}{s - s_2} + \frac{A_3}{s - s_3} \quad (4\text{-}57)$$

The quadratic can be factored in terms of the poles s_1 and s_2; thus,

$$s^2 + \omega_n^2 = (s - j\omega_n)(s + j\omega_n) = (s - s_1)(s - s_2)$$

The inverse transform of Eq. (4-57) is

$$f(t) = A_1 e^{j\omega_n t} + A_2 e^{-j\omega_n t} + A_3 e^{s_3 t} \quad (4\text{-}58)$$

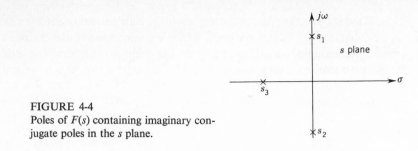

FIGURE 4-4
Poles of $F(s)$ containing imaginary conjugate poles in the s plane.

As $f(t)$ is a real quantity, the coefficients A_1 and A_2 are complex conjugates. The terms in Eq. (4-58) can be combined into the more useful form

$$f(t) = 2|A_1| \sin (\omega_n t + \phi) + A_3 e^{s_3 t} \qquad (4\text{-}59)$$

Note that since there is no damping term multiplying the sinusoid, this term represents a steady-state value. The angle

$$\phi = \text{angle of } A_1 + 90°$$

The values of A_1 and A_3, found in the conventional manner, are

$$A_1 = [(s - s_1)F(s)]_{s=s_1} \qquad A_3 = [(s - s_3)F(s)]_{s=s_3}$$

For the case where

$$F(s) = \frac{100}{(s^2 + 25)(s + 2)} = \frac{A_1}{s - j5} + \frac{A_2}{s + j5} + \frac{A_3}{s + 2} \qquad (4\text{-}60)$$

the value of the coefficients are

$$A_1 = [(s - j5)F(s)]_{s=j5} = \left[\frac{100}{(s + j5)(s + 2)}\right]_{s=j5} = 1.86\underline{/-158.2°}$$

$$A_3 = [(s + 2)F(s)]_{s=-2} = \left(\frac{100}{s^2 + 25}\right)_{s=-2} = 3.45$$

The solution is

$$f(t) = 3.72 \sin (5t - 68.2°) + 3.45 e^{-2t} \qquad (4\text{-}61)$$

Case 4   Multiple-Order Complex Poles

Multiple-order complex-conjugate poles are rare. They can be treated in much the same fashion as repeated real poles.

4-8   PARTIAL-FRACTION SHORTCUTS

Some shortcuts can be used to simplify the partial-fraction expansion procedures previously described. These new procedures are generally most useful for transform functions that have multiple poles or complex poles. With multiple poles, the evaluation of the constants by the process of repeated differentiation given by Eq. (4-48) can be very tedious, particularly when a factor containing s appears in the numerator of $F(s)$. With complex poles the residues are complex numbers, with the result that algebraic errors are easily made. The procedure is to evaluate those coefficients for real poles which can be obtained readily without differentiation. The corresponding partial fractions are then subtracted from the original function. The resultant function is easier to work with to get the additional partial fractions than the original function $F(s)$.

As the first example, consider the function $F(s)$ of Eq. (4-49), which contains a pole $s_1 = -2$ of multiplicity 3. The coefficient $A_{13} = 1$ is readily found. Subtracting the associated fraction from $F(s)$ and putting the result over a common denominator yields

$$F_x(s) = \frac{1}{(s + 2)^3(s + 3)} - \frac{1}{(s + 2)^3} = \frac{-1}{(s + 2)^2(s + 3)}$$

Note that the remainder $F_x(s)$ contains the pole $s_1 = -2$ with a multiplicity of only 2. Thus, the coefficient A_{12} is now easily obtained from $F_x(s)$ without differentiation. The procedure is repeated to obtain all the coefficients associated with the multiple pole.

As a second example, consider the function

$$F(s) = \frac{20}{(s^2 + 6s + 25)(s + 1)} = \frac{As + B}{s^2 + 6s + 25} + \frac{C}{s + 1} \qquad (4\text{-}62)$$

Instead of partial fractions for each pole, leave the quadratic factor in one fraction. Note that the numerator of the fraction containing the quadratic contains s and is one degree lower than the denominator. The residue C is $C = [(s + 1)F(s)]_{s=-1} = 1$. Subtracting this fraction from $F(s)$ gives

$$\frac{20}{(s^2 + 6s + 25)(s + 1)} - \frac{1}{s + 1} = \frac{-s^2 - 6s - 5}{(s^2 + 6s + 25)(s + 1)}$$

Since the pole at $s = -1$ has been removed, the numerator of the remainder must be divisible by $s + 1$. The simplified function must be equal to the remaining partial fraction; i.e.,

$$-\frac{s + 5}{s^2 + 6s + 25} = \frac{As + B}{s^2 + 6s + 25} = \frac{-(s + 5)}{(s + 3)^2 + 4^2}$$

The inverse transform for this fraction can be obtained directly from Appendix A, pair 26. The complete inverse transform of $F(s)$ is

$$f(t) = e^{-t} - 1.12e^{-3t} \sin(4t + 63.4°) \qquad (4\text{-}63)$$

Leaving the complete quadratic in one fraction has resulted in real constants for the numerator. The chances for error are probably less than evaluating the residues, which are complex numbers, at the complex poles.

A rule presented by Hazony and Riley[4] is very useful for evaluating the co-efficients of partial-fraction expansions. For a normalized ratio of polynomials:

1 If the denominator is one degree higher than the numerator, the sum of the residues is 1.

2 If the denominator is two or more degrees higher than the numerator, the sum of the residues is 0.

Equation (4-32) with a_w factored from the numerator is a normalized ratio of polynomials. These rules are applied to the ratio $F(s)/a_w$. It should be noted that *residue* refers only to the coefficients of terms in a partial-fraction expansion with first-degree denominators. Coefficients of terms with higher-degree denominators are referred to only as coefficients.

These rules can be used to simplify the work involved in evaluating the coefficients of partial-fraction expansions, particularly when the original function has a multiple-order pole. For example, in Eq. (4-49) it can immediately be written that $A_{11} + A_2 = 0$. Since $A_2 = -1$, the value of $A_{11} = 1$ is obtained directly. The evaluation of derivatives associated with the coefficients for multiple-order poles is therefore eliminated.

Digital-computer programs are readily available for evaluating the partial-fraction coefficients and obtaining a tabulation and plot of $f(t)$.[8]

4-9   GRAPHICAL DETERMINATION OF PARTIAL-FRACTION COEFFICIENTS[5]

The preceding sections describe the analytical evaluation of the partial-fraction coefficients. These constants are directly related to the pole-zero pattern of the function $F(s)$ and can be determined graphically, whether the poles and zeros are real or in complex-conjugate pairs. In fact, as long as $P(s)$ and $Q(s)$ are in factored form, the coefficients can be determined graphically by inspection, and a table of Laplace transforms is unnecessary.

Referring to Eq. (4-32) and rewriting with both the numerator and denominator factored and $a_w = K$ gives

$$F(s) = \frac{P(s)}{Q(s)} = \frac{K(s - z_1)(s - z_2)\cdots(s - z_m)\cdots(s - z_w)}{(s - p_1)(s - p_2)\cdots(s - p_k)\cdots(s - p_n)}$$

$$= K \frac{\displaystyle\prod_{m=1}^{w} (s - z_m)}{\displaystyle\prod_{k=1}^{n} (s - p_k)} \tag{4-64}$$

The zeros of this function, $s = z_m$, are those values of s for which the function is zero; that is, $F(z_m) = 0$. Zeros are indicated by a small circle on the s plane. The poles of this function, $s = p_k$, are those values of s for which the function is infinite; that is, $|F(p_k)| = \infty$. Poles are indicated by a small x on the s plane. Poles are also known

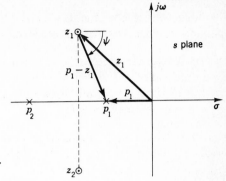

FIGURE 4-5
Directed line segments drawn on a pole-zero diagram of a function $F(s)$.

as singularities† of the function. When $F(s)$ has only simple poles (first-order poles), it can be expanded into partial fractions of the form

$$F(s) = \frac{A_1}{s - p_1} + \frac{A_2}{s - p_2} + \cdots + \frac{A_k}{s - p_k} + \cdots + \frac{A_n}{s - p_n} \qquad (4\text{-}65)$$

The coefficients are obtained from Eq. (4-38) and are given by

$$A_k = [(s - p_k)F(s)]_{s=p_k} \qquad (4\text{-}66)$$

The first coefficient is

$$A_1 = \frac{K(p_1 - z_1)(p_1 - z_2)\cdots(p_1 - z_w)}{(p_1 - p_2)(p_1 - p_3)\cdots(p_1 - p_n)} \qquad (4\text{-}67)$$

Figure 4-5 shows the poles and zeros of a function $F(s)$. The quantity $s = p_1$ is drawn as an arrow from the origin of the s plane to the pole p_1. This arrow is called a *directed line segment*. It has a magnitude equal to $|p_1|$ and an angle of 180°. Similarly, the directed line segment $s = z_1$ is drawn as an arrow from the origin to the zero z_1 and has a corresponding magnitude and angle. A directed line segment is also drawn from the zero z_1 to the pole p_1. By the rules of vector addition, it can be shown that this directed line segment is equal to $p_1 - z_1$; it has a magnitude equal to its length and an angle ψ, as shown. By referring to Eq. (4-67), it is seen that $p_1 - z_1$ appears in the numerator of A_1. While this quantity can be evaluated analytically, it can also be measured from the pole-zero diagram by means of a ruler and a protractor. The measured directed line segment is obtained in polar form, a magnitude and an angle, which is most useful in evaluating Eq. (4-67). In a similar fashion, each factor in Eq. (4-67) can be obtained graphically; the directed line segments are drawn *from* the zeros and each of the other poles *to* the pole p_1. The angle from a zero is indicated by the symbol ψ, and the angle from a pole is indicated by the symbol θ.

The general rule for evaluating the coefficients in the partial-fraction expansion is quite simple. The value of A_k is the product of K and the directed distances from each zero to the pole p_k divided by the product of the directed distances from each of

---

† A singularity of a function $F(s)$ is a point where $F(s)$ does not have a derivative.

the other poles to the pole p_k. Each of these directed distances is characterized by a magnitude and an angle. This statement can be written in equation form as

$$A_k = K \frac{\text{product of directed distances from each zero to pole } p_k}{\text{product of directed distances from all other poles to pole } p_k}$$

$$= K \frac{\prod_{m=1}^{w} |p_k - z_m| \bigg/ \sum_{m=1}^{w} \psi(p_k - z_m)}{\prod_{\substack{c=1 \\ c \neq k}}^{n} |p_k - p_c| \bigg/ \sum_{c=1}^{n} \theta(p_k - p_c)} \tag{4-68}$$

For real poles the values of A_k must be real but can be either positive or negative. A_k is positive if the total number of real poles and real zeros to the right of p_k is even, and it is negative if the total number of real poles and real zeros to the right of p_k is odd. For complex poles the values of A_k are complex. If $F(s)$ has no finite zeros, the numerator of Eq. (4-68) is equal to K. A repeated factor of the form $(s + a)^r$ in either the numerator or the denominator is included r times in Eq. (4-68), where the values of $|s + a|$ and $\underline{/s + a}$ are obtained graphically. The application of the graphical technique is illustrated by the following example.

EXAMPLE   *A function $F(s)$ with complex poles*

$$F(s) = \frac{K(s + \alpha)}{s[(s + a)^2 + b^2]} = \frac{K(s + \alpha)}{s(s + a - jb)(s + a + jb)}$$

$$= \frac{K(s - z_1)}{s(s - p_1)(s - p_2)} = \frac{A_0}{s} + \frac{A_1}{s + a - jb} + \frac{A_2}{s + a + jb} \tag{4-69}$$

The coefficient A_0 is obtained by use of the directed line segments shown in Fig. 4-6a:

$$A_0 = \frac{K(\alpha)}{(a - jb)(a + jb)} = \frac{K\alpha}{\sqrt{a^2 + b^2} \, e^{j\theta_1} \sqrt{a^2 + b^2} \, e^{j\theta_2}} = \frac{K\alpha}{a^2 + b^2} \tag{4-70}$$

Note that the angles θ_1 and θ_2 are equal in magnitude but opposite in sign. The coefficient for a real pole is therefore always a real number.

The coefficient A_1 is obtained by use of the directed line segments shown in Fig. 4-6b:

$$A_1 = \frac{K[(\alpha - a) + jb]}{(-a + jb)(j2b)} = \frac{K\sqrt{(\alpha - a)^2 + b^2} \, e^{j\psi}}{\sqrt{a^2 + b^2} \, e^{j\theta} 2b e^{j\pi/2}}$$

$$= \frac{K}{2b} \sqrt{\frac{(\alpha - a)^2 + b^2}{a^2 + b^2}} \, e^{j(\psi - \theta - \pi/2)} \tag{4-71}$$

where $\qquad \psi = \tan^{-1} \dfrac{b}{\alpha - a} \qquad$ and $\qquad \theta = \tan^{-1} \dfrac{b}{-a}$

Since the constants A_1 and A_2 are complex conjugates, it is not necessary to evaluate A_2. It should be noted that zeros in $F(s)$ affect both the magnitude and the angle of the coefficients in the time function.

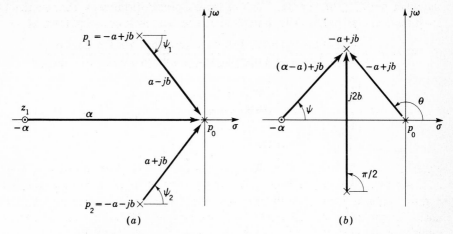

FIGURE 4-6
Directed line segments (a) for evaluating A_0 of Eq. (4-70) and (b) for evaluating A_1 of Eq. (4-71).

The response as a function of time is

$$f(t) = A_0 + 2|A_1|e^{-at} \cos\left(bt + \psi - \theta - \frac{\pi}{2} \right) \qquad (4\text{-}72)$$

The trigonometric function in the damped sinusoidal term can be changed from a cosine to a sine by a shift of $\pi/2$ in the phase angle. This gives

$$f(t) = A_0 + 2|A_1|e^{-at} \sin\left(bt + \underline{/A_1} + 90° \right)$$

$$= \frac{K\alpha}{a^2 + b^2} + \frac{K}{b} \sqrt{\frac{(\alpha - a) + b^2}{a^2 + b^2}}\, e^{-at} \sin\left(bt + \phi \right) \qquad (4\text{-}73)$$

where

$$\phi = \psi - \theta = \tan^{-1} \frac{b}{\alpha - a} - \tan^{-1} \frac{b}{-a}$$

$$= \text{angle of } A_1 + 90° \qquad (4\text{-}74)$$

This agrees with transform pair 28 in Appendix A. The evaluation of A_0 and A_1 is intended to be performed by measuring the directed line segments directly from the pole-zero diagram.

4-10 USE OF THE SPIRULE

The preceding section showed the procedure for graphically evaluating the coefficients in the partial-fraction expansion of a function $F(s)$. The directed line segments used to compute these coefficients must be measured on the pole-zero diagram. Since they

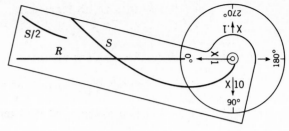

FIGURE 4-7
Outline of a Spirule.

have both a magnitude and an angle, these measurements can be made with a ruler and a protractor. A Spirule† is a convenient device for facilitating the computation of the coefficients A_k. It is essentially a protractor with a pivoted arm, as shown in Fig. 4-7. Its function is to measure and multiply or divide the lengths of the directed line segments to obtain the magnitude of A_k. It also algebraically adds the angles of the directed line segments in order to obtain the angle of A_k. Outlined below are steps of multiplication and division and the algebraic addition of angles.

A function $F(s)$ has the form

$$F(s) = \frac{K(s - z_1)(s - z_2)\cdots(s - z_w)}{(s - p_1)(s - p_2)\cdots(s - p_n)}$$

$$= \frac{A_1}{s - p_1} + \frac{A_2}{s - p_2} + \cdots + \frac{A_k}{s - p_k} + \cdots \qquad (4\text{-}75)$$

Magnitude of A_k

1 Plot the poles and zeros on rectangular coordinate graph paper. Determine the *scale factor* between the *scale* used on the graph paper and the unit length on the edge of the arm of the Spirule.
2 Line up the reference line R on the Spirule arm with the $\times 1$ arrow on the disk. Place the pivot of the Spirule over the pole p_k.
3 Rotate the arm and disk together to align the reference line R with the first zero z_1. Hold the disk stationary and turn the arm until the S curve crosses over the zero z_1. This procedure sets the magnitude of $p_k - z_1$ on the Spirule.
4 Release the disk so that the arm and disk rotate together. Repeat step 3 for all zeros of $F(s)$. This sets the product of the zero factors on the Spirule.
5 Rotate the arm and disk together so that the S curve crosses over the pole p_1. Hold the disk stationary and rotate the arm until the reference line R coincides with the pole p_1. This procedure performs the division by $p_k - p_1$.
6 Repeat step 5 for all poles of $F(s)$.

† Available from The Spirule Company, 9728 El Venado, Whittier, California 90603.
Also, see Appendix B.

7 The final *Spirule* reading is obtained from the *index arrow* on the disk and the scale on the arm. Three of the index arrows are marked ×.1, ×1, and ×10.

8 The magnitude of A_k is given by

$$A_k = K(\text{Spirule reading})(\text{index arrow marking})(\text{scale factor})^x \qquad (4\text{-}76)$$

The exponent x is equal to the number of zeros minus the number of poles of $(s - p_k)F(s)$.

Angle of A_k

1 Line up the reference line R on the Spirule arm with the $0°$ marking on the disk. Place the pivot of the Spirule over the pole p_k.
2 Rotate the arm and disk together so that the line R is pointing horizontally to the left. Hold the disk stationary and turn the arm until the R line crosses the first zero z_1. The Spirule reading is now the angle of $p_k - z_1$.
3 Release the disk so that the arm and disk rotate together, and repeat step 2 for all zeros of $F(s)$. This procedure adds the angles of the zero factors on the Spirule.
4 Rotate the arm and disk together so that the line R crosses over the pole p_1. Hold the disk stationary and turn the arm until line R is pointing horizontally to the left. This procedure subtracts the angle of $p_k - p_1$.
5 Release the disk so that the arm and disk rotate together, and repeat step 4 for all poles of $F(s)$.
6 The final reading after step 5 is the angle of A_k. The scale on the Spirule disk is from 0 to $360°$ in the positive or counterclockwise direction. If the angle is greater than $180°$, it may be more convenient to convert it to its equivalent negative angle.

With practice it is possible to develop facility in the use of the Spirule and to obtain good accuracy. The results can be checked by analytically calculating the coefficients. The Spirule may also be used in Chap. 7 for plotting the root locus.

4-11 FREQUENCY RESPONSE FROM THE POLE-ZERO DIAGRAM

The frequency response of a system is described as the steady-state response with a sine-wave forcing function for all values of frequency. This information is often presented graphically. One form of presentation requires the use of two curves. One curve shows the ratio of output amplitude to input amplitude M and the other curve the phase angle of the output α, where both are plotted as a function of frequency, often on a logarithmic scale.

Consider the input to a system as sinusoidal and given by

$$x_1(t) = X_1 \sin \omega t \qquad (4\text{-}77)$$

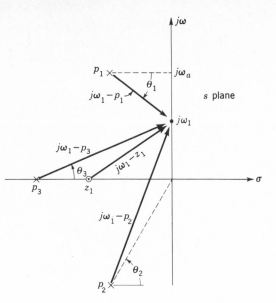

FIGURE 4-8
Pole-zero diagram of Eq. (4-79) showing
the directed line segments for evaluating
the frequency response $M/\underline{\alpha}$.

and the output is

$$x_2(t) = X_2 \sin (\omega t + \alpha) \qquad (4\text{-}78)$$

The magnitude of the frequency response is

$$M = \left| \frac{X_2(j\omega)}{X_1(j\omega)} \right| = \left| \frac{P(j\omega)}{Q(j\omega)} \right| = \left| \frac{K(j\omega - z_1)(j\omega - z_2)\cdots}{(j\omega - p_1)(j\omega - p_2)\cdots} \right| \qquad (4\text{-}79)$$

and the angle is

$$\alpha = \underline{/P(j\omega)} - \underline{/Q(j\omega)}$$
$$= \underline{/K} + \underline{/j\omega - z_1} + \underline{/j\omega - z_2} + \cdots - \underline{/j\omega - p_1} - \underline{/j\omega - p_2} - \cdots \qquad (4\text{-}80)$$

Figure 4-8 shows a possible pole-zero diagram and the directed line segments for evaluating M and α corresponding to a frequency ω_1. As ω increases, each of the directed line segments changes in both magnitude and angle. Several characteristics can be noted by observing the change in the magnitude and angle of each directed line segment as the frequency ω increases from 0 to ∞. The magnitude and angle of the frequency response can be obtained graphically from the pole-zero diagram. The Spirule can be used for this purpose.

The magnitude M and the angle α are a composite of all the effects of all the poles and zeros. In particular, for a system function which has all poles and zeros in the left half of the s plane, the following characteristics of the frequency response are noted:

1 At $\omega = 0$ the magnitude is a finite value, and the angle is $0°$.
2 If there are more poles than zeros, as $\omega \to \infty$, the magnitude of M approaches zero and the angle is $-90°$ times the difference between the number of poles and zeros.

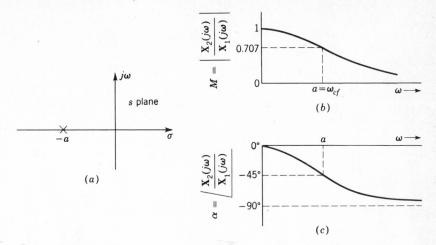

FIGURE 4-9
(a) Pole location of Eq. (4-81); (b) M vs. ω; (c) α vs. ω.

3   The magnitude can have a peak value M_m only if there are complex poles fairly close to the imaginary axis. It is shown later that the damping ratio ζ of the complex poles must be less than 0.707 to obtain a peak value for M. Of course, the presence of zeros could counteract the effect of the complex poles, and so it is possible that no peak would be present even if ζ of the poles were less than 0.707.

Typical frequency-response characteristics are illustrated for two common functions. As a first example,

$$\frac{X_2(s)}{X_1(s)} = \frac{P(s)}{Q(s)} = \frac{a}{s + a} \qquad (4\text{-}81)$$

The frequency response starts with a magnitude of unity and an angle of $0°$. As ω increases, the magnitude decreases and approaches zero while the angle approaches $-90°$. At $\omega = a$ the magnitude is 0.707 and the angle is $-45°$. The frequency $\omega = a$ is called the corner frequency ω_{cf} or the break frequency. Figure 4-9 shows the pole location, the frequency-response magnitude M, and the frequency-response angle α.

The second example is a system transform with a pair of conjugate poles:

$$\frac{X_2(s)}{X_1(s)} = \frac{\omega_n^2}{s^2 + 2\zeta\omega_n s + \omega_n^2} = \frac{\omega_n^2}{(s - p_1)(s - p_2)}$$

$$= \frac{\omega_n^2}{(s + \zeta\omega_n - j\omega_n\sqrt{1 - \zeta^2})(s + \zeta\omega_n + j\omega_n\sqrt{1 - \zeta^2})} \qquad (4\text{-}82)$$

The frequency response is given by

$$M\underline{/\alpha} = \frac{X_2(j\omega)}{X_1(j\omega)} = \frac{\omega_n^2}{(j\omega + \zeta\omega_n - j\omega_n\sqrt{1 - \zeta^2})(j\omega + \zeta\omega_n + j\omega_n\sqrt{1 - \zeta^2})} \qquad (4\text{-}83)$$

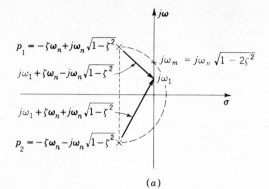

$$(a)$$

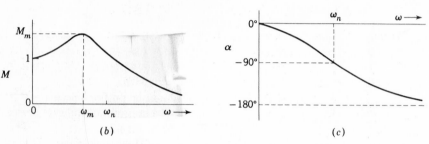

$$(b) \qquad\qquad (c)$$

FIGURE 4-10

(a) Pole-zero diagram, $\zeta < 0.707$; (b) magnitude of frequency response; (c) angle of frequency response.

The location of the poles of Eq. (4-82) and the directed line segments for $s = j\omega$ are shown in Fig. 4-10a. At $\omega = 0$ the values are $M = 1$ and $\alpha = 0°$. As ω increases, the magnitude M first increases because $j\omega - p_1$ is decreasing faster than $j\omega - p_2$ is increasing. By differentiating the magnitude of M, obtained from Eq. (4-83), with respect to ω, it can be shown (see Sec. 9-3) that the maximum value M_m and the frequency ω_m at which it occurs are given by

$$M_m = \frac{1}{2\zeta\sqrt{1 - \zeta^2}} \qquad (4\text{-}84)$$

$$\omega_m = \omega_n\sqrt{1 - 2\zeta^2} \qquad (4\text{-}85)$$

A circle drawn on the s plane with the poles p_1 and p_2 as the diameter intersects the imaginary axis at the value ω_m given by Eq. (4-85). Both Eq. (4-85) and the geometrical construction of the circle show that the M curve has a maximum value, other than at $\omega = 0$, only if $\zeta < 0.707$. The angle α becomes more negative as ω increases. At $\omega = \omega_n$ the angle is equal to $-90°$. This is the corner frequency for a quadratic factor with complex roots. As $\omega \to \infty$, the angle α approaches $-180°$. The magnitude and angle of the frequency response for $\zeta < 0.707$ are shown in Fig. 4-10. The M and α curves are continuous for all linear systems.

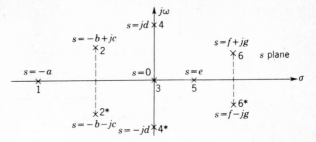

FIGURE 4-11
Location of poles in the s plane. (Numbers are used to identify the poles.)

4-12 LOCATION OF POLES AND STABILITY

The stability and the corresponding response of a system can be determined from the location of the poles of the response transform $F(s)$ in the s plane. The possible positions of the poles are shown in Fig. 4-11, and the responses are given in Table 4-1. These poles are the roots of the characteristic equation.

Poles of the response transform at the origin or on the imaginary axis that are not contributed by the forcing function result in a continuous output. These outputs are undesirable in a control system. Poles in the right-half s plane result in transient terms that increase with time. Such performance characterizes an unstable system; therefore poles in the right-half s plane are not desirable, in general.

4-13 LAPLACE TRANSFORM OF THE IMPULSE FUNCTION

Figure 4-12 shows a rectangular pulse of amplitude $1/a$ and of duration a. The analytical expression for this pulse is

$$f(t) = \frac{u_{-1}(t) - u_{-1}(t - a)}{a} \qquad (4\text{-}86)$$

and its Laplace transform is

$$F(s) = \frac{1 - e^{-as}}{as} \qquad (4\text{-}87)$$

Table 4-1 RELATION OF RESPONSE TO LOCATION OF POLES

Position of pole	Form of response	Characteristics
1	Ae^{-at}	Damped exponential
2–2*	$Ae^{-bt} \sin(ct + \phi)$	Exponentially damped sinusoid
3	A	Constant
4–4*	$A \sin(dt + \phi)$	Constant sinusoid
5	Ae^{et}	Increasing exponential (unstable)
6–6*	$Ae^{ft} \sin(gt + \phi)$	Exponentially increasing sinusoid (unstable)

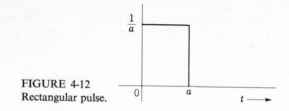

FIGURE 4-12
Rectangular pulse.

If a is decreased, the amplitude increases and the duration of the pulse decreases but the area under the pulse, and thus its strength, remains unity. The limit of $f(t)$ as $a \to 0$ is termed a unit impulse and is designated by $\delta(t) = u_0(t)$. The Laplace transform of the unit impulse, as evaluated by use of L'hopital's theorem, is $\Delta(s) = 1$.

When any function has a jump discontinuity, the derivative of the function has an impulse at the discontinuity. Some systems are subjected to shock inputs. For example, an airplane in flight may be jolted by a gust of wind of short duration. A gun that is fired has a reaction force of large magnitude and very short duration, and so does a steel ball bouncing off a steel plate. When the duration of a disturbance or input to a system is very short compared with the natural periods of the system, the response can often be well approximated by considering the input to be an impulse of proper strength. An impulse of infinite magnitude and zero duration does not occur in nature; but if the pulse duration is much smaller than the time constants of the system, a representation of the input by an impulse is a good approximation. The shape of the impulse is unimportant as long as the strength of the equivalent impulse is equal to the strength of the actual pulse.

Since the Laplace transform of an impulse is defined, the approximate response of a system to a pulse is often found more easily by this method than by the classical method.

For the RLC circuit shown in Fig. 4-13, the input voltage is an impulse. The problem is to find the voltage e_c across the capacitor as a function of time. The differential equation relating e_c to the input voltage e and the impedances is written by the node method:

$$e_c\left(CD + \frac{1}{R + LD}\right) - e\,\frac{1}{R + LD} = 0$$

By rationalizing, the equation becomes

$$e_c(LCD^2 + RCD + 1) = e \qquad (4\text{-}88)$$

Taking the Laplace transform of this equation gives

$$(s^2LC + sRC + 1)E_c(s) - sLCe_c(0) - LCDe_c(0) - RCe_c(0) = E(s)$$

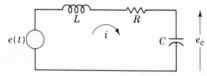

FIGURE 4-13
RLC series circuit.

Consider an initial condition $e_c(0^-) = 0$. The voltage across the capacitor is related to the current flowing in the series circuit by the expression $De_c = i/C$. Since the current cannot change instantaneously through an inductor, $De_c(0^-) = 0$. These initial conditions which exist at $t = 0^-$ are used in the Laplace transformed equation when the forcing function is an impulse. The solution for the current shows a discontinuous step change at $t = 0$.

The input voltage e is an impulse; therefore $E(s) = 1$. Solving for $E_c(s)$ gives

$$E_c(s) = \frac{1/LC}{s^2 + (R/L)s + 1/LC} = \frac{\omega_n^2}{s^2 + 2\zeta\omega_n s + \omega_n^2} \qquad (4\text{-}89)$$

Depending on the relative sizes of the parameters, the circuit may be overdamped or underdamped. In the overdamped case the voltage $e_c(t)$ rises to a peak value and then decays to zero. In the underdamped case the voltage oscillates with a frequency $\omega_d = \omega_n\sqrt{1 - \zeta^2}$ and eventually decays to zero. The form of the voltage $e_c(t)$ for several values of damping is shown in the next section.

The stability of a system is revealed by the impulse response. With an impulse input the response transform contains only poles contributed by the system parameters. The impulse response is therefore the inverse transform of the system function. The impulse response provides one method of evaluating the system function; although the method is not simple, it is possible to take the impulse response which is a function of time and convert it into a function of s.

The system response $g(t)$ to an impulse input $\delta(t)$ can also be utilized to determine the response $c(t)$ to any input $r(t)$.[6] The response $c(t)$ is determined by the use of the convolution integral

$$c(t) = \int_0^t r(\tau)g(t - \tau)\,d\tau \qquad (4\text{-}90)$$

This method is normally used for inputs which are not sinusoidal or a power series.

4-14 SECOND-ORDER SYSTEM WITH IMPULSE EXCITATION

The response of a second-order system to a step-function input is shown in Fig. 3-6. The impulse function is the derivative of the step function; therefore the response to an impulse is the derivative of the response to a step function. Figure 4-14 shows the response for several values of damping ratio ζ for the second-order function of Eq. (4-89).

The first zero of the impulse response occurs at t_p, which is the time at which the maximum value of the step-function response occurs.

For the underdamped case the impulse response is

$$f(t) = \frac{\omega_n}{\sqrt{1 - \zeta^2}}\, e^{-\zeta\omega_n t} \sin \omega_n\sqrt{1 - \zeta^2}\, t \qquad (4\text{-}91)$$

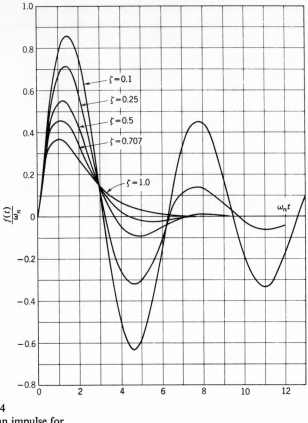

FIGURE 4-14
Response to an impulse for

$$F(s) = \frac{\omega_n^2}{s^2 + 2\zeta\omega_n s + \omega_n^2}$$

By setting the derivative of $f(t)$ with respect to time equal to zero, it can be shown that the maximum overshoot occurs at

$$t_m = \frac{\cos^{-1}\zeta}{\omega_n\sqrt{1-\zeta^2}} \qquad (4\text{-}92)$$

The maximum value of $f(t)$ is

$$f(t_m) = \omega_n \exp\left(-\frac{\zeta\cos^{-1}\zeta}{\sqrt{1-\zeta^2}}\right) \qquad (4\text{-}93)$$

4-15 ADDITIONAL MATRIX OPERATIONS AND PROPERTIES

Additional characteristics of matrices required to solve matrix state equations by using the Laplace transform are presented in this section.

Principal diagonal The principal diagonal of a square matrix $\mathbf{M} = [m_{ij}]$ consists of the elements m_{ii}.

Diagonal matrix A diagonal matrix is a square matrix in which all the elements off the principal diagonal are zero. This can be expressed as $m_{ij} = 0$ for $i \neq j$ and the principal diagonal elements $m_{ii} = m_i$ are not all zero. When the diagonal elements are all equal, the matrix is called a *scalar* matrix.

Trace The *trace* of a square matrix $\mathbf{M}$ of order n is the sum of all the elements along the principal diagonal; that is, trace $\mathbf{M} = \sum\limits_{i=1}^{n} m_{ii}$.

Determinant The determinant of a square matrix $\mathbf{M}$ of order α is the sum of all possible signed products of α elements containing one and only one element from every row and column in the matrix. The determinant of $\mathbf{M}$ may be denoted by $|\mathbf{M}|$, or det $\mathbf{M}$, or Δ_M. It is assumed that the reader knows how to evaluate determinants.

EXAMPLE 1

$$\mathbf{M} = \begin{bmatrix} 3 & 1 & 2 \\ 1 & 0 & -4 \\ 0 & 5 & 7 \end{bmatrix} \qquad |\mathbf{M}| = \begin{vmatrix} 3 & 1 & 2 \\ 1 & 0 & -4 \\ 0 & 5 & 7 \end{vmatrix} = 63$$

Characteristics of determinants are:

1 The determinant of a unit matrix is $|\mathbf{I}| = 1$.
2 The determinant of a matrix is zero if (*a*) any row or column contains all zeros or (*b*) the elements of any two rows (or columns) have a common ratio.

Singular matrix A square matrix is said to be *singular* if the value of its determinant is zero. If the value is not zero, the matrix is *nonsingular*.

Minor The minor M_{ij} of a square matrix $\mathbf{M}$ of order α is the determinant formed after the ith row and jth column are deleted from $\mathbf{M}$.

Principal minor A principal minor is a minor M_{ii} whose diagonal elements are also the diagonal elements of the matrix $\mathbf{M}$.

Cofactor A cofactor is a *signed* minor and is given by

$$C_{ij} = \Delta_{ij} = (-1)^{i+j} M_{ij} \qquad (4\text{-}94)$$

EXAMPLE 2 For the matrix $\mathbf{M}$ of Example 1, the cofactor C_{21} is obtained by deleting the second row and first column, giving

$$C_{21} = (-1)^{2+1} \begin{vmatrix} 1 & 2 \\ 5 & 7 \end{vmatrix} = +3$$

Adjoint matrix The adjoint of square matrix $\mathbf{M}$, denoted as adj $\mathbf{M}$, is the transpose of the cofactor matrix. The cofactor matrix is formed by replacing each element of $\mathbf{M}$ by its cofactor.

$$\text{adj } \mathbf{M} = [C_{ji}] = [C_{ij}]^T = \begin{bmatrix} \text{array of} \\ \text{cofactors} \end{bmatrix}^T \qquad (4\text{-}95)$$

EXAMPLE 3   The adj M for the matrix M of Example 1 is

$$\text{adj } M = \begin{bmatrix} 20 & -7 & 5 \\ 3 & 21 & -15 \\ -4 & 14 & -1 \end{bmatrix}^T = \begin{bmatrix} 20 & 3 & -4 \\ -7 & 21 & 14 \\ 5 & -15 & -1 \end{bmatrix}$$

Inverse matrix   The product of a matrix and its adjoint is a scalar matrix, as illustrated by the following example.

EXAMPLE 4   The product M adj M, using values in Examples 1 and 3, is

$$M \text{ adj } M = \begin{bmatrix} 3 & 1 & 2 \\ 1 & 0 & -4 \\ 0 & 5 & 7 \end{bmatrix} \begin{bmatrix} 20 & 3 & -4 \\ -7 & 21 & 14 \\ 5 & -15 & -1 \end{bmatrix}$$

$$= \begin{bmatrix} 63 & 0 & 0 \\ 0 & 63 & 0 \\ 0 & 0 & 63 \end{bmatrix} = 63 \begin{bmatrix} 1 & 0 & 0 \\ 0 & 1 & 0 \\ 0 & 0 & 1 \end{bmatrix} = |M|I$$

The inverse of a square matrix M is denoted by M^{-1} and has the property

$$MM^{-1} = M^{-1}M = I \qquad (4\text{-}96)$$

The inverse matrix M^{-1} is defined from the results of Example 4 as

$$M^{-1} = \frac{\text{adj } M}{|M|} \qquad (4\text{-}97)$$

The inverse exists only if $|M| \neq 0$, that is, the matrix M is nonsingular.
Product of determinants   The determinant of the product of square matrices is equal to the product of the individual determinants:

$$|ABC| = |A| \cdot |B| \cdot |C| \qquad (4\text{-}98)$$

Rank of a matrix   The rank r of a matrix M, not necessarily square, is the order of the largest square array contained in M which has a nonzero determinant.

EXAMPLE 5

$$M = \begin{bmatrix} 1 & 2 & 3 \\ 2 & 3 & 4 \\ 3 & 5 & 7 \end{bmatrix}$$

The determinant $|M| = 0$; that is, M is a singular matrix. The square array obtained by deleting the first row and second column has a nonzero determinant:

$$\begin{vmatrix} 2 & 4 \\ 3 & 7 \end{vmatrix} = 2$$

Therefore, M has a rank of 2.

Degeneracy (or nullity) of a matrix When a matrix **M** of order n has a rank r, there are $q = n - r$ rows or columns which are linear combinations of the r rows or columns. The matrix **M** is then said to be of *degeneracy q*. In Example 5 the matrix **M** is of order $n = 3$, the rank is $r = 2$, and the degeneracy is $q = 1$ (simple degeneracy). The third row of **M** is the sum of the first two rows.

Symmetric matrix A square matrix containing only real elements is symmetric if it is equal to its transpose, that is, $\mathbf{A} = \mathbf{A}^T$. The relationship between the elements is

$$a_{ij} = a_{ji}$$

A symmetric matrix is symmetrical about its principal diagonal.

Transpose of a product of matrices The transpose of a product of matrices is the product of the transposed matrices in reverse order:

$$(\mathbf{AB})^T = \mathbf{B}^T \mathbf{A}^T$$

This relationship can be described by the statement that the product of transposed matrices is equal to the transpose of the product in reverse order of the original matrices.

4-16 SOLUTION OF STATE EQUATION[7]

The homogeneous state equation is

$$\dot{\mathbf{x}} = \mathbf{Ax} \qquad (4\text{-}99)$$

The solution of this equation can be obtained by taking the Laplace transform

$$s\mathbf{X}(s) - \mathbf{x}(0) = \mathbf{AX}(s)$$

Grouping of the terms containing $\mathbf{X}(s)$ yields

$$[s\mathbf{I} - \mathbf{A}]\mathbf{X}(s) = \mathbf{x}(0)$$

The unit matrix **I** is introduced so that all terms in the equations are proper matrices. Premultiplying both sides of the equation by $[s\mathbf{I} - \mathbf{A}]^{-1}$ yields

$$\mathbf{X}(s) = [s\mathbf{I} - \mathbf{A}]^{-1}\mathbf{x}(0) \qquad (4\text{-}100)$$

The inverse Laplace transform of Eq. (4-99) gives

$$\mathbf{x}(t) = \mathscr{L}^{-1}\{[s\mathbf{I} - \mathbf{A}]^{-1}\}\mathbf{x}(0) \qquad (4\text{-}101)$$

Comparing this solution with Eq. (3-70) yields the following expression for the state transition matrix:

$$\Phi(t) = \mathscr{L}^{-1}[s\mathbf{I} - \mathbf{A}]^{-1} \qquad (4\text{-}102)$$

The *resolvent matrix* is designated by $\Phi(s)$ and is defined by

$$\Phi(s) = [s\mathbf{I} - \mathbf{A}]^{-1} \qquad (4\text{-}103)$$

EXAMPLE 1 Solve for $\mathbf{\Phi}(t)$ (see the Example of Sec. 3-13) for

$$\mathbf{A} = \begin{bmatrix} 0 & 6 \\ -1 & -5 \end{bmatrix}$$

Using Eq. (4-103) gives

$$\mathbf{\Phi}(s) = \left[\begin{bmatrix} s & 0 \\ 0 & s \end{bmatrix} - \begin{bmatrix} 0 & 6 \\ -1 & -5 \end{bmatrix} \right]^{-1} = \begin{bmatrix} s & -6 \\ 1 & s+5 \end{bmatrix}^{-1}$$

Using Eqs. (4-97)† gives

$$\mathbf{\Phi}(s) = \frac{1}{s^2 + 5s + 6} \begin{bmatrix} s+5 & 6 \\ -1 & s \end{bmatrix} = \begin{bmatrix} \dfrac{s+5}{(s+2)(s+3)} & \dfrac{6}{(s+2)(s+3)} \\ \dfrac{-1}{(s+2)(s+3)} & \dfrac{s}{(s+2)(s+3)} \end{bmatrix}$$

Using transform pairs 12 to 14 in Appendix A yields the inverse transform

$$\mathbf{\Phi}(t) = \begin{bmatrix} 3e^{-2t} - 2e^{-3t} & 6e^{-2t} - 6e^{-3t} \\ -e^{-2t} + e^{-3t} & -2e^{-2t} + 3e^{-3t} \end{bmatrix}$$

This is the same as the value obtained in Eq. (3-88) and is an alternate method of obtaining the same result. When an input is present, the state equation is

$$\dot{\mathbf{x}} = \mathbf{Ax} + \mathbf{Bu} \qquad (4\text{-}104)$$

Solving by means of the Laplace transform yields

$$\mathbf{X}(s) = [s\mathbf{I} - \mathbf{A}]^{-1}\mathbf{x}(0) + [s\mathbf{I} - \mathbf{A}]^{-1}\mathbf{BU}(s) = \mathbf{\Phi}(s)\mathbf{x}(0) + \mathbf{\Phi}(s)\mathbf{BU}(s) \qquad (4\text{-}105)$$

EXAMPLE 2 Solve for the state equation in the Example of Sec. 3-14 for a step-function input. The function $\mathbf{\Phi}(s)$ is given in Example 1 and $\mathbf{U}(s) = 1/s$. Inserting these values into Eq. (4-105) gives

$$\mathbf{X}(s) = \mathbf{\Phi}(s)\mathbf{x}(0) + \begin{bmatrix} \dfrac{6}{s(s+2)(s+3)} \\ \dfrac{1}{(s+2)(s+3)} \end{bmatrix}$$

Applying the transform pairs 10 and 11 of Appendix A yields the same result for $\mathbf{x}(t)$ as in Eq. (3-93). The Laplace transform method is a direct method of solving the state equation.

The possibility exists in the elements of $\mathbf{\Phi}(s)$ that a numerator factor of the form $s + a$ will cancel a similar term in the denominator. In that case the transient mode e^{-at} does not appear in the corresponding element of $\mathbf{\Phi}(t)$. This has a very important

---

† When A is a companion matrix, a direct algorithm for evaluating $[s\mathbf{I} - \mathbf{A}]^{-1}$ and $\mathbf{\Phi}(t)$ is presented in Ref. 9.

theoretic significance since it affects the *controllability and observability* of a control system. These properties are described in Sec. 6-9. Also, as described in Chap. 14, the ability to achieve an optimal control system requires that the system be observable and controllable.

4-17 EVALUATION OF THE TRANSFER-FUNCTION MATRIX

The transfer function matrix of a system with multiple inputs and outputs is evaluated with the system at rest, i.e., with zero initial conditions. The state and output equations for a system are

$$\dot{x} = Ax + Bu \qquad y = Cx + Du$$

Taking the Laplace transform of these equations and solving for the output $Y(s)$ in terms of the input $U(s)$ yields

$$Y(s) = C\Phi(s)x(0) + C\Phi(s)BU(s) + DU(s)$$

Since the transfer-function relationship is $Y(s) = G(s)U(s)$ and is defined for zero initial conditions, the system transfer-function matrix is given by analogy as

$$G(s) = C\Phi(s)B + D \qquad (4\text{-}106)$$

For multiple-input multiple-output systems $G(s)$ is a matrix in which the elements are the transfer functions between each output and each input of the system.

EXAMPLE Determine the transfer functions and draw a block diagram for the two-input two-output system represented by

$$\dot{x} = \begin{bmatrix} 0 & 1 \\ -2 & -3 \end{bmatrix} x + \begin{bmatrix} 1 & 1 \\ 0 & -2 \end{bmatrix} u \qquad y = \begin{bmatrix} 0 & -2 \\ 1 & 0 \end{bmatrix} x$$

The characteristic polynomial is

$$|sI - A| = (s + 1)(s + 2)$$

From Eq. (4-103), the resolvent matrix is

$$\Phi(s) = \frac{1}{(s+1)(s+2)} \begin{bmatrix} s+3 & 1 \\ -2 & s \end{bmatrix}$$

Then, from Eq. (4-106)

$$G(s) = C\Phi(s)B = \begin{bmatrix} \dfrac{4}{(s+1)(s+2)} & \dfrac{4}{s+2} \\ \dfrac{s+3}{(s+1)(s+2)} & \dfrac{1}{s+2} \end{bmatrix} = \begin{bmatrix} G_{11}(s) & G_{12}(s) \\ G_{21}(s) & G_{22}(s) \end{bmatrix} \qquad (4\text{-}107)$$

The block diagram for the system is shown in Fig. 4-15.

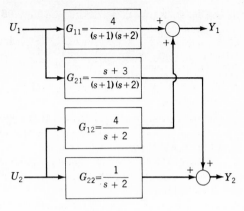

FIGURE 4-15
Block diagram for Eq. (4-107).

4-18 SUMMARY

This chapter discusses the important characteristics and the use of the Laplace transform, which is employed extensively with differential equations because it systematizes their solution. Also it is used extensively in feedback-system synthesis. The pole-zero pattern has been introduced to represent a system function. The pole-zero pattern is significant because it determines the amplitudes of all the time-response terms. The frequency response has also been shown to be a function of the pole-zero pattern. The solution of linear time-invariant state equations can be obtained by means of the Laplace transform. The procedure is one that can be adapted for the use of a digital computer. This is advantageous for multiple-input multiple-output systems.

Later chapters cover feedback-system analysis and synthesis by two methods. The first of these is the root-locus method, which locates the poles and zeros of the system in the s plane. Knowing the poles and zeros permits an exact determination of the time response. The second method is based on the frequency response. Since the frequency response is a function of the pole-zero pattern, the two methods are complementary and give equivalent information in different forms.

System stability requires that all roots of the characteristic equation be located in the left half of the s plane. They can be identified on the pole-zero diagram and are the poles of the overall transfer function. The transfer function can be obtained from the overall differential equation relating an input to an output. The transfer function can also be obtained from the state-equation formulation, as shown in this chapter.

REFERENCES

1 Churchill, R. V.: "Operational Mathematics," 3d ed., McGraw-Hill, New York, 1972.
2 Thomson, W. T.: "Laplace Transformation," 2d ed., Prentice-Hall, Englewood Cliffs, N.J., 1960.

3 Wylie, C. R., Jr.: "Advanced Engineering Mathematics," 4th ed., McGraw-Hill, New York, 1975.

4 Hazony, D., and I. Riley: Simplified Technique for Evaluating Residues in the Presence of High Order Poles, paper presented at the Western Electronic Show and Convention, August 1959.

5 Aseltine, J. A.: "Transform Method in Linear System Analysis," McGraw-Hill, New York, 1958.

6 Kinariwala, B., et al.: "Linear Circuits and Computation," Wiley, New York, 1973.

7 Ward, J. R., and R. D. Strum, "State Variable Analysis," Prentice-Hall, Englewood Cliffs, N.J., 1970.

8 Heaviside Partial Fraction Expansion and Time Response Program (PARTL), School of Engineering, Air Force Institute of Technology, Wright-Patterson Air Force Base, Ohio, 1974.

9 Taylor, F. J.: A Novel Inversion of $(s\mathbf{I} - \mathbf{A})$, *Intern. J. Systems Science*, vol. 5, no. 2, pp. 153–160, Feb. 1974.

5

SYSTEM REPRESENTATION

5-1 INTRODUCTION

This chapter introduces the basic principles of system representation. From the concepts introduced in earlier chapters, a number of systems are represented in block-diagram form. The individual functions are represented by transfer functions which are used to describe the blocks of the system. Feedback is included in these systems in order to achieve the desired performance. Also, the standard symbols and definitions are presented. These are extensively used in the technical literature on control systems and form a common basis for understanding and clarity. While block diagrams simplify the representation of functional relationships within a system, the use of signal flow graphs (SFG) provides further simplification advantageous for larger systems, which may contain many intercoupled relationships. The inclusion of initial conditions in the SFGs leads to the state-diagram representation.

To provide flexibility in the method to be used for system analysis and design, the system representation may be in either the transfer-function or the state-equation form. The procedures for converting from one representation to the other have been presented in Chaps. 3 and 4. When the state-variable format is used, it is often desirable to transform the mathematical equations so that the states are uncoupled. This simplifies the equations, leading to an **A** matrix which is diagonal. This is called the *normal* or *canonical form* and is useful for evaluating observability and controllability (see Chap. 6). It also considerably simplifies the computer programs

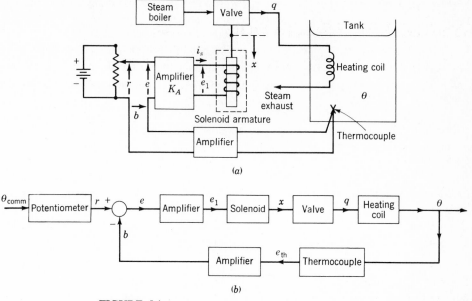

FIGURE 5-1
An industrial-process temperature-control system.

required to solve these equations. The techniques for transforming the state vector are presented.

5-2   BLOCK DIAGRAMS

The representation of physical components by blocks was shown in Chap. 2. For each block the transfer function provides the dynamic mathematical relationship between the input and output quantities. Also, Chap. 1 and Fig. 1-2 describe the concept of feedback, which is used to achieve a better response of a control system to a command input. This section presents several examples of control systems and their representation by block diagrams. The blocks represent the functions performed rather than the components of the system.

EXAMPLE 1   *A temperature-control system*   The first example is an industrial-process temperature-control system, shown in Fig. 5-1*a*. The voltage *r*, obtained from a potentiometer, is calibrated in terms of the desired temperature θ_{comm} to be produced within the tank. This voltage represents the input quantity to the control system. The actual temperature θ, the output quantity, is measured by means of a thermocouple immersed in the tank. The voltage e_{th} produced in the thermocouple is proportional to θ. The voltage e_{th} is amplified to produce a voltage *b*, which is the feedback quantity. The voltage $e = r - b$ is the actuating signal and is amplified to produce the solenoid voltage e_1. The current i_s in the solenoid, which results from applying e_1, produces a proportional force f_s which acts on the solenoid armature and valve to control the valve position x. The valve position in turn controls the flow q of hot steam into the heating coil in the tank. The resulting temperature θ of the tank is directly proportional to the steam flow with a time delay which depends on the specific heat of the fluid and the mixing rate. The block-diagram representation for this system (Fig. 5-1b) shows the functions of each unit and the signal flow through the system.

To show the operation of the system consider that an increase in the tank temperature is required. The voltage r is increased to a value that represents the value of the desired temperature. This change in r causes an increase in the actuating signal e. This increase in e, through its effect on the solenoid and valve, causes an increase in the flow of hot steam through the heating coil in the tank. The temperature θ therefore increases in proportion to the steam flow. When the output temperature rises to a value essentially equal to the desired temperature, the feedback voltage b is equal to the reference input r, $b \approx r$. The flow of steam through the heating coil is stabilized at a steady-state value that maintains θ at the desired value.

EXAMPLE 2   *Command guidance interceptor system*   A more complex system is a command guidance system which directs the flight of a missile in space in order to intercept a moving target. The target may be an enemy bomber whose aim is to drop bombs at some position. The defense uses the missile with the objective of intercepting and destroying the bomber before it launches its bombs. A sketch of a generalized command guidance interceptor system is shown in Fig. 5-2. The target-tracking radar is used first for detection and then for tracking the target. It supplies information on target range and angle and their rates of change (time derivatives). This information is continuously fed into the computer, which calculates a predicted course for the target. The missile-tracking radar supplies similar information, which is used by the computer to determine its flight path. The computer compares the two flight paths and determines the necessary change in missile flight path to produce a collision course. The necessary flight-path changes are supplied to the radio command link, which transmits this information to the missile. This electrical information containing corrections in flight path is used by a control system in the missile. The missile control system converts the error signals to mechanical displacements of the missile airframe control surfaces by means of actuators. The missile responds to the positions of the aerodynamic control surfaces to follow the prescribed flight path, which is intended to produce a collision with the target. Monitoring of the target is continuous so that changes in the missile course can be corrected up to the point of impact. A block diagram depicting the functions of this command guidance system is shown in Fig. 5-3. Within each block many individual functions are performed. Some of the components of the missile control system are shown within the block representing the missile.

EXAMPLE 3   *Aircraft control system*[1]   The feedback control system used to keep an airplane on a predetermined course or heading is necessary for the navigation of commercial airliners. Despite poor weather conditions and lack of visibility the airplane must maintain a specified heading and altitude in order to reach its destination

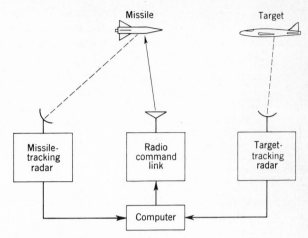

FIGURE 5-2
Command guidance interceptor system.

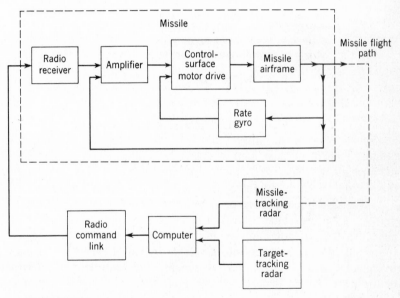

FIGURE 5-3
Block diagram of a generalized command guidance interceptor system.

safely. In addition, in spite of rough air, the trip must be made as smooth and comfortable as possible for the passengers and crew. The problem is considerably complicated by the fact that the airplane has six degrees of freedom. This makes control more difficult than the control of a ship, whose motion is limited to the surface of the water. A *flight controller* is used to control the aircraft motion.

Two typical signals to the system are the correct flight path, which is set by the pilot, and the level position of the airplane. The ultimately controlled variable is the

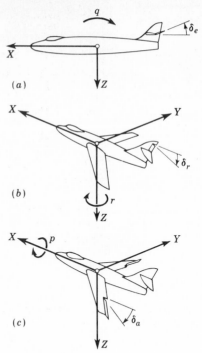

FIGURE 5-4
Airplane control surfaces: (*a*) elevator deflection produces pitching velocity q; (*b*) rudder deflection produces yawing velocity r; (*c*) aileron deflection produces rolling velocity p.

actual course and position of the airplane. The output of the control system, the controlled variable, is the aircraft heading. In conventional aircraft there are three primary control surfaces used to control the physical three-dimensional attitude of the airplane, the elevators, rudder, and ailerons. The axes used for an airplane and the motions produced by the control surfaces are shown in Fig. 5-4.

The directional gyroscope is used as the error-measuring device. Two gyros must be used to provide control of both heading and attitude (level position) of the airplane. The error that appears in the gyro as an angular displacement between the rotor and case is translated into a voltage by various methods, including the use of transducers such as potentiometers, synchros, transformers, or microsyns. Selection of the method used depends on the preference of the gyro manufacturer and the sensitivity required.

Additional stabilization for the aircraft can be provided in the control system by rate feedback. In other words, in addition to the primary feedback, which is the position of the airplane, another signal proportional to the angular rate of rotation of the airplane around the vertical axis is fed back in order to achieve a stable response. A "rate" gyro is used to supply this signal. This additional stabilization may be absolutely necessary for some of the newer high-speed aircraft.

A typical block diagram of the aircraft control system (Fig. 5-5) illustrates control of the airplane heading by controlling the rudder position. In this system the airplane heading is controlled and is the direction the airplane would travel in still air. The pilot must correct this heading, depending on the cross winds, so that the actual

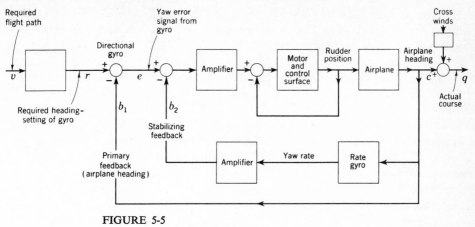

FIGURE 5-5
Airplane directional-control system.

course of the airplane coincides with the desired path. Another control is included in the complete airplane control system to keep the airplane in level flight. It controls the ailerons and elevators.

5-3 DETERMINATION OF THE OVERALL TRANSFER FUNCTION

The block diagram of a control system with negative feedback can be simplified to the form shown in Fig. 5-6, where the standard symbols and definitions used in feedback systems are indicated.

In using the block diagram to represent a linear feedback control system where the transfer functions of the components are known, the letter symbol is capitalized, indicating that it is a transformed quantity; i.e., it is a function of the operator D, the complex parameter s, or the frequency parameter $j\omega$. This applies for the transfer function, where G is used to represent $G(D)$, $G(s)$, or $\mathbf{G}(j\omega)$. It also applies to all

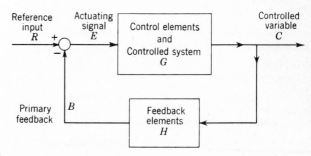

FIGURE 5-6
Block diagram of a feedback system.

variable quantities, such as C, which represents $C(D)$, $C(s)$, or $\mathbf{C}(j\omega)$. Lowercase symbols are used, as in Fig. 5-5, to represent any function in the time domain. For example, the symbol c represents $c(t)$.

The important characteristic of such a system is the *overall transfer function*, which is the ratio of the transform of the controlled variable C to the transform of the reference input R. This ratio may be expressed in operational, Laplace transform, or frequency (phasor) form. The overall transfer function is also referred to as the *control ratio*. The terms are used interchangeably in this text. The control elements and controlled system G respond to the actuating signal E to produce the controlled variable C. The feedback elements H constitute the portion of the system that responds to the controlled variable C to produce the primary feedback B. The actuating signal E is equal to the reference input R minus the primary feedback B. The transfer functions of the forward and feedback components of the system are G and H, respectively.

The equations describing this system in terms of the transform variable are

$$C(s) = G(s)E(s) \tag{5-1}$$

$$B(s) = H(s)C(s) \tag{5-2}$$

$$E(s) = R(s) - B(s) \tag{5-3}$$

Combining these equations produces the *control ratio*, or *overall transfer function*,

$$\frac{C(s)}{R(s)} = \frac{G(s)}{1 + G(s)H(s)} \tag{5-4}$$

The characteristic equation of the closed-loop system is

$$1 + G(s)H(s) = 0 \tag{5-5}$$

This is obtained from the denominator of the control ratio. The stability and response of the closed-loop system, as determined by analysis of the characteristic equation, are discussed more fully in later chapters.

For simplified systems where the feedback is unity, that is, $H(s) = 1$, the actuating signal is the error present in the system, i.e., reference input minus the controlled variable, expressed by

$$E(s) = R(s) - C(s) \tag{5-6}$$

The control ratio with unity feedback is

$$\frac{C(s)}{R(s)} = \frac{G(s)}{1 + G(s)} \tag{5-7}$$

The *open-loop transfer function* is defined as the ratio of the output of the feedback path $B(s)$ to the actuating signal $E(s)$ for any given feedback loop. In terms of Fig. 5-6, the open-loop transfer function is

$$\frac{B(s)}{E(s)} = G(s)H(s) \tag{5-8}$$

The *forward transfer function* is defined as the ratio of the controlled variable $C(s)$ to the actuating signal $E(s)$. For the system shown in Fig. 5-5 the forward transfer function is

$$\frac{C(s)}{E(s)} = G(s) \tag{5-9}$$

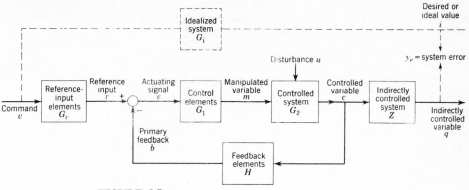

FIGURE 5-7
Block diagram of feedback control system containing all basic elements.

In the case of unity feedback, where $H(s) = 1$, the open-loop and the forward transfer functions are the same.

The forward transfer function $G(s)$ may be made up not only of elements in cascade but also may contain internal, or *minor*, feedback loops. The algebra of combining these internal feedback loops is similar to that used above. An example of a controlled system with an internal feedback loop is shown in Fig. 5-5.

It is often useful to express the actuating signal E in terms of the input R. Solving from Eqs. (5-1) to (5-3) gives

$$\frac{E(s)}{R(s)} = \frac{1}{1 + G(s)H(s)} \qquad (5\text{-}10)$$

The concept of system error y_e is important. It is defined as the ideal or desired system value minus the actual system output. The ideal value establishes the desired performance of the system. For unity-feedback systems the actuating signal is an actual measure of the error and is directly proportional to the system error (see Fig. 5-7).

5-4 STANDARD BLOCK-DIAGRAM TERMINOLOGY[2]

Figure 5-7 shows a block-diagram representation of a feedback control system containing the basic elements. Figure 5-8 shows the block diagram and symbols of a more complicated system with multiple paths. Numerical subscripts are used to distinguish between blocks of similar functions in the circuit. For example, in Fig. 5-7 the control elements are designated by G_1 and the controlled system by G_2. In Fig. 5-8 the control elements are divided into two blocks, G_1 and G_2, to aid in representing relations between parts of the system. Also, in this figure the primary feedback is represented by H_1. Additional minor feedback loops might be designated by H_2, H_3, and so forth.

The idealized system represented by the block enclosed with dashed lines in Fig. 5-7 can be understood to show the relation between the basic input to the system

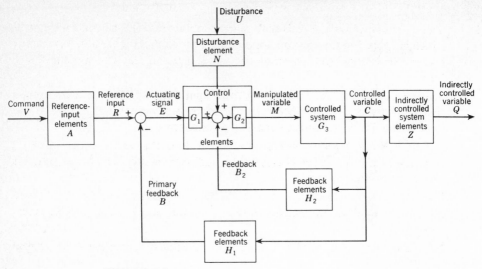

FIGURE 5-8
Block diagram of representative feedback control system showing use of multiple feedback loops (all uppercase letters to denote transformation).

and the performance of the system in terms of the desired output. This would be the system agreed upon to establish the ideal value for the output of the system. In systems where the command is actually the desired value or ideal value, the idealized system would be represented by unity. The arrows and block associated with the idealized system, the ideal value, and the system error are shown in dashed lines on the block diagram because they do not exist physically in any feedback control system. For any specific problem, it represents the system (conceived in the mind of the designer) that will give the nearest approach, when considered as a perfect system, to the desired output or ideal value.

Definitions: Variables in the System

The *command v* is the input which is established by some means external to, and independent of, the feedback control system.

The *reference input r* is derived from the command and is the actual signal input to the system.

The *controlled variable c* is the quantity that is directly measured and controlled. It is the output of the controlled system.

The *primary feedback b* is a signal which is a function of the controlled variable and which is compared with the reference input to obtain the actuating signal.

The *actuating signal e* is obtained from a comparison measuring device and is the reference input minus the primary feedback. This signal, usually at a low energy level, is the input to the control elements that produce the manipulated variable.

The *manipulated variable m* is that quantity obtained from the control elements which is applied to the controlled system. The manipulated variable is generally at a higher energy level than the actuating signal and may also be modified in form.

The *indirectly controlled variable q* is the output quantity and is related through the indirectly controlled system to the controlled variable. It is outside the closed loop and is not directly measured for control.

The *ultimately controlled variable* is a general term that refers to the indirectly controlled variable. In the absence of the indirectly controlled variable, it refers to the controlled variable.

The *ideal value i* is the value of the ultimately controlled variable that would result from an idealized system operating from the same command as the actual system.

The *system error y_e* is the ideal value minus the value of the ultimately controlled variable.

The *disturbance u* is the unwanted signal that tends to affect the controlled variable. The disturbance may be introduced into the system at many places.

Definitions: System Components

The *reference input elements G_v* produce a signal proportional to the command.

The *control elements G* produce the manipulated variable from the actuating signal.

The *controlled system G* is the device that is to be controlled. This is frequently a high-power element.

The *feedback elements H* produce the primary feedback from the controlled variable. This is generally a proportionality device but may also modify the characteristics of the controlled variable.

The *indirectly controlled system Z* relates the indirectly controlled variable to the controlled quantity. This component is outside the feedback loop.

The *idealized system G_i* is one whose performance is agreed upon to define the relationship between the ideal value and the command. In the adaptive systems this is called the *model* system.

The *disturbance element N* denotes the functional relationship between the variable representing the disturbance and its effect on the control system.

5-5   POSITION-CONTROL SYSTEM

Figure 5-9 shows a simplified block diagram of an angular-position-control system. The reference selector and the sensor, which produce the reference input $R = \theta_R$ and the controlled output position $C = \theta_o$, respectively, consist of rotational potentiometers. The combination of these units to produce a rotational comparison unit to generate the actuating signal E for the position-control system is shown in Fig. 5-10a,

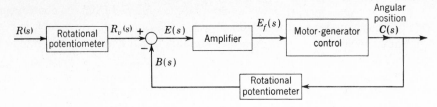

FIGURE 5-9
Position-control system.

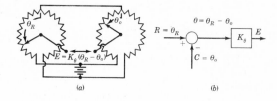

FIGURE 5-10
Rotational position comparison and its block-diagram representation.

where K_θ, in volts per radian, is the potentiometer-sensitivity constant. The symbolic comparator for this system is shown in Fig. 5-10b.

The transfer function of the motor-generator control is obtained by writing the equations for the schematic diagram shown in Fig. 5-11a. This figure shows a dc motor which has a constant field excitation and drives an inertia and friction load. The armature voltage for the motor is furnished by the generator which is driven at constant speed by a prime mover. The generator voltage e_g is determined by the voltage e_f applied to the generator field. The generator is acting as a power amplifier for the signal voltage e_f. The equations for this system are

$$e_f = L_f \, Di_f + R_f i_f \tag{5-11}$$

$$e_g = K_g i_f \tag{5-12}$$

$$e_g - e_m = (L_g + L_m) \, Di_m + (R_g + R_m)i_m \tag{5-13}$$

$$e_m = K_b \, D\theta_o \tag{5-14}$$

$$T = K_T i_m = J \, D^2\theta_o + B \, D\theta_o \tag{5-15}$$

The complete block diagram drawn in Fig. 5-11b is based on the performance represented by these equations. Starting with the input quantity e_f, Eq. (5-11) shows that a current i_f is produced. Therefore, a block is drawn with e_f as the input and i_f as the output. Equation (5-12) shows that a voltage e_g is generated as a function of the current i_f. Therefore, a second block is drawn with the current i_f as the input and e_g as the output. Equation (5-13) relates the current i_m in the motor to the difference of two voltages, $e_g - e_m$. To obtain this difference a summation point is introduced.

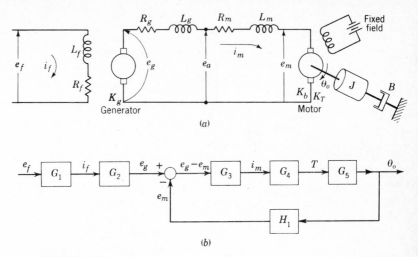

FIGURE 5-11
(a) Motor-generator control; (b) block diagram.

The quantity e_g from the previous block enters this summation point. To obtain the quantity $e_g - e_m$ there must be added the quantity e_m entering the summation point with a minus sign. Up to this point the manner in which e_m is obtained has not yet been determined. The output of this summation point is used as the input to the next block from which the current i_m is the output. In the same manner, the block with current as the input and the generated torque as the output and the block with the torque input and the resultant motor position as the output are drawn. There must be no loose ends in the complete diagram; i.e., every dependent variable must be connected through a block or blocks into the system. Therefore e_m is obtained from Eq. (5-14), and a block representing this relationship is drawn with θ_o as the input and e_m as the output. When this procedure is set up, the block diagram is completed and the functional relationships in the system are described. Note that the generator and the motor are no longer separately distinguishable. The input and the output of each block are not in the same units, as both electrical and mechanical quantities are included.

The transfer functions of each block, as determined in terms of the pertinent Laplace transforms, are

$$G_1(s) = \frac{I_f(s)}{E_f(s)} = \frac{1/R_f}{1 + (L_f/R_f)s} = \frac{1/R_f}{1 + T_f s} \tag{5-16}$$

$$G_2(s) = \frac{E_g(s)}{I_f(s)} = K_g \tag{5-17}$$

$$G_3(s) = \frac{I_m(s)}{E_g(s) - E_m(s)} = \frac{1/(R_g + R_m)}{1 + [(L_g + L_m)/(R_g + R_m)]s}$$

$$= \frac{1/R_{gm}}{1 + (L_{gm}/R_{gm})s} = \frac{1/R_{gm}}{1 + T_{gm}s} \tag{5-18}$$

$$G_4(s) = \frac{T(s)}{I_m(s)} = K_T \qquad (5\text{-}19)$$

$$G_5(s) = \frac{\Theta_o(s)}{T(s)} = \frac{1/B}{s[1 + (J/B)s]} = \frac{1/B}{s(1 + T_n s)} \qquad (5\text{-}20)$$

$$H_1(s) = \frac{E_m(s)}{\Theta_o(s)} = K_b s \qquad (5\text{-}21)$$

The block diagram can be simplified, as shown in Fig. 5-12, by combining the blocks in cascade. The block diagram is further simplified, as shown in Fig. 5-13, by evaluating an equivalent block from $E_g(s)$ to $\Theta_o(s)$, using the principle of Eq. (5-4). The final simplification results in Fig. 5-14. The overall transfer function $G_x(s)$ is

$$G_x(s) = \frac{\Theta_o(s)}{E_f(s)}$$

$$= \frac{K_g K_T / R_f B R_{gm}}{s(1 + T_f s)[(1 + K_T K_b / B R_{gm}) + (T_{gm} + T_n)s + T_{gm} T_n s^2]} \qquad (5\text{-}22)$$

This expression is the exact transfer function for the entire motor and generator combination. Certain approximations, if valid, may be made to simplify this expression. The first approximation is that the inductance of the generator and motor armatures is very small ($T_{gm} \approx 0$). With this approximation, the transfer function reduces to

$$\frac{\Theta_o(s)}{E_f(s)} = G_x(s) \approx \frac{K_x}{s(1 + T_f s)(1 + T_m s)} \qquad (5\text{-}23)$$

where

$$K_x = \frac{K_g K_T}{R_f (B R_{gm} + K_T K_b)} \qquad (5\text{-}24)$$

$$T_f = \frac{L_f}{R_f} \qquad (5\text{-}25)$$

$$T_m = \frac{J R_{gm}}{B R_{gm} + K_T K_b} \qquad (5\text{-}26)$$

If the frictional effect of the load is very small, the approximation can be made that $B \approx 0$. With this additional approximation, the transfer function remains of the same form as Eq. (5-23), but the constants are now

$$K_x = \frac{K_g}{R_f K_b} \qquad (5\text{-}27)$$

$$T_f = \frac{L_f}{R_f} \qquad (5\text{-}28)$$

$$T_m = \frac{J R_{gm}}{K_T K_b} \qquad (5\text{-}29)$$

For simple components, as in this case, an overall transfer function can often be derived more easily by combining the original system equations. This can be done

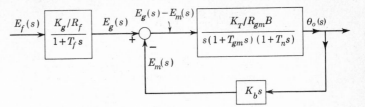

FIGURE 5-12
Simplified block diagram.

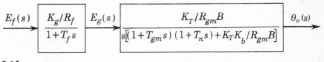

FIGURE 5-13
Reduced block diagram.

without the intermediate steps shown in this example. However, the purpose of this example is to show the representation of dynamic components by individual blocks and the combination of blocks.

A new block diagram representing the position-control system of Fig. 5-9 is redrawn in Fig. 5-15. Since the transfer function of the amplifier of Fig. 5-9 is $E_f(s)/E(s) = A$, the forward transfer function of the position-control system of Fig. 5-15 is

$$G(s) = \frac{\Theta_o(s)}{E(s)} = \frac{AK_xK_\theta}{s(1 + T_fs)(1 + T_ms)} \qquad (5\text{-}30)$$

The overall transfer function (control ratio) for this system is

$$\frac{C(s)}{R(s)} = \frac{\Theta_o(s)}{\Theta_r(s)} = \frac{G(s)}{1 + G(s)H(s)}$$

$$= \frac{AK_xK_\theta}{s(1 + T_fs)(1 + T_ms) + AK_xK_\theta} \qquad (5\text{-}31)$$

where $H(s) = 1$.

FIGURE 5-14
Overall block diagram for Fig. 5-11.

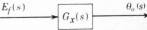

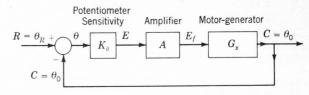

FIGURE 5-15
Equivalent representation of Fig. 5-9.

FIGURE 5-16
Signal flow graph for $x_2 = ax_1$.

$$x_1 \qquad a \qquad x_2 = ax_1$$

5-6   SIGNAL FLOW GRAPHS[3,4]

The block diagram is a useful tool for simplifying the representation of a system. The block diagrams of Fig. 5-9 and 5-15 have only one feedback loop and may be categorized as simple block diagrams. The system represented in Fig. 5-5 has a total of three feedback loops and is no longer a simple system. When intercoupling exists between feedback loops, and when a system has more than one input and one output, the control system and block diagram are more complex. Having the block diagram simplifies the analysis of a complex system. Such an analysis can be further simplified by using a signal flow graph (SFG), which looks like a simplified block diagram.

An SFG is a diagram which represents a set of simultaneous equations. It consists of a *graph* in which *nodes* are connected by directed *branches*. The nodes represent each of the system variables. A branch connected between two nodes acts as a one-way signal multiplier: the direction of signal flow is indicated by an arrow placed on the branch, and the multiplication factor (transmittance or transfer function) is indicated by a letter placed near the arrow. Thus, in Fig. 5-16, the branch transmits the signal x_1 from left to right and multiplies it by the quantity a in the process. The quantity a is the transmittance, or transfer function. It may also be indicated by $a = t_{12}$, where the subscripts show that the signal flow is from node 1 to node 2.

Flow-Graph Definitions

A *node* performs two functions:

1   Addition of the signals on all incoming branches
2   Transmission of the total node signal (the sum of all incoming signals) to all outgoing branches.

These functions are illustrated in the graph of Fig. 5-17, which represents the equations

$$w = au + bv \qquad (5\text{-}32)$$

$$x = cw \qquad (5\text{-}33)$$

$$y = dw \qquad (5\text{-}34)$$

There are three types of nodes that are of particular interest:

Source nodes (*independent nodes*)   These represent independent variables and have only outgoing branches. In Fig. 5-17, nodes u and v are source nodes.

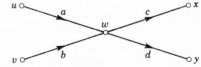

FIGURE 5-17
Signal flow graph for Eqs. (5-32) to (5-34).

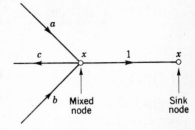

FIGURE 5-18
Mixed and sink nodes for a variable.

Sink nodes (*dependent nodes*) These represent dependent variables and have only incoming branches. In Fig. 5-17, nodes x and y are sink nodes.

Mixed nodes (*general nodes*) These have both incoming and outgoing branches. In Fig. 5-17, node w is a mixed node. A mixed node may be treated as a sink node by adding an outgoing branch of unity transmittance, as shown in Fig. 5-18.

A *path* is any connected sequence of branches whose arrows are in the same direction.

A *forward path* between two nodes is one which follows the arrows of successive branches and in which a node appears only once. In Fig. 5-17 the path uwx is a forward path between the nodes u and x.

Flow-Graph Algebra

The following rules are useful for simplifying a signal flow graph.

Series paths (*cascade nodes*) Series paths can be combined into a single path by multiplying the transmittances as shown in Fig. 5-19a.

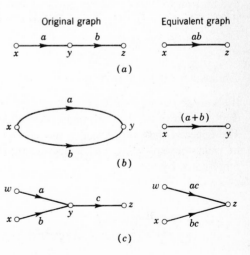

FIGURE 5-19
Flow-graph simplifications.

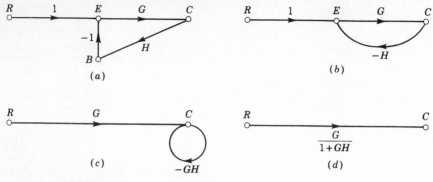

FIGURE 5-20
Successive reduction of the flow graph for the feedback system of Fig. 5-6.

Parallel paths Parallel paths can be combined by adding the transmittances as shown in Fig. 5-19b.

Node absorption A node representing a variable other than a source or sink can be eliminated as shown in Fig. 5-19c.

Feedback paths The equations for the feedback system of Fig. 5-6 are

$$C = GE \qquad (5\text{-}35)$$

$$B = HC \qquad (5\text{-}36)$$

$$E = R - B \qquad (5\text{-}37)$$

Note that an equation is written for each dependent variable. The corresponding signal flow graph is shown in Fig. 5-20a. The node B can be eliminated to produce Fig. 5-20b. The node E can be eliminated to produce Fig. 5-20c, which has a *self-loop* of value $-GH$. The final simplification is to eliminate the self-loop to produce the overall transmittance from the input R to the output C. This is obtained by summing signals at node C in Fig. 5-20c, yielding $C = GR - GHC$. Solving for C produces Fig. 5-20d.

General Flow-Graph Analysis

If all the source nodes are brought to the left and all the sink nodes are brought to the right, the SFG for an arbitrarily complex system can be represented by Fig. 5-21a. The effect of the *internal* nodes can be factored out by ordinary algebraic processes to yield the equivalent graph represented by Fig. 5-21b. This simplified graph is represented by

$$y_1 = T_a x_1 + T_d x_2 \qquad (5\text{-}38)$$

$$y_2 = T_b x_1 + T_e x_2 \qquad (5\text{-}39)$$

$$y_3 = T_c x_1 + T_f x_2 \qquad (5\text{-}40)$$

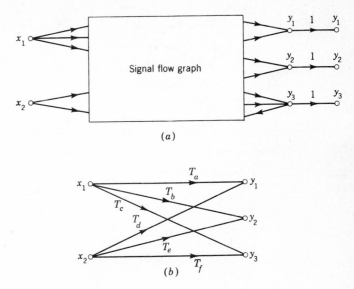

FIGURE 5-21
Equivalent signal flow graphs.

The T's, called overall graph transmittances, are the overall transmittances from a specified source node to a specified dependent node. For linear systems the principle of superposition can be used to "solve" the graph. This means that the sources can be considered one at a time. Then the output signal is equal to the sum of the contributions produced by each input.

The overall transmittances can be found by the ordinary processes of linear algebra, i.e., by the solution of the set of simultaneous equations representing the system. However, the same results can be obtained directly from the signal flow graph. The fact that they can produce answers to large sets of linear equations *by inspection* gives the signal flow graphs their power and usefulness.

The Mason Gain Rule

The overall transmittance can be obtained from the formula developed by S. J. Mason. The formula and definitions are followed by an example to show its application. The overall transmittance is given by

$$T = \frac{\sum T_n \Delta_n}{\Delta} \qquad (5\text{-}41)$$

where T_n is the transmittance of each forward path between a source and a sink node and Δ is the graph determinant found from

$$\Delta = 1 - \sum L_1 + \sum L_2 - \sum L_3 + \cdots \qquad (5\text{-}42)$$

In this equation L_1 is the transmittance of each closed path, and $\sum L_1$ is the sum of the transmittances of all closed paths in the graph. L_2 is the product of the transmittances of two nontouching loops. Loops are nontouching if they do not have any

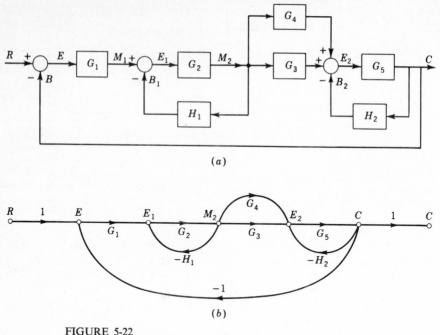

(a)

(b)

FIGURE 5-22
A block diagram and its signal flow graph.

common nodes. $\sum L_2$ is the sum of the product of transmittances of all possible combinations of nontouching loops taken two at a time. L_3 is the product of the transmittances of three nontouching loops. $\sum L_3$ is the sum of the product of transmittances of all possible combinations of nontouching loops taken three at a time.

In Eq. (5-41) Δ_n is the cofactor of T_n. It is the determinant of the remaining subgraph when the forward path which produces T_n is removed. Thus, Δ_n does not include any loops which touch the forward path in question. Δ_n is equal to unity when the forward path touches all the loops in the graph or when the graph contains no loops. Δ_n has the same form as Eq. (5-42).

EXAMPLE   Figure 5-22 shows a block diagram and its SFG. Note that not all the variables are shown. Since $E_1 = M_1 - B_1 = G_1 E - H_1 M_2$, it is not necessary to show M_1 and B_1 explicitly. Since this is a fairly complex system, the resulting equation is expected to be complex. However, the application of Mason's rule produces the resulting overall transmittance in a systematic manner. This system has four loops whose transmittances are $-G_2 H_1$, $-G_5 H_2$, $-G_1 G_2 G_3 G_5$, and $-G_1 G_2 G_4 G_5$. Therefore

$$\sum L_1 = -G_2 H_1 - G_5 H_2 - G_1 G_2 G_3 G_5 - G_1 G_2 G_4 G_5 \qquad (5\text{-}43)$$

Only two loops are nontouching; therefore

$$\sum L_2 = (-G_2 H_1)(-G_5 H_2) \qquad (5\text{-}44)$$

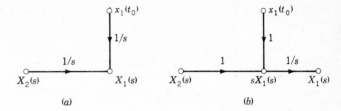

FIGURE 5-23
Representation of an integrator in the Laplace domain in a signal flow graph.

Although there are four loops, there is no set of three loops which are nontouching; therefore

$$\sum L_3 = 0 \qquad (5\text{-}45)$$

The system determinant can therefore be obtained from Eq. (5-42).

There are two forward paths between R and C. The corresponding forward transmittances are $G_1 G_2 G_3 G_5$ and $G_1 G_2 G_4 G_5$. If either path, with its corresponding nodes, is removed from the graph, the remaining subgraphs have no loops. The cofactors Δ_n are therefore both equal to unity. The complete overall transmittance from R to C, obtained from Eq. (5-41), is

$$T = \frac{G_1 G_2 G_3 G_5 + G_1 G_2 G_4 G_5}{1 + G_2 H_1 + G_5 H_2 + G_1 G_2 G_3 G_5 + G_1 G_2 G_4 G_5 + G_2 G_5 H_1 H_2} \qquad (5\text{-}46)$$

The overall transmittance has been obtained by inspection from the signal flow graph. This is much simpler than solving the five simultaneous equations which represent this system.

5-7 STATE TRANSITION SIGNAL FLOW GRAPH[5]

The state transition SFG or, more simply, the *state diagram*, is a simulation diagram for a system of equations and includes the initial conditions of the states. Since the state diagram in the Laplace domain satisfies the rules of Mason's SFG, it can be used to obtain the transfer function of the system and the state transition equation. As described in Sec. 3-15, the basic elements used in a simulation diagram are a gain, a summer, and an integrator. The signal-flow representation in the Laplace domain for an integrator is obtained as follows:

$$\dot{x}_1(t) = x_2(t)$$

$$X_1(s) = \frac{X_2(s)}{s} + \frac{x_1(t_0)}{s} \qquad (5\text{-}47)$$

Equation (5-47) may be represented either by Fig. 5-23a or Fig. 5-23b,

A differential equation which contains no derivatives of the input, as given by Eq. (3-98), is repeated here:

$$D^n y + a_{n-1} D^{n-1} y + \cdots + a_1 D y + a_0 y = u \qquad (5\text{-}48)$$

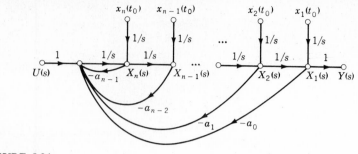

FIGURE 5-24
State diagram for Eq. (5-48).

The state diagram is drawn in Fig. 5-24 with phase variables. It is obtained by first drawing the number of integrator branches $1/s$ equal to the order of the differential equation. The outputs of the integrators are designated as the state variables, and the initial conditions of the n states are included as inputs in accordance with Fig. 5-23a. Letting $x_1 = y$, inserting the phase variables $\dot{x}_i = x_{i+1}$ into Eq. (5-48), and taking the Laplace transform yields, after dividing by s,

$$X_n(s) = \frac{1}{s}[-a_0 X_1(s) - a_1 X_2(s) - \cdots - a_{n-1}X_n(s) + x_n(t_0) + U(s)]$$

The node $X_n(s)$ in Fig. 5-24 satisfies this equation. The signal at the unlabeled node between $U(s)$ and $X_n(s)$ is equal to $sX_n(s) - x_n(t_0)$.

The overall transfer function $Y(s)/U(s)$ is defined with all initial states equal to zero. Using this condition and applying the Mason gain formula given by Eq. (5-41) yields

$$G(s) = \frac{Y(s)}{U(s)} = \frac{s^{-n}}{\Delta(s)} = \frac{s^{-n}}{1 + a_{n-1}s^{-1} + \cdots + a_1 s^{-(n-1)} + a_0 s^{-n}}$$

$$= \frac{1}{s^n + a_{n-1}s^{n-1} + \cdots + a_1 s + a_0} \tag{5-49}$$

The state transition equation of the system can be obtained from the state diagram by applying the Mason gain formula and considering each initial condition as a source. This is illustrated by the following example.

EXAMPLE For

$$\ddot{y} + \frac{R}{L}\dot{y} + \frac{1}{LC}y = \frac{1}{LC}u \tag{3-94}$$

(a) draw the state diagram and (b) determine the state transition equation.

(a) The state diagram, Fig. 5-25, includes two integrators since this is a second-order equation. The state variables are selected as the phase variables which are the outputs of the integrators, that is, $x_1 = y$ and $x_2 = \dot{x}_1$.

SYSTEM REPRESENTATION

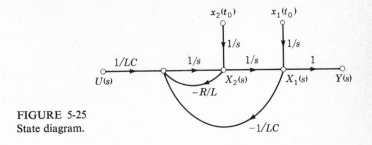

FIGURE 5-25
State diagram.

(b) The state transition equations are obtained by applying the Mason gain formula:

$$X_1(s) = \frac{s^{-1}(1 + s^{-1}R/L)}{\Delta(s)} x_1(t_0) + \frac{s^{-2}}{\Delta(s)} x_2(t_0) + \frac{s^{-2}/LC}{\Delta(s)} U(s)$$

$$X_2(s) = \frac{-s^{-2}/LC}{\Delta(s)} x_1(t_0) + \frac{s^{-1}}{\Delta(s)} x_2(t_0) + \frac{s^{-1}/LC}{\Delta(s)} U(s)$$

$$\Delta(s) = 1 + \frac{s^{-1}R}{L} + \frac{s^{-2}}{LC}$$

After simplification these equations become

$$\mathbf{X}(s) = \begin{bmatrix} X_1(s) \\ X_2(s) \end{bmatrix}$$

$$= \frac{1}{s^2 + (R/L)s + 1/LC} \left(\begin{bmatrix} s + \dfrac{R}{L} & 1 \\ -\dfrac{1}{LC} & s \end{bmatrix} \begin{bmatrix} x_1(t_0) \\ x_2(t_0) \end{bmatrix} + \begin{bmatrix} \dfrac{1}{LC} \\ \dfrac{s}{LC} \end{bmatrix} U(s) \right)$$

This equation can be recognized as being of the form given by Eq. (4-105), i.e.,

$$\mathbf{X}(s) = \mathbf{\Phi}(s)\mathbf{x}(t_0) + \mathbf{\Phi}(s)\mathbf{B}U(s)$$

Thus the resolvent matrix $\mathbf{\Phi}(s)$ is readily identified and by use of the inverse Laplace transform yields the state transition matrix $\mathbf{\Phi}(t)$. The elements of the resolvent matrix can be obtained directly from the SFG with $U(s) = 0$ and only one $x(t_0)$ considered at a time, i.e., all other initial conditions set to zero. Each element of $\mathbf{\Phi}(s)$ is

$$\phi_{ij}(s) = \frac{X_i(s)}{x_j(t_0)}$$

For example,

$$\phi_{11}(s) = \frac{X_1(s)}{x_1(t_0)} = \frac{s^{-1}(1 + s^{-1}R/L)}{\Delta(s)} = \frac{s + R/L}{s^2 \Delta(s)}$$

The complete state transition equation $\mathbf{x}(t)$ is obtainable through use of the state diagram. Therefore, $\mathbf{\Phi}(s)$ is obtained without performing the inverse operation $[s\mathbf{I} - \mathbf{A}]^{-1}$. Since $\mathbf{x}(t)$ represents phase variables, the system output is $y(t) = x_1(t)$.

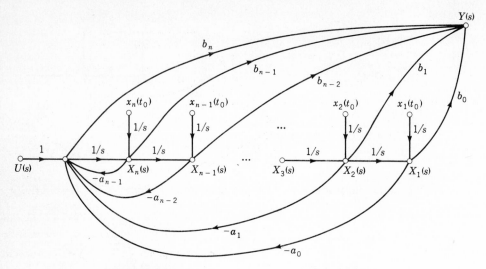

FIGURE 5-26
State diagram for Eq. (5-50) using phase variables, when $w = n$.

A differential equation containing derivatives of the input, given by Eq. (3-101), is repeated below.

$$(D^n + a_{n-1}D^{n-1} + \cdots + a_1 D + a_0)y$$
$$= (b_w D^w + b_{w-1}D^{w-1} + \cdots + b_1 D + b_0)u \qquad w \le n \qquad (5\text{-}50)$$

Phase variables can be specified as the state variables, provided the output is identified by Eq. (3-102) as

$$y = (b_w D^w + b_{w-1}D^{w-1} + \cdots + b_1 D + b_0)x_1 \qquad (5\text{-}51)$$

Equation (5-50) then reduces to

$$(D^n + a_{n-1}D^{n-1} + \cdots + a_1 D + a_0)x_1 = u \qquad (5\text{-}52)$$

The resulting state diagram for Eqs. (5-51) and (5-52) is shown in Fig. 5-26 for $w = n$. The state equations obtained from the state diagram are given by Eq. (3-99). The output equation in matrix form is readily obtained from Fig. 5-26 in terms of the transformed variables $X(s)$ and the input $U(s)$ as

$$Y(s) = [(b_0 - a_0 b_n)\ (b_1 - a_1 b_n)\ \cdots$$
$$(b_{n-2} - a_{n-2}b_n)\ (b_{n-1} - a_{n-1}b_n)]X(s) + b_n U(s) \qquad (5\text{-}53)$$

Equation (5-53) is expressed in terms of the state $X(s)$ and the input $U(s)$. In order to obtain Eq. (5-53) from Fig. 5-26, delete all the branches having a transmittance $1/s$ and all the initial conditions in Fig. 5-26. The Mason gain formula is then used to obtain the output in terms of the state variables and the input. Note that there are two forward paths from each state variable to the output.

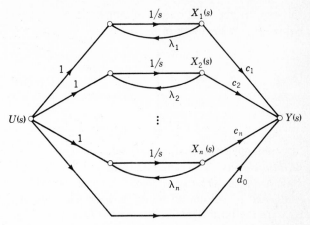

FIGURE 5-27
Simulation of Eq. (5-54) by parallel decomposition for $w = n$ (Jordan diagram for distinct roots).

5-8   PARALLEL STATE DIAGRAMS FROM TRANSFER FUNCTIONS

A single-input single-output system represented by the differential equation (5-50) may also be represented by an overall transfer function of the form

$$\frac{Y(s)}{U(s)} = G(s) = \frac{b_w s^w + b_{w-1} s^{w-1} + \cdots + b_1 s + b_0}{s^n + a_{n-1} s^{n-1} + \cdots + a_1 s + a_0} \qquad w \leq n \qquad (5\text{-}54)$$

An alternate method of determining a simulation diagram and a set of state variables is to factor the denominator and to express $G(s)$ in partial fractions. When there are no repeated roots and $w = n$, the form is

$$G(s) = d_0 + \frac{c_1}{s - \lambda_1} + \frac{c_2}{s - \lambda_2} + \cdots + \frac{c_n}{s - \lambda_n} = d_0 + \sum_{i=1}^{n} G_i(s)$$

where
$$G_i(s) = c_i X_i(s)/U(s) = c_i/(s - \lambda_i)$$

The output therefore is

$$Y(s) = d_0 U(s) + \frac{c_1 U(s)}{s - \lambda_1} + \frac{c_2 U(s)}{s - \lambda_2} + \cdots$$

$$= d_0 U(s) + c_1 X_1(s) + c_2 X_2(s) + \cdots \qquad (5\text{-}55)$$

The state variables $X_i(s)$ are selected to satisfy this equation. Each fraction represents a first-order differential equation of the form

$$\dot{x}_i - \lambda_i x_i = u \qquad (5\text{-}56)$$

This can be simulated by an integrator with a feedback path of gain equal to λ_i, followed by a gain c_i. Therefore the complete simulation diagram is drawn in Fig. 5-27. The term $d_0 U(s)$ in Eq. (5-55) is satisfied in Fig. 5-27 by the feedforward path of gain d_0. This term appears only when the numerator and denominator of $G(s)$ have the same degree. The output of each integrator is defined as a state variable. The state equations are therefore of the form

$$\dot{\mathbf{x}} = \begin{bmatrix} \lambda_1 & & & 0 \\ & \lambda_2 & & \\ & & \ddots & \\ 0 & & & \lambda_n \end{bmatrix} \mathbf{x} + \begin{bmatrix} 1 \\ 1 \\ \vdots \\ 1 \end{bmatrix} u = \mathbf{\Lambda}\mathbf{x} + \mathbf{b}_n u \qquad (5\text{-}57)$$

$$y = \begin{bmatrix} c_1 & c_2 & \cdots & c_n \end{bmatrix}\mathbf{x} + d_0 u = \mathbf{c}_n{}^T\mathbf{x} + \mathbf{d}_n u \qquad (5\text{-}58)$$

An important feature of these equations is that the $\mathbf{A}$ matrix appears in diagonal or *normal form*. This diagonal form is indicated by $\mathbf{\Lambda}$ (or $\mathbf{A}^*$), and the corresponding state variables are often called *canonical variables*. The elements of the $\mathbf{B}$ vector (indicated by $\mathbf{b}_n$) are all unity, and the elements of the $\mathbf{C}$ row matrix (indicated by $\mathbf{c}_n{}^T$) are the coefficients of the partial fractions. The state diagram of Fig. 5-27 and the state equation of Eq. (5-57) represent the Jordan form for the system with distinct eigenvalues.

The diagonal matrix $\mathbf{A} = \mathbf{\Lambda}$ means that each state equation is *uncoupled*; i.e., each state x_i can be solved independently of the other states. This simplifies the procedure for finding the state transition matrix $\mathbf{\Phi}(t)$. This form is also useful for studying the observability and controllability of a system, as discussed in Chap. 6. When the state equations are expressed in normal form, the state variables are often denoted by z_i. Thus, the state and output equations for a multiple-input multiple-output system can be expressed as

$$\dot{\mathbf{z}} = \mathbf{\Lambda}\mathbf{z} + \mathbf{B}_n\mathbf{u} \qquad (5\text{-}59)$$

$$\mathbf{y} = \mathbf{C}_n\mathbf{z} + \mathbf{D}_n\mathbf{u} \qquad (5\text{-}60)$$

Transfer functions with repeated roots are not encountered very frequently and are not considered in this text. If they should occur, the reader is referred to more extensive books in linear algebra.

EXAMPLE Draw the parallel state diagram and determine the state equation. A partial fraction expansion is performed on $G(s)$, and the state diagram is drawn in Fig. 5-28.

$$G(s) = \frac{4s^2 + 15s + 13}{s^2 + 3s + 2} = 4 + \frac{1}{s + 2} + \frac{2}{s + 1}$$

$$\dot{\mathbf{x}}(t) = \begin{bmatrix} -2 & 0 \\ 0 & -1 \end{bmatrix}\mathbf{x}(t) + \begin{bmatrix} 1 \\ 1 \end{bmatrix} u(t)$$

$$\mathbf{y}(t) = \begin{bmatrix} 1 & 2 \end{bmatrix}\mathbf{x}(t) + 4u(t)$$

5-9 DIAGONALIZING THE A MATRIX[6,7]

In Sec. 5-8 the method of partial-fraction expansion was shown to lead to the desirable normal form of the state equation in which the $\mathbf{A}$ matrix is diagonal. The partial-fraction method is not convenient for multiple-input multiple-output systems or when

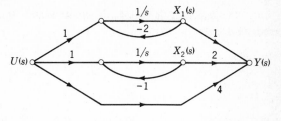

FIGURE 5-28
Simulation diagram for example of Sec. 5-8.

the system equations are already given in state form. Therefore, this section presents a more general method for converting the state equation by means of a linear *similarity transformation*. Since the state variables are not unique, the intention is to transform the state vector $\mathbf{x}$ to a new state vector $\mathbf{z}$ by means of a constant, square, nonsingular transformation matrix $\mathbf{T}$ so that

$$\mathbf{x} = \mathbf{Tz} \qquad (5\text{-}61)$$

Since $\mathbf{T}$ is a constant matrix, the differentiation of this equation yields

$$\dot{\mathbf{x}} = \mathbf{T}\dot{\mathbf{z}} \qquad (5\text{-}62)$$

Substituting these values into the state equation $\dot{\mathbf{x}} = \mathbf{Ax} + \mathbf{Bu}$ produces

$$\mathbf{T}\dot{\mathbf{z}} = \mathbf{ATz} + \mathbf{Bu} \qquad (5\text{-}63)$$

Premultiplying by $\mathbf{T}^{-1}$ gives

$$\dot{\mathbf{z}} = \mathbf{T}^{-1}\mathbf{ATz} + \mathbf{T}^{-1}\mathbf{Bu} \qquad (5\text{-}64)$$

The corresponding output equation is

$$\mathbf{y} = \mathbf{CTz} + \mathbf{Du} \qquad (5\text{-}65)$$

It is easily shown below that the eigenvalues are the same for the original and for the transformed equations. From Eq. (5-64) the new characteristic equation is

$$|\lambda\mathbf{I} - \mathbf{T}^{-1}\mathbf{AT}| = 0 \qquad (5\text{-}66)$$

In this equation the unit matrix can be replaced by $\mathbf{I} = \mathbf{T}^{-1}\mathbf{T}$. Then, after prefactoring $\mathbf{T}^{-1}$ and postfactoring $\mathbf{T}$, the characteristic equation becomes

$$|\mathbf{T}^{-1}(\lambda\mathbf{I} - \mathbf{A})\mathbf{T}| = 0 \qquad (5\text{-}67)$$

From the property that the determinant of the product of matrices is equal to the product of the determinants of the individual matrices, Eq. (5-67) is equal to

$$|\mathbf{T}^{-1}| \cdot |\lambda\mathbf{I} - \mathbf{A}| \cdot |\mathbf{T}| = |\mathbf{T}^{-1}| \cdot |\mathbf{T}| \cdot |\lambda\mathbf{I} - \mathbf{A}| = 0$$

Since $|\mathbf{I}| = |\mathbf{T}^{-1}| \cdot |\mathbf{T}| = 1$, the characteristic equation of the transformed equation as given by Eq. (5-66) is equal to the characteristic equation of the original system, i.e.,

$$|\lambda\mathbf{I} - \mathbf{A}| = 0$$

Therefore it is concluded that the eigenvalues are invariant in a linear transformation given by Eq. (5-61).

The matrix $\mathbf{T}$ is called the *modal* matrix when it is selected so that $\mathbf{T}^{-1}\mathbf{A}\mathbf{T}$ is diagonal, i.e.,

$$\mathbf{T}^{-1}\mathbf{A}\mathbf{T} = \mathbf{\Lambda} = \begin{bmatrix} \lambda_1 & & & 0 \\ & \lambda_2 & & \\ & & \ddots & \\ 0 & & & \lambda_n \end{bmatrix} \tag{5-68}$$

This supposes that the eigenvalues are distinct. With this transformation the system equations are

$$\dot{\mathbf{z}} = \mathbf{\Lambda}\mathbf{z} + \mathbf{B}'\mathbf{u} \tag{5-69}$$

$$\mathbf{y} = \mathbf{C}'\mathbf{z} + \mathbf{D}'\mathbf{u} \tag{5-70}$$

where $\qquad \mathbf{\Lambda} = \mathbf{T}^{-1}\mathbf{A}\mathbf{T} \qquad \mathbf{B}' = \mathbf{T}^{-1}\mathbf{B} \qquad \mathbf{C}' = \mathbf{C}\mathbf{T} \qquad \mathbf{D}' = \mathbf{D}$

When $\mathbf{T}^{-1}\mathbf{A}\mathbf{T}$ is used to transform the equations to any other state-variable representation $\mathbf{A}'$, $\mathbf{T}$ is just called a *transformation matrix*, that is, $\mathbf{T}^{-1}\mathbf{A}\mathbf{T} = \mathbf{A}'$.

There are three methods for obtaining the modal matrix $\mathbf{T}$ for the case of distinct eigenvalues.

Method 1   When there are distinct eigenvalues $\lambda_1, \lambda_2, \ldots, \lambda_n$ for the matrix $\mathbf{A}$ and it is in companion form, the Vandermonde matrix, which is easily obtained, is the modal matrix. The Vandermonde matrix is defined by

$$\mathbf{T} = \begin{bmatrix} 1 & 1 & \cdots & 1 \\ \lambda_1 & \lambda_2 & \cdots & \lambda_n \\ \lambda_1^2 & \lambda_2^2 & \cdots & \lambda_n^2 \\ \cdots\cdots\cdots\cdots\cdots\cdots\cdots \\ \lambda_1^{n-1} & \lambda_2^{n-1} & \cdots & \lambda_n^{n-1} \end{bmatrix} \tag{5-71}$$

EXAMPLE 1   Transform the state variables in the following equations in order to uncouple the states:

$$\dot{\mathbf{x}} = \begin{bmatrix} 0 & 1 & 0 \\ 0 & 0 & 1 \\ -24 & -26 & -9 \end{bmatrix} \mathbf{x} + \begin{bmatrix} 1 \\ 0 \\ 2 \end{bmatrix} u \tag{5-72}$$

$$y = \begin{bmatrix} 3 & 3 & 1 \end{bmatrix} \mathbf{x} \tag{5-73}$$

The characteristic equation $|\lambda\mathbf{I} - \mathbf{A}| = 0$ yields the roots $\lambda_1 = -2$, $\lambda_2 = -3$, and $\lambda_3 = -4$. Since these eigenvalues are distinct and the $\mathbf{A}$ matrix is in companion form, the Vandermonde matrix can be used as the modal matrix. Using Eq. (5-71) gives

$$\mathbf{T} = \begin{bmatrix} 1 & 1 & 1 \\ -2 & -3 & -4 \\ 4 & 9 & 16 \end{bmatrix}$$

The inverse is

$$\mathbf{T}^{-1} = \frac{1}{2}\begin{bmatrix} 12 & 7 & 1 \\ -16 & -12 & -2 \\ 6 & 5 & 1 \end{bmatrix}$$

In order to show that this does uncouple the states, the matrix $\mathbf{T}^{-1}\mathbf{AT}$ is evaluated and found to be

$$\mathbf{T}^{-1}\mathbf{AT} = \begin{bmatrix} -2 & 0 & 0 \\ 0 & -3 & 0 \\ 0 & 0 & -4 \end{bmatrix} = \Lambda$$

The matrix system equations, in terms of the new state vector $\mathbf{z}$, are obtained from Eqs. (5-64) and (5-65) as

$$\dot{\mathbf{z}} = \begin{bmatrix} -2 & 0 & 0 \\ 0 & -3 & 0 \\ 0 & 0 & -4 \end{bmatrix} \mathbf{z} + \begin{bmatrix} 7 \\ -10 \\ 4 \end{bmatrix} u \qquad (5\text{-}74)$$

$$y = \begin{bmatrix} 1 & 3 & 7 \end{bmatrix}\mathbf{z} \qquad (5\text{-}75)$$

Method 2 The second method does not require the $\mathbf{A}$ matrix to be in companion form. Premultiplying Eq. (5-68) by $\mathbf{T}$, where the elements of the modal matrix are defined by $\mathbf{T} = [v_{ij}]$, yields

$$\mathbf{AT} = \mathbf{T\Lambda} = \begin{bmatrix} \lambda_1 v_{11} & \lambda_2 v_{12} & \cdots & \lambda_n v_{1n} \\ \lambda_1 v_{21} & \lambda_2 v_{22} & \cdots & \lambda_n v_{2n} \\ \cdots\cdots\cdots\cdots\cdots\cdots\cdots\cdots \\ \lambda_1 v_{n1} & \lambda_2 v_{n2} & \cdots & \lambda_n v_{nn} \end{bmatrix} \qquad (5\text{-}76)$$

For convenience the columns of the modal matrix $\mathbf{T}$ are designated by the column vectors $\mathbf{v}_i$ and are called *eigenvectors*. The development in the early part of this section involves the use of $\mathbf{T}^{-1}$. A necessary condition, therefore, is that $\mathbf{T}$ not be singular. This is satisfied if $\mathbf{v}_i$ are linearly independent. Equating columns on the left and right sides of Eq. (5-76) yields

$$\mathbf{Av}_i = \lambda_i \mathbf{v}_i \qquad (5\text{-}77)$$

This equation can be put in the form

$$[\lambda_i \mathbf{I} - \mathbf{A}]\mathbf{v}_i = \mathbf{0} \qquad (5\text{-}78)$$

This is a homogeneous equation whose rank is $n - 1$. Therefore, there is one independent solution which is proportional to any nonzero column of adj $[\lambda_i \mathbf{I} - \mathbf{A}]$.†
The columns of adj $[\lambda_i \mathbf{I} - \mathbf{A}]$ are linearly related; thus each value of λ_i yields only one column $\mathbf{v}_i$ of the modal matrix.

EXAMPLE 2   Use adj $[\lambda_i \mathbf{I} - \mathbf{A}]$ to obtain the modal matrix for the equations

$$\dot{\mathbf{x}} = \begin{bmatrix} -9 & 1 & 0 \\ -26 & 0 & 1 \\ -24 & 0 & 0 \end{bmatrix} \mathbf{x} + \begin{bmatrix} 2 \\ 5 \\ 0 \end{bmatrix} u \qquad (5\text{-}79)$$

$$y = \begin{bmatrix} 1 & 2 & -1 \end{bmatrix}\mathbf{x} \qquad (5\text{-}80)$$

† This result is given here without proof. A derivation is contained in "State Variables for Engineers," P. M. DeRusso, R. J. Roy, and C. M. Close, sec. 4.7, Wiley, New York, 1965.

The characteristic equation $|\lambda\mathbf{I} - \mathbf{A}| = 0$ yields the roots $\lambda_1 = -2$, $\lambda_2 = -3$, and $\lambda_3 = -4$. These eigenvalues are distinct, but $\mathbf{A}$ is not in companion form. Therefore, the Vandermonde matrix is not the modal matrix. Using the adjoint method gives

$$\text{adj}\,[\lambda\mathbf{I} - \mathbf{A}] = \text{adj}\begin{bmatrix} \lambda + 9 & -1 & 0 \\ 26 & \lambda & -1 \\ 24 & 0 & \lambda \end{bmatrix}$$

$$= \begin{bmatrix} \lambda^2 & \lambda & 1 \\ -26\lambda - 24 & \lambda^2 + 9\lambda & \lambda + 9 \\ -24\lambda & -24 & \lambda^2 + 9\lambda + 26 \end{bmatrix}$$

For $\lambda_1 = -2$:

$$\text{adj}\,[-2\mathbf{I} - \mathbf{A}] = \begin{bmatrix} 4 & -2 & 1 \\ 28 & -14 & 7 \\ 48 & -24 & 12 \end{bmatrix} \qquad \mathbf{v}_1 = \begin{bmatrix} 1 \\ 7 \\ 12 \end{bmatrix}$$

For $\lambda_2 = -3$:

$$\text{adj}\,[-3\mathbf{I} - \mathbf{A}] = \begin{bmatrix} 9 & -3 & 1 \\ 54 & -18 & 6 \\ 72 & -24 & 8 \end{bmatrix} \qquad \mathbf{v}_2 = \begin{bmatrix} 1 \\ 6 \\ 8 \end{bmatrix}$$

For $\lambda_3 = -4$:

$$\text{adj}\,[-4\mathbf{I} - \mathbf{A}] = \begin{bmatrix} 16 & -4 & 1 \\ 80 & -20 & 5 \\ 96 & -24 & 6 \end{bmatrix} \qquad \mathbf{v}_3 = \begin{bmatrix} 1 \\ 5 \\ 6 \end{bmatrix}$$

In each case the columns of adj $|\lambda_i\mathbf{I} - \mathbf{A}|$ are linearly related; i.e., they are proportional. The $\mathbf{v}_i$ may be multiplied by a constant and are selected to contain the smallest integers; often the leading term is reduced to 1. In practice, it is necessary to calculate only one column of the adjoint matrix. The modal matrix is

$$\mathbf{T} = [\mathbf{v}_1 \quad \mathbf{v}_2 \quad \mathbf{v}_3] = \begin{bmatrix} 1 & 1 & 1 \\ 7 & 6 & 5 \\ 12 & 8 & 6 \end{bmatrix}$$

The inverse is

$$\mathbf{T}^{-1} = \frac{-1}{2}\begin{bmatrix} -4 & 2 & -1 \\ 18 & -6 & 2 \\ -16 & 4 & -1 \end{bmatrix}$$

The matrix equations, in terms of the new state vector $\mathbf{z}$, are

$$\dot{\mathbf{z}} = \begin{bmatrix} -2 & 0 & 0 \\ 0 & -3 & 0 \\ 0 & 0 & -4 \end{bmatrix}\mathbf{z} + \begin{bmatrix} -1 \\ -3 \\ 6 \end{bmatrix}u = \mathbf{\Lambda}\mathbf{z} + \mathbf{B}'u \qquad (5\text{-}81)$$

$$y = [3 \quad 5 \quad 5]\mathbf{z} = \mathbf{C}'\mathbf{z} \qquad (5\text{-}82)$$

Once each eigenvalue is assigned the designation λ_1, λ_2, and λ_3, they appear in this order along the diagonal of $\mathbf{\Lambda}$. Note that the systems of Examples 1 and 2 have the same eigenvalues, but different modal matrices are required to uncouple the states.

Method 3 An alternate method of evaluating the n elements of each $\mathbf{v}_i$ is to form a set of n equations from the matrix equation, Eq. (5-77). This is illustrated by the following example.

EXAMPLE 3 Rework Example 2 using Eq. (5-77). The matrices formed by using the **A** matrix of Eq. (5-79) are

$$\begin{bmatrix} -9 & 1 & 0 \\ -26 & 0 & 1 \\ -24 & 0 & 0 \end{bmatrix} \begin{bmatrix} v_{1i} \\ v_{2i} \\ v_{3i} \end{bmatrix} = \lambda_i \begin{bmatrix} v_{1i} \\ v_{2i} \\ v_{3i} \end{bmatrix}$$

Performing the multiplication yields

$$\begin{bmatrix} -9v_{1i} + v_{2i} \\ -26v_{1i} + v_{3i} \\ -24v_{1i} \end{bmatrix} = \begin{bmatrix} \lambda_i v_{1i} \\ \lambda_i v_{2i} \\ \lambda_i v_{3i} \end{bmatrix}$$

Each value of λ_i is inserted in this matrix, and the corresponding elements are equated to form three equations. For $\lambda_1 = -2$, the equations are

$$-9v_{11} + v_{21} = -2v_{11}$$
$$-26v_{11} + v_{31} = -2v_{21}$$
$$-24v_{11} = -2v_{31}$$

Only two of these equations are independent. This can be demonstrated by inserting v_{21} from the first equation and v_{31} from the third equation into the second equation. The result is the identity $v_{11} = v_{11}$. This merely confirms the fact that $[\lambda_i \mathbf{I} - \mathbf{A}]$ is of rank $n - 1 = 2$. Therefore the procedure is to arbitrarily select one of the elements equal to unity. After letting $v_{11} = 1$, these equations yield

$$\mathbf{v}_1 = \begin{bmatrix} v_{11} \\ v_{21} \\ v_{31} \end{bmatrix} = \begin{bmatrix} v_{11} \\ 7v_{11} \\ 12v_{11} \end{bmatrix} = \begin{bmatrix} 1 \\ 7 \\ 12 \end{bmatrix}$$

This is the same result as obtained in Example 2. The procedure is repeated to evaluate $\mathbf{v}_2$ and $\mathbf{v}_3$.

The need to select one of the elements in each column vector v_i arbitrarily can be eliminated by specifying the desired $\mathbf{B}_n$ matrix. In Sec. 5-8 the partial-fraction expansion of the transfer function led to a $\mathbf{b}_n$ matrix containing all 1's, that is,

$$\mathbf{b}_n = \begin{bmatrix} 1 & 1 & \cdots & 1 \end{bmatrix}^T$$

In order to achieve this form for $\mathbf{b}_n$ there is imposed the restriction that $\mathbf{b} = \mathbf{Tb}_n$. This produces n equations which are added to the n^2 equations formed by equating the elements of $\mathbf{AT} = \mathbf{T\Lambda}$. Only n^2 of these equations are independent, and they can be solved simultaneously for the n^2 elements of $\mathbf{T}$. The restriction that $\mathbf{b} = \mathbf{Tb}_n$ can be satisfied only if the system is controllable. The determination and definition of controllability are presented in Sec. 6-9.

The matrix $\mathbf{T}$ (not a modal matrix) can be used to transform a state equation from physical variables $\mathbf{x}_p$ to phase variables $\mathbf{x}$ by using $\mathbf{x}_p = \mathbf{Tx}$. Then the state equation $\dot{\mathbf{x}}_p = \mathbf{A}_p\mathbf{x}_p + \mathbf{b}_p u$ is transformed to $\dot{\mathbf{x}} = \mathbf{Ax} + \mathbf{b}_0 u$, where $\mathbf{A}$ is in companion form and $\mathbf{b}_0{}^T = \begin{bmatrix} 0 & \cdots & 0 & b_n \end{bmatrix}$. The matrix $\mathbf{T}$ is evaluated by equating the elements of $\mathbf{A}_p\mathbf{T} = \mathbf{TA}$ and $\mathbf{b}_p = \mathbf{Tb}_0$. Examples are contained in Secs. 14-9 and 14-12.

5-10 TRANSFORMING THE A MATRIX WITH COMPLEX EIGENVALUES

The transformation of variables described in Sec. 5-9 simplifies the system representation mathematically by removing any coupling between the system modes. This diagonalizing of the **A** matrix produces the type of system representation shown in Fig. 5-27. Each state equation corresponding to the normal form of the state equation, Eq. (5-69), has the form

$$\dot{z}_i(t) = \lambda_i z_i(t) + b_i u \qquad (5\text{-}83)$$

Since only one state variable z_i appears in this equation, it can be solved independently of all the other states.

When the eigenvalues are complex, the matrix **T** which produces the diagonal matrix **Λ** contains complex numbers. This is illustrated for the differential equation

$$\ddot{x} - 2\sigma\dot{x} + (\sigma^2 + \omega_d^2)x = u(t) \qquad (5\text{-}84)$$

where $\omega_n^2 = \sigma^2 + \omega_d^2$ and the eigenvalues are $\lambda_{1,2} = \sigma \pm j\omega_d$. Using the phase variables $x_1 = x$ and $x_2 = \dot{x}_1$, the state and output equations are

$$\dot{\mathbf{x}} = \begin{bmatrix} 0 & 1 \\ -\omega_n^2 & 2\sigma \end{bmatrix} \mathbf{x} + \begin{bmatrix} 0 \\ 1 \end{bmatrix} u \qquad (5\text{-}85)$$

$$y = \begin{bmatrix} 1 & 0 \end{bmatrix} \mathbf{x} \qquad (5\text{-}86)$$

The modes can be uncoupled, but this is undesirable, as shown below. Since the eigenvalues are distinct and **A** is a companion matrix, the modal matrix **T** is the Vandermonde matrix

$$\mathbf{T} = \begin{bmatrix} 1 & 1 \\ \sigma + j\omega_d & \sigma - j\omega_d \end{bmatrix} = \begin{bmatrix} \mathbf{v}_1 & \mathbf{v}_2 \end{bmatrix} \qquad (5\text{-}87)$$

When the transformation $\mathbf{x} = \mathbf{T}\mathbf{z}$ is used,[8] the normal form of the state and output equations obtained from Eqs. (5-69) and (5-70) are

$$\dot{\mathbf{z}} = \begin{bmatrix} \sigma + j\omega_d & 0 \\ 0 & \sigma - j\omega_d \end{bmatrix} \mathbf{z} + \begin{bmatrix} \dfrac{1}{j2\omega_d} \\ -\dfrac{1}{j2\omega_d} \end{bmatrix} u \qquad (5\text{-}88)$$

$$y = \begin{bmatrix} 1 & 1 \end{bmatrix} \mathbf{z} \qquad (5\text{-}89)$$

The simulation diagram based on these equations is shown in Fig. 5-29. The parameters in this diagram are complex quantities, which increases the difficulty when obtaining the mathematical solution. Thus, it may be desirable to perform another transformation in order to obtain a simulation diagram which contains only real quantities. The pair of complex conjugate eigenvalues λ_1 and λ_2 jointly contribute to one transient mode which has the form of a damped sinusoid. The additional transformation is $\mathbf{z} = \mathbf{Q}\mathbf{w}$, where

$$\mathbf{Q} = \begin{bmatrix} \dfrac{1}{2} & -\dfrac{j}{2} \\ \dfrac{1}{2} & \dfrac{j}{2} \end{bmatrix} \qquad (5\text{-}90)$$

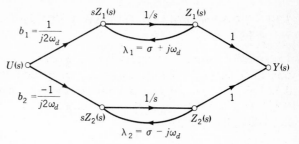

FIGURE 5-29
Simulation diagram for Eqs. (5-88) and (5-89).

The new state and output equations are

$$\dot{\mathbf{w}} = \mathbf{Q}^{-1}\mathbf{\Lambda}\mathbf{Q}\mathbf{w} + \mathbf{Q}^{-1}\mathbf{T}^{-1}\mathbf{B}u = \mathbf{\Lambda}_m\mathbf{w} + \mathbf{B}_m u \qquad (5\text{-}91)$$

$$y = \mathbf{C}\mathbf{T}\mathbf{Q}\mathbf{w} + \mathbf{D}u = \mathbf{C}_m\mathbf{w} + \mathbf{D}_m u \qquad (5\text{-}92)$$

For this example these equations† are

$$\dot{\mathbf{w}} = \begin{bmatrix} \sigma & \omega_d \\ -\omega_d & \sigma \end{bmatrix} \mathbf{w} + \begin{bmatrix} 0 \\ \dfrac{1}{\omega_d} \end{bmatrix} u \qquad (5\text{-}93)$$

$$y = \begin{bmatrix} 1 & 0 \end{bmatrix}\mathbf{w} \qquad (5\text{-}94)$$

The simulation diagram for these equations is shown in Fig. 5-30. Note that only real quantities appear. Also, the two modes have not been isolated, but this is an advantage for complex-conjugate roots.

The effect of the two transformations $\mathbf{x} = \mathbf{T}\mathbf{z}$ and $\mathbf{z} = \mathbf{Q}\mathbf{w}$ can be considered as one transformation given by

$$\mathbf{x} = \mathbf{T}\mathbf{Q}\mathbf{w} = \begin{bmatrix} 1 & 0 \\ \sigma & \omega_d \end{bmatrix} \mathbf{w} = \mathbf{T}_m\mathbf{w} \qquad (5\text{-}95)$$

Comparison of the modified modal matrix $\mathbf{T}_m$ in Eq. (5-95) with the original matrix in Eq. (5-87) shows that

$$\mathbf{T}_m = [\text{Re } (\mathbf{v}_1) \quad \text{Im } (\mathbf{v}_1)] \qquad (5\text{-}96)$$

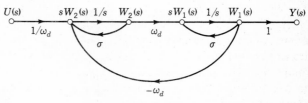

FIGURE 5-30
Simulation for Eqs. (5-93) and (5-94).

† Although the trajectories in the state space are not drawn here, the result of this transformation produces trajectories which are symmetrical with respect to the $\mathbf{w}$ axes; see Sec. 13-2.

Therefore $\mathbf{T}_m$ can be obtained directly from $\mathbf{T}$ without using the $\mathbf{Q}$ transformation.

In the general case containing both complex and real eigenvalues the modified matrix $\mathbf{\Lambda}_m$ and the resulting matrices $\mathbf{B}_m$ and $\mathbf{C}_m$ have the form

$$\mathbf{\Lambda}_m = \begin{bmatrix} \sigma_1 & \omega_1 & & & & & \\ -\omega_1 & \sigma_1 & & & \mathbf{0} & & \\ & & \sigma_2 & \omega_2 & & & \\ & & -\omega_2 & \sigma_2 & & & \\ & & & & \lambda_5 & & \\ & \mathbf{0} & & & & \ddots & \\ & & & & & & \lambda_n \end{bmatrix} \qquad \mathbf{B}_m = \mathbf{T}_m^{-1}\mathbf{B} \qquad (5\text{-}97)$$

$$\mathbf{C}_m = \mathbf{C}\mathbf{T}_m \qquad (5\text{-}98)$$

With this modified matrix $\mathbf{\Lambda}_m$ the two oscillating modes produced by the two sets of complex eigenvalues $\sigma_1 \pm j\omega_1$ and $\sigma_2 \pm j\omega_2$ are uncoupled. The modified matrix $\mathbf{\Lambda}_m$ can be achieved by using the modified modal matrix $\mathbf{T}_m = \mathbf{T}\mathbf{Q}$, where

$$\mathbf{Q} = \begin{bmatrix} \dfrac{1}{2} & \dfrac{-j}{2} & & & & & \\ \dfrac{1}{2} & \dfrac{j}{2} & & & \mathbf{0} & & \\ & & \dfrac{1}{2} & \dfrac{-j}{2} & & & \\ & & \dfrac{1}{2} & \dfrac{j}{2} & & & \\ & & & & 1 & & \mathbf{0} \\ & & & & & 1 & \\ & \mathbf{0} & & & & & \ddots \\ & & & & \mathbf{0} & & 1 \end{bmatrix} \qquad (5\text{-}99)$$

The modal matrix $\mathbf{T}$ is evaluated by any of the methods described in Sec. 5-9. The column vectors contained in $\mathbf{T}$ for the conjugate eigenvalues are also conjugates. Thus

$$\mathbf{T} = \begin{bmatrix} \mathbf{v}_1 & \mathbf{v}_1^* & \mathbf{v}_3 & \mathbf{v}_3^* & \mathbf{v}_5 & \cdots & \mathbf{v}_n \end{bmatrix} \qquad (5\text{-}100)$$

The modified matrix $\mathbf{T}_m$ can be obtained directly by using

$$\mathbf{T}_m = \begin{bmatrix} \text{Re}\,(\mathbf{v}_1) & \text{Im}\,(\mathbf{v}_1) & \text{Re}\,(\mathbf{v}_3) & \text{Im}\,(\mathbf{v}_3) & \mathbf{v}_5 & \cdots & \mathbf{v}_n \end{bmatrix} \qquad (5\text{-}101)$$

5-11 SUMMARY

The foundation has been completed for analyzing the performance of feedback control systems. The analysis of such systems and the adjustments of parameters to achieve the desired performance is covered in the following chapters. In order to provide clarity in understanding the principles and to avoid ambiguity, the systems considered are restricted to those which have a single input and a single output. The procedures can be extended to multiple-input multiple-output systems. The block-diagram representation containing a single feedback loop is the basic unit used to develop the various techniques of analysis. This point in preparation for overall system analysis has been reached by writing the differential equation in Chap. 2, obtaining the solution of the differential equations in Chap. 3, introducing the Laplace transform in Chap. 4, and incorporating this into the representation of a complete system in Chap. 5.

REFERENCES

1 Etkin, B.: "Dynamics of Atmospheric Flight," Wiley, New York, 1972.
2 "IEEE Standard Dictionary of Electrical and Electronics Terms," Wiley-Interscience, New York, 1972.
3 Mason, S. J.: Feedback Theory: Further Properties of Signal Flow Graphs, *Proc. IRE*, vol. 44, no. 7, pp. 920–926, July 1956.
4 Mason, S. J., and H. J. Zimmerman: "Electronic Circuits, Signals, and Systems," M.I.T. Press, Cambridge, Mass., 1960.
5 Kuo, B. C.: "Linear Networks and Systems," McGraw-Hill, New York, 1967.
6 Trimmer, J. D.: "Response of Physical Systems," p. 17, Wiley, New York, 1950.
7 Langholz, G., and S. Frankenthal: Reduction to Normal Form of a State Equation in the Presence of Input Derivatives, *Intern. J. Systems Science*, vol. 5, no .7, July 1974.
8 Crossley, T. R., and B. Porter: Inversion of Complex Matrices, *Electronics Letters*, vol. 6, no. 4, Feb. 1970.

6

CONTROL-SYSTEM CHARACTERISTICS

6-1 INTRODUCTION

In Chaps. 2 and 4 the transfer functions of both open- and closed-loop systems were developed for various types of controlled quantities. These transfer functions have certain basic characteristics that permit transient and steady-state analyses of the feedback-controlled system. Five factors of prime importance in feedback-controlled systems are *stability, the existence and magnitude of the steady-state error, controllability, observability, and parameter sensitivity*. The stability characteristic of a linear time-invariant system is determined from the system's characteristic equation. Routh's stability criterion provides a means for determining stability without evaluating the roots of this equation. For unity-feedback systems the steady-state characteristics are obtainable from the forward transfer function, yielding figures of merit and a ready means for classifying systems. These properties are developed in this chapter. The concepts of controllability and observability are essential considerations in every practical result of modern control theory. These concepts are defined in this chapter. The sensitivity function is an important system figure of merit. It is descriptive of the variation of the system output due to the variation of a system parameter.

6-2 ROUTH'S STABILITY CRITERION[1]

The *response transform* $X_2(s)$ has the general form given by Eq. (4-32), which is repeated here in slightly modified form. $X_1(s)$ is the driving transform

$$X_2(s) = \frac{P(s)}{Q(s)} X_1(s) = \frac{P(s)X_1(s)}{b_n s^n + b_{n-1} s^{n-1} + b_{n-2} s^{n-2} + \cdots + b_1 s + b_0} \qquad (6\text{-}1)$$

Sections 4-6 to 4-10 describe the methods used to evaluate the inverse transform $\mathscr{L}^{-1}[F(s)] = f(t)$. However, before the inverse transformation can be performed, the polynomial $Q(s)$ must be factored. Computer programs are readily available for obtaining the roots of a polynomial. Section 4-12 shows that stability of the response $x_2(t)$ requires that all zeros of $Q(s)$ have negative real parts. Since it is usually not necessary to find the exact solution when the response is unstable, a simple procedure to determine the existence of zeros with positive real parts is needed. If such zeros of $Q(s)$ with positive real parts are found, the system must be modified. Routh's criterion is a simple method of determining the number of zeros with positive real parts without actually solving for the zeros of $Q(s)$. It should be noted that zeros of $Q(s)$ are poles of $X_2(s)$.

The characteristic equation is

$$Q(s) = b_n s^n + b_{n-1} s^{n-1} + b_{n-2} s^{n-2} + \cdots + b_1 s + b_0 = 0 \qquad (6\text{-}2)$$

If the b_0 term is zero, divide by s to obtain the equation in the form of Eq. (6-2). The b's are real coefficients, and all powers of s from s^n to s^0 must be present in the characteristic equation. If any coefficients other than b_0 are zero, or if the coefficients do not all have the same sign, then there are pure imaginary roots or roots with positive real parts and the system is unstable. It is therefore unnecessary to continue if only stability or instability is to be determined. In special situations it may be necessary to determine the actual number of roots in the right-half s plane. For these situations the procedure described in this section can be used.

The coefficients of the characteristic equation are arranged in the pattern shown in the following Routhian array. These coefficients are then used to evaluate the rest of the constants to complete the array.

$$
\begin{array}{c|ccccc}
s^n & b_n & b_{n-2} & b_{n-4} & b_{n-6} & \cdots \\
s^{n-1} & b_{n-1} & b_{n-3} & b_{n-5} & b_{n-7} & \cdots \\
s^{n-2} & c_1 & c_2 & c_3 & \cdots \\
s^{n-3} & d_1 & d_2 & \cdots \\
\cdots & \cdots \cdots \cdots \cdots \cdots \\
s^1 & j_1 \\
s^0 & k_1
\end{array}
$$

The constants c_1, c_2, c_3, etc., in the third row are evaluated as follows:

$$c_1 = \frac{b_{n-1}b_{n-2} - b_n b_{n-3}}{b_{n-1}} \qquad (6\text{-}3)$$

$$c_2 = \frac{b_{n-1}b_{n-4} - b_n b_{n-5}}{b_{n-1}} \qquad (6\text{-}4)$$

$$c_3 = \frac{b_{n-1}b_{n-6} - b_n b_{n-7}}{b_{n-1}} \qquad (6\text{-}5)$$

This pattern is continued until the rest of the c's are all equal to zero. Then the d row is formed by using the s^{n-1} and s^{n-2} row. The constants are

$$d_1 = \frac{c_1 b_{n-3} - b_{n-1} c_2}{c_1} \qquad (6\text{-}6)$$

$$d_2 = \frac{c_1 b_{n-5} - b_{n-1} c_3}{c_1} \qquad (6\text{-}7)$$

$$d_3 = \frac{c_1 b_{n-7} - b_{n-1} c_4}{c_1} \qquad (6\text{-}8)$$

This is continued until no more d terms are present. The rest of the rows are formed in this way down to the s^0 row. The complete array is triangular, ending with the s^0 row. Notice that s^1 and s^0 rows contain only one term each. Once the array has been found, *Routh's criterion states that the number of roots of the characteristic equation with positive real parts is equal to the number of changes of sign of the coefficients in the first column.* Therefore the system is stable if all terms in the first column have the same sign.

The following example illustrates this criterion:

$$Q(s) = s^5 + s^4 + 10s^3 + 72s^2 + 152s + 240 \qquad (6\text{-}9)$$

The Routhian array is formed by using the procedure described above:

s^5	1	10	152
s^4	1	72	240
s^3	-62	-88	
s^2	70.6	240	
s^1	122.6		
s^0	240		

In the first column there are two changes of sign, from 1 to -62 and from -62 to 70.6; therefore $Q(s)$ has two roots in the right-half s plane. Note that this criterion gives the number of roots with positive real parts but does not tell the values of the roots. If Eq. (6-9) is factored, the roots are $s_1 = -3$, $s_{2,3} = -1 \pm j\sqrt{3}$, and $s_{4,5} = +2 \pm j4$. This confirms that there are two roots with positive real parts. The Routh criterion does not distinguish between real and complex roots.

Theorem 1   Division of a row   The coefficients of any row may be multiplied or divided by a positive number without changing the signs of the first column. The labor of evaluating the coefficients in Routh's array can be reduced by multiplying or dividing any row by a constant. This may result, for example, in reducing the size of the coefficients and therefore simplifying the evaluation of the remaining coefficients. The following example illustrates this theorem:

$$Q(s) = s^6 + 3s^5 + 2s^4 + 9s^3 + 5s^2 + 12s + 20 \qquad (6\text{-}10)$$

The Routhian array is

$$
\begin{array}{c|cccc}
s^6 & 1 & 2 & 5 & 20 \\
s^5 & \cancel{3} & \cancel{9} & \cancel{12} & \\
    & 1 & 3 & 4 & \text{(after dividing by 3)} \\
s^4 & -1 & 1 & 20 & \\
s^3 & \cancel{4} & \cancel{24} & & \\
    & 1 & 6 & & \text{(after dividing by 4)} \\
s^2 & 7 & 20 & & \\
s^1 & 22 & & & \text{(after multiplying by 7)} \\
s^0 & 20 & & &
\end{array}
$$

Notice that the size of the numbers has been reduced by dividing the s^5 row by 3 and the s^3 row by 4. The result is unchanged; i.e., there are two changes of signs in the first column and therefore there are two roots with positive real parts.

Theorem 2   A zero coefficient in the first column When the first term in a row is zero but not all the other terms are zero, the following methods can be used:

 1 Substitute in the original equation $s = 1/x$; then solve for the roots of x with positive real parts. The number of roots x with positive real parts will be the same as the number of s roots with positive real parts.
 2 Multiply the original polynomial by the factor $(s + 1)$ which introduces an additional negative root. Then form the Routhian array for the new polynomial.

Both these methods are illustrated in the following example:

$$Q(s) = s^4 + s^3 + 2s^2 + 2s + 5 \qquad (6\text{-}11)$$

The Routhian array is:

$$
\begin{array}{c|ccc}
s^4 & 1 & 2 & 5 \\
s^3 & 1 & 2 & \\
s^2 & 0 & 5 &
\end{array}
$$

The zero in the first column prevents completion of the array. The following methods overcome this problem.

Method 1   Letting $s = 1/x$ and rearranging the polynomial gives

$$Q(x) = 5x^4 + 2x^3 + 2x^2 + x + 1 \qquad (6\text{-}12)$$

The new Routhian array is

$$
\begin{array}{c|ccc}
x^4 & 5 & 2 & 1 \\
x^3 & 2 & 1 & \\
x^2 & -1 & 2 & \\
x^1 & 5 & & \\
x^0 & 2 & &
\end{array}
$$

There are two changes of sign; therefore there are two roots of x in the right-half s plane. The number of roots s with positive real parts is also two. This method does not work when the coefficients of $Q(s)$ and of $Q(x)$ are identical.

Method 2

$$Q_1(s) = Q(s)(s + 1) = s^5 + 2s^4 + 3s^3 + 4s^2 + 7s + 5$$

s^5	1	3	7
s^4	2	4	5
s^3	2	9	
s^2	-10	10	
s^1	9		
s^0	10		

The same result is obtained by both methods. There are two changes of sign in the first column, so there are two zeros of $Q(s)$ with positive real parts. An additional method is described in Ref. 9.

Theorem 3   A zero row   When all the coefficients of one row are zero, the procedure is as follows:

1 The auxiliary equation can be formed from the preceding row, as shown below.
2 The Routhian array can be completed by replacing the all-zero row with the coefficients obtained by differentiating the auxiliary equation.
3 The roots of the auxiliary equation are also roots of the original equation. These roots occur in pairs and are the negatives of each other. Therefore these roots may be imaginary (complex conjugates) or real (one positive and one negative), may lie in quadruplets (two pairs of complex-conjugate roots), etc.

Consider the system which has the characteristic equation

$$Q(s) = s^4 + 2s^3 + 11s^2 + 18s + 18 = 0 \qquad (6\text{-}13)$$

The Routhian array is

s^4	1	11	18
s^3	$\not2$	$\not1\not8$	
	1	9	(after dividing by 2)
s^2	$\not2$	$\not1\not8$	
	1	9	(after dividing by 2)
s^1	0		

The presence of a zero row indicates that there are roots that are the negatives of each other. The next step is to form the auxiliary equation from the preceding row, which is the s^2 row. The highest power of s is s^2, and only even powers of s appear. Therefore the auxiliary equation is

$$s^2 + 9 = 0 \qquad (6\text{-}14)$$

The roots of this equation are

$$s = \pm j3$$

These are also roots of the original equation. The presence of imaginary roots indicates that the output includes a sinusoidally oscillating component.

To complete the Routhian array, the auxiliary equation is differentiated and is

$$2s + 0 = 0 \qquad (6\text{-}15)$$

The coefficients of this equation are inserted in the s^1 row, and the array is then completed:

$$
\begin{array}{c|c}
s^1 & 2 \\
s^0 & 9
\end{array}
$$

Since there are no changes of sign in the first column, there are no roots with positive real parts.

In feedback systems, covered in detail in the following chapters, the ratio of the output to the input does not have an explicitly factored denominator. An example of such a function is

$$\frac{X_2(s)}{X_1(s)} = \frac{P(s)}{Q(s)} = \frac{K(s + 2)}{s(s + 5)(s^2 + 2s + 5) + K(s + 2)} \qquad (6\text{-}16)$$

The value K is an adjustable parameter in the system and may be positive or negative. The value of K determines the location of the poles and therefore the stability of the system. It is important to know the range of values of K for which the system is stable. This information must be obtained from the characteristic equation, which is

$$Q(s) = s^4 + 7s^3 + 15s^2 + (25 + K)s + 2K = 0 \qquad (6\text{-}17)$$

The coefficients must all be positive in order for the zeros of $Q(s)$ to lie in the left half of the s plane, but this is not a sufficient condition for stability. The Routhian array permits evaluation of precise boundaries for K:

$$
\begin{array}{c|ccc}
s^4 & 1 & 15 & 2K \\
s^3 & 7 & 25 + K & \\
s^2 & 80 - K & 14K & \\
s^1 & \dfrac{(80 - K)(25 + K) - 98K}{80 - K} & & \\
s^0 & 14K & &
\end{array}
$$

The term $80 - K$ from the s^2 row imposes the restriction $K < 80$, and the s^0 row requires $K > 0$. The numerator of the first term in the s^1 row is equal to $-K^2 - 43K + 2{,}000$. By use of the quadratic formula the zeros of this function are $K = 28.1$ and $K = -71.1$. The combined restrictions on K for stability of the system are therefore $0 < K < 28.1$. For the value $K = 28.1$ the characteristic equation has imaginary roots which can be evaluated by applying Theorem 3. Also, for $K = 0$, it can be seen from Eq. (6-16) that there is no output. The methods for selecting the "best" value of K in the range between 0 and 28.1 are contained in later chapters. It is important to note that the Routh criterion has provided useful but restricted information.

Another method of determining stability is Hurwitz's criterion,[1,8] which establishes the necessary conditions in terms of the system determinants. This method is more amenable to digital-computer system-stability evaluation.

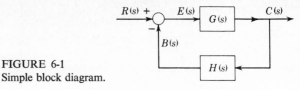

FIGURE 6-1
Simple block diagram.

6-3 MATHEMATICAL AND PHYSICAL FORMS

In various systems the controlled variable C shown in Fig. 6-1 may have the physical form of position or speed, temperature, rate of change of temperature, voltage, rate of flow, pressure, etc. To generalize the study of feedback systems the controlled variable has been labeled C. Once the blocks in the diagram are related to transfer functions, it is immaterial to the analysis of the system what the physical form of the controlled variable may be.

Generally, the important quantities are the controlled quantity c, its rate of change Dc, and its second derivative D^2c, that is, the first several derivatives of c, including the zeroth derivative. For any specific control system each of these "mathematical" functions has a definite "physical" meaning. For example, if the controlled variable c is position, Dc is velocity and D^2c is acceleration. As a second example, if the controlled variable c is velocity, then Dc is acceleration and D^2c is the rate of change of acceleration.

In the analysis of a given system the mathematical response is important. For example, the input signal to a system may have the irregular form shown in Fig. 6-2, which cannot be expressed by any simple equation. This prevents a straightforward analysis of system response. One notes, though, that the signal form shown in Fig. 6-2 may be considered to be composed of three basic forms of known types of input signals, i.e., a step in the region cde, a ramp in the region $0b$, and a parabola in the region ef. This leads to the conclusion that if the given linear system is analyzed separately for each of these types of input signals, there is then established a fair measure of performance with the irregular input. Also, use of these standard inputs provides a means for comparing the performance of different systems.

Consider that the system shown in Fig. 6-1 is a position-control system and that the input position signal is of the form shown in Fig. 6-2. In general, feedback control systems are analyzed on the basis of a unit step input signal (see Fig. 6-3). This system can be analyzed on the basis that the unit step input signal $r(t)$ represents position. Since this gives only a limited idea of how the system responds to the actual input signal, one can then analyze the system on the basis that the unit step signal represents a constant velocity $Dr(t) = u_{-1}(t)$. This in reality gives an input position signal of the

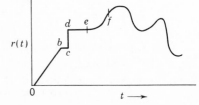

FIGURE 6-2
Input signal to a system.

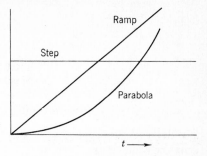

FIGURE 6-3
Graphical forms of step, ramp, and parabolic input functions.

form of a ramp (Fig. 6-3) and thus a closer idea of how the system responds to the actual input signal.

In the same manner one can consider that the unit step input signal represents a constant acceleration, $D^2 r(t) = u_{-1}(t)$, to obtain the system's performance to a parabolic position input signal. The curves shown in Fig. 6-3 then represent acceleration, velocity, and position.

6-4 TYPES OF FEEDBACK SYSTEMS

A simple closed-loop feedback system with unity feedback is shown in Fig. 6-4. The open-loop transfer function for this system is $G(s) = C(s)/E(s)$, which is determined by the components of the actual control system. Several examples are derived in the preceding chapter. Generally the transfer function has one of the following mathematical forms:

$$G(s) = \frac{K_0(1 + T_1 s)(1 + T_2 s)\cdots}{(1 + T_a s)(1 + T_b s)\cdots} \qquad (6\text{-}18)$$

$$G(s) = \frac{K_1(1 + T_1 s)(1 + T_2 s)\cdots}{s(1 + T_a s)(1 + T_b s)\cdots} \qquad (6\text{-}19)$$

$$G(s) = \frac{K_2(1 + T_1 s)(1 + T_2 s)\cdots}{s^2(1 + T_a s)(1 + T_b s)\cdots} \qquad (6\text{-}20)$$

The above equations are expressed in a more generalized manner by defining the standard form of the transfer function as

$$G(s) = \frac{K_m(1 + b_1 s + b_2 s^2 + \cdots + b_w s^w)}{s^m(1 + a_1 s + a_2 s^2 + \cdots + a_u s^u)} = K_m G'(s) \qquad (6\text{-}21)$$

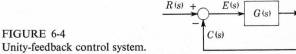

FIGURE 6-4
Unity-feedback control system.

where

$a_1, a_2, \ldots$ = constant coefficients

$b_1, b_2, \ldots$ = constant coefficients

K_m = gain constant of the transfer function $G(s)$

$m = 0, 1, 2, \ldots$ denotes *type* of transfer function

$G'(s)$ = forward transfer function with unity gain

The degree of the denominator is $n = m + u$. For a unity-feedback system, E and C have the same units. Therefore, K_0 is nondimensional, K_1 has the units of seconds^{-1}, and K_2 has the units of seconds^{-2}.

Once a physical system has been expressed mathematically, the analysis that must follow is independent of the nature of the physical system. It is immaterial whether the system is electrical, mechanical, hydraulic, thermal, or a combination of these. The most common types of feedback control systems fall into the three categories expressed by Eqs. (6-18) to (6-20). It is important to analyze each category thoroughly and to relate it as closely as possible to its transient and steady-state solution. In order to analyze best each control system, a "type" designation is introduced. The designation is based upon the order of the exponent m of s in Eq. (6-21). Thus, when $m = 0$, the system represented by this equation is called a Type 0 system; when $m = 1$ it is called a Type 1 system; when $m = 2$ it is called a Type 2 system; etc.

The various types exhibit the following properties:

Type 0   A constant actuating signal results in a constant value for the controlled variable.

Type 1   A constant actuating signal results in a constant rate of change (constant velocity) of the controlled variable.

Type 2   A constant actuating signal results in a constant second derivative (constant acceleration) of the controlled variable.

Type 3   A constant actuating signal results in a constant rate of change of acceleration of the controlled variable.

These classifications lend themselves to definition in terms of the differential equations of the system and to identification in terms of the forward transfer function. For all classifications the degree of the denominator of the $G(s)H(s)$ function usually is equal to or greater than the degree of the numerator because of the physical nature of feedback control systems. That is, in every physical system there are energy-storage and dissipative elements such that there can be no instantaneous transfer of energy from the input to the output. However, exceptions do occur.

6-5   ANALYSIS OF SYSTEM TYPES

The properties presented in the preceding section are now examined in detail for each type of $G(s)$ appearing in stable systems. First, remember that

Final-value theorem:

$$\lim_{t \to \infty} f(t) = \lim_{s \to 0} sF(s) \qquad (6\text{-}22)$$

Differentiation theorem:

$$\mathcal{L}[D^m c(t)] = s^m C(s) \qquad \text{when all initial conditions are zero} \qquad (6\text{-}23)$$

It should also be remembered that the steady-state output of a stable closed-loop system that has *unity feedback* has the same form as the input when the input is a polynomial. Therefore, if the input is a ramp function, the steady-state output must also include a ramp function plus a constant, and so forth.

From the preceding section it is seen that the forward transfer function defines the system type. In deriving the transfer function it is generally in the factored form

$$G(s) = \frac{C(s)}{E(s)} = \frac{K_m(1 + T_1 s)(1 + T_2 s) \cdots}{s^m (1 + T_a s)(1 + T_b s)(1 + T_c s) \cdots} \qquad (6\text{-}24)$$

Solving this equation for $E(s)$ yields

$$E(s) = \frac{(1 + T_a s)(1 + T_b s)(1 + T_c s) \cdots}{K_m(1 + T_1 s)(1 + T_2 s) \cdots} s^m C(s) \qquad (6\text{-}25)$$

Thus, applying the final-value theorem gives

$$e(t)_{ss} = \lim_{s \to 0} [sE(s)] = \lim_{s \to 0} \left[\frac{s(1 + T_a s)(1 + T_b s)(1 + T_c s) \cdots}{K_m(1 + T_1 s)(1 + T_2 s) \cdots} s^m C(s) \right]$$

$$= \lim_{s \to 0} \frac{s[s^m C(s)]}{K_m} \qquad (6\text{-}26)$$

But applying the final-value theorem to Eq. (6-23) gives

$$\lim_{s \to 0} s[s^m C(s)] = D^m c(t)_{ss} \qquad (6\text{-}27)$$

Therefore, Eq. (6-26) can be written as

$$e(t)_{ss} = \frac{D^m c(t)_{ss}}{K_m} \qquad (6\text{-}28)$$

or

$$K_m e(t)_{ss} = D^m c(t)_{ss} \qquad (6\text{-}29)$$

This equation relates a derivative of the output to the error; it is most useful for the case where $D^m c(t)_{ss}$ = constant. Then $e(t)_{ss}$ must also equal a constant, that is, $e(t)_{ss} = E_0$, and Eq. (6-29) may be expressed as

$$K_m E_0 = D^m c(t)_{ss} = \text{constant} = C_m \qquad (6\text{-}30)$$

Note that $C(s)$ has the form

$$C(s) = \frac{G(s)}{1 + G(s)} R(s)$$

$$= \frac{K_m[(1 + T_1 s)(1 + T_2 s) \cdots]}{s^m (1 + T_a s)(1 + T_b s) \cdots + K_m(1 + T_1 s)(1 + T_2 s) \cdots} R(s) \qquad (6\text{-}31)$$

The expression for $E(s)$ in terms of the input $R(s)$ is obtained as follows:

$$E(s) = \frac{C(s)}{G(s)} = \frac{1}{G(s)} \frac{G(s)R(s)}{1 + G(s)H(s)} = \frac{R(s)}{1 + G(s)H(s)} \qquad (6\text{-}32)$$

For unity feedback $[H(s) = 1]$, and with $G(s)$ given by Eq. (6-24) the expression for $E(s)$ is

$$E(s) = \frac{s^m(1 + T_a s)(1 + T_b s) \cdots R(s)}{s^m(1 + T_a s)(1 + T_b s) \cdots + K_m(1 + T_1 s)(1 + T_2 s) \cdots} \qquad (6\text{-}33)$$

Applying the final-value theorem to Eq. (6-33) yields

$$e(t)_{ss} = \lim_{s \to 0} s \left[\frac{s^m(1 + T_a s)(1 + T_b s) \cdots R(s)}{s^m(1 + T_a s)(1 + T_b s) \cdots + K_m(1 + T_1 s)(1 + T_2 s) \cdots} \right] \qquad (6\text{-}34)$$

Equation (6-34) relates the steady-state error to the input; it is now analyzed for various system types and for step, ramp, and parabolic inputs.

Case 1 $m = 0$ (Type 0 System)

For a step input $r(t) = R_0 u_{-1}(t)$, $R(s) = R_0/s$. From Eq. (6-34)

$$e(t)_{ss} = \frac{R_0}{1 + K_0} = \text{constant} = E_0 \neq 0 \qquad (6\text{-}35)$$

From Eq. (6-35) it is seen that a Type 0 system with a constant input produces a constant value of the output and a constant actuating signal. The same results can be obtained by applying Eq. (6-30). This means that in a Type 0 system a fixed error E_0 is required to produce a desired constant output C_0; that is, $K_0 E_0 = c(t)_{ss} = \text{constant} = C_0$. For steady-state conditions

$$e(t)_{ss} = r(t)_{ss} - c(t)_{ss} = R_0 - C_0 = E_0 \qquad (6\text{-}36)$$

Differentiating the above equation yields $Dr(t)_{ss} = Dc(t)_{ss} = 0$. Figure 6-5 illustrates the results obtained above.

For a ramp input $r(t) = R_1 t u_{-1}(t)$, $R(s) = R_1/s^2$. From Eq. (6-34)

$$e(t)_{ss} = \infty$$

Also, by the use of the Heaviside partial-fraction expansion, the particular solution of $e(t)$ can be shown, by use of Eq. (6-33), to contain the term $[R_1/(1 + K_0)]t$. Therefore the conclusion is that a Type 0 system with a ramp-function input produces a ramp output with a smaller slope; thus there is an error which increases with time and approaches a value of infinity. This means that a Type 0 system cannot follow a ramp input.

In the same manner, it can be shown that a Type 0 system cannot follow a parabolic input

$$r(t) = \frac{R_2 t^2}{2} u_{-1}(t) \qquad (6\text{-}37)$$

that is, $e(t)_{ss} = r(t)_{ss} - c(t)_{ss}$ approaches a value of infinity.

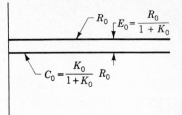

FIGURE 6-5
Steady-state response of a Type 0 system with a step input.

Case 2 $m = 1$ (Type 1 System)

Step input $R(s) = R_0/s$ From Eq. (6-34)

$$e(t)_{ss} = 0 \qquad (6\text{-}38)$$

Therefore it can be stated that a Type 1 system with a constant input produces a steady-state constant output of value identical with the input. This means that for a Type 1 system there is zero steady-state error between the output and input for a step input, i.e.,

$$e(t)_{ss} = r(t)_{ss} - c(t)_{ss} = 0 \qquad (6\text{-}39)$$

The above analysis is in agreement with Eq. (6-30). That is, for a constant input $r(t) = R_0 u_{-1}(t)$, the steady-state output must be a constant $c(t) = C_0 u_{-1}(t)$ so that

$$Dc(t)_{ss} = 0 = K_1 E_0 \qquad (6\text{-}40)$$

or

$$E_0 = 0 \qquad (6\text{-}41)$$

Ramp input $R(s) = R_1/s^2$ From Eq. (6-34)

$$e(t)_{ss} = \frac{R_1}{K_1} = \text{constant} = E_0 \neq 0 \qquad (6\text{-}42)$$

From Eq. (6-42) it is seen that a Type 1 system with a ramp input produces a ramp output with a constant actuating signal. That is, in a Type 1 system a fixed error E_0 is required to produce a ramp output. This result can also be obtained from Eq. (6-30), that is, $K_1 E_0 = Dc(t)_{ss} = \text{constant} = C_1$. For steady-state conditions

$$e(t)_{ss} = r(t)_{ss} - c(t)_{ss} = E_0 \qquad (6\text{-}43)$$

For the ramp input

$$r(t) = R_1 t u_{-1}(t) \qquad (6\text{-}44)$$

the output has the form of a power series

$$c(t)_{ss} = C_0 + C_1 t \qquad (6\text{-}45)$$

Substituting Eqs. (6-44) and (6-45) into Eq. (6-43) yields

$$E_0 = R_1 t - C_0 - C_1 t \qquad (6\text{-}46)$$

This can occur only with

$$R_1 = C_1 \qquad (6\text{-}47)$$

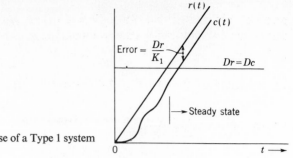

FIGURE 6-6
Steady-state response of a Type 1 system
with a ramp input.

This result signifies that the slopes of the ramp input and the ramp output are equal. This, of course, is a necessary condition if the difference between input and output is a constant. Figure 6-6 illustrates the results obtained above.

Parabolic input $r(t) = (R_2 t^2/2)u_{-1}(t)$, $R(s) = R_2/s^3$ From Eq. (6-34), $e(t)_{ss} = \infty$. Also, by use of the partial-fraction expansion, the particular solution of $e(t)$ obtained from Eq. (6-33) can be shown to contain the term

$$e(t) = \frac{R_2 t}{K_1} \qquad (6\text{-}48)$$

Therefore, it can be stated that a Type 1 system with a parabolic input produces a parabolic output but with an error which increases with time and approaches a value of infinity. This means that a Type 1 system cannot follow a parabolic input.

Case 3 $m = 2$ (Type 2 System)

Step input $R(s) = R_0/s$ From Eq. (6-34)

$$e(t)_{ss} = 0 \qquad (6\text{-}49)$$

Therefore, it can be stated that a Type 2 system with a constant input produces a constant output of value identical with the input. This means that for a Type 2 system, zero error exists in the steady state between the output and input for a step input, i.e.,

$$e(t)_{ss} = r(t)_{ss} - c(t)_{ss} = 0 \qquad (6\text{-}50)$$

The above analysis is in agreement with Eq. (6-30). That is, for a constant input the output is also a constant and

$$D^2 c(t)_{ss} = 0 = K_2 E_0 \qquad (6\text{-}51)$$

or

$$E_0 = 0 \qquad (6\text{-}52)$$

Ramp input $R(s) = R_1/s^2$ From Eq. (6-34)

$$e(t)_{ss} = 0 \qquad (6\text{-}53)$$

Therefore, it can be stated that a Type 2 system with a ramp input produces a ramp output of identical slope with the input after steady-state conditions have been reached. This means that, for a Type 2 system, zero error exists between the steady-state output and input for a ramp input, i.e.,

$$e(t)_{ss} = r(t)_{ss} - c(t)_{ss} = 0 \qquad (6\text{-}54)$$

Substituting Eqs. (6-44) and (6-45) into Eq. (6-54) yields

$$R_1 t - C_0 - C_1 t = 0 \qquad (6\text{-}55)$$

Differentiating the above gives

$$R_1 = C_1 \qquad (6\text{-}56)$$

Substituting this result into Eq. (6-55) yields

$$C_0 = 0 \qquad (6\text{-}57)$$

and

$$R_1 t = C_1 t \qquad (6\text{-}58)$$

Thus, the steady-state slopes are identical. The above analysis is in agreement with Eq. (6-30); that is, for a ramp input the output has the form $c(t)_{ss} = C_1 t + C_0$ and

$$D^2 c(t)_{ss} = 0 = K_2 E_0 \qquad (6\text{-}59)$$

or

$$E_0 = 0 \qquad (6\text{-}60)$$

Parabolic input $R(s) = R_2/s^3$ From Eq. (6-34)

$$e(t)_{ss} = \frac{R_2}{K_2} = \text{constant} = E_0 \neq 0 \qquad (6\text{-}61)$$

From Eq. (6-61) it is seen that a Type 2 system with a parabolic input produces a parabolic output with a constant actuating signal. That is, in a Type 2 system a fixed error E_0 is required to produce a parabolic output. This result is further confirmed by applying Eq. (6-30), i.e.,

$$K_2 E_0 = D^2 c(t)_{ss} = \text{constant} = C_2$$

Thus, for steady-state conditions

$$e(t)_{ss} = r(t)_{ss} - c(t)_{ss} = E_0 \qquad (6\text{-}62)$$

For a parabolic input given by

$$r(t) = \frac{R_2 t^2}{2} u_{-1}(t) \qquad (6\text{-}63)$$

the particular solution of the output must be given by

$$c(t)_{ss} = \frac{C_2}{2} t^2 + C_1 t + C_0 \qquad (6\text{-}64)$$

Substituting Eqs. (6-63) and (6-64) into Eq. (6-62) yields

$$E_0 = \frac{R_2 t^2}{2} - \frac{C_2 t^2}{2} - C_1 t - C_0 \qquad (6\text{-}65)$$

Differentiating the above equation twice results in

$$R_2 t - C_2 t - C_1 = 0 \qquad (6\text{-}66)$$

$$R_2 - C_2 = 0 \qquad (6\text{-}67)$$

Table 6-1 STEADY-STATE RESPONSE CHARACTERISTICS FOR STABLE UNITY-FEEDBACK SYSTEMS

System type m	$r(t)_{ss}$	$c(t)_{ss}$	$e(t)_{ss}$	$e(\infty)$	Derivatives
0	$R_0 u_{-1}(t)$	$\dfrac{K_0}{1+K_0} R_0$	$\dfrac{R_0}{1+K_0}$	$\dfrac{R_0}{1+K_0}$	$Dr = Dc = 0$
	$R_1 t u_{-1}(t)$	$\dfrac{K_0 R_1}{1+K_0} t + C_0$	$\dfrac{R_1}{1+K_0} t - C_0$	∞	$Dr \neq Dc$
	$\dfrac{R_2 t^2}{2} u_{-1}(t)$	$\dfrac{K_0 R_2}{2(1+K_0)} t^2 + C_1 t + C_0$	$\dfrac{R_2}{2(1+K_0)} t^2 - C_1 t - C_0$	∞	$Dr \neq Dc$
1	$R_0 u_{-1}(t)$	R_0	0	0	$Dr = Dc = 0$
	$R_1 t u_{-1}(t)$	$R_1 t - \dfrac{R_1}{K_1}$	$\dfrac{R_1}{K_1}$	$\dfrac{R_1}{K_1}$	$Dr = Dc = R_1$
	$\dfrac{R_2 t^2}{2} u_{-1}(t)$	$\dfrac{R_2 t^2}{2} + C_1 t + C_0$	$-C_1 t - C_0$	∞	$Dr \neq Dc$
2	$R_0 u_{-1}(t)$	R_0	0	0	$Dr = Dc = 0$
	$R_1 t u_{-1}(t)$	$R_1 t$	0	0	$Dr = Dc = R_1$
	$\dfrac{R_2 t^2}{2} u_{-1}(t)$	$\dfrac{R_2 t^2}{2} - \dfrac{R_2}{K_2}$	$\dfrac{R_2}{K_2}$	$\dfrac{R_2}{K_2}$	$D^2 r = D^2 c = R_2$ $\quad Dr = Dc = R_2 t$

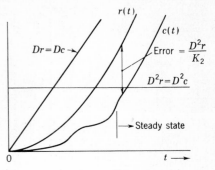

FIGURE 6-7
Steady-state response of a Type 2 system
with a parabolic input.

Therefore, $C_1 = 0$, $R_2 = C_2$, and $E_0 = -C_0$. Thus the input and output curves have the same shape but are displaced by a constant error. Figure 6-7 illustrates the results obtained above.

The results determined in this section verify the properties stated in the previous section for the system types. The steady-state response characteristics for Type 0, 1, and 2 unity-feedback systems are given in Table 6-1 and apply only to stable systems.

6-6   EXAMPLE OF A TYPE 2 SYSTEM

A Type 2 system is one in which the second derivative of the output is maintained constant by a constant actuating signal. In Fig. 6-8 is shown a positioning system which is a Type 2 system, where

K_b = motor back emf constant, V/(rad/s)
K_x = potentiometer constant, V/rad
K_T = motor torque constant, lb-ft/A
A = integrator amplifier gain, s^{-1}
K_g = generator constant, V/A
$T_m = JR_{gm}/(K_bK_T + R_{gm}B)$ = motor mechanical constant, s
T_f = generator field constant, s
$K_M = K_T/(BR_{gm} + K_TK_b)$ = overall motor constant, rad/V-s

The motor-generator control unit is described in Secs. 2-13 and 2-14. It is assumed that the inductance L_{gm} of the generator and motor armatures is negligible. The transfer function of this unit, with the inductance neglected, is given by Eq. (2-148). Figure 6-9 is the block-diagram representation of the Type 2 position-control system shown in Fig. 6-8. Thus the open-loop transfer function is

$$G(s) = \frac{\theta_o(s)}{E(s)} = \frac{K_xAK_gK_M/R_f}{s^2(1 + T_fs)(1 + T_ms)} \qquad (6\text{-}68)$$

or

$$G(s) = \frac{K_2}{s^2(1 + T_fs)(1 + T_ms)} \qquad (6\text{-}69)$$

where K_2 has the units of seconds^{-2}.

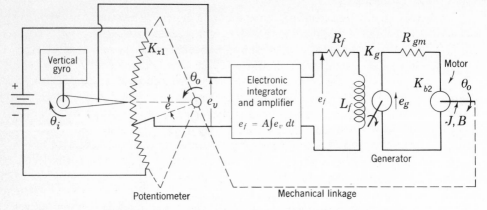

FIGURE 6-8
Position control of space-vehicle camera (Type 2 system).

This Type 2 control system is unstable. As shown in Chap. 10, an appropriate cascade compensator can be added to the forward path to produce a stable system. The new transfer function has the form

$$G_0(s) = G_c(s)G(s) = \frac{K_2'(1 + T_1 s)}{s^2(1 + T_f s)(1 + T_m s)(1 + T_2 s)} \qquad (6\text{-}70)$$

For a constant input $r(t) = R_0 u_{-1}(t)$ the steady-state value of $c(t)_{ss}$ is

$$c(t)_{ss} = \lim_{s \to 0}\left[s\,\frac{C(s)}{R(s)}\,\frac{R_0}{s}\right] = \lim_{s \to 0}\left[\frac{G_0(s)}{1 + G_0(s)}\,R_0\right] = R_0 \qquad (6\text{-}71)$$

Thus
$$E_0 = r(t)_{ss} - c(t)_{ss} = R_0 - R_0 = 0 \qquad (6\text{-}72)$$

Therefore, as expected, a Type 2 system follows a step-function input with no error.

6-7 STEADY-STATE ERROR COEFFICIENTS[2]

In the preceding sections of this chapter, system types are defined. This is the first step toward establishing a set of standard characteristics that permit the engineer to obtain as much information as possible about a given system with a minimum amount of calculation. Also, these standard characteristics must point the direction in which a given system must be modified to meet a given set of performance specifications.

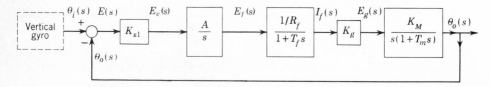

FIGURE 6-9
Block-diagram representation of the system of Fig. 6-8.

Another item of importance is the ability of a system to maintain the output at the desired value with a minimum error. Thus, in this section are defined system error coefficients that are a measure of a unity-feedback control system's steady-state accuracy for a given desired output that is relatively constant or slowly varying.

In Eq. (6-30) it is shown that when the derivative of the output is constant, a constant actuating signal exists. This derivative is proportional to the actuating signal E_0 and to a constant K_m, which is the gain of the forward transfer function. The conventional names for these constants for the Type 0, 1, and 2 systems are *position*, *velocity*, and *acceleration error coefficients*, respectively. Since the conventional names were originally selected for application to position-control systems (servomechanisms), these names referred to the actual physical form of $c(t)$ or $r(t)$, which is position, as well as to the mathematical forms of $c(t)$, that is, c, Dc, and D^2c. It is the belief of the authors that these names are ambiguous when the analysis is extended to cover control of temperature, velocity, etc. To avoid this ambiguity of terminology and to define general terms that are universally applicable, the authors have selected the terminology *step*, *ramp*, and *parabolic steady-state error coefficients*. Table 6-2 lists the conventional and the authors' designation of the error coefficients.

The *following derivations of the error coefficients are independent of the system type. They apply to any system type and are defined for specific forms of the input, i.e., for a step, ramp, or parabolic input. These error coefficients are useful only for stable* unity-feedback *systems*. The results are summarized in Table 6-3.

Steady-State Step Error Coefficient

The step error coefficient is defined as

$$\text{Step error coefficient} = \frac{\text{steady-state value of output } c(t)_{ss}}{\text{steady-state actuating signal } e(t)_{ss}} = K_p \qquad (6\text{-}73)$$

and applies only for a step input $r(t) = R_0 u_{-1}(t)$. The steady-state value of the output is obtained by applying the final-value theorem to Eq. (6-31):

$$c(t)_{ss} = \lim_{s \to 0} sC(s) = \lim_{s \to 0} \left[\frac{sG(s)}{1 + G(s)} \frac{R_0}{s} \right] = \lim_{s \to 0} \left[\frac{G(s)}{1 + G(s)} R_0 \right] \qquad (6\text{-}74a)$$

Table 6-2 CORRESPONDENCE BETWEEN THE CONVENTIONAL AND THE AUTHORS' DESIGNATION OF STEADY-STATE ERROR COEFFICIENTS

Symbol	Conventional designation of error coefficients	Authors' designation of error coefficients
K_p	Position	Step
K_v	Velocity	Ramp
K_a	Acceleration	Parabolic

Similarly, from Eq. (6-32), for a unity-feedback system

$$e(t)_{ss} = \lim_{s \to 0} \left[s \frac{1}{1 + G(s)} \frac{R_0}{s} \right] = \lim_{s \to 0} \left[\frac{1}{1 + G(s)} R_0 \right] \qquad (6\text{-}74b)$$

Substituting Eqs. (6-74a) and (6-74b) into Eq. (6-73) yields

$$\text{Step error coefficient} = \frac{\lim_{s \to 0} \left[\dfrac{G(s)}{1 + G(s)} R_0 \right]}{\lim_{s \to 0} \left[\dfrac{1}{1 + G(s)} R_0 \right]} \qquad (6\text{-}75)$$

Since both the numerator and the denominator of Eq. (6-75) in the limit can never be zero or infinity simultaneously, where $K_m \neq 0$, the indeterminate forms 0/0 and ∞/∞ never occur. Thus, this equation reduces to

$$\text{Step error coefficient} = \lim_{s \to 0} G(s) = K_p \qquad (6\text{-}76)$$

Therefore,

For Type 0 *system:*

$$K_p = \lim_{s \to 0} \frac{K_0(1 + T_1 s)(1 + T_2 s) \cdots}{(1 + T_a s)(1 + T_b s)(1 + T_c s) \cdots} = K_0 \qquad (6\text{-}77)$$

For Type 1 *system:*

$$K_p = \infty \qquad (6\text{-}78)$$

For Type 2 *system:*

$$K_p = \infty \qquad (6\text{-}79)$$

Steady-State Ramp Error Coefficient

The ramp error coefficient is defined as

Ramp error coefficient

$$= \frac{\text{steady-state value of derivative of output } (Dc)_{ss}}{\text{steady-state actuating signal } e(t)_{ss}}$$

$$= K_v \qquad (6\text{-}80)$$

Table 6-3   DEFINITIONS OF STEADY-STATE ERROR
COEFFICIENTS FOR STABLE UNITY-FEEDBACK
SYSTEMS

Error coefficient	Definition of error coefficient	Value of error coefficient	Form of input signal $r(t)$
Step	$\dfrac{c(t)_{ss}}{e(t)_{ss}}$	$\lim_{s \to 0} G(s)$	$R_0 u_{-1}(t)$
Ramp	$\dfrac{(Dc)_{ss}}{e(t)_{ss}}$	$\lim_{s \to 0} sG(s)$	$R_1 t u_{-1}(t)$
Parabolic	$\dfrac{(D^2 c)_{ss}}{e(t)_{ss}}$	$\lim_{s \to 0} s^2 G(s)$	$(R_2 t^2/2) u_{-1}(t)$

and applies only for a ramp input $r(t) = R_1 t u_{-1}(t)$. The first derivative of the output is given by

$$\mathscr{L}[Dc] = sC(s) = \frac{sG(s)}{1 + G(s)} R(s) \qquad (6\text{-}81)$$

The steady-state value of the derivative of the output is obtained by using the final-value theorem:

$$(Dc)_{ss} = \lim_{s \to 0} s[sC(s)] = \lim_{s \to 0} \left[\frac{s^2 G(s)}{1 + G(s)} \frac{R_1}{s^2} \right] = \lim_{s \to 0} \left[\frac{G(s)}{1 + G(s)} R_1 \right] \qquad (6\text{-}82)$$

Similarly, from Eq. (6-32), for a unity-feedback system

$$e(t)_{ss} = \lim_{s \to 0} \left[s \frac{1}{1 + G(s)} \frac{R_1}{s^2} \right] = \lim_{s \to 0} \left[\frac{1}{1 + G(s)} \frac{R_1}{s} \right] \qquad (6\text{-}83)$$

Substituting Eqs. (6-82) and (6-83) into Eq. (6-80) yields

$$\text{Ramp error coefficient} = \frac{\lim\limits_{s \to 0} \left[\dfrac{G(s)}{1 + G(s)} R_1 \right]}{\lim\limits_{s \to 0} \left[\dfrac{1}{1 + G(s)} \dfrac{R_1}{s} \right]} \qquad (6\text{-}84)$$

Since the above equation never has the indeterminate form $0/0$ or ∞/∞, it can be simplified to

$$\text{Ramp error coefficient} = \lim_{s \to 0} sG(s) = K_v \qquad (6\text{-}85)$$

Therefore,

For a Type 0 system:

$$K_v = \lim_{s \to 0} \frac{sK_0(1 + T_1 s)(1 + T_2 s) \cdots}{(1 + T_a s)(1 + T_b s)(1 + T_c s) \cdots} = 0 \qquad (6\text{-}86)$$

For a Type 1 system:

$$K_v = K_1 \qquad (6\text{-}87)$$

For a Type 2 system:

$$K_v = \infty \qquad (6\text{-}88)$$

Steady-State Parabolic Error Coefficient

The parabolic error coefficient is defined as

Parabolic error coefficient

$$= \frac{\text{steady-state value of second derivative of output } (D^2 c)_{ss}}{\text{steady-state actuating signal } e(t)_{ss}} = K_a \qquad (6\text{-}89)$$

and applies only for a parabolic input $r(t) = (R_2 t^2/2)u_{-1}(t)$. The second derivative of the output is given by

$$\mathscr{L}[D^2 c] = s^2 C(s) = \frac{s^2 G(s)}{1 + G(s)} R(s) \qquad (6\text{-}90)$$

The steady-state value of the second derivative of the output is obtained by using the final-value theorem:

$$(D^2c)_{ss} = \lim_{s\to 0} s[s^2C(s)] = \lim_{s\to 0}\left[\frac{s^3G(s)}{1 + G(s)}\frac{R_2}{s^3}\right] = \lim_{s\to 0}\left[\frac{G(s)}{1 + G(s)}R_2\right] \qquad (6\text{-}91)$$

Similarly, from Eq. (6-32), for a unity-feedback system

$$e(t)_{ss} = \lim_{s\to 0}\left[s\frac{1}{1 + G(s)}\frac{R_2}{s^3}\right] = \lim_{s\to 0}\left[\frac{1}{1 + G(s)}\frac{R_2}{s^2}\right] \qquad (6\text{-}92)$$

Substituting Eqs. (6-91) and (6-92) into Eq. (6-89) yields

$$\text{Parabolic error coefficient} = \frac{\displaystyle\lim_{s\to 0}\left[\frac{G(s)}{1 + G(s)}R_2\right]}{\displaystyle\lim_{s\to 0}\left[\frac{1}{1 + G(s)}\frac{R_2}{s^2}\right]} \qquad (6\text{-}93)$$

Since the above equation never has the indeterminate form $0/0$ or ∞/∞, it can be simplified to

$$\text{Parabolic error coefficient} = \lim_{s\to 0} s^2G(s) = K_a \qquad (6\text{-}94)$$

Therefore,

For a Type 0 system:

$$K_a = \lim_{s\to 0}\frac{s^2K_0(1 + T_1s)(1 + T_2s)\cdots}{(1 + T_as)(1 + T_bs)(1 + T_cs)\cdots} = 0 \qquad (6\text{-}95)$$

For a Type 1 system:

$$K_a = 0 \qquad (6\text{-}96)$$

For a Type 2 system:

$$K_a = K_2 \qquad (6\text{-}97)$$

6-8   USE OF STEADY-STATE ERROR COEFFICIENTS

The use of steady-state error coefficients is discussed for unity-feedback systems, as illustrated in Fig. 6-10.

Type 1 System

For a Type 1 system considered at steady state, the value of $(Dc)_{ss}$ is

$$(Dc)_{ss} = K_1E_0 \qquad (6\text{-}98)$$

Thus, the larger K_1, the smaller the size of the actuating signal necessary to maintain a constant rate of change of the output. From the standpoint of trying to maintain $c(t) = r(t)$ at all times, a larger K_1 results in a more sensitive system. In other words, a larger K_1 results in a greater speed of response of the system to a given actuating signal $e(t)$. Therefore, K_1 is another standard characteristic of a system's perform-

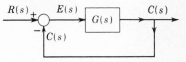

FIGURE 6-10
Simple block diagram.

ance. The maximum value of K_1 is limited by stability considerations and is discussed in later chapters.

For the Type 1 system the step-error coefficient is equal to infinity, and the steady-state error is zero for a step input. Therefore, the steady-state output $c(t)_{ss}$ for a Type 1 system is equal to the input when $r(t) = $ constant.

Consider now a ramp input $r(t) = R_1 tu_{-1}(t)$. The steady-state value of $(Dc)_{ss}$ is found by using the final-value theorem:

$$(Dc)_{ss} = \lim_{s \to 0} s[sC(s)] = \lim_{s \to 0} \left[s \frac{sG(s)}{1 + G(s)} R(s) \right] \qquad (6\text{-}99)$$

where $R(s) = R_1/s^2$ and

$$G(s) = \frac{K_1(1 + T_1 s)(1 + T_2 s) \cdots (1 + T_w s)}{s(1 + T_a s)(1 + T_b s) \cdots (1 + T_u s)}$$

Inserting these values in Eq. (6-99) gives

$$(Dc)_{ss} = \lim_{s \to 0} \left[s \frac{s K_1(1 + T_1 s)(1 + T_2 s) \cdots (1 + T_w s)}{s(1 + T_a s)(1 + T_b s) \cdots (1 + T_u s)} \frac{R_1}{s^2} \right] = R_1$$
$$+ K_1(1 + T_1 s)(1 + T_2 s) \cdots (1 + T_w s)$$

Therefore,
$$(Dc)_{ss} = (Dr)_{ss} \qquad (6\text{-}100)$$

The magnitude of the steady-state error is found by using the ramp error coefficient. From Eq. (6-85),

$$K_v = \lim_{s \to 0} sG(s) = K_1$$

From the definition of ramp error coefficient, the steady-state error is

$$e(t)_{ss} = \frac{(Dc)_{ss}}{K_1} \qquad (6\text{-}101)$$

Since $(Dc)_{ss} = Dr$,

$$e(t)_{ss} = \frac{Dr}{K_1} = \frac{R_1}{K_1} = E_0 \qquad (6\text{-}102)$$

Therefore a Type 1 system follows a ramp input with a constant error E_0. Figure 6-11 illustrates these conditions graphically.

Table of Steady-State Error Coefficients

Table 6-4 gives the values of the error coefficients for the Type 0, 1, and 2 systems. These values are determined from Table 6-3. The reader should be able to make ready use of Table 6-4 for evaluating the appropriate error coefficient. The error coefficient

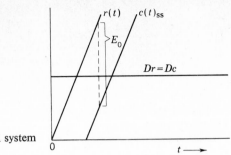

FIGURE 6-11
Steady-state response of a Type 1 system
for Dr = constant.

is then used with the definitions given in Table 6-3 to evaluate the magnitude of the steady-state error.

Polynomial Input: t^{m+1}

Note that a Type m system can follow an input of the form t^{m-1} with zero steady-state error. It can follow an input t^m, but there is a constant steady-state error. It cannot follow an input t^{m+1} because the steady-state error approaches infinity. However, for this case the input may be present only for a finite length of time. Thus the error is also finite. Then the error can be evaluated by taking the inverse Laplace transform of Eq. (6-33) and inserting the value of time. The maximum permissible error limits the time $(0 < t < t_1)$ that an input t^{m+1} can be applied to a control system.

6-9   CONTROLLABILITY AND OBSERVABILITY[3]

An important objective of modern control theory is the design of systems that have an optimum performance. The optimal control is based on the optimization of some specific performance criterion. The achievement of such optimal linear control systems is governed by the *controllability* and *observability* properties of the system. Further, these properties establish the conditions for complete equivalence between the state-variable and transfer-function representations. A study of these properties, presented below, provides a basis for consideration of the optimal control problem, contained in Chap. 14.

Table 6-4   STEADY-STATE ERROR COEFFICIENTS FOR
STABLE SYSTEMS

System type	Step error coefficient K_p	Ramp error coefficient K_v	Parabolic error coefficient K_a
0	K_0	0	0
1	∞	K_1	0
2	∞	∞	K_2

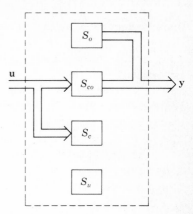

FIGURE 6-12
The four possible subdivisions
of a system.

Controllability A system is said to be completely *state-controllable* if, for any t_0, each initial state $\mathbf{x}(t_0)$ can be transferred to any final state $\mathbf{x}(t_f)$ in a finite time, $t_f > t_0$, by means of an unconstrained control input vector $\mathbf{u}(t)$. An unconstrained control vector has no limit on the amplitudes of $\mathbf{u}(t)$. This definition implies that $\mathbf{u}(t)$ is able to affect each state variable in

$$\mathbf{x}(t) = \mathbf{\Phi}(t - t_0)\mathbf{x}(t_0) + \int_{t_0}^{t} \mathbf{\Phi}(t - \tau)\mathbf{Bu}(\tau)\,d\tau \qquad (6\text{-}103)$$

Observability A system is said to be completely *observable* if every initial state $\mathbf{x}(t_0)$ can be exactly determined from the measurements of the output $\mathbf{y}(t)$ over the finite interval of time $t_0 \le t \le t_f$. This definition implies that every state of $\mathbf{x}(t)$ affects the output $\mathbf{y}(t)$:

$$\mathbf{y}(t) = \mathbf{Cx}(t) = \mathbf{C\Phi}(t - t_0)\mathbf{x}(t_0) + \mathbf{C}\int_{t_0}^{t} \mathbf{\Phi}(t - \tau)\mathbf{Bu}(\tau)\,d\tau \qquad (6\text{-}104)$$

where the initial state $\mathbf{x}(t_0)$ is the result of control inputs prior to t_0.

The concepts of controllability and observability[4] can be illustrated graphically by the block diagram of Fig. 6-12. By the proper selection of the state variables it is possible to divide a system into the four subdivisions shown in Fig. 6-12, where

S_{co} = completely controllable and completely observable subsystem
S_o = completely observable but uncontrollable subsystem
S_c = completely controllable but unobservable subsystem
S_u = uncontrollable and unobservable subsystem

An inspection of this figure readily reveals that only the controllable and observable subsystem S_{co} satisfies the definition of a transfer-function matrix, that is,

$$\mathbf{G}(s)\mathbf{U}(s) = \mathbf{Y}(s)$$

If the entire system is completely controllable and completely observable, then the state-variable and transfer-function matrix representations of a system are equivalent and accurately represent the system. In that case the resulting transfer function does carry all the information characterizing the dynamic performance of the system. As mentioned in Chap. 2, there is no unique method of selecting the state variables to be used in representing a system, but the transfer-function matrix is completely and uniquely specified once the state-variable representation of the system is known. In the transfer-function approach the state vector is suppressed, i.e., the transfer-function method is concerned only with the system's input-output characteristics. The state-variable method also includes a description of the system's internal behavior.

The determination of controllability and observability of a system by subdividing into its four possible subdivisions is not easy. A simpler method for determining these system characteristics has been developed. Utilizing Eq. (6-103), with $t_0 = 0$, and defining the final state vector $x(t_f) = 0$ yields

$$0 = e^{At_f}x(0) + \int_0^{t_f} e^{A(t_f - \tau)}Bu(\tau)\,d\tau$$

or
$$x(0) = -\int_0^{t_f} e^{-A\tau}Bu(\tau)\,d\tau \qquad (6\text{-}105)$$

The Cayley-Hamilton method presented in Sec. 3-13 shows that exp $(-A\tau)$ can be expressed as a polynomial of order $n - 1$ as follows:

$$e^{-A\tau} = \sum_{k=0}^{n-1} \alpha_k(\tau)A^k \qquad (6\text{-}106)$$

Inserting this equation into Eq. (6-105) yields

$$x(0) = -\sum_{k=0}^{n-1} A^k B \int_0^{t_f} \alpha_k(\tau)u(\tau)\,d\tau \qquad (6\text{-}107)$$

With the input $u(t)$ of dimension r, the integral in Eq. (6-107) can be evaluated, and the result is

$$\int_0^{t_f} \alpha_k(\tau)u(\tau)\,d\tau = \beta_k$$

Equation (6-107) can now be expressed as

$$x(0) = -\sum_{k=0}^{n-1} A^k B\beta_k = -[B \mid AB \mid \cdots \mid A^{n-1}B]\begin{bmatrix} \beta_0 \\ \beta_1 \\ \vdots \\ \beta_{n-1} \end{bmatrix} \qquad (6\text{-}108)$$

According to the definition of controllability, each initial state $x_i(0)$ must be influenced by the input $u(t)$. This requires that

$$\text{Rank of } [B \mid AB \mid \cdots \mid A^{n-1}B] = n \qquad (6\text{-}109)$$

For a single-input system the matrix B reduces to the vector b, and Eq. (6-109) represents an $n \times n$ matrix.

A simpler alternate method for determining controllability when the system has nonrepeated eigenvalues is to use the transformation $x = Tz$ to convert the A matrix into diagonal form, as described in Sec. 5-9. The resulting state equation is

$$\dot{z} = T^{-1}ATz + T^{-1}Bu = \Lambda z + B'u \qquad (6\text{-}110)$$

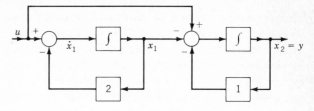

FIGURE 6-13
Block-diagram representation of the system of Eq. (6-116).

According to the definition of controllability, each transformed state z_i can be influenced by the input $\mathbf{u}(t)$ only if $\mathbf{B}' = \mathbf{T}^{-1}\mathbf{B}$ has no zero row. Therefore, for non-repeated eigenvalues, *a system is controllable if $\mathbf{B}'$ has no zero row.* For repeated eigenvalues see Refs. 4 and 5. $\mathbf{B}'$ is called the *mode controllability* matrix.

In a similar manner the condition for observability is derived from the homogeneous system equations:

$$\dot{\mathbf{x}} = \mathbf{A}\mathbf{x} \qquad (6\text{-}111)$$

$$\mathbf{y} = \mathbf{C}\mathbf{x} \qquad (6\text{-}112)$$

When Eq. (6-106) is used, the output vector $\mathbf{y}(t)$ can be expressed as

$$\mathbf{y}(t) = \mathbf{C}e^{\mathbf{A}t}\mathbf{x}(0) = \sum_{k=0}^{n-1} \alpha_k(t)\mathbf{C}\mathbf{A}^k\mathbf{x}(0) \qquad (6\text{-}113)$$

For observability each output must be influenced by each state x_i. This imposes restrictions on $\mathbf{C}\mathbf{A}^k$. It can be shown that the system is completely observable iff the following matrix has the property

$$\text{Rank of } [\mathbf{C}^T \mid \mathbf{A}^T\mathbf{C}^T \mid (\mathbf{A}^T)^2\mathbf{C}^T \mid \cdots \mid (\mathbf{A}^T)^{(n-1)}\mathbf{C}^T] = n \qquad (6\text{-}114)$$

For a single-output system the matrix $\mathbf{C}^T$ is a column matrix which can be represented by $\mathbf{c}$.

A simpler procedure for determining observability can be used when the system has nonrepeated eigenvalues and the equations are put in canonical form. The output equation is then

$$\mathbf{y} = \mathbf{C}\mathbf{T}\mathbf{z} = \mathbf{C}'\mathbf{z} \qquad (6\text{-}115)$$

If a column of $\mathbf{C}'$ has all zeros, then one mode is not coupled to *any* of the outputs and the system is unobservable. *The condition for observability, for the case of non-repeated eigenvalues, is that $\mathbf{C}'$ have no zero columns.*

EXAMPLE Figure 6-13 shows the block-diagram representation of a system whose state and output equations are

$$\dot{\mathbf{x}} = \begin{bmatrix} -2 & 0 \\ -1 & -1 \end{bmatrix} \mathbf{x} + \begin{bmatrix} 1 \\ 1 \end{bmatrix} u \qquad (6\text{-}116)$$

$$y = \begin{bmatrix} 0 & 1 \end{bmatrix}\mathbf{x} \qquad (6\text{-}117)$$

Determine (a) the eigenvalues from the state equation, (b) the transfer function $Y(s)/U(s)$, (c) whether the system is controllable and/or observable, and (d) the transformation matrix $\mathbf{T}$ that transforms the general state-variable representation of Eqs. (6-116) and (6-117) into the canonical-form representation.

(a)
$$|s\mathbf{I} - \mathbf{A}| = \begin{vmatrix} s+2 & 0 \\ 1 & s+1 \end{vmatrix} = (s+2)(s+1)$$

The eigenvalues are $\lambda_1 = -2$ and $\lambda_2 = -1$

(b)
$$\Phi(s) = [s\mathbf{I} - \mathbf{A}]^{-1} = \frac{\begin{bmatrix} s+1 & 0 \\ -1 & s+2 \end{bmatrix}}{(s+1)(s+2)}$$

$$G(s) = \mathbf{c}^T\Phi(s)\mathbf{b}$$

$$= \left([0 \quad 1]\begin{bmatrix} s+1 & 0 \\ -1 & s+2 \end{bmatrix}\begin{bmatrix} 1 \\ 1 \end{bmatrix}\right)\frac{1}{(s+1)(s+2)}$$

$$= \frac{s+1}{(s+1)(s+2)} = \frac{1}{s+2}$$

Note that there is a cancellation of the factor $s+1$, and only the $\lambda = -2$ mode is controllable and observable.

(c)
$$\text{Rank of } [\mathbf{b} \mid \mathbf{Ab}] = \text{Rank of } \begin{bmatrix} 1 & -2 \\ 1 & -2 \end{bmatrix} = 1$$

This is a singular matrix; thus the system is *not* controllable.

$$\text{Rank of } [\mathbf{c} \mid \mathbf{A}^T\mathbf{c}] = \text{Rank of } \begin{bmatrix} 0 & -1 \\ 1 & -1 \end{bmatrix} = 2$$

This is a nonsingular matrix; thus the system is observable. Note that the $\lambda = -1$ mode, which is canceled in the determination of the transfer function, is observable but not controllable.

(d) Using the third method of Sec. 5-9, with $\mathbf{x} = \mathbf{Tz}$, gives

$$\mathbf{AT} = \mathbf{T\Lambda}$$

$$\begin{bmatrix} -2 & 0 \\ -1 & -1 \end{bmatrix}\begin{bmatrix} t_{11} & t_{12} \\ t_{21} & t_{22} \end{bmatrix} = \begin{bmatrix} t_{11} & t_{12} \\ t_{21} & t_{22} \end{bmatrix}\begin{bmatrix} -2 & 0 \\ 0 & -1 \end{bmatrix}$$

$$\begin{bmatrix} -2t_{11} & -2t_{12} \\ -t_{11}-t_{21} & -t_{12}-t_{22} \end{bmatrix} = \begin{bmatrix} -2t_{11} & -t_{12} \\ -2t_{21} & -t_{22} \end{bmatrix}$$

Equating elements on both sides of this equation yields

$$-2t_{11} = -2t_{11} \qquad -2t_{12} = -t_{12}$$
$$-t_{11}-t_{21} = -2t_{21} \qquad -t_{12}-t_{22} = -t_{22}$$

Since there are only two independent equations, assume that $t_{11} = t_{22} = 1$. Then $t_{21} = 1$ and $t_{12} = 0$,

$$\mathbf{b}' = \mathbf{T}^{-1}\mathbf{b} = \begin{bmatrix} 1 & 0 \\ -1 & 1 \end{bmatrix}\begin{bmatrix} 1 \\ 1 \end{bmatrix} = \begin{bmatrix} 1 \\ 0 \end{bmatrix}$$

$$\mathbf{c}'^T = \mathbf{c}^T\mathbf{T} = [0 \quad 1]\begin{bmatrix} 1 & 0 \\ 1 & 1 \end{bmatrix} = [1 \quad 1]$$

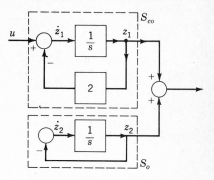

FIGURE 6-14
Simulation diagram for Eqs. (6-118)
and (6-119).

Since $\mathbf{b}'$ contains a zero element, the system is uncontrollable, and since $\mathbf{c}'$ contains all nonzero elements, the system is observable. These results agree with the results obtained by the method used in part (c). The canonical equations for this system are

$$\dot{\mathbf{z}} = \begin{bmatrix} -2 & 0 \\ 0 & -1 \end{bmatrix} \mathbf{z} + \begin{bmatrix} 1 \\ 0 \end{bmatrix} u \qquad (6\text{-}118)$$

$$y = \begin{bmatrix} 1 & 1 \end{bmatrix} \mathbf{z} \qquad (6\text{-}119)$$

The individual equations are

$$\dot{z}_1 = -2z_1 + u \qquad (6\text{-}120)$$

$$\dot{z}_2 = -z_2 \qquad (6\text{-}121)$$

$$y = z_1 + z_2 \qquad (6\text{-}122)$$

Equation (6-120) satisfies the requirement of controllability; i.e., for any t_0 the initial state $z_1(t_0)$ can be transferred to any final state $\dot{z}_1(t_f)$ in a finite time $t_f \geq t_0$ by means of the unconstrained control vector u. Since u does not appear in Eq. (6-121), it is impossible to control the state z_2. Thus the system is uncontrollable. Equation (6-122) satisfies the definition of observability; i.e., the states $z_1(t_0)$ and $z_2(t_0)$ can be determined from the measurements of the output $y(t)$ over the finite interval of time $t_0 \leq t \leq t_f$.

From Eqs. (6-118) and (6-119) the simulation diagram of Fig. 6-14 is synthesized. The advantage of the canonical form is that the modes $\lambda_1 = -2$ and $\lambda_2 = -1$ are completely decoupled. This feature permits the establishment of the existence of some or all of the four subdivisions of Fig. 6-12 and the application of the definitions of controllability and observability.

When deriving the transfer function from the system differential equations, cancellation of a pole by a zero is not permitted in developing the state representation because observability and controllability of the system are determined from the state equations. Cancellation of a pole by a zero in the transfer function may hide some of the dynamics of the entire system. This eliminates information pertaining to S_o, S_c, and S_u, illustrated in the last example.

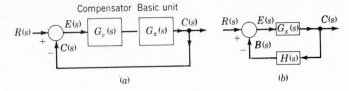

FIGURE 6-15
Improving system performance by means of (a) cascade compensation and
(b) feedback compensation.

6-10   SENSITIVITY[6,7]

A basic closed-loop control system is one composed of the minimum amount of equipment G_x necessary to perform the control function and the necessary sensors and
comparators required to provide feedback. The following chapters present detailed
methods of performing the system analysis. If the performance of the basic system is
not satisfactory, the desired performance can be achieved by inserting a compensator
in cascade with G_x, as shown in Fig. 6-15a, or in the feedback path, as shown in Fig.
6-15b. Note that in Fig. 6-15b the input signal is being compared with a modified
form of the output signal and not directly with it. The compensator inserted in the
feedback or in the cascade path can produce the same overall response. The design
methods are presented in Chapters 10 and 11.

The following factors must be considered when making a choice between cascade
and feedback compensation.

1. The design procedures for a cascade compensator are more direct than those
for a feedback compensator. The application of feedback compensators is sometimes
more laborious.

2. Because of the physical form of the control system, i.e., whether it is electrical,
hydraulic, mechanical, etc., a cascade or feedback compensator may not exist or be
practical.

3. The type of signal input to the compensator must be considered. For example, if the system utilizes a 400-Hz carrier, the design of a feedback compensator
may be more difficult than that of a cascade compensator.

4. The economics in the use of either technique for a given control system involves items such as the size, weight, and cost of components and amplifiers. In the
forward path the signal goes from a low- to a high-energy level, whereas the reverse is
true in the feedback loop. Thus, generally an amplifier may not be necessary in the
feedback path. The cascade path generally requires an amplifier for gain and/or
isolation. Also, the size and weight may be different for the cascade and the feedback
compensators. These items are of great importance in aircraft, both commercial and
military, and in spacecraft, where minimum size and weight of equipment are essential.

5. The environmental conditions in which the feedback control system is to be
utilized affect the accuracy and stability of the controlled quantity. This is a serious
problem in an airplane or space vehicle, which is subjected to rapid changes in altitude
and temperature. It is shown in Chaps. 11 to 12 that control-system performance
can be improved by the use of a feedback compensator or by state-variable feedback.

6. The problem of noise within a control system may determine the choice of compensator. The noise problem is accentuated in situations where a greater amplifier gain is required with a forward compensator than by the use of feedback networks. Also, the frequency characteristics of the compensator needed to give the desired system improvement may be such that it attenuates the high-frequency portion of the noise content.

7. The time of response desired for a control system is a determining factor. Often a faster time of response can be achieved by the use of feedback compensation.

8. Some systems require "tight-loop" stabilization to isolate the dynamics of one portion of a control system from other portions of the complete system. This can be accomplished by introducing an inner feedback loop around the portion to be isolated.

9. Besides all these factors, the available components and the designer's experience and preferences influence the choice between a cascade and feedback compensator.

The environmental conditions to which a control system is subjected affect the accuracy and stability of the system. The performance characteristics of most components are affected by their environment and by aging. Thus any change in the component characteristics causes a change in the transfer function and therefore in the controlled quantity. The effect of a parameter change on system performance can be expressed in terms of a *sensitivity function*. This sensitivity function $S_\delta{}^M$ is a measure of the sensitivity of the system's response to a system parameter variation and is given by

$$S_\delta{}^M = \frac{d(\ln M)}{d(\ln \delta)} = \frac{d(\ln M)}{d\delta} \frac{d\delta}{d(\ln \delta)} \qquad (6\text{-}123)$$

where

  $\ln$ = logarithm to base e
  M = system's output response
  δ = system parameter that varies

Now

$$\frac{d(\ln M)}{d\delta} = \frac{1}{M}\frac{dM}{d\delta} \qquad (6\text{-}124)$$

Accordingly, Eq. (6-123) can be written

$$S_\delta{}^M \bigg|_{\substack{M=M_o \\ \delta=\delta_o}} = \frac{dM/M_o}{d\delta/\delta_o} = \frac{\text{fractional change in output}}{\text{fractional change in system parameter}} \qquad (6\text{-}125)$$

where M_o and δ_o represent the nominal values of M and δ. When M is a function of more than one parameter, say $\delta_1, \delta_2, \ldots, \delta_k$, the corresponding formulas for the sensitivity entails partial derivatives. For a small change in δ from δ_o, M changes from M_o, and the sensitivity can be written as

$$S_\delta{}^M \bigg|_{\substack{M=M_o \\ \delta=\delta_o}} \approx \frac{\Delta M/M_o}{\Delta\delta/\delta_o} \qquad (6\text{-}126)$$

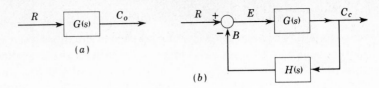

FIGURE 6-16
Control systems: (*a*) open loop; (*b*) closed loop.

To illustrate the effect of changes in the transfer function, four cases are considered for which the input signal $r(t)$ and its transform $R(s)$ are fixed.

Case 1   Open-Loop System of Fig. 6-16*a*

The effect of a change in $G(s)$, for a fixed $r(t)$ and thus a fixed $R(s)$, can be determined
by differentiating the output expression

$$C_o(s) = R(s)G(s) \qquad (6\text{-}127)$$

giving

$$dC_o(s) = R(s)\,dG(s) \qquad (6\text{-}128)$$

Combining these two equations gives

$$dC_o(s) = \frac{dG(s)}{G(s)}\,C_o(s) \rightarrow S_{G(s)}{}^{C(s)}(s) = \frac{dC_o(s)/C_o(s)}{dG(s)/G(s)} = 1 \qquad (6\text{-}129)$$

Therefore a change in $G(s)$ causes a proportional change in the transform of the output
$C_o(s)$. This requires that the performance specifications of $G(s)$ be such that any variation still results in the degree of accuracy within the prescribed limits. In Eq. (6-129)
the varying function in the system is the transfer function $G(s)$.

Case 2   Closed-Loop Unity Feedback System of Fig. 6-16*b* [$H(s) = 1$]

Proceeding in the same manner as for case 1 leads to

$$C_c(s) = R(s)\,\frac{G(s)}{1 + G(s)} \qquad (6\text{-}130)$$

$$dC_c(s) = R(s)\,\frac{dG(s)}{[1 + G(s)]^2} \qquad (6\text{-}131)$$

$$dC_c(s) = \frac{dG(s)}{G(s)[1 + G(s)]}\,C_c(s) = \frac{1}{1 + G(s)}\frac{dG(s)}{G(s)}C_c(s) \qquad (6\text{-}132)$$

Comparing Eq. (6-132) with Eq. (6-129) readily reveals that the effect of changes of
$G(s)$ upon the transform of the output of the closed-loop control is reduced by the
factor $1/|1 + G(s)|$ compared to the open-loop control. *This is an important reason
why feedback systems are used.*

Case 3   Closed-Loop Nonunity Feedback System of Fig. 6-16*b* [Feedback Function $H(s)$ Fixed and $G(s)$ Variable]

Proceeding as before gives

$$C_c(s) = R(s)\frac{G(s)}{1 + G(s)H(s)} \tag{6-133}$$

$$dC_c(s) = R(s)\frac{dG(s)}{[1 + G(s)H(s)]^2} \tag{6-134}$$

$$dC_c(s) = \frac{dG(s)}{G(s)[1 + G(s)H(s)]}C_c(s) = \frac{1}{1 + G(s)H(s)}\frac{dG(s)}{G(s)}C_c(s) \tag{6-135}$$

Comparing Eqs. (6-129) and (6-135) shows that the closed-loop variation is reduced by the factor $1/|1 + G(s)H(s)|$. In comparing Eqs. (6-132) and (6-135), if the term $|1 + G(s)H(s)|$ is larger than the term $|1 + G(s)|$, there is an advantage to using a non-unity-feedback system. Further, $H(s)$ may be introduced both to provide an improvement in system performance and to reduce the effect of parameter variations within $G(s)$.

Case 4   Closed-Loop Nonunity Feedback System of Fig. 6-16*b* [Feedback Function $H(s)$ Variable and $G(s)$ Fixed]

From Eq. (6-133),

$$dC_c(s) = R(s)\frac{-G(s)^2\,dH(s)}{[1 + G(s)H(s)]^2} \tag{6-136}$$

Multiplying and dividing Eq. (6-136) by $H(s)$ and also dividing by Eq. (6-133) results in

$$dC_c(s) = -\left[\frac{G(s)H(s)}{1 + G(s)H(s)}\frac{dH(s)}{H(s)}\right]C_c(s) \approx -\frac{dH(s)}{H(s)}C_c(s) \tag{6-137}$$

The approximation applies for those cases where $|G(s)H(s)| \gg 1$. When Eq. (6-137) is compared with Eq. (6-129), it is seen that a variation in the feedback function has approximately a direct effect upon the output, the same as for the open-loop case. Thus, the components of $H(s)$ must be selected as precision elements to maintain the desired degree of accuracy and stability in the transform $C(s)$.

The two situations of cases 3 and 4 serve to point out the advantage of feedback compensation from the standpoint of parameter changes. Since feedback compensation minimizes the effect of variations in the components of $G(s)$, prime consideration can be given to obtaining the necessary power requirement in the forward loop rather than to accuracy and stability. $H(s)$ can be designed as a precision device so that the transform $C(s)$ of the output $c(t)$ has the desired accuracy and stability. *In other words, by use of feedback compensation the performance of the system can be made to depend more on the feedback term than on the forward term.*

Applying the sensitivity equation (6-125) to each of the four cases, where $M(s) = C(s)/R(s)$, where $\delta = G(s)$ (for cases 1, 2, and 3), and $\delta = H(s)$ (for case 4), yields the results shown in Table 6-5. The table reveals that S_δ^M *never exceeds a magnitude of* 1, and the smaller this value, the less sensitive the system is to a variation in the transfer function. For an increase in the variable function, a positive value of the sensitivity function means that the output increases from its nominal response. Similarly, a negative value of the sensitivity function means that the output decreases from its nominal response. It must be realized that the results presented in this table are based upon a functional analysis, i.e., the "variations" considered are in $G(s)$ and $H(s)$. The results are easily interpreted only when $G(s)$ and $H(s)$ are real numbers. When they are not real numbers, and where δ represents the parameter that varies within $G(s)$ or $H(s)$, the interpretation can be made as a function of frequency.

The analysis in this section so far has considered variations in the transfer functions $G(s)$ and $H(s)$. Take next the case when $r(t)$ is sinusoidal. Then the input can be represented by the phasor $\mathbf{R}(j\omega)$ and the output by the phasor $\mathbf{C}(j\omega)$. The system is now represented by the frequency transfer functions $\mathbf{G}(j\omega)$ and $\mathbf{H}(j\omega)$. All the formulas developed earlier in this section are the same in form, but the arguments are $j\omega$ instead of s. As parameters vary within $\mathbf{G}(j\omega)$ and $\mathbf{H}(j\omega)$, the magnitude of the sensitivity function can be plotted as a function of frequency. The magnitude $|S_\delta^M(j\omega)|$ does not have the limits of 0 to 1 given in Table 6-5 but can vary from 0 to any large magnitude. In Chap. 12 the sensitivity to parameter variation is investigated in detail. That analysis shows that the sensitivity function can be considerably reduced with appropriate feedback included in the system structure.

6-11   SUMMARY

System stability requires that all poles be located in the left half of the s plane. Routh's criterion is a straightforward method for determining system stability. It identifies the necessary restrictions on any adjustable parameter in order to maintain stability.

Table 6-5   SENSITIVITY FUNCTIONS

Case	System variable parameter	S_δ^M
1	$G(s)$	$\dfrac{dC_o/C_o}{dG/G} = 1$
2	$G(s)$	$\dfrac{dC_c/C_c}{dG/G} = \dfrac{1}{1 + G}$
3	$G(s)$	$\dfrac{dC_c/C_c}{dG/G} = \dfrac{1}{1 + GH}$
4	$H(s)$	$\dfrac{dC_c/C_c}{dH/H} = -1$

In this chapter an attempt is made to clarify the distinction between the physical forms that the reference input and the controlled variable may have and the mathematical formulations of these quantities. In a position-control system the error always represents position regardless of whether the input is $r(t) = u_{-1}(t)$, $u_{-2}(t)$, or $u_{-3}(t)$. A system that controls another physical quantity, such as temperature, the error represents temperature or whatever is being controlled. Since, in general, the forward transfer functions of most feedback control systems fall into three categories, they can be identified as Type 0, 1, and 2 systems, with the corresponding definitions of the steady-state error coefficients. These error coefficients are indicative of a system's steady-state performance. Thus, a start has been made in developing a set of standard characteristics.

The definitions and meaning of controllability and observability are presented in this chapter. These definitions play an important role in applying optimization techniques to control systems, as described in Chaps. 14 and 15.

The best location and design of a compensator to improve performance of a system, whose parameters may vary, can be based upon a sensitivity analysis. The sensitivity function analysis is applied to a specific example in Chap. 12, which takes into consideration the fact that the transfer functions are frequency-sensitive.

REFERENCES

1 Guillemin, E. A.: "The Mathematics of Circuit Analysis," Wiley, New York, 1949.
2 Chestnut, H., and R. W. Mayer: "Servomechanisms and Regulating System Design," 2d ed., vol. 1, Wiley, New York, 1959.
3 Athans, M., and P. L. Falb: "Optimal Control," McGraw-Hill, New York, 1966.
4 Gilbert, E. G.: Controllability and Observability in Multi-Variable Control Systems, *J. Soc. Ind. Appl. Math.*, ser. A, *Control*, vol. 1, no. 2, pp. 128–151, 1963.
5 Chen, C. T., and C. A. Desoer: "Proof of Controllability of Jordan Form State Equations," *Trans. IEEE*, vol. AC-13, pp. 195–196, 1968.
6 Sensitivity and Modal Response for Single-Loop and Multiloop Systems, *Flight Control Lab.*, *ASD, AFSC Tech. Doc. Rep.* ASD-TDR-62-812, Wright-Patterson Air Force Base, Ohio, January 1963.
7 Cruz, J. B., Jr.: "Feedback Systems," McGraw-Hill, New York, 1972.
8 Porter, Brian: "Stability Criteria for Linear Dynamical Systems," Academic, New York, 1968.
9 Gantmacher, F. R.: "Applications of the Theory of Matrices," Interscience Division, Wiley, New York, 1959.

7
ROOT LOCUS

7-1 INTRODUCTION

A designer can determine whether his design of a control system meets the specifications if he knows the desired time response of the controlled variable. By deriving the differential equations for the control system and solving them, an accurate solution of the system's performance can be obtained, but this approach is not feasible for other than simple systems. If the response does not meet the specifications, it is not easy to determine from this solution just what physical parameters in the system should be changed to improve the response.

A designer wishes to be able to predict a system's performance by an analysis that does not require the actual solution of the differential equations. Also, he would like this analysis to indicate readily the manner or method by which this system must be adjusted or compensated to produce the desired performance characteristics.

The first thing that a designer wants to know about a given system is whether or not it is stable. This can be determined by examining the roots obtained from the characteristic equation $1 + G(s)H(s) = 0$. Since the work involved in determining the roots can become tedious, a simpler approach is desirable. By applying Routh's criterion to the characteristic equation it is possible in short order to determine whether the system is stable or unstable. Yet this does not satisfy the designer because it does not indicate the degree of stability of the system, i.e., the amount of overshoot and the settling time of the controlled variable. Not only must the system

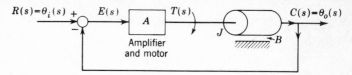

FIGURE 7-1
A position-control system.

be stable, but the overshoot must be maintained within prescribed limits and transients must die out in a sufficiently short time. The graphical methods to be described in this text not only indicate whether a system is stable or unstable but, for a stable system, also show the degree of stability.

There are two basic methods available to a designer. He can choose to analyze and interpret the steady-state sinusoidal response of the transfer function of the system to obtain an idea of the system's response. This method is based upon the interpretation of a Nyquist plot, discussed in more detail in Chaps. 8 to 9. Although this frequency-response approach does not yield an exact quantitative prediction of the system's performance, i.e., the poles of the control ratio $C(s)/R(s)$ cannot be determined, enough information can be obtained to indicate whether the system needs to be adjusted or compensated. Also, the analysis indicates how the system should be compensated.

This chapter deals with the second method, *the root-locus method*,[1,2] devised by W. R. Evans, which incorporates the more desirable features of both the classical method and the frequency-response method. *The root locus is a plot of the roots of the characteristic equation of the closed-loop system as a function of the gain.* This graphical approach yields a clear indication of the effect of gain adjustment with relatively small effort compared with other methods. The underlying principle is that the poles of $C(s)/R(s)$ (transient-response modes) are related to the zeros and poles of the open-loop transfer function $G(s)H(s)$ and also to the gain. An important advantage of the root-locus method is that the roots of the characteristic equation of the system can be obtained directly; this results in a complete and accurate solution of the transient and steady-state response of the controlled variable. Another important feature is that an approximate solution can be obtained with a reduction of the work required. As with any other design technique, a person who has obtained sufficient experience with this method is able to apply it and to synthesize a compensating network, if one is required, with relative ease.

7-2 PLOTTING ROOTS OF A CHARACTERISTIC EQUATION

To give a better insight into the root-locus plots, consider the position-control system shown in Fig. 7-1. Its forward transfer function is

$$G(s) = \frac{\theta_o(s)}{E(s)} = \frac{A/J}{s(s + B/J)} = \frac{K}{s(s + a)} \qquad (7\text{-}1)$$

where $K = A/J$ and $a = B/J$. Assume that $a = 2$. Thus

$$G(s) = \frac{C(s)}{E(s)} = \frac{K}{s(s + 2)} \qquad (7\text{-}2)$$

When the transfer function is expressed with the coefficients of the highest powers of s in both the numerator and the denominator equal to unity, the value of K is defined as the *loop sensitivity*. The control ratio (closed-loop transfer function) is

$$\cdot \quad \frac{C(s)}{R(s)} = \frac{K}{s(s + 2) + K} = \frac{K}{s^2 + 2s + K} = \frac{\omega_n{}^2}{s^2 + 2\zeta\omega_n s + \omega_n{}^2} \qquad (7\text{-}3)$$

where $\omega_n = \sqrt{K}$, $\zeta = 1/\sqrt{K}$, and K is considered to be adjustable from zero to an infinite value.

The problem is to determine the roots of the characteristic equation for all values of K and to plot these roots in the s plane. The roots of the characteristic equation are given by

$$s_{1,2} = -1 \pm \sqrt{1 - K} = -\zeta\omega_n \pm \omega_n\sqrt{\zeta^2 - 1} \qquad (7\text{-}4)$$

For $K = 0$, the roots are $s_1 = 0$ and $s_2 = -2$, which also are the poles of the open-loop transfer function given by Eq. (7-2). When $K = 1$, then $s_{1,2} = -1$. Thus when $0 < K < 1$, the roots $s_{1,2}$ are real and lie on the negative real axis of the s plane between -2 to -1 and -1 to 0, respectively. For the case where $K > 1$, the roots are complex and are given by

$$s_{1,2} = \sigma \pm j\omega_d = -\zeta\omega_n \pm j\omega_n\sqrt{1 - \zeta^2} = -1 \pm j\sqrt{K - 1} \qquad (7\text{-}5)$$

Note that the real part of all the roots is constant for values of $K > 1$.

The roots of the characteristic equation $s^2 + 2s + K = 0$ are determined for a number of values of K (see Table 7-1) and are plotted in Fig. 7-2. Curves are drawn through these plotted points. On these curves, containing two branches, lie all possible roots of the characteristic equation for all values of K from zero to infinity. Note that each branch is calibrated with K as a parameter and the values of K at points on the locus are underlined; the arrows show the direction of increasing values of K.

Table 7-1 LOCATION OF ROOTS FOR THE
CHARACTERISTIC EQUATION
$s^2 + 2s + K = 0$

K	s_1	s_2
0	$-0 + j0$	$-2.0 - j0$
0.5	$-0.293 + j0$	$-1.707 - j0$
0.75	$-0.5 + j0$	$-1.5 - j0$
1.0	$-1.0 + j0$	$-1.0 - j0$
2.0	$-1.0 + j1.0$	$-1.0 - j1.0$
3.0	$-1.0 + j1.414$	$-1.0 - j1.414$

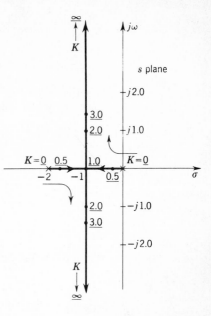

FIGURE 7-2
Plot of all roots of the characteristic
equation $s^2 + 2s + K = 0$ for $0 \le$
$K < \infty$. Values of K are underlined.

These curves are defined as the root-locus plot of Eq. (7-3). Once this plot is obtained,
the roots that best fit the system performance specifications can be selected. Corresponding to the selected roots there is a required value of K which can be determined
from the plot. When the roots have been selected, the time response can be obtained.
Since this process of finding the root locus by calculating the roots for various values
of K becomes tedious for characteristic equations of order higher than second, a
simpler method of obtaining the root locus is desired. The graphical methods for
determining the root-locus plot are the subject of the rest of this chapter.

The value of K is normally considered to be positive. However, it is possible
for K to be negative. For the example in this section, if the value of K is negative,
Eq. (7-4) gives only real roots. Thus the entire locus lies on the real axis, that is,
$0 \le s_1 < +\infty$ and $-2 \ge s_2 > -\infty$ for $0 \ge K > -\infty$. For any negative value
of K there is a root in the right half of the s plane, and the system is unstable.

Once the root locus has been obtained for a control system, it is possible to
determine the variation in system performance with respect to a variation in sensitivity
K. For the example of Fig. 7-1, the control ratio is written in terms of its roots, for
$K > 1$, as

$$\frac{C(s)}{R(s)} = \frac{K}{(s + \sigma - j\omega_d)(s + \sigma + j\omega_d)} \tag{7-6}$$

Note, as defined in Fig. 4-3, that a root with a damping ratio ζ lies on a line making
the angle $\eta = \cos^{-1} \zeta$ with the negative real axis. The damping ratio of several
roots is indicated in Fig. 7-3. Analysis of the root locus reveals the following characteristics for an increase in the gain of the system:

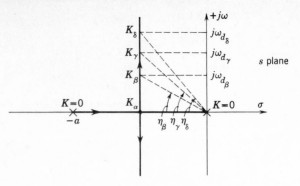

FIGURE 7-3
Root-locus plot of the position-control system of Fig. 7-1.

1 A decrease in the damping ratio ζ. This increases the overshoot of the time response.

2 An increase in the undamped natural frequency ω_n. The value of ω_n is the distance from the origin to the complex root.

3 An increase in the damped natural frequency ω_d. The value of ω_d is the imaginary component of the complex root.

4 No effect on the rate of decay σ; that is, it remains constant for all values of gain equal to or greater than K_α. For more complex systems this will not be the case.

5 The root locus is a vertical line for $K \geq K_\alpha$, and $\sigma = -\zeta\omega_n$ is constant. This means that no matter how much the gain is increased in a linear *simple* second-order system, the system can never become unstable. The time response of this system with a step-function input, for $\zeta < 1$, is of the form

$$c(t) = A_0 + A_1 e^{-\zeta\omega_n t} \sin(\omega_d t + \phi)$$

The root locus of each control system can be analyzed in a similar manner to obtain an idea of the variation in its time response which results from a variation in its loop sensitivity K.

7-3 QUALITATIVE ANALYSIS OF THE ROOT LOCUS

A zero is added to the simple second-order system of the preceding section so that the transfer function is

$$G(s) = \frac{K(s + 1/T_2)}{s(s + 1/T_1)} \qquad (7\text{-}7)$$

The root locus of the control system having this transfer function is shown in Fig. 7-4*b*. When this root locus is compared with that of the original system, shown in Fig. 7-4*a*, it is seen that the branches have been "pulled" to the left, or farther from the imaginary axis. For values of static loop sensitivity greater than K_α, the roots are

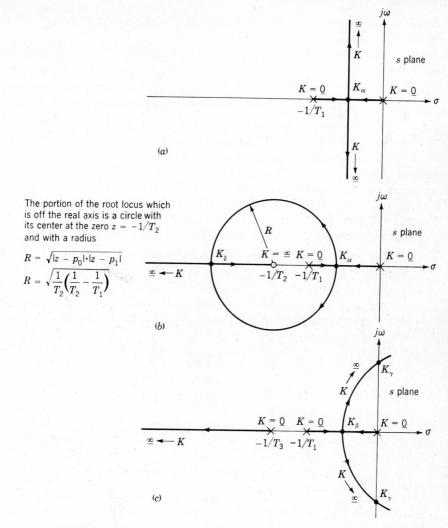

The portion of the root locus which is off the real axis is a circle with its center at the zero $z = -1/T_2$ and with a radius

$$R = \sqrt{|z - p_0| \cdot |z - p_1|}$$

$$R = \sqrt{\frac{1}{T_2}\left(\frac{1}{T_2} - \frac{1}{T_1}\right)}$$

FIGURE 7-4
Various root-locus configurations.
(a) Root locus of basic transfer function:

$$G(s) = \frac{K}{s(s + 1/T_1)} \qquad H(s) = 1$$

(b) Root locus with additional zero:

$$G(s) = \frac{K(s + 1/T_2)}{s(s + 1/T_1)} \qquad H(s) = 1$$

(c) Root locus with additional pole:

$$G(s) = \frac{K}{s(s + 1/T_1)(s + 1/T_3)} \qquad H(s) = 1$$

farther to the left than for the original system. Therefore the transients will decay faster, yielding a more stable system.

If a pole, instead of a zero, is added to the original system the resulting transfer function is

$$G(s) = \frac{K}{s(s + 1/T_1)(s + 1/T_3)} \qquad (7\text{-}8)$$

Figure 7-4c shows the root locus of the control system having this transfer function. Note that the addition of a pole has pulled the locus to the right so that two branches cross the imaginary axis. For values of static loop sensitivity greater than K_β, the roots are closer to the imaginary axis than for the original system. Therefore the transients will decay more slowly, yielding a less stable system. Also for values of $K > K_\gamma$ two of the three roots lie in the right half of the s plane, resulting in an unstable system. The addition of the pole has resulted in a less stable system, compared with the original system. Thus the following general conclusions can be drawn.

1 The addition of a zero to a system has the effect of pulling its root locus to the left, tending to make it a more stable and faster-responding (shorter T_s) system.

2 The addition of a pole to a system has the effect of pulling the root locus to the right, tending to make it a less stable and slower-responding system.

Figure 7-5 illustrates the root-locus configurations for negative-feedback control systems having the following transfer functions:

$$G(s)H(s) = \frac{K(s + 1/T_2)(s + 1/T_4)}{s(s + 1/T_1)(s + 1/T_3)} \qquad (7\text{-}9)$$

$$G(s)H(s) = \frac{K(s + 1/T_2)(s + 1/T_4)(s + 1/T_6)}{(s + 1/T_1)(s + 1/T_3)(s + 1/T_5)} \qquad (7\text{-}10)$$

$$G(s)H(s) = \frac{K(s + 1/T_2)}{s(s - 1/T)(s^2 + 2\zeta\omega_n s + \omega_n^2)} \qquad (7\text{-}11)$$

Note that the third system contains a pole in the right half of the s plane. It represents the performance of an airplane with an autopilot in the longitudinal mode.

The root-locus method is a graphical technique for readily determining the location of all possible roots of a characteristic equation as the gain is varied from zero to infinity. Also, how the locus should be altered in order to improve the system's performance can be readily determined, based upon the knowledge of the effect of the addition of poles or zeros.

7-4 PROCEDURE OUTLINE

To help the reader visualize the order of the root-locus approach, the procedure to be followed in applying this method is first outlined.

Step 1 Derive the open-loop transfer function $G(s)H(s)$ of the system.

Step 2 Factor the numerator and denominator of the transfer function into linear factors of the form $s + a$, where a may be real or complex.

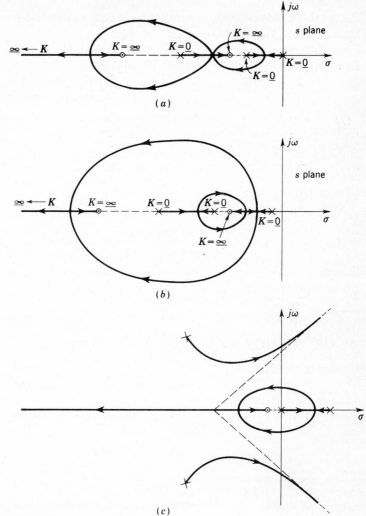

FIGURE 7-5
Various root-locus configurations:

(a) $G(s)H(s) = \dfrac{K(s + 1/T_2)(s + 1/T_4)}{s(s + 1/T_1)(s + 1/T_4)}$

(b) $G(s)H(s) = \dfrac{K(s + 1/T_2)(s + 1/T_4)(s + 1/T_6)}{(s + 1/T_1)(s + 1/T_3)(s + 1/T_5)}$

(c) $G(s)H(s) = \dfrac{K(s + 1/T_2)}{s(s - 1/T)(s^2 + 2\xi\omega_n s + \omega_n^2)}$

Note: These figures are not drawn to scale. Several other root-locus shapes are possible for a given pole-zero arrangement, depending on the specific values of the poles and zeros. Interesting variations of the possible root-locus plots for a given pole-zero arrangement are shown in V. C. M. Yeh, *The Study of Transients in Linear Feedback Systems by Conformal Mapping and Root-Locus Method, Trans. ASME*, vol. 76, pp. 349–361, 1954.

Step 3 Plot the zeros and poles of the open-loop transfer function in the $s = \sigma + j\omega$ plane.

Step 4 The plotted zeros and poles of the open-loop function determine the roots of the characteristic equation of the closed-loop system $[1 + G(s)H(s) = 0]$. By use of the geometrical shortcuts and the Spirule† or a digital-computer program,[3] determine the locus that describes the roots of the closed-loop characteristic equation.

Step 5 Calibrate the locus in terms of the loop sensitivity K. If the gain of the open-loop system is predetermined, the location of the exact roots of $1 + G(s)H(s)$ is immediately known. If the location of the roots is specified, the required value of K can be determined.

Step 6 Once the roots have been found, in step 5, the equation of the system's response can be calculated by taking the inverse Laplace transform, either by analytical or by graphical calculations. Computer programs are also available.[7]

Step 7 If the response does not meet the desired specifications, determine the shape that the root locus must have to meet these specifications.

Step 8 Synthesize the network that must be inserted into the system, if other than gain adjustment is required, to make the required modification on the original locus. This process, called *compensation*, is described in later chapters.

7-5 OPEN-LOOP TRANSFER FUNCTION

In securing the open-loop transfer function, keep the terms in the factored form of $s + a$ or $s^2 + 2\zeta\omega_n s + \omega_n^2$. For unity feedback the open-loop function is equal to the forward transfer function $G(s)$. For nonunity feedback it also includes the transfer function of the feedback path. This open-loop transfer function is of the form

$$G(s)H(s) = \frac{K(s + a_1) \cdots (s + a_h) \cdots (s + a_w)}{s^m(s + b_1)(s + b_2) \cdots (s + b_c) \cdots (s + b_u)} \qquad (7\text{-}12)$$

where a_h and b_c may be real or complex numbers and may lie in either the left-half or right-half s plane. The value of K may be either positive or negative. For example, consider

$$G(s)H(s) = \frac{K(s + a_1)}{s(s + b_1)(s + b_2)} \qquad (7\text{-}13)$$

When the transfer function is in this form (with the coefficients of s all equal to unity), the K is defined as the *loop sensitivity*. By inspection it can be seen that for this example a zero of the open-loop transfer function exists at $s = -a_1$ and the poles are at $s = 0$, $s = -b_1$, and $s = -b_2$. Now let the zeros and poles be denoted by the letters z and p, respectively, i.e.,

$$z_1 = -a_1 \quad z_2 = -a_2 \quad \cdots \quad z_w = -a_w$$
$$p_1 = -b_1 \quad p_2 = -b_2 \quad \cdots \quad p_u = -b_u$$

† Appendix B describes the construction and use of the Spirule for obtaining points on the root locus.

Then Eq. (7-12) can be rewritten as

$$G(s)H(s) = \frac{K(s - z_1) \cdots (s - z_w)}{s^m(s - p_1) \cdots (s - p_u)} = \frac{K \prod\limits_{h=1}^{w} (s - z_h)}{s^m \prod\limits_{c=1}^{u} (s - p_c)} \tag{7-14}$$

where $\prod$ indicates a product of terms. The degree of the numerator is w, and that of the denominator is $m + u = n$.

7-6 POLES OF THE CONTROL RATIO $C(s)/R(s)$

The underlying principle of the root-locus method is that the poles of the control ratio $C(s)/R(s)$ are related to the zeros and poles of the $G(s)H(s)$ function and to the loop sensitivity K. This can be shown as follows. Let

$$G(s) = \frac{N_1(s)}{D_1(s)} \tag{7-15}$$

and

$$H(s) = \frac{N_2(s)}{D_2(s)} \tag{7-16}$$

Then

$$G(s)H(s) = \frac{N_1 N_2}{D_1 D_2} \tag{7-17}$$

Thus

$$\frac{C(s)}{R(s)} = M(s) = \frac{A(s)}{B(s)} = \frac{G(s)}{1 + G(s)H(s)} = \frac{N_1/D_1}{1 + N_1 N_2/D_1 D_2} \tag{7-18}$$

where

$$B(s) \equiv 1 + G(s)H(s) = 1 + \frac{N_1 N_2}{D_1 D_2} = \frac{D_1 D_2 + N_1 N_2}{D_1 D_2} \tag{7-19}$$

Rationalizing Eq. (7-18) gives

$$\frac{N}{D+N} = \frac{C(s)}{R(s)} = M(s) = \frac{N_1 D_2}{D_1 D_2 + N_1 N_2} = \frac{P(s)}{Q(s)} \tag{7-20}$$

From Eqs. (7-19) and (7-20) it is seen that the zeros of $B(s)$ are the poles of $M(s)$ and determine the form of the system's transient response. In terms of $G(s)H(s)$, given by Eq. (7-14), the degree of $B(s)$ is equal to $m + u$; therefore $B(s)$ has $n = m + u$ finite zeros. As shown in Chap. 4, Laplace Transforms, the factors of $Q(s)$ produce transient components of $c(t)$ which fall into the categories shown in Table 7-2. The numerator $P(s)$ of Eq. (7-20) merely modifies the constant multiplier of these transient components. The roots of $B(s) = 0$, which is the characteristic equation of the system, can be determined as follows:

$$B(s) \equiv 1 + G(s)H(s) = 0 \tag{7-21}$$

or

$$G(s)H(s) = \frac{K(s - z_1) \cdots (s - z_w)}{s^m(s - p_1) \cdots (s - p_u)} = -1 \tag{7-22}$$

Thus, as the loop sensitivity K assumes values from zero to infinity, the transfer function $G(s)H(s)$ must always be equal to -1. The corresponding values of s, for any value of K, which satisfy Eq. (7-22) are the poles of $M(s)$. The plots of these values of s are defined as the root locus of $M(s)$.

Conditions that determine the root locus for *positive* values of loop sensitivity are now determined. The general form of $G(s)H(s)$ for any value of s is

$$G(s)H(s) = Fe^{-j\beta}$$

The right side of Eq. (7-22), -1, can be written as

$$-1 = e^{j(1+2h)\pi} \qquad h = 0, \pm 1, \pm 2, \ldots$$

Equation (7-22) is satisfied *only* for those values of s for which

$$Fe^{-j\beta} = e^{j(1+2h)\pi}$$

where

$$F = |G(s)H(s)| = 1$$

and

$$-\beta = (1 + 2h)\pi \qquad (7\text{-}23)$$

From the above it can be concluded that the magnitude of $G(s)H(s)$, a function of the complex variable s, must always be unity and its phase angle must be an odd multiple of π if the particular value of s is to be a zero of $B(s) = 1 + G(s)H(s)$. Consequently, the following two conditions are formalized for the root locus for all positive values of K from zero to infinity:

Magnitude condition:

$$|G(s)H(s)| = 1 \qquad (7\text{-}24)$$

Angle condition:

$$\underline{/G(s)H(s)} = (1 + 2h)180° \qquad \text{for } h = 0, \pm 1, \pm 2, \ldots \qquad (7\text{-}25)$$

$$\mathbf{K > 0}$$

In a similar manner, the conditions for *negative* values of loop sensitivity $(-\infty < K < 0)$ can be determined. [This corresponds to positive feedback,

Table 7-2

Denominator factor of $C(s)$	Corresponding inverse	Form
s	$u_{-1}(t)$	Step function
$s + \dfrac{1}{T}$	$e^{-t/T}$	Decaying exponential
$s^2 + 2\zeta\omega_n s + \omega_n{}^2$	$e^{-\zeta\omega_n t} \sin(\omega_n\sqrt{1 - \zeta^2}\, t + \phi)$ where $\zeta < 1$	Damped sinusoid

$e(t) = r(t) + b(t)$, and positive values of K.] The root locus must satisfy the conditions

Magnitude condition:

$$|G(s)H(s)| = 1 \qquad (7\text{-}26)$$

Angle condition:

$$\underline{/G(s)H(s)} = h360° \qquad \text{for } h = 0, \pm1, \pm2, \ldots \qquad (7\text{-}27)$$

$$\mathbf{K < 0}$$

Thus the root-locus method provides a plot of the variation of each of the poles of $C(s)/R(s)$ in the complex s plane as the loop sensitivity is varied from $K = 0$ to $K = \pm\infty$.

7-7 APPLICATION OF THE MAGNITUDE AND ANGLE CONDITIONS

Once the open-loop transfer function $G(s)H(s)$ has been determined and put into the proper form, the poles and zeros of this function are plotted in the $s = \sigma + j\omega$ plane. As an example, consider

$$G(s)H(s) = \frac{K(s + 1/T_1)^2}{s(s + 1/T_2)(s^2 + 2\zeta\omega_n s + \omega_n{}^2)} = \frac{K(s - z_1)^2}{s(s - p_1)(s - p_2)(s - p_3)} \qquad (7\text{-}28)$$

For the quadratic factor $s^2 + 2\zeta\omega_n s + \omega_n{}^2$ with the damping ratio $\zeta < 1$, the complex-conjugate poles of Eq. (7-28) are

$$p_{2,3} = -\zeta\omega_n \pm j\omega_n\sqrt{1 - \zeta^2} = \sigma \pm j\omega_d$$

The plot of the poles and zeros of Eq. (7-28) is shown in Fig. 7-6. Remember that complex poles or zeros always occur in conjugate pairs, that σ is the damping constant, and ω_d is the damped natural frequency of oscillation. A multiple pole or zero is indicated on the pole-zero diagram by $x]_q$ or $\odot]_q$, where $q = 1, 2, 3, \ldots$ is the order of the pole or zero. With the open-loop poles and zeros plotted, they are now used in the graphical construction of the locus of the poles (the roots of the characteristic equation) of the closed-loop control ratio.

For any particular value (real or complex) of s the terms $s, s - p_1, s - p_2, s - z_1, \ldots$ are complex numbers designating directed line segments. For example, if $s = -4 + j4$ and $p_1 = -1$, then $s - p_1 = -3 + j4$ or

$$|s - p_1| = 5$$

and
$$\phi_1 = \underline{/s - p_1} = 126.8° \qquad \text{(see Fig. 7-6)}$$

In the preceding section it is shown that the roots of the characteristic equation $1 + G(s)H(s) = 0$ are all values of s which satisfy the conditions

$$|G(s)H(s)| = 1 \qquad (7\text{-}29)$$

$$\underline{/G(s)H(s)} = \begin{cases} (1 + 2h)180° & \text{for } K > 0 \\ h360° & \text{for } K < 0 \end{cases}$$

$$\text{for } h = 0, \pm1, \pm2, \ldots \qquad (7\text{-}30)$$

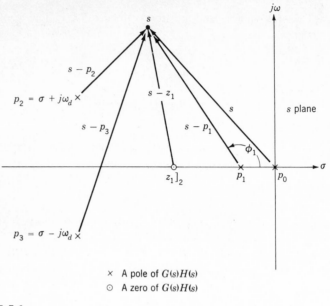

× A pole of $G(s)H(s)$
○ A zero of $G(s)H(s)$

FIGURE 7-6
Pole-zero diagram for Eq. (7-28).

These are labeled as the magnitude and angle conditions, respectively. Therefore, applying these two conditions to the general equation (7-14) results in

$$\frac{|K| \cdot |s - z_1| \cdots |s - z_w|}{|s^m| \cdot |s - p_1| \cdot |s - p_2| \cdots |s - p_u|} = 1 \qquad (7\text{-}31)$$

and

$$-\beta = \underline{/s - z_1} + \cdots + \underline{/s - z_w} - m\underline{/s} - \underline{/s - p_1} - \cdots - \underline{/s - p_u}$$

$$= \begin{cases} (1 + 2h)180° & \text{for } K > 0 \\ h360° & \text{for } K < 0 \end{cases} \qquad (7\text{-}32)$$

where the number of values used for h is $n - w$.
Rewriting these equations gives

$$|K| = \frac{|s^m| \cdot |s - p_1| \cdot |s - p_2| \cdots |s - p_u|}{|s - z_1| \cdots |s - z_w|} = \text{loop sensitivity} \qquad (7\text{-}33)$$

and

$$-\beta = \Sigma(\text{angles of numerator terms}) - \Sigma(\text{angles of denominator terms})$$

$$= \begin{cases} (1 + 2h)180° & \text{for } K > 0 \\ h360° & \text{for } K < 0 \end{cases} \qquad (7\text{-}34)$$

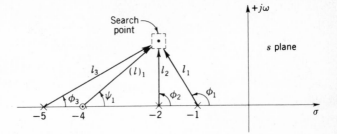

FIGURE 7-7
Construction of the root locus.

All angles are considered positive, measured in the counterclockwise sense. Since $G(s)H(s)$ usually has more poles than zeros, it is convenient to multiply Eq. (7-34) by -1. Since h may be positive or negative, rearranging terms yields

$$\beta = \underline{/\text{denominator}} - \underline{/\text{numerator}}$$

$$= \begin{cases} (1 + 2h)180° & \text{for } K > 0 \\ h360° & \text{for } K < 0 \end{cases} \tag{7-35}$$

Equations (7-33) and (7-35) are in the form generally used in the graphical construction of the root locus. In other words, there are particular values of s for which $G(s)H(s)$ satisfies the angle condition. For a given loop sensitivity only certain of these values of s simultaneously satisfy the magnitude condition. Those values of s which satisfy both the angle and the magnitude conditions are the roots of the characteristic equation and are $n = m + u$ in number. Thus, corresponding to step 4 in Sec. 7-4, the locus of all possible roots is obtained by applying the angle condition. This root locus can be calibrated in terms of the loop sensitivity K by using the magnitude condition.

EXAMPLE

$$G(s)H(s) = \frac{K_0(1 + 0.25s)}{(1 + s)(1 + 0.5s)(1 + 0.2s)} \tag{7-36}$$

Determine the locus of all possible closed-loop poles of $C(s)/R(s)$;

$$G(s)H(s) = \frac{K_0(0.25)}{(0.5)(0.2)} \frac{s + 4}{(s + 1)(s + 2)(s + 5)} \tag{7-37}$$

$$G(s)H(s) = \frac{K(s + 4)}{(s + 1)(s + 2)(s + 5)} \qquad K = 2.5K_0$$

Step 1 The poles and zeros are plotted in Fig. 7-7.
Step 2 In Fig. 7-7 the ϕ's are denominator angles and ψ's are numerator angles. Also, the l's are the lengths of the directed segments stemming from the denominator

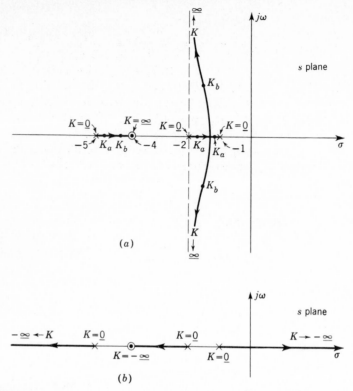

FIGURE 7-8
The complete root locus of Eq. (7-36): (a) for $K > 0$; (b) for $K < 0$.

factors, and (l)'s are the lengths of the directed segments stemming from the numerator factors. After plotting the poles and zeros of the open-loop transfer function, arbitrarily choose a search point. To this point, draw directed line segments from all the open-loop poles and zeros and label as indicated. For this search point to be a point on the locus, the following angle condition must be true:

$$\beta = \phi_1 + \phi_2 + \phi_3 - \psi_1 = \begin{cases} (1 + 2h)180° & \text{for } K > 0 \\ h360° & \text{for } K < 0 \end{cases} \qquad (7\text{-}38)$$

If this equation is not satisfied, select another search point until it is satisfied. Locate a sufficient number of points in the s plane that satisfy the angle condition. In the next section additional information is given that lessens the work involved in this trial-and-error approach.

Step 3 Once the complete locus has been determined by this trial-and-error method, the locus can be calibrated in terms of the loop sensitivity for any root s_1 as follows:

$$|K| = \frac{l_1 l_2 l_3}{(l)_1} \qquad (7\text{-}39)$$

where

$$l_1 = |s_1 + 1| \qquad l_2 = |s_1 + 2| \qquad l_3 = |s_1 + 5| \qquad (l)_1 = |s_1 + 4|$$

In other words, the values of l_1, l_2, l_3, and $(l)_1$ are known for a given point s_1 that satisfies the angle condition; thus the value of $|K|$ for this point can be calculated. The appropriate sign must be given to the magnitude of K compatible with the particular angle condition which was utilized to obtain the root locus. Note that since complex roots must occur in conjugate pairs, the locus is symmetrical about the real axis. Thus the bottom half of the locus can be drawn once the locus above the real axis has been determined. The root locus for this system is shown in Fig. 7-8.

$W(s)$ Plane

From Eq. (7-37)

$$W(s) = u_x + jv_y = \frac{(s + 1)(s + 2)(s + 5)}{s + 4} = -K \qquad (7\text{-}40)$$

The line $u_x = -K$ in the $W(s)$ plane maps into the curves indicated in Fig. 7-8. That is, for each value of u_x in the $W(s)$ plane there is a particular value or a set of values of s in the s plane.

7-8 GEOMETRICAL PROPERTIES (CONSTRUCTION RULES)

To facilitate the application of the root-locus method, the following rules are established for $K > 0$. These rules are based upon the interpretation of the angle condition and the analysis of the characteristic equation. The reader should be able to extend these rules for the case where $K < 0$. The rules for both $K > 0$ and $K < 0$ are listed in Sec. 7-14 for easy reference.

The rules presented aid in obtaining the root locus by expediting the manual plotting of the locus. The root locus can also be obtained by various automatic methods using the analog or digital computer.[3] For automatic plotting these rules provide *checkpoints* to ensure that the computer solution is correct.

Rule 1 Number of branches of the locus The characteristic equation $B(s) = 0$ is of degree $n = m + u$; therefore there are n roots, which are continuous functions of the open-loop sensitivity K. As K is varied from zero to infinity, each root traces a continuous curve. Since there are n roots, there are the same number of curves or branches in the complete root locus. Since the degree of the polynomial $B(s)$ is determined by the poles of the open-loop transfer function, the *number of branches of the root locus is equal to the number of poles of the open-loop transfer function.*

Rule 2 Real-axis locus In Fig. 7-9 are shown a number of open-loop poles and zeros. If the angle condition is applied to any search point such as s_1 on the real axis,

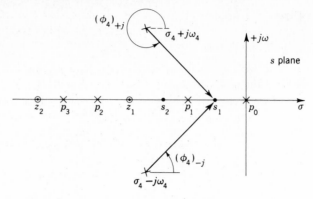

FIGURE 7-9
Determination of the real-axis locus.

the angular contribution of all the poles and zeros on the real axis to the left of this point is zero. The angular contribution of the complex-conjugate poles to this point is 360°. (This is also true for complex-conjugate zeros.) Finally, the poles and zeros on the real axis to the right of this point each contribute 180° (with the appropriate sign included). From Eq. (7-35) the angle of $G(s)H(s)$ to the point s_1 is given by

$$\phi_0 + \phi_1 + \phi_2 + \phi_3 + (\phi_4)_{+j} + (\phi_4)_{-j} - (\psi_1 + \psi_2) = (1 + 2h)180° \qquad (7\text{-}41)$$

or

$$180° + 360° = (1 + 2h)180° \qquad (7\text{-}42)$$

Therefore s_1 is a point on a locus. Similarly, it can be shown that the point s_2 is not a point on the locus. It can be seen that the poles and zeros to the left of the s point and the 360° contributed by the complex-conjugate poles or zeros do not affect the odd-multiple-of-180° requirement. Thus, *if the total number of real poles and zeros to the right of a search point s on the real axis is odd, this point lies on the locus.* In Fig. 7-9 the root locus exists on the real axis from 0 to p_1, z_1 to p_2, and p_3 to z_2.

All points on the real axis between z_1 and p_2 in Fig. 7-9 satisfy the angle condition and are therefore points on the root locus. However, there is no guarantee that this section of the real axis is part of just one branch. Figure 7-5b and Prob. 7-4 illustrate the situation where the part of the real axis between a pole and a zero is divided into three sections which are parts of three different branches.

Rule 3 Locus end points The magnitude of the loop sensitivity which satisfies the magnitude condition is given by Eq. (7-33) and has the general form

$$|W(s)| = K = \frac{\displaystyle\prod_{c=1}^{n} |s - p_c|}{\displaystyle\prod_{h=1}^{w} |s - z_h|} \qquad (7\text{-}43)$$

When it is remembered that the numerator and denominator factors of Eq. (7-43) are the poles and zeros, respectively, of the open-loop transfer function, the following conclusions can be drawn:

1 When $s = p_c$, the loop sensitivity K is zero.

2 When $s = z_h$, the loop sensitivity K is infinite. When the numerator of Eq. (7-43) is of higher order than the denominator, $s = \infty$ also makes K infinite, thus being equivalent in effect to a zero.

Thus it can be said that the locus starting points ($K = 0$) are at the open-loop poles and that the locus ending points ($K = \infty$) are at the open-loop zeros (the point at infinity being considered as an equivalent zero of multiplicity equal to $n - w$).

Rule 4 Asymptotes of locus as *s* approaches infinity Plotting the locus is greatly facilitated if one can determine the asymptotes approached by the various branches as s takes on large values. Taking the limit of $G(s)H(s)$ as s approaches infinity, based on Eq. (7-15), yields

$$\lim_{s \to \infty} G(s)H(s) = \lim_{s \to \infty} \left[K \frac{\prod_{h=1}^{w} (s - z_h)}{\prod_{c=1}^{n} (s - p_c)} \right] = \frac{K}{s^{n-w}} = -1 \qquad (7\text{-}44)$$

It must be remembered that K in Eq. (7-44) is still a variable in the manner prescribed previously, thus allowing the magnitude condition to be met. Therefore,

$$-K = s^{n-w} \qquad (7\text{-}45)$$

$$|-K| = |s^{n-w}| \qquad \text{magnitude condition} \qquad (7\text{-}46)$$

$$\underline{/-K} = \underline{/s^{n-w}} = (1 + 2h)180° \qquad \text{angle condition} \qquad (7\text{-}47)$$

Rewriting Eq. (7-47) gives

$$(n - w) \underline{/s} = (1 + 2h)180°$$

or
$$\gamma = \frac{(1 + 2h)180°}{n - w} \qquad \text{as } s \to \infty \qquad (7\text{-}48)$$

There are $n - w$ asymptotes of the root locus and their angles are given by

$$\gamma = \frac{(1 + 2h)180°}{[\text{number of poles of } G(s)H(s)] - [\text{number of zeros of } G(s)H(s)]} \qquad (7\text{-}49)$$

Equation (7-49) reveals that, no matter what magnitude s may have, after a sufficiently large value has been reached, the argument (angle) of s on the root locus remains constant. For a search point that has a sufficiently large magnitude the open-loop poles and zeros appear to it as if they had collapsed into a single point. Therefore the branches are asymptotic to straight lines whose slopes and directions are given by Eq. (7-49) (see Fig. 7-10). These asymptotes usually do not go through the origin. The correct real-axis intercept of the asymptotes is obtained from Rule 5.

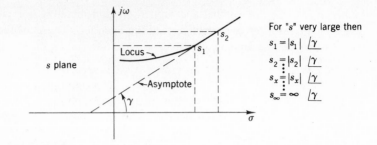

FIGURE 7-10
Asymptotic condition for large values of s.

Rule 5   Real-axis intercept of the asymptotes   The real-axis crossing σ_o of the asymptotes can be obtained by applying the theory of equations. The result is

$$\sigma_o = \frac{\sum\limits_{c=1}^{n} \text{Re}\,(p_c) - \sum\limits_{h=1}^{w} \text{Re}\,(z_h)}{n - w} \tag{7-50}$$

The asymptotes are not dividing lines, and a locus may cross its asymptote. It may be valuable to know from which side the root locus approaches its asymptote. Lorens and Titsworth[4] present a method for obtaining this information which shows that the locus lies exactly along the asymptote if the pole-zero pattern is symmetric about the asymptote line extended through the point σ_o.

Rule 6   Breakaway point on the real axis   It has been shown that the locus starts at the poles where $K = 0$ and ends at the zeros which are finite or at $s = \infty$. Consider now the case where the locus has branches on the real axis between two poles (between p_0 and p_1 in Fig. 7-11a and between p_2 and p_3 in Fig. 7-11b). There must be a point at which the two branches break away from the real axis and enter the complex region of the s plane in order to approach zeros or the point at infinity. For two finite zeros (or one finite zero and one at infinity) (see Fig. 7-11) the branches are coming from the complex region and enter the real axis. In Fig. 7-11 between two poles there is a point s_1 for which the loop sensitivity K_z is greater than for points on either side of s_1 on the real axis. In other words, since K starts with a value of zero at the poles and increases in value as the locus moves away from the poles, there is a point somewhere in between where the K's for the two branches simultaneously reach a maximum value. This point is called the *breakaway point*. Plots of K vs. σ utilizing Eq. (7-33) are shown in Fig. 7-11 for the portions of the root locus which exist on the real axis for $K > 0$. The point s_2 for which the value of K is a minimum between two zeros is called the *break-in point*. The breakaway and break-in points can easily be calculated for an open-loop pole-zero combination for which the derivative of $W(s) = -K$ is of the second order. As an example, if

$$G(s)H(s) = \frac{K}{s(s + 1)(s + 2)} \tag{7-51}$$

then $$W(s) = s(s + 1)(s + 2) = -K$$

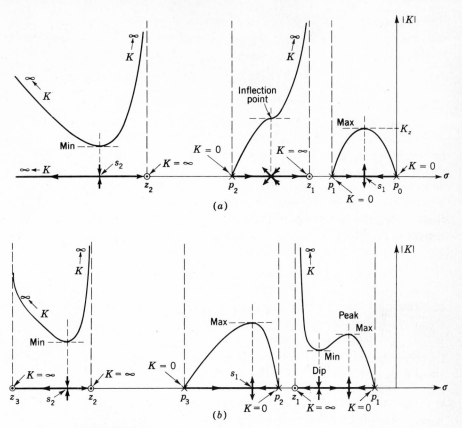

FIGURE 7-11
The plots of K and the corresponding real-axis locus for (a) Fig. 7-5a and (b) Fig. 7-5b.

Multiplying the factors together gives

$$W(s) = s^3 + 3s^2 + 2s = -K \qquad (7\text{-}52)$$

When $s^3 + 3s^2 + 2s$ is a minimum, $-K$ is a minimum and K is a maximum. Thus by taking the derivative of this function and setting it equal to zero, the points s_1 can be determined:

$$\frac{dW(s)}{ds} = 3s^2 + 6s + 2 = 0 \qquad (7\text{-}53)$$

or $\qquad\qquad s_{1,2} = -1 \pm 0.5743 = -0.4257, \, -1.5743$

Since the breakaway point for $K > 0$ must lie between $s = 0$ and $s = -1$ in order to satisfy the angle condition, the value of s_1 is

$$s_1 = -0.4257 \qquad (7\text{-}54)$$

(The other point, $s_2 = -1.5743$, is the break-in point on the root locus for $K < 0$.) Substituting $s_1 = -0.4257$ into Eq. (7-52) gives the value of K at the breakaway point for $K > 0$ as

$$-K = (-0.426)^3 + (3)(-0.426)^2 + (2)(-0.426) \qquad (7\text{-}55)$$

$$K = 0.385$$

When the derivative of $W(s)$ is of higher order than 2, a digital-computer program can be used to calculate the roots of the numerator polynomial of $dW(s)/ds$; these roots locate the breakaway and break-in points. An alternate method for finding the breakaway and break-in points is the "hills-and-dales" method, i.e., plotting the value of K vs. real values of s. This plot permits the determination of the maxima (breakaway) and minima (break-in) without differentiation. Note that it is also possible to have both a breakaway and a break-in point between a pole and zero (finite or infinite) on the real axis, as shown in Figs. 7-5 and 7-11. The plot of K vs. σ for a locus between a pole and zero falls into one of the following categories:

1 The plot clearly indicates a peak and a dip, as illustrated between p_1 and z_1 in Fig. 7-11b. The peak represents a "maximum" value of K that satisfies the condition for a breakaway point. The dip represents a "minimum" value of K that satisfies the condition for a break-in point.

2 The plot does not indicate a dip-and-peak combination but contains an inflection point. An inflection point occurs when the breakaway and break-in points coincide, as is the case between p_2 and z_1 in Fig. 7-11a.

3 The plot does not indicate a dip-and-peak combination and clearly indicates the absence of any possibility of the existence of an inflection point. For this situation there are no break-in or breakaway points.

Rule 7 Complex pole (or zero): angle of departure The next geometrical shortcut is the rapid determination of the direction in which the locus leaves a complex pole or enters a complex zero. Although in Fig. 7-12 a complex pole is considered, the results also hold for a complex zero.

In Fig. 7-12a, an area about p_2 is chosen so that l_2 is very much smaller than l_0, l_1, l_3, and $(l)_1$. For illustrative purposes, this area has been enlarged many times in Fig. 7-12b. Under these conditions the angular contributions from all the other poles and zeros, except p_2, to a search point anywhere in this area are approximately constant. They can be considered to have values determined as if the search point were right at p_2. When the angle condition is applied to this small area, the angle equation is

$$\phi_0 + \phi_1 + \phi_2 + \phi_3 - \psi_1 = (1 + 2h)180° \qquad (7\text{-}56)$$

or

$$\phi_2 = (1 + 2h)180° - (\phi_0 + \phi_1 + 90° - \psi_1)$$

In a similar manner the angle of approach to a complex zero can be determined. For an open-loop transfer function having the pole-zero arrangement shown in Fig. 7-13, the angle of approach ψ_1 to z_1 is given by

$$\psi_1 = (\phi_0 + \phi_1 + \phi_2 - 90°) - (1 + 2h)180° \qquad (7\text{-}57)$$

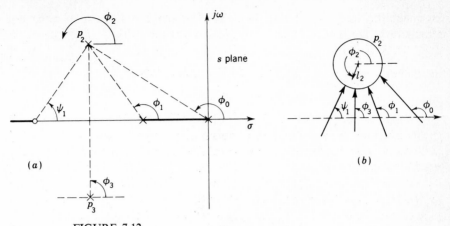

FIGURE 7-12
Angular condition in the vicinity of a complex pole.

In other words, the direction of the locus as it leaves a pole or zero can be determined by adding up, according to the angle condition, all the angles of all vectors from all the other poles and zeros to the pole or zero in question. Subtracting this sum from $(1 + 2h)180°$ gives the required direction.

Rule 8 Imaginary-axis crossing point In cases where the locus crosses the imaginary axis into the right-half s plane, the crossover point can usually be determined by Routh's method or by similar means. For example, if the closed-loop characteristic equation $D_1 D_2 + N_1 N_2 = 0$ is of the form

$$s^3 + bs^2 + cs + Kd = 0$$

the Routhian array is

$$
\begin{array}{c|cc}
s^3 & 1 & c \\
s^2 & b & Kd \\
s^1 & (bc - Kd)/b \\
s^0 & Kd \\
\end{array}
$$

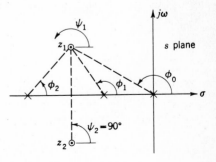

FIGURE 7-13
Angular condition in the vicinity of a complex zero.

An undamped oscillation may exist if the s^1 row in the table equals zero. For this condition the auxiliary equation obtained from the s^2 row is

$$bs^2 + Kd = 0 \qquad (7\text{-}58)$$

and its roots are

$$s_{1,2} = \pm j \sqrt{\frac{Kd}{b}} = \pm j\omega_n \qquad (7\text{-}59)$$

The loop sensitivity term K is determined by setting the s^1 row to zero:

$$K = \frac{bc}{d} \qquad (7\text{-}60)$$

For $K > 0$, Eq. (7-59) gives the natural frequency of the undamped oscillation. This corresponds to the point on the imaginary axis where the locus crosses over into the right-half s plane. The imaginary axis divides the s plane into stable and unstable regions. Also, the value of K from Eq. (7-60) determines the value of the loop sensitivity at the crossover point. For values of $K < 0$ the term in the s^0 row is negative, thus characterizing an unstable response. The limiting values for a stable response are therefore

$$0 < K < \frac{bc}{d} \qquad (7\text{-}61)$$

In like manner, the crossover point can be determined for higher-order characteristic equations. For these higher-order systems care must be exercised in *analyzing all terms in the first column that contain the term K* in order to obtain the correct limiting values of gain.

Rule 9   Nonintersection or intersection of root-locus branches[5]   By utilizing the theory of complex variables, the following properties are evolved:

1   A value of s which satisfies the angle condition of Eq. (7-35) is a point on the root locus. If $dW(s)/ds \neq 0$ at this point, there is one and only one branch of the root locus through the point. Thus it can be said that there are no root-locus intersections at this point.

2   If the first $y - 1$ derivatives of $W(s)$ vanish at a given point on the root locus, there will be y branches approaching and y branches leaving this point. Thus, it can be said that there are root-locus intersections at this point. The angle between two adjacent *approaching* branches is given by

$$\lambda_y = \pm \frac{360°}{y} \qquad (7\text{-}62)$$

Also, the angle between a branch *leaving* and an adjacent branch that is *approaching* the same point is given by

$$\theta_y = \pm \frac{180°}{y} \qquad (7\text{-}63)$$

Figure 7-14 illustrates these angles.

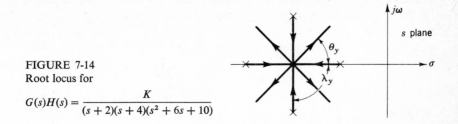

FIGURE 7-14
Root locus for

$$G(s)H(s) = \frac{K}{(s+2)(s+4)(s^2+6s+10)}$$

Rule 10 Conservation of the sum of the system roots[6] The technique described by this rule aids in the determination of the general shape of the root locus. Consider the general open-loop transfer function in the form

$$G(s)H(s) = \frac{K \prod\limits_{h=1}^{w} (s - z_h)}{s^m \prod\limits_{c=1}^{u} (s - p_c)} \qquad (7\text{-}64)$$

By recalling that for physical systems $w \le n = u + m$, the denominator of $C(s)/R(s)$ can be written

$$B(s) = 1 + G(s)H(s) = \frac{\prod\limits_{j=1}^{n} (s - r_j)}{s^m \prod\limits_{c=1}^{u} (s - p_c)} \qquad (7\text{-}65)$$

where r_j are the roots described by the root locus.

Substituting from Eq. (7-64) into Eq. (7-65) and equating numerators on each side of the resulting equation yields

$$s^m \prod\limits_{c=1}^{u} (s - p_c) + K \prod\limits_{h=1}^{w} (s - z_h) = \prod\limits_{j=1}^{n} (s - r_j) \qquad (7\text{-}66)$$

Expanding both sides of this equation gives

$$\left(s^n - \sum\limits_{c=1}^{u} p_c s^{n-1} + \cdots \right) + K \left(s^w - \sum\limits_{h=1}^{w} z_h s^{w-1} + \cdots \right)$$

$$= s^n - \sum\limits_{j=1}^{n} r_j s^{n-1} + \cdots \qquad (7\text{-}67)$$

For those open-loop transfer functions which satisfy the condition $w \le n - 2$, the following is obtained by equating the coefficients of s^{n-1} of Eq. (7-67):

$$\sum\limits_{c=1}^{u} p_c = \sum\limits_{j=1}^{n} r_j$$

Since m open-loop poles have values of zero, this equation can also be written

$$\sum\limits_{j=1}^{n} p_j = \sum\limits_{j=1}^{n} r_j \qquad (7\text{-}68)$$

where p_j now represents all the open-loop poles, including those at the origin, and r_j are the roots of the characteristic equation. This equation reveals that as the system gain is varied from zero to infinity, the sum of the system roots is constant. In other words, the sum of the system roots is conserved and is independent of K. When a system has several root-locus branches which go to infinity (as $K \rightarrow \infty$), the directions of the branches are such that the sum of the roots is constant. A branch going to the right therefore requires that there will be a branch going to the left. The root locus of Fig. 7-2 satisfies the conservancy law for the root locus. The sum of the roots is a constant for all values of K.

Rule 11   Determination of roots on the root locus   After the root locus has been plotted, the specifications for system performance are used to determine the dominant roots (the roots closest to the imaginary axis). For the dominant roots the required loop sensitivity can be determined by applying the magnitude condition, as shown in Eq. (7-33). The remaining roots on each of the other branches can be determined by any of the following methods:

Method 1   Trial-and-error search for a point on each branch of the locus that satisfies the loop sensitivity for the dominant roots.

Method 2   If all except one real or a complex pair of roots are known, either of the following procedures can be used.

Procedure 1   Divide the characteristic equation by the factors representing the known roots. The remainder gives the remaining roots. One should not expect an exact divison because of the inaccuracies introduced in graphically determining the poles and the loop sensitivity.

Procedure 2   Equation (7-68), known as *Grant's rule*, can be used to find some of the roots. A necessary condition is that the denominator of $G(s)H(s)$ be at least of degree 2 higher than the numerator. If all the roots except one real root are known, application of Eq. (7-68) yields directly the value of the real root. However, for complex roots of the form $r = \sigma \pm j\omega_d$ it yields only the value of its real component σ. The magnitude of these roots is determined from the relationship

$$|r|^2 = K_m(-1)^m \frac{\prod\limits_{c=1}^{u} (p_c)}{\prod\limits_{j=1}^{n-2} (r_j)} \qquad m \geq 1 \qquad (7\text{-}69)$$

where K_m = steady-state error coefficient (gain constant)
p_c = poles of $G(s)H(s)$, excluding poles at origin
r_j = known roots of characteristic equation
The imaginary part of the roots can now be determined from

$$\omega_d = (|r|^2 - \sigma^2)^{1/2} \qquad (7\text{-}70)$$

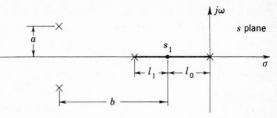

FIGURE 7-15
Location of the breakaway point.

Digital-computer root-locus programs[3] are available that yield the loop sensitivity and the roots of the characteristic equation of the system. When the damping ratio is specified for the dominant roots, these roots determine the value of K and the remaining roots.

7-9 EXAMPLE

Given:

$$G(s) = \frac{K_1}{s(s^2/2{,}600 + s/26 + 1)} \quad \text{and} \quad H(s) = \frac{1}{0.04s + 1}$$

Rearranging gives

$$G(s) = \frac{2{,}600K_1}{s(s^2 + 100s + 2{,}600)} = \frac{N_1}{D_1} \quad \text{and} \quad H(s) = \frac{25}{s + 25} = \frac{N_2}{D_2}$$

Thus

$$G(s)H(s) = \frac{65{,}000K_1}{s(s + 25)(s^2 + 100s + 2{,}600)}$$

Find $C(s)/R(s)$ with $\zeta = 0.5$ for the dominant roots (roots closest to the imaginary axis).

1. The poles of $G(s)H(s)$ are plotted on the s plane in Fig. 7-15; the values of these poles are $s = 0, -25, -50 + j10, -50 - j10$. The system is completely unstable for $K < 0$. Therefore this example is solved only for the condition $K > 0$.
2. There are four branches of the root locus.
3. The locus exists on the real axis between 0 and -25.
4. The angles of the asymptotes are

$$\gamma = \frac{(1 + 2h)180°}{4} = \pm 45°, \pm 135°$$

5. The real-axis intercept of the asymptotes is

$$\sigma_o = \frac{0 - 25 - 50 - 50}{4} = -31.25$$

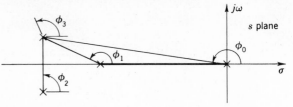

FIGURE 7-16
Determination of the departure angle.

6. The breakaway point s_1 (see Fig. 7-15) on the real axis between 0 and -25 is found by using a Spirule to evaluate the maximum K [see Eq. (7-33)]:

$$s_1 = 9.0$$

7. The angle of departure ϕ_3 (see Fig. 7-16) from the pole $-50 + j10$ is obtained from

$$\phi_0 + \phi_1 + \phi_2 + \phi_3 = (1 + 2h)180°$$
$$168.7° + 158.2° + 90° + \phi_3 = (1 + 2h)180°$$
$$\phi_3 = 123.1°$$

Similarly, the angle of departure from the pole $-50 - j10$ is $-123.1°$.

8. The imaginary-axis intercepts are obtained from

$$\frac{C(s)}{R(s)} = \frac{2{,}600K_1(s + 25)}{s^4 + 125s^3 + 5{,}100s^2 + 65{,}000s + 65{,}000K_1} \qquad (7\text{-}71)$$

The Routhian array for the denominator of $C(s)/R(s)$, which is the characteristic polynomial, is

s^4	1	5,100	$65{,}000K_1$
s^3	1	520 (after division by 125)	
s^2	1	$14.2K_1$ (after division by 4,580)	
s^1	$520 - 14.2K_1$		
s^0	$14.2K_1$		

Pure imaginary roots exist when the s^1 row is zero. This occurs when $K_1 = 520/14.2 = 36.7$. The auxiliary equation is

$$s^2 + 14.2K_1 = 0$$

and the imaginary roots are

$$s = \pm j\sqrt{14.2K_1} = \pm j\sqrt{520} = \pm j22.8$$

9. Additional points on the root locus are found by use of the Spirule, locating points that satisfy the angle condition

$$\underline{/s} + \underline{/s + 25} + \underline{/s + 50 - j10} + \underline{/s + 50 + j10} = (1 + 2m)180°$$

The locus is shown in Fig. 7-17.

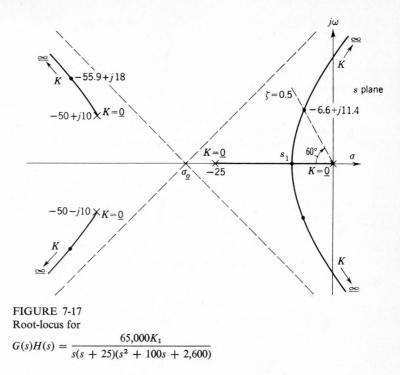

FIGURE 7-17
Root-locus for

$$G(s)H(s) = \frac{65,000K_1}{s(s + 25)(s^2 + 100s + 2,600)}$$

10. The radial line for $\zeta = 0.5$ is drawn on the graph of Fig. 7-17 at the angle (see Fig. 4-3 for definition of η)

$$\eta = \cos^{-1} 0.5 = 60°$$

The dominant roots obtained from the graph are

$$s = -6.6 \pm j11.4$$

11. The gain is obtained graphically from the expression

$$K = 65,000K_1 = |s| \cdot |s + 25| \cdot |s + 50 - j10| \cdot |s + 50 + j10|$$

For $s = -6.6 + j11.4$,

$$K = 65,000K_1 = 600,900$$

$$K_1 = 9.24$$

12. The other roots are evaluated to satisfy the magnitude condition $K = 600,900$. By trial and error the other roots of the characteristic equation are

$$s = -55.9 \pm j18.0$$

12. *Alternative methods* The other roots can also be obtained by dividing the characteristic equation by the quadratic factor representing the two dominant roots. With $K_1 = 9.24$ the characteristic equation is

$$s^4 + 125s^3 + 5,100s^2 + 65,000s + 600,900 = 0$$

The quadratic factor representing the dominant roots is

$$(s + 6.6 - j11.4)(s + 6.6 + j11.4) = s^2 + 13.2s + 173.5$$

Dividing the characteristic equation by this quadratic factor leaves the quadratic representing the other roots as

$$s^2 + 112s + 3,450$$

There is a remainder when this division is performed which is due to the graphical inaccuracy. The other roots obtained from this factor are

$$s = -56 \pm j18$$

The values of the other roots obtained by both methods are almost equal.

The additional roots can also be determined by using Grant's rule. From Eq. (7-68),

$$0 - 25 + (-50 + j10) + (-50 - j10)$$
$$= (-6.6 + j11.4) + (-6.6 - j11.4) + (\sigma + j\omega_d) + (\sigma - j\omega_d)$$

This gives

$$\sigma = -55.9$$

By using this value, the roots can be determined from the root locus as $-55.9 \pm j18.0$. If these branches of the locus have not been drawn, proceed as shown below.

From Eq. (7-69),

$$|r|^2 = (9.24)(-1)^1 \frac{(-25)(-50 + j10)(-50 - j10)}{(-6.6 + j11.4)(-6.6 - j11.4)} = 3,525$$

From Eq. (7-70),

$$\omega_d = (3,525 - 3,140)^{1/2} = 19.6$$

This gives the roots as $-55.9 \pm j19.6$, which are within the expected graphical accuracy.

13. The control ratio, using values of the roots obtained in steps 10 and 12, is

$$\frac{C(s)}{R(s)} = \frac{N_1 D_2}{\text{factors as determined from root locus}}$$

$$= \frac{24,040(s + 25)}{(s + 6.6 + j11.4)(s + 6.6 - j11.4)}$$
$$\times (s + 55.9 + j18)(s + 55.9 - j18)$$

$$= \frac{24,040(s + 25)}{(s^2 + 13.2s + 173.5)(s^2 + 111.8s + 3,450)}$$

14. The response $c(t)$ for a unit step input is found from[7]

$$C(s) = \frac{24,040(s + 25)}{s(s^2 + 13.2s + 173.5)(s^2 + 111.8s + 3,450)}$$

$$= \frac{A_0}{s} + \frac{A_1}{s + 6.6 - j11.4} + \frac{A_2}{s + 6.6 + j11.4}$$

$$+ \frac{A_3}{s + 55.9 - j18} + \frac{A_4}{s + 55.9 + j18}$$

The constants are

$$A_0 = 1.0 \qquad A_1 = 0.604 \,\underline{/-201.7^\circ} \qquad A_3 = 0.14 \,\underline{/-63.9^\circ}$$

Note that A_0 must be exactly 1 since $G(s)$ is Type 1 and the gain of $H(s)$ is unity. It can also be obtained from Eq. (7-71) for $C(s)/R(s)$ in step 8. Inserting $R(s) = 1/s$ and finding the final value gives $c(t)_{ss} = 1.0$. The response $c(t)$ is

$$c(t) = 1 + 1.21^{-6.6t} \sin (11.4t - 111.7^\circ) + 0.28e^{-55.9t} \sin (18t + 26.1^\circ) \qquad (7\text{-}72)$$

A plot of $c(t)$ is shown in Fig. 7-23.

7-10 FREQUENCY RESPONSE

Once the root locus has been determined and the system gain has been set for the desired performance, it is very easy to determine the steady-state frequency response. The frequency response gives the ratio of phasor output to phasor input for sinusoidal inputs over a band of frequencies. The plots of the magnitude M and the angle α of $C(j\omega)/R(j\omega)$ vs. the frequency ω define the frequency response of a control system. These curves are very useful in control-system design:

    1 They enable a designer to minimize the effect of any undesirable noise within the system.

    2 They present a qualitative picture of the system's transient response. The correlation between the transient and frequency responses is discussed in detail in Chap. 9.

For the problem of Sec. 7-9 the closed-loop control ratio is

$$\frac{C(s)}{R(s)} = \frac{24{,}040(s + 25)}{(s + 6.6 - j11.4)(s + 6.6 + j11.4)(s + 55.9 - j18)(s + 55.9 + j18)}$$

$$(7\text{-}73)$$

To obtain the steady-state frequency response, let s assume values equal to $j\omega$. The control ratio as a function of frequency is

$$\frac{C(j\omega)}{R(j\omega)} = \frac{24{,}040(j\omega + 25)}{\begin{array}{c}(j\omega + 6.6 - j11.4)(j\omega + 6.6 + j11.4) \\ \times (j\omega + 55.9 - j18)(j\omega + 55.9 + j18)\end{array}} \qquad (7\text{-}74)$$

For any frequency ω_1 each of the factors of Eq. (7-74) is a directed line segment and can be measured on the pole-zero plot, as shown in Fig. 7-18. When the magnitudes and angles for each term of Eq. (7-74) have been obtained, the value of $C(j\omega)/R(j\omega)$ can be determined. Note that the angles are measured as shown in Fig. 7-18, where clockwise angles are negative and counterclockwise angles are positive. This procedure is repeated for a sufficient number of values of frequency to draw a plot of $C(j\omega)/R(j\omega)$ vs. ω. Figure 7-19 shows a plot of the magnitude $|C(j\omega)/R(j\omega)|$ vs. ω obtained from the pole-zero plot of Fig. 7-18. The angle curve of $C(j\omega)/R(j\omega)$ is also evaluated graphically but is not shown here. The maximum value of the magnitude

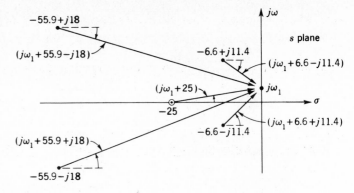

FIGURE 7-18
Frequency response from the plot, in the s plane, of the closed-loop poles and zeros.

curve is labeled M_m, and the frequency at which this occurs is labeled ω_m. In Chap. 9 these quantities are related to the time response.

7-11 PERFORMANCE CHARACTERISTICS

As pointed out early in this chapter, the root-locus method incorporates the more desirable features of both the classical method and the steady-state sinusoidal phasor analysis. In the example of Sec. 7-9 a direct relationship is noted between the root locus and the time solution. This section is devoted to strengthening this relationship to enable the designer to synthesize and/or compensate a system.

General Introduction

In review, consider a simple second-order system whose control ratio is

$$\frac{C(s)}{R(s)} = \frac{K}{s^2 + 2\zeta\omega_n s + \omega_n^2} \tag{7-75}$$

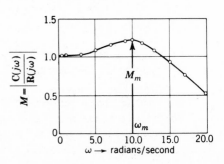

FIGURE 7-19
Closed-loop frequency response for the system of Sec. 7-10.

and whose transient component of the response to a step input is

$$c(t)_t = C_1 e^{s_1 t} + C_2 e^{s_2 t} \qquad (7\text{-}76)$$

where
$$s_1 = -\zeta\omega_n + j\omega_n\sqrt{1 - \zeta^2} = \sigma + j\omega_d \qquad \zeta < 1$$
$$s_2 = -\zeta\omega_n - j\omega_n\sqrt{1 - \zeta^2} = \sigma - j\omega_d \qquad (7\text{-}77)$$

Thus
$$c(t)_t = Ae^{\sigma t} \sin(\omega_d t + \phi) \qquad (7\text{-}78)$$

Consider now a plot of the roots in the s plane and their correlation with the transient solution in the time domain for a step input. In Fig. 7-20 are illustrated six cases for different values of damping, showing both the locations of the roots and the corresponding transient plots.

In the case of $\zeta = 0$ the roots lie on the $\pm j\omega$ axis. For this case there are sustained oscillations. Those portions of the locus which yield roots in the right-half s plane result in unstable operation. Thus, the desirable roots are on that portion of the locus in the left-half s plane. (As shown in Chap. 9, the $\pm j\omega$ axis of the s plane corresponds to the $-1 + j0$ point of the Nyquist plot.)

Table 4-1 summarizes the information available from Fig. 7-20, i.e., the correlation between the location of the closed-loop poles and the corresponding transient component of the response. Thus the value of the root-locus method is that it is possible to determine all the forms of the transient component of the response that a control system may have.

Plot of Characteristic Roots for $0 < \zeta < 1$

The important desired roots of the time solution lie in the region in which $0 < \zeta < 1$ (generally between 0.4 and 0.8). In Fig. 7-20a, the radius r from the origin to the root s_1 is

$$r = \sqrt{\omega_{d_1}{}^2 + \sigma_{1,2}{}^2} = \sqrt{\omega_n{}^2(1 - \zeta^2) + \omega_n{}^2\zeta^2} = \omega_n \qquad (7\text{-}79)$$

and
$$\cos\eta = \left|\frac{-\sigma_{1,2}}{r}\right| = \frac{\zeta\omega_n}{\omega_n} = \zeta \qquad (7\text{-}80)$$

or
$$\eta = \cos^{-1}\zeta \qquad (7\text{-}81)$$

From the above equations the constant-parameter loci are drawn in Fig. 7-21.

From Figs. 7-20 and 7-21 and Eqs. (7-79) to (7-81) the following conclusions are drawn pertaining to the s plane:

1 Horizontal lines represent lines of constant damped natural frequency ω_d. The closer these lines are to the real axis, the lower the value of ω_d. For roots lying on the real axis ($\omega_d = 0$) there is no oscillation.

2 Vertical lines represent lines of constant damping or constant rate of decay of the transient. The closer these lines (or the characteristic roots) are to the imaginary axis, the longer it takes for the transient response to die out.

3 Circles about the origin are circles of constant undamped natural angular

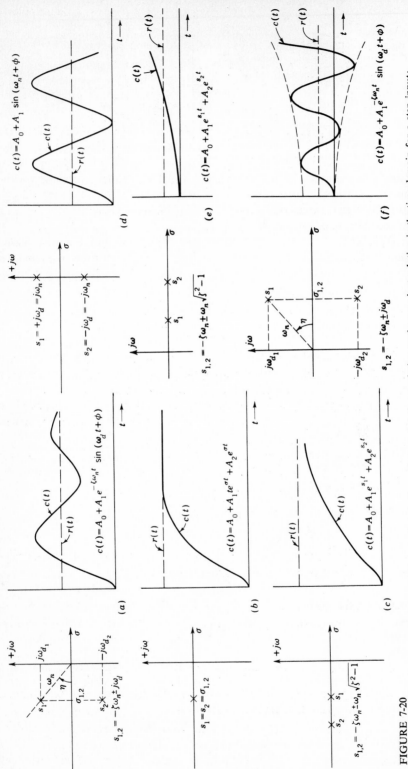

FIGURE 7-20

Plot of roots of a second-order characteristic equation in the s plane and their correlation to the transient solution in the time domain for a step input: (a) underdamped, stable, $0 < \zeta < 1$; (b) critically damped, stable, $\zeta = 1$; (c) overdamped, stable, $\zeta > 1$; (d) undamped, sustained oscillations, $\zeta = 0$; (e) real roots, unstable; (f) underdamped, unstable, $0 > \zeta > -1$.

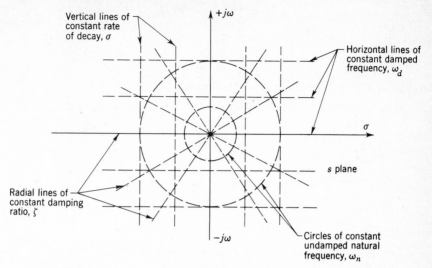

FIGURE 7-21
Constant-parameter curves on the s plane.

frequency ω_n. Since $\sigma^2 + \omega_d{}^2 = \omega_n{}^2$, the locus of the roots of constant ω_n is a circle in the s plane; that is, $|s_1| = |s_2| = \omega_n$. The smaller the circles, the lower ω_n.

4 Radial lines passing through the origin with the angle η are lines of constant damping ratio ζ. The angle η is measured clockwise from the negative real axis for positive ζ, as shown in Fig. 7-20a.

Variation of Roots with ζ

Note in Fig. 7-22 the following:

For $\zeta > 1$, the roots $s_{1,2} = -\zeta\omega_n \pm \omega_n\sqrt{\zeta^2 - 1}$ are real.

For $\zeta = 1$, the roots $s_{1,2} = -\zeta\omega_n$ are real and equal.

For $\zeta < 1$, the roots $s_{1,2} = -\zeta\omega_n \pm j\omega_n\sqrt{1 - \zeta^2}$ are complex conjugates.

For $\zeta = 0$, the roots $s_{1,2} = \pm j\omega_n$ are imaginary.

Higher-Order Systems

The control ratio of a system of order n is given by

$$\frac{C(s)}{R(s)} = \frac{P(s)}{s^n + a_{n-1}s^{n-1} + \cdots + a_0} \qquad (7\text{-}82)$$

In a system having one or more sets of complex-conjugate roots in the characteristic equation, each quadratic factor is of the form

$$s^2 + 2\zeta\omega_n s + \omega_n{}^2 \qquad (7\text{-}83)$$

and the roots are

$$s_{1,2} = \sigma \pm j\omega_d \qquad (7\text{-}84)$$

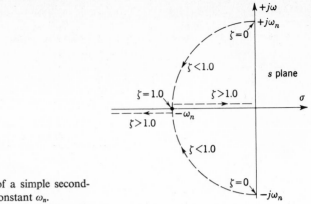

FIGURE 7-22
Variation of roots of a simple second-order system for a constant ω_n.

The relationships developed for ω_n and ζ earlier in this section apply equally well for each complex-conjugate pair of roots of an nth-order system. The distinction is that the dominant ζ and ω_n apply for that pair of complex-conjugate roots which lie closest to the imaginary axis. These values of ζ and ω_n are dominant because the corresponding transient term has the longest settling time and the largest magnitude. Thus the dominant values of ζ and ω_n are selected for the desired response. It must be remembered that in setting the values of the dominant ζ and ω_n the other ζ's and ω_n's are automatically set. Depending on the location of the other roots, they modify the solution obtained from the dominant roots.

In the example of Sec. 7-9 the response $c(t)$ with a step input given by Eq. (7-72) is

$$c(t) = 1 + 1.21e^{-6.6t} \sin(11.4t - 111.7°) + 0.28e^{-55.9t} \sin(18t + 26.1°) \qquad (7\text{-}85)$$

Note that the transient term due to the roots of $s = -55.9 \pm j18$ dies out in approximately one-tenth the time of the transient term due to the dominant roots $s = -6.6 \pm j11.4$. Therefore, the solution can be approximated by the simplified equation

$$c(t) \approx 1 + 1.21e^{-6.6t} \sin(11.4t - 111.7°) \qquad (7\text{-}86)$$

This expression is valid except for a short initial period of time while the other transient term dies out. Equation (7-86) suffices if an approximate solution is desired, i.e., for determining M_p, t_p, and T_s, since the neglected term does not appreciably affect these three quantities. In general, the designer can judge from the location of the roots which ones may be neglected. If an exact solution is desired, all roots must be considered. Plots of $c(t)$ corresponding to Eqs. (7-85) and (7-86) are shown in Fig. 7-23. Since there is little difference between the two curves, the system can be approximated as a second-order system. The significant characteristics of the time response are $t_p = 0.26$ s, $M_p = 1.19$, $t_s = 0.61$ s, and $\omega_d \approx 11.4$ rad/s.

The transient response of any complex system, with the effect of all the roots taken into account, often can be considered to be the result of an equivalent second-order system. On this basis, it is possible to define effective (or equivalent) values of ζ and ω_n. Realize that in order to alter either effective quantity, the location of one or more of the roots must be altered.

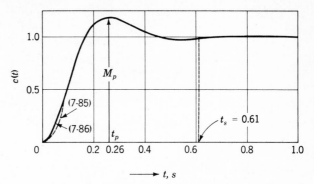

FIGURE 7-23
Plot of $c(t)$ vs. t for Eqs. (7-85) and (7-86).

7-12 TRANSPORT LAG[8]

Some elements in control systems are characterized by dead time or transport lag. This appears as a dead interval for which the output is delayed in response to an input. Figure 7-24a shows the block diagram of a transport-lag element. Figure 7-24b shows a typical resultant output $e_o(t) = e_i(t - \tau)u_{-1}(t - \tau)$, which has the same form as the input but is delayed by a time interval τ. Dead time is a characteristic which can be represented precisely in terms of the Laplace transform (see Sec. 4-4) as a transcendental function. Thus

$$E_o(s) = e^{-\tau s}E_i(s) \qquad (7\text{-}87)$$

The transfer function of the transport-lag element is

$$G_\tau(s) = \frac{E_o(s)}{E_i(s)} = e^{-\tau s} = e^{-\sigma\tau}\big/\!\underline{-\omega\tau} \qquad (7\text{-}88)$$

It has a negative angle which is directly proportional to the frequency.

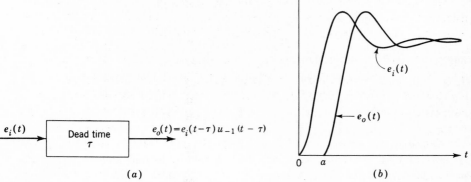

FIGURE 7-24
(a) Transport lag τ; (b) time characteristic.

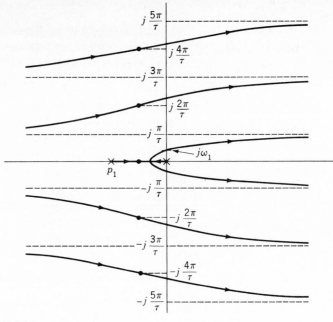

FIGURE 7-25
Root locus for

$$G(s)H(s) = \frac{Ke^{-\tau s}}{s(s - p_1)}$$

The root locus for a system containing a transport lag is illustrated by the following example:[8]

$$G(s)H(s) = \frac{Ke^{-\tau s}}{s(s - p_1)} = \frac{Ke^{-\sigma\gamma}e^{-j\omega\gamma}}{s(s - p_1)} \qquad (7\text{-}89)$$

For $K > 0$ the angle condition is

$$\underline{/s} + \underline{/s - p_1} + \underline{/\omega\tau} = (1 + 2h)180° \qquad (7\text{-}90)$$

The root locus for the system with transport lag is drawn in Fig. 7-25. The magnitude condition used to calibrate the root locus is

$$K = |s| \cdot |s - p_1|e^{\sigma\tau} \qquad (7\text{-}91)$$

The same system without transport lag has only two branches and is stable for all $K > 0$ (see Fig. 7-2). The system with transport lag has an infinite number of branches. The values of K at the points where the branches cross the imaginary axis are

$$K = \omega\sqrt{p_1^2 + \omega^2} \qquad (7\text{-}92)$$

The two branches closest to the origin thus have the largest influence on system stability and are therefore the principal branches. The maximum value of loop sensitivity K given by Eq. (7-92) is therefore determined by the frequency ω_1 (see Fig. 7-25) which satisfies the angle condition $\omega_1\tau = \tan^{-1}|p_1|/\omega_1$.

7-13   SYNTHESIS

The root-locus method lends itself very readily to the problem of synthesis because of the direct relationship between the frequency and the time domains to the s domain. The desired response can be achieved by keeping in mind the following five points.

1   First plot the locus; then from the known information the location of the dominant closed-loop poles must be specified.

2   Alternately, if the gain has been specified, then by applying the magnitude condition to each branch of the locus all the roots can be determined.

3   If the gain is not known but must be determined, enough information must be known about the desired response to proceed with the synthesis.

   a   Since the desired time response, in general, has a peak value M_p between 1.0 and 1.4, the locus has at least one set of complex-conjugate poles. Often there is a pair of complex poles near the imaginary axis which dominate the time response of the system; if so, these poles are referred to as the *dominant-pole pair*.

   b   To set the system gain, any of the following items must be specified for the dominant-pole pair: the damping ratio ζ, the settling time T_s (for 2 percent: $T_s = 4/\zeta\omega_n$), the undamped natural frequency ω_n, or the damped natural frequency ω_d. As previously determined, each of these factors corresponds to either a line or a circle in the s domain.

   c   When the line or circle corresponding to the given information has been drawn, as stated in point 3*b*, its intersection with the locus corresponding to the dominant-pole pair fixes the value of the system gain. By applying the magnitude condition to this point the value of the gain required can be determined. Setting the value of the gain to this value results in the transient-response terms, corresponding to the dominant-pole pair, having the desired characterizing parameter.

4   Once the value of the gain on the dominant branch is fixed, thus fixing the dominant poles, the locations of all remaining roots on the other branches are also fixed.

5   If the root locus does not yield the desired response, the locus must be altered by compensation to achieve the desired results. The subject of compensation in the s domain is discussed in later chapters.

7-14   SUMMARY OF ROOT-LOCUS CONSTRUCTION RULES FOR NEGATIVE FEEDBACK

Rule 1   The number of branches of the root locus is equal to the number of poles of the open-loop transfer function.

Rule 2   For *positive* values of K, the root locus exists on those portions of the real axis for which the sum of the poles and zeros to the right is an odd number. For *negative* values of K, the root locus exists on those portions of the real axis for which the sum of the poles and zeros to the right is an even number (including zero).

Rule 3 The root locus starts ($K = 0$) at the open-loop poles and ends ($K = \pm\infty$) at the open-loop zeros or at infinity.

Rule 4 The angles of the asymptotes of the root locus that end at infinity are determined by

$$\gamma = \frac{(1 + 2h)180° \text{ for } K > 0 \quad \text{ or } \quad h360° \text{ for } K < 0}{[\text{number of poles of } G(s)H(s)] - [\text{number of zeros of } G(s)H(s)]}$$

Rule 5 The real-axis intercept of the asymptotes is

$$\sigma_o = \frac{\sum\limits_{c=1}^{n} \text{Re}\,(p_c) - \sum\limits_{h=1}^{w} \text{Re}\,(z_h)}{n - w}$$

Rule 6 The breakaway point for the locus between two poles on the real axis (or the break-in point for the locus between two zeros on the real axis) can be determined by taking the derivative of the loop sensitivity K with respect to s.   Equate this derivative to zero and find the roots of the resulting equation.   The root that occurs between the poles (or the zeros) is the breakaway (or break-in) point.

Rule 7 For $K > 0$ the angle of departure from a complex pole is equal to 180° minus the sum of the angles from the other poles plus the sum of the angles from the zeros. Any of these angles may be positive or negative. For $K < 0$ the departure angle is 180° from that obtained for $K > 0$.

For $K > 0$ the angle of approach to a complex zero is equal to the sum of the angles from the poles minus the sum of the angles from the other zeros minus 180°. For $K < 0$ the approach angle is 180° from that obtained for $K > 0$.

Rule 8 The imaginary-axis crossing of the root locus can be determined by setting up the Routhian array from the closed-loop characteristic equation. Equate the s^1 row to zero and form the auxiliary equation from the s^2 row. The roots of the auxiliary equation are the imaginary-axis crossover points.

Rule 9 The selection of the dominant roots of the characteristic equation is based on the specifications that give the required system performance. The loop sensitivity for these roots is determined by means of the magnitude condition. The remaining roots are then determined to satisfy the same magnitude condition.

Rule 10 For those open-loop transfer functions for which $w \le n - 2$, Grant's rule states that the sum of the closed-loop roots is equal to the sum of the open-loop poles.

Rule 11 Once the dominant roots have been located, Grant's rule

$$\sum_{j=1}^{n} p_j = \sum_{j=1}^{n} r_j$$

may be used to find one real or two complex roots. Factoring known roots from the characteristic equation can simplify the work of finding the remaining roots. A trial-and-error procedure can also be used to locate all the roots for a specified value of K.

A root-locus digital-computer program[3] will produce an accurate calibrated root locus. This considerably simplifies the work required for the system design.

7-15 SUMMARY

In this chapter the root-locus method is developed to solve graphically for the roots of the characteristic equation. This method can be extended to solving for the roots of any polynomial. Any polynomial can be rearranged and put into the mathematical form of a ratio of factored polynomials which is equal to plus or minus unity, i.e.,

$$\frac{N(s)}{D(s)} = \pm 1$$

Once this form has been obtained, the procedures given in this chapter can be utilized to locate the roots of the polynomial.

The root locus permits the analysis of the performance of feedback systems and provides a basis for selecting the gain in order to best meet the performance specifications. Since the closed-loop poles are obtained explicitly, the form of the time response is directly available. A computer program is available for obtaining $c(t)$ vs. t.[7] If the performance specifications cannot be met, the root locus can be analyzed to determine the appropriate compensation to yield the desired results. This is covered in Chap. 10.

REFERENCES

1 Truxal, J. G.: "Automatic Feedback Control System Synthesis," McGraw-Hill, New York, 1955.
2 Evans, W. R.: "Control-System Dynamics," McGraw-Hill, New York, 1954.
3 "User's Manual for a Digital Computer Routine to Calculate the Root Locus (ROOTL)," School of Engineering, Air Force Institute of Technology, Wright-Patterson Air Force Base, Ohio, 1974.
4 Lorens, C. S., and R. C. Titsworth: Properties of Root Locus Asymptotes, letter in *IRE Trans. Autom. Control*, vol. AC-5, pp. 71–72, January 1960.
5 Wilts, C. H.: "Principles of Feedback Control," Addison-Wesley, Reading, Mass., 1960.
6 Grant, A. J., North American Aviation, Inc.: "The Conservation of the Sum of the System Roots as Applied to the Root Locus Method," unpublished paper, Apr. 10, 1953.
7 Heaviside Partial Fraction Expansion and Time Response Program (PARTL), School of Engineering, Air Force Institute of Technology, Wright-Patterson Air Force Base, Ohio, 1974.
8 Chang, C. S.: "Analytical Method for Obtaining the Root Locus with Positive and Negative Gain," *IEEE Trans. Autom. Control*, vol. AC-10, pp. 92–94, 1965.

8

FREQUENCY RESPONSE

8-1 INTRODUCTION

In conventional control-system analysis there are two basic methods for predicting and adjusting a system's performance without resorting to the solution of the system's differential equation. One of these, the root-locus method, is discussed in detail in Chap. 7. The frequency-response method, to which the Nyquist stability criterion may be applied, is developed in this chapter.[1,2] For the comprehensive study of a system by conventional methods it is necessary to use both methods of analysis. The principal advantage of the root-locus method is that the actual time response is easily obtained by means of the inverse Laplace transform because the precise root locations are known. However, it is sometimes necessary to have performance requirements in terms of the frequency response. Also, the noise which is always present in any system can result in poor overall performance. The frequency response of a system permits analysis with respect to both these items. The design of a passband for the system response may exclude the noise and therefore improve the system performance as long as the dynamic performance specifications are met. The frequency response is also useful in situations for which the transfer functions of some or all of the blocks in a block diagram are unknown. The frequency response can be determined experimentally for these situations. Then an approximate expression for the transfer function can be obtained from the graphical plot of the experimental data. Thus, no particular representation can be judged superior to the rest. Each has its particular use and advantage in a particular situation. As the reader becomes fully acquainted with all methods, he will become aware of the potentialities of each method and will know when each should be used.

Design of control systems by modern control-theory techniques is based upon achieving an optimum performance according to a specified performance index PI; for example, minimizing the integral of squared error (ISE): PI $= \int_0^\infty e(t)^2\, dt$, where $e \equiv r - c$. Both frequency-response and root-locus methods are valuable complementary tools for many of the techniques of modern control theory.

In this chapter two graphical representations of transfer functions are presented, the logarithmic and the polar plots. These plots are used to develop Nyquist's stability criterion.

8-2   CORRELATION OF THE SINUSOIDAL AND TIME RESPONSES[2]

Earlier in the text it is pointed out that solving for $c(t)$ by the classical method is laborious and impractical for synthesis purposes, especially when the input is not a simple analytical function. The use of Laplace transform theory lessens the work involved and permits the engineer to synthesize and improve a system. Chapter 7 on the root locus illustrates this fact. The advantages of the graphical representations in the frequency domain of the transfer functions are developed in the following pages.

Once the frequency response of a system has been determined, the time response can be determined by inverting the corresponding Fourier transform. The behavior in the frequency domain for a given driving function $r(t)$ can be determined by the Fourier transform as

$$\mathbf{R}(j\omega) = \int_{-\infty}^{\infty} r(t)e^{-j\omega t}\, dt \qquad (8\text{-}1)$$

For a given control system the frequency response of the controlled variable is

$$\mathbf{C}(j\omega) = \frac{\mathbf{G}(j\omega)}{1 + \mathbf{G}(j\omega)\mathbf{H}(j\omega)}\, \mathbf{R}(j\omega) \qquad (8\text{-}2)$$

By use of the inverse Fourier transform the controlled variable as a function of time is

$$c(t) = \frac{1}{2\pi} \int_{-\infty}^{\infty} \mathbf{C}(j\omega)e^{j\omega t}\, d\omega \qquad (8\text{-}3)$$

This approach is much used in practice. If the design engineer cannot evaluate Eq. (8-3) by reference to a table of definite integrals, this equation can be evaluated by numerical or graphical integration. This is necessary if $\mathbf{C}(j\omega)$ is available only as a curve and cannot be simply expressed in analytical form, as is often the case. The procedure is described in several books.[10] In addition, methods have been developed based on the Fourier transform and a step input signal, relating $\mathbf{C}(j\omega)$ qualitatively to the time solution without actually taking the inverse Fourier transform. These methods permit the engineer to make an approximate determination of the response of his system through the interpretation of graphical plots in the frequency domain. This makes the design and improvement of feedback systems possible with a minimum effort.

Sections 4-11 and 7-10 show that the frequency response is a function of the pole-zero pattern in the s plane. It is therefore related to the time response of the system. Two features of the frequency response are the maximum value M_m and the resonant frequency ω_m. Section 9-3 describes the qualitative relationship between the time response and the values M_m and ω_m. Since the location of the poles can be determined from the root locus, there is a direct relationship between the root-locus and frequency-response methods.

8-3 FREQUENCY-RESPONSE CURVES

From experience one finds that if analytical solutions are tedious, graphical solutions may be easier to use. The plots in the frequency domain that have found great use in graphical analysis in the field of feedback control systems belong to two categories. The first category is the plot of the magnitude of the output-input ratio vs. frequency in rectangular coordinates, as illustrated in Secs. 4-11 and 7-10. In logarithmic coordinates these are known as *Bode plots*. Associated with this plot is a second plot of the corresponding phase angle vs. frequency. In the second category the output-input ratio may be plotted in polar coordinates with frequency as a parameter. For this category there are two types of polar plots, direct and inverse. Polar plots are generally used only for the open-loop response and are commonly referred to as *Nyquist plots*.[3] The plots can readily be obtained experimentally or by computer.[4] When a computer is not available, the Bode plots are easily obtained by a graphical procedure. The other plots can then be obtained from the Bode plots.

For a given sinusoidal input signal, the input and steady-state output are of the following forms:

$$r(t) = R \sin \omega t \tag{8-4}$$

$$c(t) = C \sin (\omega t + \alpha) \tag{8-5}$$

The closed-loop frequency response

$$\frac{C(j\omega)}{R(j\omega)} = \frac{G(j\omega)}{1 + G(j\omega)H(j\omega)} = M(\omega)/\underline{\alpha(\omega)} \tag{8-6}$$

can be determined for any given value of frequency. For each value of frequency, Eq. (8-6) yields a phasor quantity whose magnitude is M and whose phase angle α is the angle between $C(j\omega)$ and $R(j\omega)$.

An ideal system may be defined as one for which $\alpha = 0^\circ$ and $R = C$ for $0 < \omega < \infty$. Curves 1 in Fig. 8-1 represent the characteristics of that ideal system. However, this definition implies an instantaneous transfer of energy from the input to the output. Such a transfer cannot be achieved in practice since any physical system has some energy dissipation and some energy-storage elements. Curves 2 and 3 in Fig. 8-1 represent the frequency responses of practical control systems. The passband,

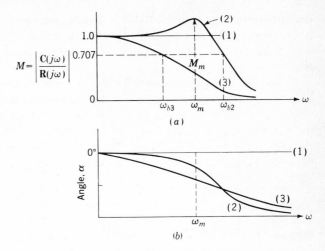

FIGURE 8-1
Frequency-response characteristics of $C(j\omega)/R(j\omega)$ in rectangular coordinates.

or bandwidth, of the frequency response is defined as the range of frequencies from 0 to the frequency ω_b, where $M = 0.707$ of the value at $\omega = 0$. The frequency ω_m is more easily obtained than ω_b. The values M_m and ω_m are often used as figures of merit.

In any system the input signal may contain spurious noise signals in addition to the true signal input, or there may be sources of noise within the closed-loop system. This noise is generally in a band of frequencies above the dominant frequency band of the true signal. Thus, in order to reproduce the true signal and attenuate the noise, feedback control systems are designed to have a definite passband. In certain cases the noise frequency may exist in the same frequency band as the true signal. When this occurs, the problem of eliminating it becomes complicated. Therefore, even if the ideal system were possible, it would not be desirable.

8-4 BODE PLOTS (LOGARITHMIC PLOTS)

The plotting of the frequency transfer function can be systematized and simplified by using logarithmic plots. The use of semilog paper eliminates the need to take logarithms of very many numbers and also expands the low-frequency range, which is of primary importance. The advantages of logarithmic plots are that (1) the mathematical operations of multiplication and division are transformed to addition and subtraction and (2) the work of obtaining the transfer function is largely graphical instead of analytical. The basic factors of the transfer function fall into three categories, and these can easily be plotted by means of straight-line asymptotic approximations.

In preliminary design studies the straight-line approximations are used to obtain approximate performance characteristics very quickly or to check values obtained from the computer. As the design becomes more firmly specified, the straight-line curves can be corrected for greater accuracy. From these logarithmic plots enough data in the frequency range of concern can readily be obtained to determine the corresponding polar plots.

Some basic definitions of logarithmic terms follow.

Logarithm   The logarithm of a complex number is itself a complex number. The abbreviation log is used to indicate the logarithm to the base 10:

$$\log |\mathbf{G}(j\omega)|e^{j\phi(\omega)} = \log |\mathbf{G}(j\omega)| + \log e^{j\phi(\omega)}$$
$$= \log |\mathbf{G}(j\omega)| + j0.434\phi(\omega) \qquad (8\text{-}7)$$

The real part is equal to the logarithm of the magnitude, $\log |\mathbf{G}(j\omega)|$, and the imaginary part is proportional to the angle, $0.434\phi(\omega)$. In the rest of this book the factor 0.434 is dropped, and only the angle $\phi(\omega)$ is used.

Decibel   In feedback-system work the unit commonly used for the logarithm of the magnitude is the *decibel*. Logarithms of transfer functions are used, where the transfer function is the ratio of output to input of a block. The variables are not necessarily in the same units; e.g., the output may be speed in radians per second, and the input may be voltage in volts.

Log magnitude   The logarithm of the magnitude of a transfer function $\mathbf{G}(j\omega)$ expressed in decibels is

$$20 \log |\mathbf{G}(j\omega)| \qquad \text{dB}$$

This quantity is called the *log magnitude*, abbreviated Lm. Thus

$$\text{Lm } \mathbf{G}(j\omega) = 20 \log |\mathbf{G}(j\omega)| \qquad \text{dB} \qquad (8\text{-}8)$$

Since the transfer function is a function of frequency, the log magnitude is also a function of frequency.

Octave and decade   Two units used to express frequency bands or frequency ratios are the octave and the decade. An octave is a frequency band from f_1 to f_2, where $f_2/f_1 = 2$. Thus, the frequency band from 1 to 2 Hz is 1 octave in width, and the frequency band from 17.4 to 34.8 Hz is also 1 octave in width. Note that 1 octave is not a fixed frequency bandwidth but depends on the frequency range being considered. The number of octaves in the frequency range from f_1 to f_2 is

$$\frac{\log (f_2/f_1)}{\log 2} = 3.32 \log \frac{f_2}{f_1} \qquad \text{octaves} \qquad (8\text{-}9)$$

There is an increase of 1 decade Hz from f_1 to f_2 when $f_2/f_1 = 10$. The frequency band from 1 to 10 Hz or from 2.5 to 25 Hz is 1 decade in width. The number of decades from f_1 to f_2 is given by

$$\log \frac{f_2}{f_1} \qquad \text{decades} \qquad (8\text{-}10)$$

The decibel values of some common numbers are given in Table 8-1. Note that the reciprocals of numbers differ only in sign. Thus, the decibel value of 2 is $+6$ dB and the decibel value of $\frac{1}{2}$ is -6 dB. Two properties are illustrated in Table 8-1.

Property 1 As a number doubles, the decibel value increases by 6 dB. The number 2.70 is twice as big as 1.35, and its decibel value is 6 dB more. The number 200 is twice as big as 100, and its decibel value is 6 dB greater.

Property 2 As a number increases by a factor of 10, the decibel value increases by 20. The number 100 is ten times as large as the number 10, and its decibel value is 20 dB more. The number 200 is one hundred times larger than the number 2, and its decibel value is 40 dB greater.

8-5 GENERAL FREQUENCY-TRANSFER-FUNCTION RELATIONSHIPS

The frequency-transfer function can be written in generalized form as the ratio of polynomials:

$$\mathbf{G}(j\omega) = \frac{K_m(1 + j\omega T_1)(1 + j\omega T_2)^r \cdots}{(j\omega)^m(1 + j\omega T_a)[1 + (2\zeta/\omega_n)j\omega + (1/\omega_n^2)(j\omega)^2] \cdots} = K_m \mathbf{G}'(j\omega) \quad (8\text{-}11)$$

where K_m is the gain constant. The logarithm of the transfer function is a complex quantity; the real portion is proportional to the log of the magnitude, and the complex portion is proportional to the angle. Two separate equations are written, one for the log magnitude and one for the angle:

$$\text{Lm } \mathbf{G}(j\omega) = \text{Lm } K_m + \text{Lm } (1 + j\omega T_1)$$
$$+ r \text{ Lm } (1 + j\omega T_2) + \cdots - m \text{ Lm } j\omega$$
$$- \text{Lm } (1 + j\omega T_a) - \text{Lm} \left[1 + \frac{2\zeta}{\omega_n} j\omega + \frac{1}{\omega_n^2} (j\omega)^2 \right] - \cdots \quad (8\text{-}12)$$

$$\underline{/\mathbf{G}(j\omega)} = \underline{/K_m} + \underline{/1 + j\omega T_1} + r\underline{/1 + j\omega T_2} + \cdots - m\underline{/j\omega}$$
$$- \underline{/1 + j\omega T_a} - \underline{\left| 1 + \frac{2\zeta}{\omega_n} j\omega + \frac{1}{\omega_n^2} (j\omega)^2 \right.} - \cdots \quad (8\text{-}13)$$

Table 8-1 DECIBEL VALUES OF SOME COMMON NUMBERS

Number	Decibels
0.01	−40
0.1	−20
0.5	−6
1.0	0
2.0	6
10.0	20
100.0	40
200.0	46

The angle equation may be rewritten as

$$\underline{/\mathbf{G}(j\omega)} = \underline{/K_m} + \tan^{-1}\omega T_1 + r\tan^{-1}\omega T_2 + \cdots - m90°$$

$$- \tan^{-1}\omega T_a - \tan^{-1}\frac{2\zeta\omega/\omega_n}{1 - \omega^2/\omega_n^2} - \cdots \qquad (8\text{-}14)$$

The gain K_m is a real number but may be positive or negative; therefore its angle is correspondingly 0 or 180°. Unless otherwise indicated, a positive value of gain is assumed in this book. Both the log magnitude and the angle given by these equations are functions of frequency. When the log magnitude and the angle are plotted as functions of the log of frequency, the resulting curves are referred to as the Bode plots or the *log magnitude diagram and the phase diagram*. Equations (8-12) and (8-13) show that the resultant curves are obtained by the addition and subtraction of the corresponding individual terms in the transfer-function equation. The two curves can be combined into a single curve of log magnitude vs. angle with frequency as a parameter. This curve, called the Nichols plot or the *log magnitude–angle diagram*, corresponds to the Nyquist polar plot and is used for the quantitative design of the feedback system to meet specifications of required performance.

8-6 DRAWING THE BODE PLOTS

The generalized form of the transfer function as given by Eq. (8-11) shows that the numerator and denominator have four basic types of factors:

$$K_m \qquad (8\text{-}15)$$

$$(j\omega)^{\pm m} \qquad (8\text{-}16)$$

$$(1 + j\omega T)^{\pm r} \qquad (8\text{-}17)$$

$$\left[1 + \frac{2\zeta}{\omega_n} j\omega + \frac{1}{\omega_n^2}(j\omega)^2 \right]^{\pm p} \qquad (8\text{-}18)$$

Each of these terms except K_m may appear raised to an integral power other than 1. The curves of log magnitude and angle vs. the log frequency can easily be drawn for each factor. Then these curves for each factor can be added together graphically to get the curves for the complete transfer function. The procedure can be further simplified by using asymptotic approximations to these curves, as shown in the following pages.

Constants

Since the constant K_m is frequency-invariant, the plot of

$$\text{Lm } K_m = 20 \log K_m \qquad \text{dB}$$

is a horizontal straight line. The constant raises or lowers the log magnitude curve of the complete transfer function by a fixed amount. The angle, of course, is zero as long as K_m is positive.

$j\omega$ Factors

The factor $j\omega$ appearing in the denominator has a log magnitude

$$\text{Lm} \frac{1}{j\omega} = 20 \log \left| \frac{1}{j\omega} \right| = -20 \log \omega \qquad (8\text{-}19)$$

When plotted against $\log \omega$, this curve is a straight line with a negative slope of 6 dB/octave or 20 dB/decade. Values of this function can be obtained from Table 8-1 for several values of ω. The angle is constant and equal to $-90°$.

When the factor $j\omega$ appears in the numerator, the log magnitude is

$$\text{Lm} (j\omega) = 20 \log |j\omega| = 20 \log \omega \qquad (8\text{-}20)$$

This curve is a straight line with a positive slope of 6 dB/octave or 20 dB/decade. The angle is constant and equal to $+90°$. Notice that the only difference between the curves for $j\omega$ and for $1/j\omega$ is a change in the sign of the slope of the log magnitude and a change in the sign of the angle. Both curves go through the point 0 dB at $\omega = 1$.

For the factor $(j\omega)^{\pm m}$ the log magnitude curve has a slope of $\pm 6m$ dB/octave or $\pm 20m$ dB/decade, and the angle is constant and equal to $\pm m90°$.

$1 + j\omega T$ Factors

The factor $1 + j\omega T$ appearing in the denominator has a log magnitude

$$\text{Lm} \frac{1}{1 + j\omega T} = 20 \log \left| \frac{1}{1 + j\omega T} \right| = -20 \log \sqrt{1 + \omega^2 T^2} \qquad (8\text{-}21)$$

For very small values of ω, that is, $\omega T \ll 1$,

$$\text{Lm} \frac{1}{1 + j\omega T} \approx \log 1 = 0 \text{ dB} \qquad (8\text{-}22)$$

The plot of the log magnitude at small frequencies is the 0-dB line. For very large values of ω, that is, $\omega T \gg 1$,

$$\text{Lm} \frac{1}{1 + j\omega T} \approx 20 \log \left| \frac{1}{j\omega T} \right| = -20 \log \omega T \qquad (8\text{-}23)$$

The value of Eq. (8-23) at $\omega = 1/T$ is 0. For values of $\omega > 1/T$ this function is a straight line with a negative slope of 6 dB/octave. The asymptotes of the plot of $\text{Lm} [1/(1 + j\omega T)]$ are two straight lines, one of zero slope below $\omega = 1/T$ and one of -6 dB/octave slope above $\omega = 1/T$. These asymptotes are drawn in Fig. 8-2. *The frequency at which the asymptotes to the log magnitude curve intersect is defined as the corner frequency* ω_{cf}. The value $\omega_{cf} = 1/T$ is the corner frequency for the function $(1 + j\omega T)^{\pm r} = (1 + j\omega/\omega_{cf})^{\pm r}$.

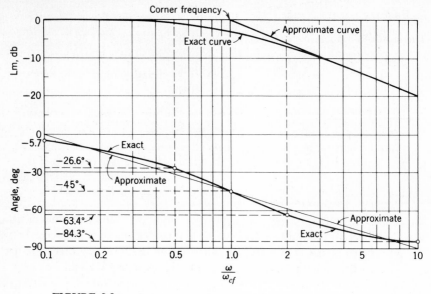

FIGURE 8-2
Log magnitude and phase diagram for

$$(1 + j\omega T)^{-1} = \left(1 + j\frac{\omega}{\omega_{cf}}\right)^{-1}$$

The exact values of Lm $(1 + j\omega T)^{-1}$ are given in Table 8-2 for several frequencies in the range a decade above and below the corner frequency. The exact curve is also drawn in Fig. 8-2. The error, in decibels, between the exact curve and the asymptotes is approximately as follows:

1 At the corner frequency: 3 dB
2 One octave above and below the corner frequency: 1 dB
3 Two octaves from the corner frequency: 0.26 dB

Frequently the preliminary design studies are made by using the asymptotes only. The correction to the straight-line approximation to yield the true log magnitude curve is shown in Fig. 8-3. For more exact studies the corrections are put in at the corner frequency and at 1 octave above and below the corner frequency, and the new curve is drawn with a french curve. If a more accurate curve is desired, computer-calculated values should be used.

The phase curve for this function is plotted in Fig. 8-2. At zero frequency the angle is 0°; at the corner frequency $\omega = \omega_{cf}$ the angle is $-45°$; and at infinite frequency the angle is $-90°$. The angle curve is symmetrical about the corner-frequency value when plotted against log (ω/ω_{cf}) or log ω. Since the abscissa of the curves in Fig. 8-2 is ω/ω_{cf}, the shapes of the angle and log magnitude curves are independent of the time constant T. Thus when the curves are plotted with the abscissa in terms of ω, changing T just "slides" the log magnitude and the angle curves left or right so that the -3 dB and the $-45°$ points occur at the frequency $\omega = \omega_{cf}$.

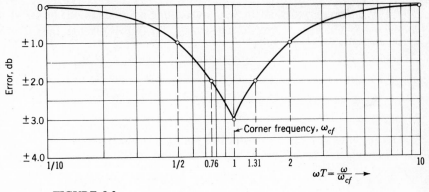

FIGURE 8-3
Log magnitude correction for $(1 + j\omega T)^{\pm 1}$.

Templates of the log magnitude and of the angle can be drawn to the same scale as the semilog paper used and can then be used for all functions of the form $1 + j\omega T$. When templates are not available, fairly accurate log magnitude and phase curves are obtained by drawing smooth curves through the key points given in Tables 8-2 and 8-3.

Table 8-2   VALUES OF Lm $(1 + j\omega T)^{-1}$ FOR SEVERAL FREQUENCIES

$\dfrac{\omega}{\omega_{cf}}$	Exact value, dB	Value of the asymptote, dB	Error, dB
0.1	0.04	0	−0.04
0.25	−0.26	0	−0.26
0.5	−0.97	0	−0.97
0.76	−2.00	0	−2.00
1	−3.01	0	−3.01
1.31	−4.35	−2.35	−2.00
2	−6.99	−6.02	−0.97
4	−12.30	−12.04	−0.26
10	−20.04	−20.0	−0.04

Table 8-3   ANGLES OF $(1 + j\omega/\omega_{cf})^{-1}$ FOR KEY FREQUENCY POINTS

$\dfrac{\omega}{\omega_{cf}}$	Angle, deg
0.1	−5.7
0.5	−26.6
1.0	−45.0
2.0	−63.4
10.0	−84.3

For preliminary design studies, straight-line approximations of the phase curve can be used. The approximation is a straight line drawn through the following three points:

ω/ω_{cf}	0.1	1.0	10
Angle	$0°$	$-45°$	$-90°$

The maximum error resulting from this approximation is about $\pm 6°$.

The factor $1 + j\omega T$ appearing in the numerator has the log magnitude

$$\text{Lm } (1 + j\omega T) = 20 \log \sqrt{1 + \omega^2 T^2}$$

This is the same function as its inverse $\text{Lm } [1/(1 + j\omega T)]$ except that it is positive. The corner frequency is the same, and the angle varies from 0 to 90° as the frequency increases from zero to infinity. The log magnitude and angle curves for the function $1 + j\omega T$ are symmetrical about the abscissa to the curves for $1/(1 + j\omega T)$. Therefore the same templates can be used for both.

Quadratic Factors

Quadratic factors of the form

$$\left[1 + \frac{2\zeta}{\omega_n} j\omega + \frac{1}{\omega_n^2} (j\omega)^2\right]^{-1} \quad (8\text{-}24)$$

often occur in feedback-system transfer functions. For $\zeta > 1$ the quadratic can be factored into two first-order factors with real zeros which can be plotted in the manner shown previously. But for $\zeta < 1$ the factors are conjugate-complex factors, and the entire quadratic is plotted without factoring:

$$\text{Lm}\left[1 + \frac{2\zeta}{\omega_n} j\omega + \frac{1}{\omega_n^2} (j\omega)^2\right]^{-1} = -20 \log\left[\left(1 - \frac{\omega^2}{\omega_n^2}\right)^2 + \left(\frac{2\zeta\omega}{\omega_n}\right)^2\right]^{1/2} \quad (8\text{-}25)$$

$$\text{Angle}\left[1 + \frac{2\zeta}{\omega_n} j\omega + \frac{1}{\omega_n^2} (j\omega)^2\right]^{-1} = -\tan^{-1} \frac{2\zeta\omega/\omega_n}{1 - \omega^2/\omega_n^2} \quad (8\text{-}26)$$

From Eq. (8-25) it is seen that for very small values of ω the low-frequency asymptote is represented by log magnitude $= 0$ dB. For very high values of frequency, the log magnitude is approximately

$$-20 \log \frac{\omega^2}{\omega_n^2} = -40 \log \frac{\omega}{\omega_n}$$

and the high-frequency asymptote has a slope of -40 dB/decade. The asymptotes cross at the corner frequency $\omega_{cf} = \omega_n$.

From Eq. (8-25) it can be seen that a resonant condition, i.e., the peak value of the $\text{Lm} > 0$ dB, is exhibited in the vicinity of $\omega = \omega_n$. Therefore there may be a substantial deviation of the log magnitude curve from the straight-line asymptotes, depending on the value of ζ. A family of curves for several values of $\zeta < 1$ is plotted

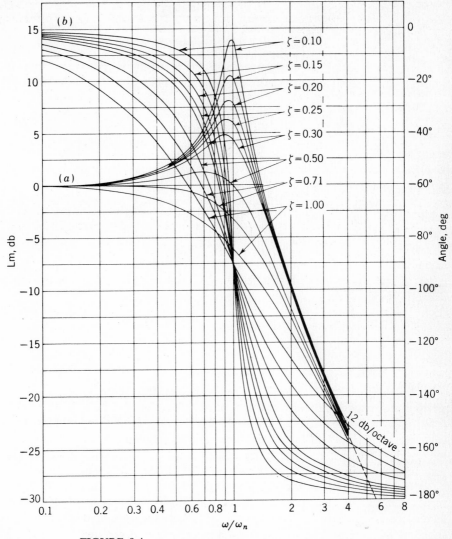

FIGURE 8-4
Log magnitude and phase diagram for

$$\left[1 + \frac{j2\zeta\omega}{\omega_n} + \left(\frac{j\omega}{\omega_n}\right)^2\right]^{-1}$$

in Fig. 8-4. If these curves are used often enough, it is worthwhile to make templates for several values of ζ. Then the appropriate correct curve can be traced directly on the graph being drawn. If templates are not used, sufficient points can be picked from Fig. 8-4 to draw the curve.

The phase-angle curve for this function also varies with ζ. At zero frequency the angle is $0°$, at the corner frequency the angle is $-90°$, and at infinite frequency the

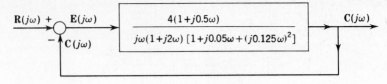

FIGURE 8-5
Block diagram of a control system with unity feedback.

angle is $-180°$. A family of curves for various values of $\zeta < 1$ is plotted in Fig. 8-4. Templates can be made of these curves, or enough values to draw the appropriate curve can be taken from Fig. 8-4.

When the quadratic factor appears in the numerator, the magnitudes of the log magnitude and phase angle are the same as those in Fig. 8-4 except that they are changed in sign.

The Lm $[1 + j2\zeta\omega/\omega_n + (j\omega/\omega_n)^2]^{-1}$ with $\zeta < 1$ has a peak value. The magnitude of this peak value and the frequency at which it occurs are important terms. These values, given earlier in Sec. 4-11 and derived in Sec. 9-3, are repeated here:

$$M_m = \frac{1}{2\zeta\sqrt{1 - \zeta^2}} \qquad (8\text{-}27)$$

$$\omega_m = \omega_n\sqrt{1 - 2\zeta^2} \qquad (8\text{-}28)$$

Note that the peak value M_m depends only on the damping ratio ζ. Equation (8-28) is meaningful only for real values of ω_m. Therefore the curve of M vs. ω has a peak value greater than unity only for $\zeta < 0.707$. The frequency at which the peak value occurs depends on both the damping ratio ζ and the undamped natural frequency ω_n. This information is used when adjusting a control system for good response characteristics. These characteristics are discussed in Chap. 9.

The discussion thus far has dealt with minimum-phase factors (all poles and zeros are in the left-half s plane). The log magnitude curves for non-minimum-phase factors (poles and zeros in the right-half s plane) are the same as those for the corresponding minimum-phase factors. However, the angle curves are different. For the factor $1 - j\omega T$ the angle varies from 0 to $-90°$ as ω varies from zero to infinity. If ζ is negative, the quadratic factor of expression (8-24) becomes a non-minimum-phase term. Its angle varies from $-360°$ at $\omega = 0$ to $-180°$ at $\omega = \infty$. This information can be obtained from the pole-zero diagram discussed in Sec. 4-11.

8-7 EXAMPLE OF DRAWING A BODE PLOT

Figure 8-5 shows the block diagram of a feedback control system with unity feedback. The log magnitude and phase diagram is drawn for the open-loop transfer function of this system. The log magnitude curve is drawn both for the straight-line approximation and for the exact curve. Table 8-4 lists the pertinent characteristics for each factor. The log magnitude asymptotes and angle curves for each factor are shown in Figs. 8-6 and 8-7, respectively. They are added algebraically to obtain the composite curve. The composite log magnitude curve using straight-line approximations is drawn directly, as outlined below.

Step 1 At frequencies less than ω_1, the first corner frequency, only the factors Lm 4 and Lm $(j\omega)^{-1}$ are effective. All the other factors have zero value. At ω_1, Lm 4 = 12 dB, and Lm $(j\omega_1)^{-1}$ = 6 dB; thus at this frequency the composite curve has the value of 18 dB. Below ω_1 the composite curve has a slope of -20 dB/decade because of the Lm $(j\omega)^{-1}$ term.

Step 2 Above ω_1, the factor Lm $(1 + j2\omega)^{-1}$ has a slope of -20 dB/decade and must be added to the terms in step 1. When the slopes are added, the composite curve has a total slope of -40 dB/decade in the frequency band from ω_1 to ω_2. Since this bandwidth is 2 octaves, the value of the composite curve at ω_2 is -6 dB.

Step 3 Above ω_2, the factor Lm $(1 + j0.5\omega)$ is effective. This factor has a slope of $+20$ dB/decade above ω_2 and must be added to obtain the composite curve. The composite curve now has a total slope of -20 dB/decade in the frequency band from ω_2 to ω_3. The bandwidth from ω_2 to ω_3 is 2 octaves; therefore the value of the composite curve at ω_3 is -18 dB.

Step 4 Above ω_3 the last term Lm $[1 + j0.05\omega + (j0.125\omega)^2]^{-1}$ must be added. This factor has a slope of -40 dB/decade; therefore the total slope of the composite curve above ω_3 is -60 dB/decade.

Table 8-4 CHARACTERISTICS OF LOG MAGNITUDE–ANGLE DIAGRAM FOR VARIOUS FACTORS

Factor	Corner frequency ω_{cf}	Log magnitude	Angle characteristics
4	None	Constant magnitude of $+12$ dB	Constant $0°$
$(j\omega)^{-1}$	None	Constant slope of -20 dB/decade	Constant $-90°$
$(1 + j2\omega)^{-1}$	$\omega_1 = 0.5$	0 slope below corner frequency; -20 dB/decade slope above corner frequency	Varies from 0 to $-90°$
$1 + j0.5\omega$	$\omega_2 = 2.0$	0 slope below corner frequency; $+20$ dB/decade slope above corner frequency	Varies from 0 to $+90°$
$[1 + j0.05\omega + (j0.125\omega)^2]^{-1}$ $\zeta = 0.2$ $\omega_n = 8$	$\omega_3 = 8.0$	0 slope below corner frequency; -40 dB/decade slope above corner frequency	Varies from 0 to $-180°$

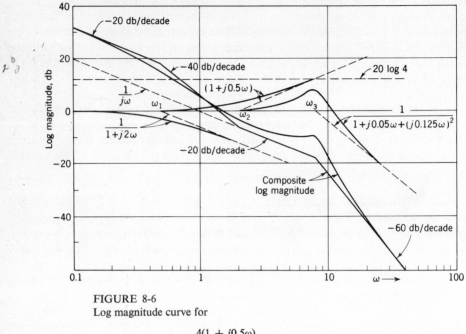

FIGURE 8-6
Log magnitude curve for

$$G(j\omega) = \frac{4(1 + j0.5\omega)}{j\omega(1 + j2\omega)[1 + j0.05\omega + (j0.125\omega)^2]}$$

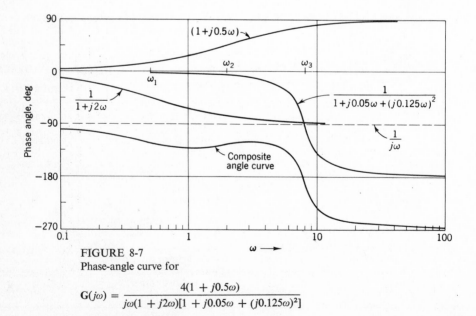

FIGURE 8-7
Phase-angle curve for

$$G(j\omega) = \frac{4(1 + j0.5\omega)}{j\omega(1 + j2\omega)[1 + j0.05\omega + (j0.125\omega)^2]}$$

Step 5 Once the asymptotic plot of the log magnitude of $\mathbf{G}(j\omega)$ has been drawn, the corrections can be added if desired. The corrections at each corner frequency and at an octave above and below the corner frequency are usually sufficient. For first-order terms the corrections are ± 3 dB at the corner frequencies and ± 1 dB at an octave above and below the corner frequency. The values for quadratic terms can be obtained from Fig. 8-4 since they are a function of the damping ratio. If the correction curve for the quadratic factor is not available, the correction at the frequencies $\omega = \omega_n$ and $\omega = 0.707\omega_n$ can easily be calculated from Eq. (8-25).

The corrected log magnitude curves for each factor and for the composite curve are shown in Fig. 8-6.

Determination of the phase-angle curve of $\mathbf{G}(j\omega)$ can be simplified by using the following procedure.

1 For the $(j\omega)^{-m}$ term, draw a line at the angle of $(-m)90°$.

2 For each $(1 + j\omega T)^{\pm 1}$ term, locate the angles from Table 8-4 at the corner frequency, an octave above and an octave below the corner frequency and a decade above and below the corner frequency. Then draw a curve through these points for each $(1 + j\omega T)^{\pm 1}$ term.

3 For each $1 + j2\zeta\omega/\omega_n + (j\omega/\omega_n)^2$ term:

   a Locate the $\pm 90°$ point at the corner frequency.

   b From Fig. 8-4, for the respective ζ, obtain a few points to draw the phase plot for each term with the aid of a french curve. The angle at $\omega = 0.707\omega_n$ may be sufficient and can be evaluated easily from Eq. (8-26).

4 Once the phase plot of each term of $\mathbf{G}(j\omega)$ has been drawn, the composite phase plot of $\mathbf{G}(j\omega)$ is determined by adding the individual phase curves.

   a Use the line representing the angle equal to

$$\left/ \lim_{\omega \to 0} \mathbf{G}(j\omega) = (-m)90° \right.$$

as the base line, where m is the system type. Add or subtract the angles of each factor from this reference line.

   b At a particular frequency on the graph, measure the angle for each single-order and quadratic factor. Add and/or subtract them from the base line until all terms have been accounted for. The number of frequency points used is determined by the desired accuracy of the phase plots.

   c At $\omega = \infty$ the phase is $90°$ times the difference of the orders of the numerator and denominator $[-(n - w)90°]$.

8-8 SYSTEM TYPE AND GAIN AS RELATED TO LOG MAGNITUDE CURVES

The steady-state error of a closed-loop system depends on the system type and the gain. The system error coefficients are determined by these two characteristics, as noted in Chap. 6. For any given log magnitude curve the system type and gain can be

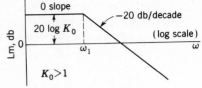

FIGURE 8-8
Log magnitude plot for
$G(j\omega) = K_0/(1 + j\omega T_a)$.

determined. Also, with the transfer function given so that the system type and gain are known, they can expedite drawing the log magnitude curve. This is described for Type 0, 1, and 2 systems.

Type 0 System

A Type 0 system has a transfer function of the form

$$G(j\omega) = \frac{K_0}{1 + j\omega T_a}$$

At low frequencies, $\omega < 1/T_a$, Lm $G(j\omega) = 20 \log K_0$, which is a constant. The slope of the log magnitude curve is zero below the corner frequency $\omega_1 = 1/T_a$ and -20 dB/decade above the corner frequency. The log magnitude curve is shown in Fig. 8-8.

For a Type 0 system the characteristics are as follows:

1 The slope at low frequencies is zero.
2 The magnitude at low frequencies is $20 \log K_0$.
3 The gain K_0 is the steady-state step-error coefficient.

Type 1 System

A Type 1 system has a transfer function of the form

$$G(j\omega) = \frac{K_1}{j\omega(1 + j\omega T_a)}$$

At low frequencies $(\omega < 1/T_a)$ Lm $G(j\omega) = $ Lm $(K_1/j\omega) = $ Lm $K_1 - $ Lm $j\omega$, which has a slope of -20 dB/decade. At $\omega = K_1$, Lm $(K_1/j\omega) = 0$. If the corner frequency $\omega_1 = 1/T_a$ is greater than K_1, the low-frequency portion of the curve of slope -20 dB/decade crosses the 0-dB axis at a value of $\omega_x = K_1$, as shown in Fig. 8-9a. If the corner frequency is less than K_1, the low-frequency portion of the curve of slope -20 dB/decade may be extended until it does cross the 0-dB axis. The value of the frequency at which the extension crosses the 0-dB axis is $\omega_x = K_1$. In other words, the plot Lm $(K_1/j\omega)$ crosses the 0-dB value at $\omega_x = K_1$, as illustrated in Fig. 8-9b.

At $\omega = 1$, Lm $j\omega = 0$; therefore Lm $(K_1/j\omega)_{\omega=1} = 20 \log K_1$. For $T_a < 1$ this value is a point on the slope of -20 dB/decade. For $T_a > 1$ this value is a point on the extension of the initial slope, as shown in Fig. 8-9b. The frequency ω_x is smaller or larger than unity according as K_1 is smaller or larger than unity.

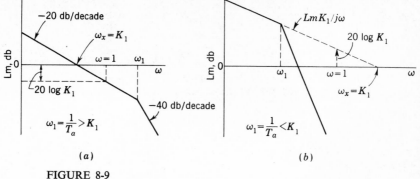

FIGURE 8-9

Log magnitude plot for $G(j\omega) = K_1/j\omega(1 + j\omega T_a)$.

For a Type 1 system the characteristics are as follows:

1 The slope at low frequencies is -20 dB/decade.
2 The intercept of the low-frequency slope of -20 dB/decade (or its extension) with the 0-dB axis occurs at a frequency ω_x, where $\omega_x = K_1$.
3 The value on the low-frequency slope of -20 dB/decade (or its extension) at the frequency $\omega = 1$ is equal to 20 log K_1.
4 The gain K_1 is the steady-state ramp error coefficient.

Type 2 System

A Type 2 system has a transfer function of the form

$$G(j\omega) = \frac{K_2}{(j\omega)^2(1 + j\omega T_a)}$$

At low frequencies, $\omega < 1/T_a$, the Lm $G(j\omega) = $ Lm $[K_2/(j\omega)^2] = $ Lm $K_2 - $ Lm $(j\omega)^2$, for which the slope is -40 dB/decade. At $\omega^2 = K_2$, Lm $[K_2/(j\omega)^2] = 0$; therefore the intercept of the initial slope of -40 dB/decade (or its extension, if necessary) with the 0-dB axis occurs at a frequency ω_y so that $\omega_y{}^2 = K_2$.

At $\omega = 1$, Lm $(j\omega)^2 = 0$; therefore, Lm $G(j\omega)_{\omega=1} = 20$ log K_2. This point occurs on the initial slope or on its extension, according as $\omega_1 = 1/T_a$ is larger or smaller than $\sqrt{K_2}$. If $K_2 > 1$, the quantity 20 log K_2 is positive, and if $K_2 < 1$, the quantity 20 log K_2 is negative.

The log magnitude curve for a Type 2 transfer function is shown in Fig. 8-10. The determination of gain K_2 from the graph is shown.

For a Type 2 system the characteristics are as follows:

1 The slope at low frequencies is -40 dB/decade.
2 The intercept of the low-frequency slope of -40 dB/decade (or its extension, if necessary) with the 0-dB axis occurs at a frequency ω_y, where $\omega_y{}^2 = K_2$.
3 The value on the low-frequency slope of -40 dB/decade (or its extension) at the frequency $\omega = 1$ is equal to 20 log K_2.
4 The gain K_2 is the steady-state parabolic error coefficient.

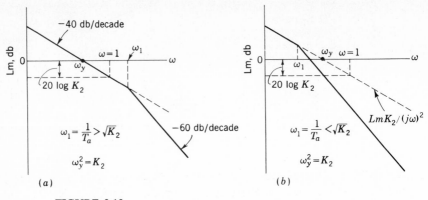

FIGURE 8-10
Log magnitude plot for $G(j\omega) = K_2/(j\omega)^2(1 + j\omega T_a)$.

8-9 EXPERIMENTAL DETERMINATION OF TRANSFER FUNCTIONS[5,9]

The log magnitude–phase-angle diagram is of great value when the mathematical expression for the transfer function of a given system is not known. The magnitude and angle of the ratio of the output to the input can be obtained experimentally for a steady-state sinusoidal input signal for a number of frequencies. These data are used to obtain the exact log magnitude and angle diagram. Asymptotes are drawn on the exact log magnitude curve by utilizing the fact that they must be multiples of ± 20 dB/decade. From these asymptotes the system type and the approximate time constants are determined. Thus, in this manner, the transfer function of the system can be synthesized.

Care must be exercised in determining whether any poles or zeros of the transfer function are in the right-half s plane. A system that has no open-loop poles or zeros in the right-half s plane is defined as a *minimum-phase* system,[6] and all factors are of the form $1 + Ts$ and/or $1 + As + Bs^2$. A system that has open-loop poles or zeros in the right-half s plane is a *non-minimum-phase* system. For this situation, one or more terms in the transfer function have the form $1 - Ts$ and/or $1 \pm As \pm Bs^2$. As an example, consider the functions $1 + j\omega T$ and $1 - j\omega T$. The log magnitude plots of these functions are identical, but the angle diagram for the former goes from 0 to 90° whereas for the latter it goes from 0 to $-90°$. Therefore care must be exercised in interpreting the angle plot to determine whether any factors of the transfer function lie in the right-half s plane. Many practical systems are in the minimum-phase category.

8-10 DIRECT POLAR PLOTS

In the earlier sections of this chapter a simple graphical method is presented for plotting the characteristic curves of transfer functions. These curves are the log magnitude and the angle of $G(j\omega)$ vs. ω, plotted on semilog graph paper. The reason for

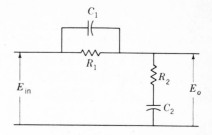

FIGURE 8-11
A complex RC circuit with an effective
voltage E_{in} applied.

presenting this method first is that these curves can be constructed easily and rapidly. Frequently, the polar plot of the transfer function is desired. The magnitude and angle of $\mathbf{G}(j\omega)$, for sufficient frequency points, are readily obtainable from the Lm $\mathbf{G}(j\omega)$ and $\underline{/\mathbf{G}(j\omega)}$ vs. log ω curves. This approach often takes less time than calculating $|\mathbf{G}(j\omega)|$ and $\phi(\omega)$ analytically for each frequency point desired, unless a digital computer is available.[4]

The data for drawing the polar plot of the frequency response can also be obtained from the pole-zero diagram, as described in Sec. 4-11. It is possible to visualize the complete shape of the frequency-response curve from the pole-zero diagram because the angular contribution of each pole and zero is readily apparent. The Spirule can also be used to evaluate the magnitude and angle of the frequency response for any value of frequency. The polar plot of $\mathbf{G}(j\omega)$ is called the *direct polar plot*, and the polar plot of $[\mathbf{G}(j\omega)]^{-1}$ is called the *inverse polar plot*. The shape of these curves is related to the system type.

Complex RC Network (Lag-Lead Compensator)

The circuit of Fig. 8-11 is used in later chapters as a compensator:

$$G(s) = \frac{E_o(s)}{E_{in}(s)} = \frac{1 + (T_1 + T_2)s + T_1 T_2 s^2}{1 + (T_1 + T_2 + T_{12})s + T_1 T_2 s^2} \qquad (8\text{-}29)$$

where the time constants are

$$T_1 = R_1 C_1 \qquad T_2 = R_2 C_2 \qquad T_{12} = R_1 C_2$$

As a function of frequency, the transfer function is

$$\mathbf{G}(j\omega) = \frac{\mathbf{E}_o(j\omega)}{\mathbf{E}_{in}(j\omega)} = \frac{(1 - \omega^2 T_1 T_2) + j\omega(T_1 + T_2)}{(1 - \omega^2 T_1 T_2) + j\omega(T_1 + T_2 + T_{12})} \qquad (8\text{-}30)$$

By the proper choice of the time constants, the circuit acts as a lag network in the lower-frequency range of 0 to ω_x and as a lead network in the higher-frequency range of ω_x to ∞. This means that the steady-state sinusoidal output E_o lags or leads the sinusoidal input E_{in}, according as ω is smaller than or larger than ω_x. The polar plot of this transfer function is a circle with its center on the real axis and lying in the first and fourth quadrants. Figure 8-12 illustrates the polar plot in nondimensionalized

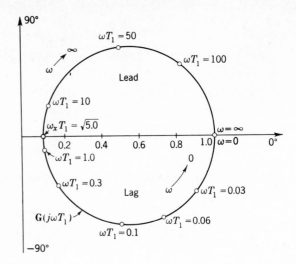

FIGURE 8-12

$$G(j\omega T_1) = \frac{(1 + j\omega T_1)(1 + j0.2\omega T_1)}{(1 + j11.1\omega T_1)(1 + j0.0179\omega T_1)}$$

for $T_2 = 0.2T_1$, $T_{12} = 10.0T_1$.

form for a typical circuit. When T_2 and T_{12} in Eq. (8-30) are expressed in terms of T_1, the expression given in the title of Fig. 8-12 results. Its properties are:

1 $\lim\limits_{\omega \to 0} \ \mathbf{G}(j\omega T_1) \to 1\underline{/0°}$.

2 $\lim\limits_{\omega \to \infty} \ \mathbf{G}(j\omega T_1) \to 1\underline{/0°}$.

3 At the frequency $\omega = \omega_x$, for which $\omega_x^2 T_1 T_2 = 1$, Eq. (8-30) becomes

$$\mathbf{G}(j\omega_x T_1) = \frac{T_1 + T_2}{T_1 + T_2 + T_{12}} = | \ (j\omega_x T_1)|\underline{/0°} \qquad (8\text{-}31)$$

Note that Eq. (8-31) represents the minimum value of the transfer function in the whole frequency spectrum. From Fig. 8-12 it is seen that for frequencies below ω_x the transfer function has a negative or lag angle. For frequencies above ω_x it has a positive or lead angle. The applications and advantages of this circuit are discussed in later chapters on compensation.

Type 0 Feedback Control System

The field-controlled servomotor described by Eq. (2-152) illustrates a typical Type 0 device. It has the transfer function

$$\mathbf{G}(j\omega) = \frac{\mathbf{C}(j\omega)}{\mathbf{E}(j\omega)} = \frac{K_0}{(1 + j\omega T_f)(1 + j\omega T_m)} \qquad (8\text{-}32)$$

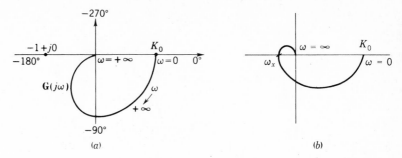

FIGURE 8-13
Polar plot for typical Type 0 transfer functions: (a) Eq. (8-32); (b) Eq. (8-33).

Note from Eq. (8-32) that

when $\omega = 0$: $\qquad\qquad\qquad\qquad\qquad$ $G(j\omega) = K_0\underline{/0°}$

when $\omega \to \infty$: $\qquad\qquad\qquad\qquad\quad$ $G(j\omega) \to 0\underline{/-180°}$

Also, for each term in the denominator the angular contribution to $G(j\omega)$, as ω goes from 0 to ∞, goes from 0 to $-90°$. Thus the polar plot of this transfer function must start at $G(j\omega) = K_0\underline{/0°}$ for $\omega = 0$ and proceed first through the fourth and then through the third quadrants to $\lim\limits_{\omega \to \infty} G(j\omega) = 0\underline{/-180°}$ as the frequency approaches infinity. In other words, the angular variation of $G(j\omega)$ is continuously decreasing, going in a clockwise direction from 0 to $-180°$. The exact shape of this plot is determined by the particular values of the time constants.

If Eq. (8-32) had another term of the form $1 + j\omega T$ in the denominator, the transfer function would be

$$G(j\omega) = \frac{K_0}{(1 + j\omega T_f)(1 + j\omega T_m)(1 + j\omega T)} \qquad (8-33)$$

The $G(j\omega)|_{\omega=\infty}$ point rotates clockwise by an additional 90°. In other words, when $\omega \to \infty$, $G(j\omega) \to 0\underline{/-270°}$. In this case the curve crosses the real axis at a frequency ω_x for which the imaginary part of the transfer function is zero.

When terms of the form $1 + j\omega T$ appear in the numerator, each results in an angular variation of 0 to 90° (a counterclockwise rotation) as the frequency is varied from 0 to ∞. Thus the angular variation $G(j\omega)$ may not continuously change in one direction. Also, the resultant polar plot may not be as smooth as the one shown in Fig. 8-13. As an example, consider the transfer function

$$G(j\omega) = \frac{K_0(1 + j\omega T_1)^2}{(1 + j\omega T_2)(1 + j\omega T_3)(1 + j\omega T_4)^2} \qquad (8-34)$$

whose polar plot is indented as shown in Fig. 8-14. The time constants T_2 and T_3 are greater than T_1, and T_1 is greater than T_4. From the angular contribution of each factor and from the analysis above, it can be surmised that the polar plot has the

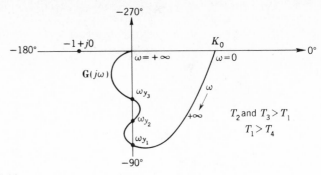

FIGURE 8-14
Polar plot for a more complex Type 0 transfer function [Eq. (8-34)].

general shape shown in the figure. In the event that T_1 is smaller than all the others, its polar plot is similar in shape to the one shown in Fig. 8-13a.

In the same manner, a quadratic in either the numerator or the denominator of a transfer function results in an angular contribution of 0 to $\pm 180°$, respectively. The polar plot of $\mathbf{G}(j\omega)$ is affected accordingly. It can be seen from the examples that the polar plot of a Type 0 system always starts at a value K_0 (step error coefficient) on the positive real axis for $\omega = 0$ and ends at zero magnitude (for $n > w$) and tangent to one of the major axes at $\omega = \infty$. The final angle is $-90°$ times the order of the denominator minus the order of the numerator of $\mathbf{G}(j\omega)$.

Type 1 Feedback Control System

A typical Type 1 system is

$$\mathbf{G}(j\omega) = \frac{\mathbf{C}(j\omega)}{\mathbf{E}(j\omega)} = \frac{K_1}{j\omega(1 + j\omega T_m)(1 + j\omega T_c)(1 + j\omega T_q)} \tag{8-35}$$

For Eq. (8-35):

$$\omega \to 0^+: \qquad\qquad\qquad \mathbf{G}(j\omega) \to \infty\underline{/-90°} \tag{8-36}$$

$$\omega \to \infty: \qquad\qquad\qquad \mathbf{G}(j\omega) \to 0\underline{/-360°} \tag{8-37}$$

Note that the $j\omega$ term in the denominator contributes the angle $-90°$ to the total angle of $\mathbf{G}(j\omega)$ for all frequencies. Thus the basic difference between Eqs. (8-33) and (8-35) is the presence of the term $j\omega$ in the denominator of the latter equation. Since all the $1 + j\omega T$ terms of Eq. (8-35) appear in the denominator, its polar plot, as shown in Fig. 8-15, has no dents. From the remarks of this and previous sections, it can be seen that the angular variation of $\mathbf{G}(j\omega)$ decreases continuously in the same direction from -90 to $-360°$ as ω increases from 0 to ∞. The presence of any frequency-dependent factor in the numerator has the same general effect on the polar plot as that described previously.

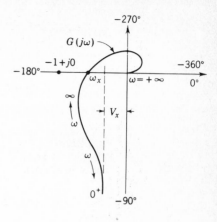

FIGURE 8-15
Polar plot for a typical Type 1 transfer
function [Eq. (8-35)].

It is seen from Eq. (8-36) that the magnitude of the function $\mathbf{G}(j\omega)$ approaches infinity as the value of ω approaches zero. This equation does not indicate whether the function approaches infinity asymptotically to the $-90°$ axis or to some line parallel to it. The true asymptote is determined by finding the value of the real part of $\mathbf{G}(j\omega)$ as ω approaches zero. Thus

$$V_x = \lim_{\omega \to 0} \operatorname{Re}\left[\mathbf{G}(j\omega)\right] \qquad (8\text{-}38)$$

or, for this particular transfer function,

$$V_x = -K_1(T_q + T_c + T_m) \qquad (8\text{-}39)$$

Equation (8-39) shows that the magnitude of $\mathbf{G}(j\omega)$ approaches infinity asymptotically to a vertical line whose real-axis intercept equals V_x, as illustrated in Fig. 8-15. Note that the value of V_x is a direct function of the ramp error coefficient.

The frequency of the crossing point on the negative real axis of the $\mathbf{G}(j\omega)$ function is that value of frequency ω_x for which the imaginary part of $\mathbf{G}(j\omega)$ is equal to zero. Thus

$$\operatorname{Im}\left[\mathbf{G}(j\omega_x)\right] = 0 \qquad (8\text{-}40)$$

or, for this particular transfer function,

$$\omega_x = (T_c T_q + T_q T_m + T_m T_c)^{-1/2} \qquad (8\text{-}41)$$

The significance of the real-axis crossing point is pointed out in later sections dealing with system stability.

Type 2 Feedback Control System

The transfer function of the Type 2 system illustrated in Sec. 6-6, prior to its modification, is

$$\mathbf{G}(j\omega) = \frac{\mathbf{C}(j\omega)}{\mathbf{E}(j\omega)} = \frac{K_2}{(j\omega)^2(1 + j\omega T_f)(1 + j\omega T_m)} \qquad (8\text{-}42)$$

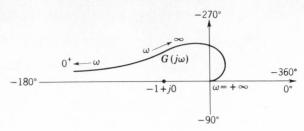

FIGURE 8-16
Polar plot for a typical Type 2 transfer function, resulting in an unstable feedback control system [Eq. (8-42)].

Its properties are:

$\omega \to 0^+$: $\qquad\qquad\qquad\qquad\qquad\qquad$ $\mathbf{G}(j\omega) \to \infty \underline{/-180°}$ $\qquad$ (8-43)

$\omega \to +\infty$: $\qquad\qquad\qquad\qquad\qquad\qquad$ $\mathbf{G}(j\omega) \to 0 \underline{/-360°}$ $\qquad$ (8-44)

The presence of the $(j\omega)^2$ term in the denominator contributes a constant $-180°$ to the total angle of $\mathbf{G}(j\omega)$ for all frequencies. For the transfer function of Eq. (8-42) the polar plot (Fig. 8-16) is a smooth curve whose angle $\phi(\omega)$ decreases continuously from -180 to $-360°$.

The introduction of an additional pole and a zero can alter the shape of the polar plot. Consider the transfer function

$$\mathbf{G}_o(j\omega) = \frac{K_2'(1 + j\omega T_1)}{(j\omega)^2(1 + j\omega T_f)(1 + j\omega T_m)(1 + j\omega T_2)} \qquad (8-45)$$

where $T_1 > T_2$. The polar plot can be obtained from the pole-zero diagram shown in Fig. 8-17. At $s = j\omega = j0^+$ the angle of each factor is zero except for the double pole at the origin. The angle at $\omega = 0^+$ is therefore $-180°$, as given by Eq. (8-43), which is still applicable. As ω increases from zero, the angle of $j\omega + 1/T_1$ increases faster than the angles of the other poles. In fact, at low frequencies the angle due to the zero is larger than the sum of the angles due to the poles located to the left of the zero. This is shown qualitatively at the frequency ω_1 in Fig. 8-17. Therefore, the angle of $\mathbf{G}(j\omega)$ at low frequencies is greater than $-180°$. As the frequency in-

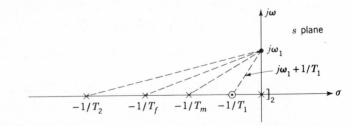

FIGURE 8-17
Pole-zero diagram for

$$G_o(s) = \frac{K_2' T_1}{T_f T_m T_2} \frac{s + 1/T_1}{s^2(s + 1/T_f)(s + 1/T_m)(s + 1/T_2)}$$

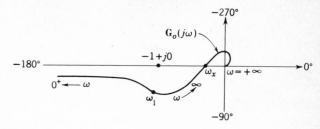

FIGURE 8-18
Polar plot for a typical Type 2 transfer function, resulting in a stable system [Eq. (8-45)].

creases to a value ω_x, the sum of the component angles of $\mathbf{G}(j\omega)$ is $-180°$ and the polar plot crosses the real axis, as shown in Fig. 8-18. As ω increases further, the angle of $j\omega + 1/T_1$ shows only a small increase, but the angles from the poles increase rapidly. In the limit, as $\omega \to \infty$, the angles of $j\omega + 1/T_1$ and $j\omega + 1/T_2$ are equal and opposite in sign, so the angle of $\mathbf{G}(j\omega)$ approaches $-360°$, as given by Eq. (8-44).

Figure 8-18 shows the complete polar plot of $\mathbf{G}(j\omega)$. A comparison of Figs. 8-16 and 8-18 shows that both curves approach $-180°$ at $\omega = 0^+$, which is typical of Type 2 systems. As $\omega \to \infty$, the angle approaches $-360°$ since both $\mathbf{G}(j\omega)$ and $\mathbf{G}_o(j\omega)$ have the same degree, $n - w = 4$. When the Nyquist stability criterion described in Secs. 8-13 and 8-14 is used, the feedback system containing $\mathbf{G}(j\omega)$ can be shown to be unstable, while the system containing $\mathbf{G}_o(j\omega)$ is stable.

It can be shown that as $\omega \to 0^+$, the polar plot for a Type 2 system is below the real axis if

$$\Sigma(T_{\text{numerator}}) - \Sigma(T_{\text{denominator}})$$

is a positive value and is above the real axis if it is a negative value. Thus for this example the necessary condition is $T_1 > T_f + T_m + T_2$.

8-11 SUMMARY OF DIRECT POLAR PLOTS

To obtain the direct polar plot of a system's forward transfer function, the following criteria are used to determine the key parts of the curve.

Step 1 The forward transfer function has the general form

$$\mathbf{G}(j\omega) = \frac{K_m(1 + j\omega T_a)(1 + j\omega T_b) \cdots (1 + j\omega T_w)}{(j\omega)^m(1 + j\omega T_1)(1 + j\omega T_2) \cdots (1 + j\omega T_u)} \qquad (8\text{-}46)$$

For this transfer function the system type is determined. Then the portion of the polar plot representing the $\lim_{\omega \to 0} \mathbf{G}(j\omega)$ is approximately located. The low-frequency polar-plot characteristics (as $\omega \to 0$) of the different system types are summarized in Fig. 8-19. The angle at $\omega = 0$ is $m(-90°)$.

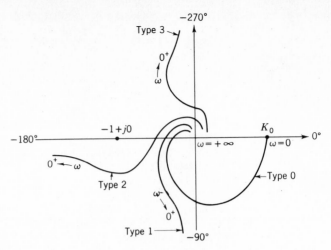

FIGURE 8-19
A summary of direct polar plots for different types of systems.

Step 2   The high-frequency end of the polar plot can be determined as follows:

$$\lim_{\omega \to +\infty} \mathbf{G}(j\omega) = 0/\underline{(w - m - u)90°} \qquad (8\text{-}47)$$

Note that since the degree of the denominator of Eq. (8-46) is always greater than the degree of the numerator, the high-frequency point ($\omega = \infty$) is approached (i.e., the angular condition) in the clockwise sense. The plot ends at the origin tangent to the axis determined by Eq. (8-47). Tangency may occur on either side of the axis.

Step 3   The asymptote that the low-frequency end approaches, for a Type 1 system, is determined by taking the limit as $\omega \to 0$ of the real part of the transfer function.

Step 4   The frequencies at the points of intersection of the polar plot with the negative real axis and the imaginary axis are determined, respectively, by setting

$$\text{Im}\,[\mathbf{G}(j\omega)] = 0 \qquad (8\text{-}48)$$

$$\text{Re}\,[\mathbf{G}(j\omega)] = 0 \qquad (8\text{-}49)$$

Step 5   If there are no time constants in the numerator of the transfer function, the curve is a smooth one in which the angle of $\mathbf{G}(j\omega)$ continuously decreases as ω goes from 0 to ∞. With time constants in the numerator, and depending upon their values, the angle may not continuously vary in the same direction, thus creating "dents" in the polar plot.

Step 6   As is seen later in this chapter, it is important to know the exact shape of the polar plot in the vicinity of the $-1 + j0$ point. Enough points of $\mathbf{G}(j\omega)$ should be accurately determined in this area.

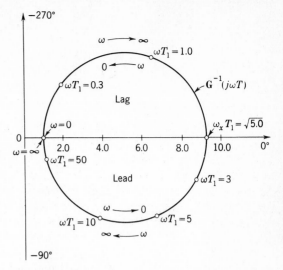

FIGURE 8-20
The inverse polar plot of a lag-lead network for $T_2 = 0.2T_1$, $T_{12} = 10.0T_1$:

$$G^{-1}(j\omega T_1) = \frac{(1 + j11.1\omega T_1)(1 + j0.0179\omega T_1)}{(1 + j\omega T_1)(1 + j0.2\omega T_1)}$$

8-12 INVERSE POLAR PLOTS

The direct polar plots have certain drawbacks when they are used for systems that utilize elements in the feedback path. In these cases it is found that the graphical analysis of the inverse polar plots is much simpler and therefore advantageous. The inverse polar plot is obtained by plotting the phasor quantity

$$G^{-1}(j\omega) = \frac{1}{G(j\omega)} = \frac{E(j\omega)}{C(j\omega)} \qquad (8\text{-}50)$$

as a function of frequency.

Compensators

The inverse plot of the lag-lead compensator discussed in Sec. 8-10 is illustrated in Fig. 8-20.

Type 0 Feedback Control System

The inverse of the transfer function used in Sec. 8-10 for the Type 0 system of Eq. (8-32) is

$$G^{-1}(j\omega) = \frac{(1 + j\omega T_f)(1 + j\omega T_m)}{K_0} \qquad (8\text{-}51)$$

The two limiting points are

$$\lim_{\omega \to 0^+} G^{-1}(j\omega) = \frac{1}{K_0}$$

$$\lim_{\omega \to +\infty} G^{-1}(j\omega) = \lim_{\omega \to +\infty} (j\omega)^2 T_f T_{ma} = \infty/\underline{180°}$$

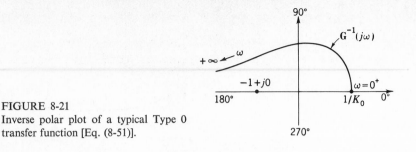

FIGURE 8-21
Inverse polar plot of a typical Type 0
transfer function [Eq. (8-51)].

As ω goes from 0 to ∞, the angle of each term in the numerator of $\mathbf{G}^{-1}(j\omega)$ goes
from 0 to 90°. Since there are no frequency-dependent terms in the denominator,
the angular condition of $\mathbf{G}^{-1}(j\omega)$ is continuously increasing from 0 to 180° as ω
goes from 0 to ∞. The inverse plot of this Type 0 system is shown in Fig. 8-21. Each
additional $1 + j\omega T$ factor in the numerator of $\mathbf{G}^{-1}(j\omega)$ rotates the high-frequency
portion of the plot counterclockwise by 90°.

Type 1 Feedback Control System

The inverse of the transfer function used in Sec. 8-10 for the Type 1 system of Eq.
(8-35) is

$$\mathbf{G}^{-1}(j\omega) = \frac{j\omega(1 + j\omega T_m)(1 + j\omega T_c)(1 + j\omega T_q)}{K_1} \qquad (8\text{-}52)$$

The two limiting points are

$$\lim_{\omega \to 0^+} \mathbf{G}^{-1}(j\omega) = 0\underline{/90°} \qquad \lim_{\omega \to +\infty} \mathbf{G}^{-1}(j\omega) = \infty\underline{/360°}$$

With these end points and previous analyses, the inverse polar plot of Eq. (8-52)
is a smooth curve, as shown in Fig. 8-22. Additional terms in the numerator or the
denominator alter the location of the end points and the shape of the plot accordingly.

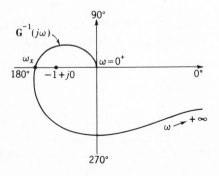

FIGURE 8-22
Inverse polar plot of a typical Type 1
transfer function [Eq. (8-52)].

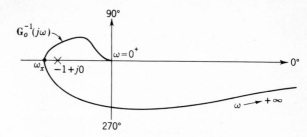

FIGURE 8-23
Inverse polar plot of a typical Type 2 transfer function [Eq. (8-53)].

Type 2 Feedback Control System

In the same manner, the inverse polar plot for the modified Type 2 system of Eq. (8-45) can be obtained. The inverse of the transfer function of Eq. (8-45) for this system is

$$G_o^{-1}(j\omega) = \frac{(j\omega)^2(1 + j\omega T_f)(1 + j\omega T_m)(1 + j\omega T_2)}{K_2'(1 + j\omega T_1)} \qquad (8\text{-}53)$$

and its polar plot is illustrated in Fig. 8-23.

The inverse polar plots of a forward transfer function are obtained with the aid of the following criteria, which determine the key parts of the curve.

Step 1   Once the system type is determined from the forward transfer function, the portion of the inverse polar plot representing the $\lim\limits_{\omega \to 0^+} G^{-1}(j\omega)$ is located. Note that in taking the limit the zero-frequency point is approached in the clockwise sense.

Step 2   By using the general form of the inverse forward transfer function as given by

$$G^{-1}(s) = \frac{s^m(1 + T_1 s)(1 + T_2 s) \cdots (1 + T_u s)}{K_m(1 + T_a s) \cdots (1 + T_w s)} \qquad (8\text{-}54)$$

the high-frequency end of the polar plot is determined as follows:

$$\lim_{\omega \to +\infty} \mathbf{G}^{-1}(j\omega) = \infty \underline{/(m + u - w)90^\circ} \qquad (8\text{-}55)$$

Step 3   The frequencies at the points of intersection of the polar plot with the negative real axis and the imaginary axis are determined, respectively, by setting

$$\text{Im}\,[\mathbf{G}^{-1}(j\omega)] = 0 \qquad (8\text{-}56)$$
$$\text{Re}\,[\mathbf{G}^{-1}(j\omega)] = 0 \qquad (8\text{-}57)$$

Step 4   The exact shape of the polar plot in the vicinity of the negative real axis crossing is important, and points on $\mathbf{G}^{-1}(j\omega)$ should be accurately determined in this area with the aid of step 3 or a computer.[4]

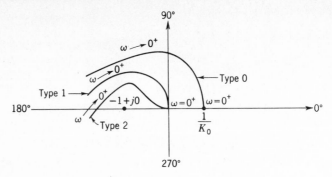

FIGURE 8-24
A summary of inverse polar plots for different types of systems.

A summary of the inverse polar plots for the different types of systems is made in Fig. 8-24. Just as for the direct polar plots, the low-frequency characteristic (as $\omega \to 0^+$) distinguishes the different system types.

8-13 NYQUIST'S STABILITY CRITERION

A system designer must be sure that the closed-loop system he designs is stable. The Nyquist stability criterion[2,3,7] provides a simple graphical procedure for determining closed-loop stability from the frequency-response curves of the open-loop transfer function $G(j\omega)H(j\omega)$. The application of this method in terms of the polar plot is covered in this section; application in terms of the log magnitude–angle (Nichols) diagram is covered in Sec. 8-19.

For a stable system the roots of the characteristic equation

$$B(s) = 1 + G(s)H(s) = 0 \qquad (8\text{-}58)$$

cannot be permitted to lie in the right-half s plane or on the $j\omega$ axis, as shown in Fig. 8-25. In terms of $G = N_1/D_1$ and $H = N_2/D_2$, this becomes

$$B(s) = 1 + \frac{N_1 N_2}{D_1 D_2} = \frac{D_1 D_2 + N_1 N_2}{D_1 D_2} \qquad (8\text{-}59)$$

It is seen that *the poles of the open-loop transfer function $G(s)H(s)$ are the poles of $B(s)$.* The condition for stability may be restated: for a stable system none of the zeros of $B(s)$ can lie in the right-half s plane or on the imaginary axis. Briefly, Nyquist's stability criterion relates the number of zeros and poles of $B(s)$ that lie in the right-half s plane to the polar plot of $G(s)H(s)$.

Limitations

In this analysis it is assumed that all the control systems are inherently linear or that their limits of operation are confined to give a linear operation. This yields a set of

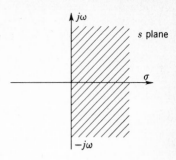

FIGURE 8-25
Prohibited region in the s plane.

linear differential equations which describe the dynamic performance of the systems. Because of the physical nature of feedback control systems, the order of the denominator D_1D_2 is equal to or greater than the order of the numerator N_1N_2 of the open-loop transfer function $G(s)H(s)$. Mathematically, this means that $\lim_{s\to\infty} G(s)H(s) \to 0$ or a constant. These two factors satisfy the necessary limitations to the generalized Nyquist stability criterion.

Mathematical Basis for Nyquist's Stability Criterion

A rigorous mathematical derivation of Nyquist's stability criterion involves complex-variable theory, and a rigorous derivation is available in the literature. Fortunately, the result of the derivation is simple and is readily applied. However, a complete knowledge of its derivation ensures that it will be used with greater facility and sureness. A qualitative approach is now presented for the special case that $B(s)$ is a rational fraction. The function $B(s)$ given by Eq. (8-59) is rationalized, factored, and then written in the form

$$B(s) = \frac{(s - Z_1)(s - Z_2)\cdots(s - Z_n)}{(s - p_1)(s - p_2)\cdots(s - p_n)} \qquad (8\text{-}60)$$

where $Z_1, Z_2, \ldots, Z_n$ are the zeros of the characteristic function and $p_1, p_2, \ldots, p_n$ are the poles of the characteristic function. Remember that z_i is used to denote a zero of $G(s)H(s)$. The poles p_i are the same as the poles of the open-loop transfer function $G(s)H(s)$ and include the s term for which $p = 0$, if it is present.

In Fig. 8-26 some of the poles and zeros of a generalized function $B(s)$ are arbitrarily drawn on the s plane. Also, an arbitrary *closed* curve Q' is drawn which encloses the zero Z_1. To the point O' on Q', whose coordinates are $s = \sigma + j\omega$, are drawn directed line segments from all the poles and zeros. The lengths of these directed line segments are given by $s - Z_1, s - Z_2, s - p_1, s - p_2$, etc. Not all the directed segments from the poles and zeros are indicated in the figure, as they are not necessary to proceed with this development. *As the point O' is rotated clockwise once around the closed curve Q', the length $s - Z_1$ rotates through a net angle of* 360°. *All the other directed* segments have rotated through a *net angle of* 0°. Thus, by referring to Eq. (8-60) it is seen that the clockwise rotation of 360° for the length $s - Z_1$ *must simultaneously be realized* by the function $B(s)$ for the enclosure of the zero Z_1 by the path Q'.

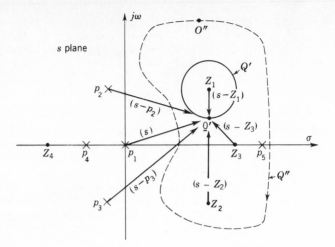

FIGURE 8-26
A plot of some poles and zeros of Eq. (8-60).

Consider now a larger closed contour Q'' which includes the zeros Z_1, Z_2, Z_3 and the pole p_5. *As a point O'' is rotated clockwise once around the closed curve Q'', each of the directed line segments from the enclosed pole and zeros rotates through a net angle of* 360°. Since the angular rotation of the pole is experienced by the characteristic function in its denominator, the net angular rotation realized by Eq. (8-60) must be equal to the net angular rotations due to the pole p_5 minus the net angular rotations due to the zeros Z_1, Z_2, and Z_3. In other words, the net angular rotation experienced by $1 + G(s)H(s)$ is 360° − (3)(360°) = −720°. Therefore, for this case, the net number of rotations N experienced by $1 + G(s)H(s)$ by the clockwise movement of point O'' *once* about the closed contour Q'' is equal to −2; that is,

(Number of poles enclosed) − (number of zeros enclosed) = 1 − 3 = −2

where the minus sign denotes clockwise (cw) rotation. Note that if the contour Q'' includes only the pole p_5, $B(s)$ experiences one counterclockwise (ccw) rotation as the point O'' is moved around the contour. Also, for *any closed path* that may be chosen, all the poles and zeros that lie outside the closed path each contribute a net angular rotation of 0° to $B(s)$ as a point is moved once around this contour.

Generalizing Nyquist's Stability Criterion

Consider now a closed contour Q such that the whole right-half s plane is encircled (see Fig. 8-27), thus encircling all zeros and poles of $B(s)$ that may have positive real parts. As a consequence of the theory of complex variables used in the derivation, it is mandatory that the contour Q not pass *through any poles or zeros of $B(s)$*. When the results previously discussed are applied to the contour Q, the following conclusions are reached:

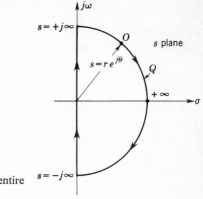

FIGURE 8-27
The contour that encloses the entire right-half s plane.

1 The total number of clockwise rotations of $B(s)$ due to its zeros is equal to the total number of zeros Z_R in the right-half s plane.
2 The total number of counterclockwise rotations of $B(s)$ due to its poles is equal to the total number of poles P_R in the right-half s plane.
3 The *net* number of rotations N of $B(s) = 1 + G(s)H(s)$ about the origin is equal to its total number of poles P_R minus its total number of zeros Z_R in the right-half s plane. N may be positive (ccw), negative (cw), or zero.

The essence of these three conclusions can be represented by the equation

$$N = \frac{\text{change in phase of } [1 + G(s)H(s)]}{2\pi} = P_R - Z_R \qquad (8\text{-}61)$$

where counterclockwise rotation is defined as being positive and clockwise rotation is negative. In order for the characteristic function $B(s)$ to realize a net rotation N, the directed line segment representing $B(s)$ (see Fig. 8-28) must rotate about the origin $360N$ degrees, or N complete revolutions.

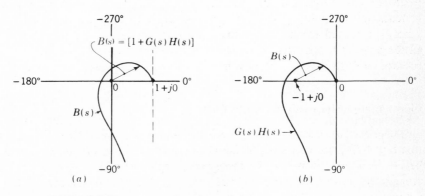

FIGURE 8-28
A change of reference for $B(s)$.

In a stable system $B(s)$ can have no zeros Z_R in the right-half s plane; therefore it can be concluded that, *for a stable system, the net number of rotations of $B(s)$ about the origin must be counterclockwise and equal to the number of poles P_R that lie in the right-half s plane; i.e.,*

$$N = P_R \qquad (8\text{-}62)$$

In other words, if $B(s)$ experiences a net clockwise rotation, this indicates that $Z_R > P_R$, where $P_R \geq 0$, and the system is unstable. If there are zero net rotations, then $Z_R = P_R$ and the system may or may not be stable, according as $P_R = 0$ or $P_R > 0$.

Figure 8-28a and b shows a plot of $B(s)$ and a plot of $G(s)H(s)$. By moving the origin of Fig. 8-28b to the $-1 + j0$ point, the curve is now equal to $1 + G(s)H(s)$, which is $B(s)$. Since $G(s)H(s)$ is known, this function is plotted and then the origin is moved to the -1 point to obtain $B(s)$.

In general, the open-loop transfer functions of many physical systems do not have any poles P_R in the right-half s plane. In this case, $N = Z_R$. *Thus for a stable system the net number of rotations about the $-1 + j0$ point must be zero when there are no poles of $G(s)H(s)$ in the right-half s plane.*

In the event that the function $G(s)H(s)$ has some poles in the right-half s plane and the denominator is not in factored form, the number P_R can be determined by applying Routh's criterion to $D_1 D_2$. The Routhian array gives the number of roots in the right-half s plane by the number of sign changes in the first column.

Analysis of Path Q

In applying Nyquist's criterion, the whole right-half s plane must be encircled to ensure the inclusion of all poles or zeros in this portion of the plane. In Fig. 8-27 the entire right-half s plane is included by considering the closed path Q to be composed of the following two segments:

1 One segment is the imaginary axis from $-j\infty$ to $+j\infty$.
2 The other segment is a semicircle of infinite radius that encircles the entire right-half s plane.

The portion of the path along the imaginary axis is represented mathematically by $s = j\omega$. Thus, replacing s by $j\omega$ in Eq. (8-60) and letting ω take on all values from $-\infty$ to $+\infty$ gives the portion of the $B(s)$ plot corresponding to that portion of the closed contour Q on the imaginary axis.

One of the limitations of the Nyquist criterion is that $\lim\limits_{s \to \infty} G(s)H(s) \to 0$ or a constant. Thus $\lim\limits_{s \to \infty} B(s) = \lim\limits_{s \to \infty} [1 + G(s)H(s)] \to 1$ or 1 plus the constant. As a consequence, as the point O moves along the segment of the closed contour represented by the semicircle of infinite radius, the corresponding portion of the $B(s)$ plot is a fixed point. As a result, the movement of point O along only the imaginary axis from $-j\infty$ to $+j\infty$ results in giving the same net rotation of $B(s)$ as if the whole contour Q were considered. *In other words, all the rotation of $B(s)$ occurs while the point O goes from $-j\infty$ to $+j\infty$ along the imaginary axis if the denominator of $B(s)$ is at least one degree greater than the degree of the numerator.* More generally, this

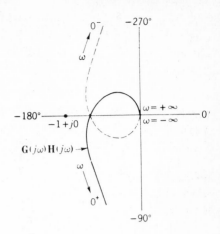

FIGURE 8-29
A plot of the transfer function
$G(j\omega)H(j\omega)$ for Eq. (8-63).

statement applies only to those transfer functions $G(s)H(s)$ that conform to the limitations stated earlier in this section.†

Effect of Poles at the Origin on the Rotation of $B(s)$

Some transfer functions $G(s)H(s)$ have an s^m in the denominator. Since no poles or zeros can lie on the contour Q, the contour shown in Fig. 8-27 must be modified. For these cases the manner in which the $\omega = 0^-$ and $\omega = 0^+$ portions of the plot are joined is now investigated. This can best be done by taking an example. Consider the transfer function

$$G(s)H(s) = \frac{K_1}{s(1 + T_1 s)(1 + T_2 s)} \qquad (8\text{-}63)$$

To obtain the direct polar plot $G(j\omega)H(j\omega)$ of this function, substitute $s = j\omega$ into Eq. (8-63). Figure 8-29 represents the plot of $G(s)H(s)$ for values of s along the imaginary axis; that is, $j0^+ < j\omega < j\infty$ and $-j\infty < j\omega < j0^-$. The plot is drawn for both positive and negative frequency values. *The polar plot drawn for negative frequencies is the conjugate of the plot drawn for positive frequencies.* This means that the curve for negative frequencies is symmetrical to the curve for positive frequencies, with the real axis as the axis of symmetry.

To determine the system stability, the closed contour Q of Fig. 8-27, in the vicinity of $s = 0$, is modified to avoid passing through the origin, as shown in Fig. 8-30a. In other words, the point O is moved along the negative imaginary axis from $s = -j\infty$ to a point where $s = -j\omega = 0^-\underline{/-\pi/2}$ becomes very small; that is, $s = -j\varepsilon$. Then the point O moves along a semicircular path of radius $s = \varepsilon e^{j\theta}$ in the right-half s plane with a very small radius ε until it reaches the positive imaginary axis at $s = +j\omega = j0^+ = 0^+\underline{/\pi/2}$. From here the point O proceeds along the positive imaginary axis to $s = +j\infty$. Letting the radius approach zero, $\varepsilon \to 0$, for

---

† A transfer function that does not conform to these limitations and to which the above italicized statement does not apply is $G(s)H(s) = s$.

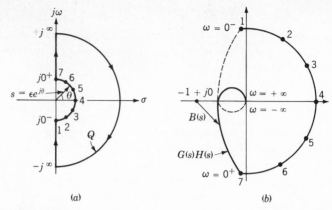

FIGURE 8-30
(a) The contour Q which encircles the right-half s plane.
(b) Complete plot for Eq. (8-63).

the semicircle around the origin ensures the inclusion of all poles and zeros in the right-half s plane. To complete the plot of $B(s)$ the effect of moving point O on this semicircle around the origin must be investigated.

For the semicircular portion of the path Q represented by $s = \varepsilon e^{j\theta}$, where $\varepsilon \to 0$ and $-\pi/2 \le \theta \le \pi/2$, Eq. (8-63) becomes

$$G(s)H(s) = \frac{K_1}{s} = \frac{K_1}{\varepsilon e^{j\theta}} = \frac{K_1}{\varepsilon} e^{-j\theta} = \frac{K_1}{\varepsilon} e^{j\psi} \qquad (8\text{-}64)$$

where $K_1/\varepsilon \to \infty$ as $\varepsilon \to 0$, and $\psi = -\theta$ goes from $\pi/2$ to $-\pi/2$ as the directed segment s goes from $\varepsilon\underline{/-\pi/2}$ to $\varepsilon\underline{/+\pi/2}$. Thus, in Fig. 8-29, the end points from $\omega \to 0^-$ and $\omega \to 0^+$ are joined by a semicircle of infinite radius in the first and fourth quadrants. Figure 8-30 illustrates the above procedure and shows the completed contour of $G(s)H(s)$ as the point O moves along the modified contour Q in the s plane in the clockwise direction. When the origin is moved to the $-1 + j0$ point, the curve becomes $B(s)$.

The plot of $B(s)$ in Fig. 8-30b does not encircle the $-1 + j0$ point; therefore the encirclement N is zero. From Eq. (8-63) there are no poles within Q; that is, $P_R = 0$. Thus, $Z_R = 0$ and the system is stable.

Transfer functions that have more than one s factor in the denominator have the general form, as $\varepsilon \to 0$,

$$G(s)H(s) = \frac{K_m}{s^m} = \frac{K_m}{(\varepsilon^m)e^{jm\theta}} = \frac{K_m}{\varepsilon^m} e^{-jm\theta} = \frac{K_m}{\varepsilon^m} e^{jm\psi} \qquad (8\text{-}64a)$$

where $m = 1, 2, 3, 4, \ldots$. With the reasoning used in the preceding example, it is seen from Eq. (8-64a) that, as s moves from 0^- to 0^+, the plot of $G(s)H(s)$ traces m clockwise semicircles of infinite radius about the origin. If $m = 2$, then, as θ goes from $-\pi/2$ to $\pi/2$ in the s plane, $G(s)H(s)$ experiences a net rotation, in the vicinity of $s = 0$, of $(2)(180°)$, or $360°$.

Since the polar plots are symmetrical about the real axis, it is only necessary to determine the shape of the plot of $G(s)H(s)$ for a range of values of $0 < \omega < +\infty$. The net rotation of the plot for the range of $-\infty < \omega < +\infty$ is twice that of the plot for the range of $0 < \omega < +\infty$.

When $G(j\omega)H(j\omega)$ Passes Through the Point $-1 + j0$

When the curve of $G(j\omega)H(j\omega)$ passes through the $-1 + j0$ point, the number of encirclements N is indeterminate. This corresponds to the condition where $B(s)$ has zeros on the imaginary axis. A necessary condition for the Nyquist criterion is that the path encircling the specified area must not pass through any poles or zeros of $B(s)$. When this condition is violated, the value for N becomes indeterminate and the Nyquist stability criterion cannot be applied. Simple imaginary zeros of $B(s)$ mean that the closed-loop system will have a continuous steady-state sinusoidal component in its output which is independent of the form of the input. Unless otherwise stated, this condition is considered unstable.

8-14 EXAMPLES OF NYQUIST'S CRITERION USING DIRECT POLAR PLOT

Several polar plots are illustrated in this section. These plots can be obtained with the aid of pole-zero diagrams. In other words, the angular variation of each term of a $G(s)H(s)$ function, as the contour Q is traversed, is readily determined from its pole-zero diagram. Both minimum- and non-minimum-phase systems are illustrated in the following examples.

EXAMPLE 1   *Type 0*   The direct polar plot is shown in Fig. 8-31 for the following Type 0 transfer function:

$$G(s)H(s) = \frac{K_0}{(1 + T_1 s)(1 + T_2 s)} \qquad (8\text{-}65)$$

From this plot it is seen that $N = 0$, with both T_1 and T_2 positive; Eq. (8-65) yields the value $P_R = 0$. Therefore, the value of Z_R is $Z_R = P_R - N = 0$ and the system represented by Eq. (8-65) is stable. Thus the criterion indicates that no matter how much the gain K_0 is increased, this system is always stable.

EXAMPLE 2   *Type 1*   The example in Fig. 8-30 for the transfer function of the form

$$G(s)H(s) = \frac{K_1}{s(1 + T_1 s)(1 + T_2 s)} \qquad (8\text{-}66)$$

is shown in the preceding section to be stable. If the gain K_1 is increased, the system is made unstable, as seen from Fig. 8-32. Note that the rotation of $B(s)$, in the direction

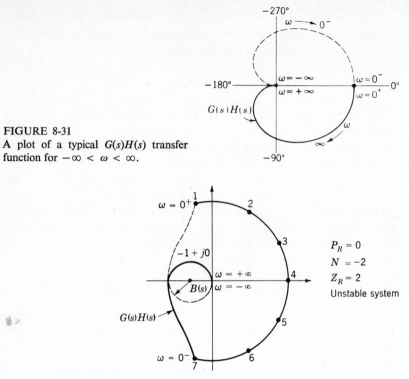

FIGURE 8-31
A plot of a typical $G(s)H(s)$ transfer function for $-\infty < \omega < \infty$.

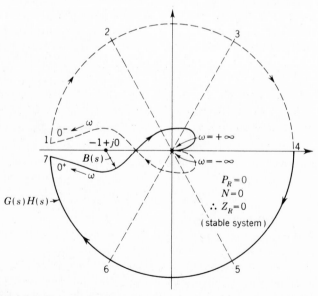

$P_R = 0$
$N = -2$
$Z_R = 2$
Unstable system

FIGURE 8-32
Polar plot of Fig. 8-30 with increased gain.

$P_R = 0$
$N = 0$
$\therefore Z_R = 0$
(stable system)

FIGURE 8-33
The complete polar plot of Eq. (8-67).

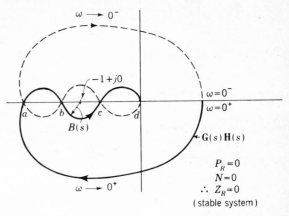

FIGURE 8-34
The complete polar plot of Eq. (8-68).

of increasing frequency, produces a net angular rotation of $-720°$, or two complete clockwise rotations ($N = -2$).

EXAMPLE 3   *Type 2*

$$G(s)H(s) = \frac{K_2(1 + T_4 s)}{s^2(1 + T_1 s)(1 + T_2 s)(1 + T_3 s)} \qquad (8\text{-}67)$$

where $T_4 > T_1$, T_2, and T_3. Figure 8-33 shows the mapping of $G(s)H(s)$ for the contour Q of the s plane. The word *mapping*, as used here, means that for a given point in the s plane there corresponds a given point in the $G(s)H(s)$ or $B(s)$ plane.

As pointed out in the preceding section, the presence of the s^2 term in the denominator of Eq. (8-67) results in a net rotation of $360°$ in the vicinity of $\omega = 0$, as shown in Fig. 8-33. For the complete range of frequencies the net rotation is zero; thus with $P_R = 0$ the system is stable. Like the previous example, this system can be made unstable by increasing the gain sufficiently for the $G(s)H(s)$ plot to cross the negative real axis to the left of the $-1 + j0$ point.

EXAMPLE 4   *Conditionally stable system*

$$G(s)H(s) = \frac{K_0(1 + T_1 s)^2}{(1 + T_2 s)(1 + T_3 s)(1 + T_4 s)(1\ T + {}_5 s)^2} \qquad (8\text{-}68)$$

where $T_5 < T_1 < T_2$, T_3, and T_4. The complete polar plot of Eq. (8-68) is illustrated in Fig. 8-34 for a particular value of gain.

In this example it is seen that the system can be made unstable not only by increasing the gain but also by decreasing the gain. If the gain is increased sufficiently for the $-1 + j0$ point to lie between the points c and d of the polar plot, the net clockwise rotation is equal to 2. Therefore $Z_R = 2$ and the system is unstable. On

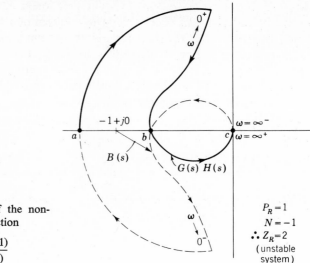

FIGURE 8-35
The complete polar plot of the non-minimum-phase transfer function

$$G(s)H(s) = \frac{K_1(T_2s + 1)}{s(T_1s - 1)}$$

$P_R = 1$
$N = -1$
$\therefore Z_R = 2$
(unstable system)

the other hand, if the gain is decreased so that the $-1 + j0$ point lies between the points a and b of the polar plot, the net clockwise rotation is again 2 and the system is unstable. The gain can be further decreased so that the $-1 + j0$ point lies to the left of point a of the polar plot, resulting in a stable system. *This system is therefore conditionally stable.* A conditionally stable system is stable for a given range of values of gain but becomes unstable if the gain is either reduced or increased sufficiently. Such a system places a greater restriction on the stability and drift of amplifier-gain characteristics. In addition, an effective gain reduction occurs in amplifiers that reach saturation with large input signals. This, in turn, may result in an unstable operation for the conditionally stable system.

EXAMPLE 5 *Nonminimum phase*

$$G(s)H(s) = \frac{K_1(T_2s + 1)}{s(T_1s - 1)} \qquad (8\text{-}69)$$

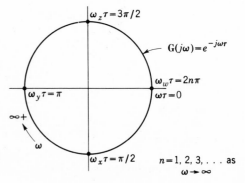

FIGURE 8-36
Polar-plot characteristic for transport lag.

The complete polar plot for this equation is illustrated in Fig. 8-35 for a particular value of gain. In this example it is seen that the system is unstable for low values of gain $(0 < K_1 < K_{1x})$ and that the system is stable for large values of gain $(K_{1x} < K_1 < \infty)$. For the range $0 < K_1 < K_{1x}$ the $-1 + j0$ point is located, as shown in Fig. 8-35, between the points a and b. For the range $K_{1x} < K_1 < \infty$ the $-1 + j0$ point lies between the points b and c, which yields $N = +1$, thus resulting in $Z_R = 0$ and a stable system.

8-15 NYQUIST'S STABILITY CRITERION APPLIED TO SYSTEMS HAVING DEAD TIME

Transport lag, as described in Sec. 7-12, is represented by the transfer function $G_\tau(s) = e^{-\tau s}$. The frequency transfer function is

$$\mathbf{G}(j\omega) = e^{-j\omega\tau} = 1\underline{/-\omega\tau} \qquad (8\text{-}70)$$

It has a magnitude of unity and a negative angle whose magnitude increases directly in proportion to frequency. The polar plot of Eq. (8-70) is a unit circle which is traced indefinitely, as shown in Fig. 8-36. The log magnitude and phase-angle diagram shows a constant value of 0 dB and a phase angle which decreases with frequency.

An example can best illustrate the application of the Nyquist criterion to a control system having dead time, or transport lag. Figure 8-37a illustrates the complete polar plot of a stable system having a specified value of gain and the transfer function

$$G_x(s)H(s) = \frac{K_1}{s(1 + T_1 s)(1 + T_2 s)} \qquad (8\text{-}71)$$

If dead time is added to this system, its transfer function becomes

$$G(s)H(s) = \frac{K_1 e^{-\tau s}}{s(1 + T_1 s)(1 + T_2 s)} \qquad (8\text{-}72)$$

and the resulting complete polar plot is shown in Fig. 8-37b. When the contour Q is traversed and the polar-plot characteristic of dead time, shown in Fig. 8-36, is included, the effects on the complete polar plot are as follows:

1 In traversing the imaginary axis of the contour Q between $0^+ < \omega < +\infty$, the polar plot of $\mathbf{G}(j\omega)\mathbf{H}(j\omega)$ in the third quadrant is shifted clockwise, closer to the $-1 + j0$ point. Thus, if the dead time is increased sufficiently, the $-1 + j0$ point will be enclosed by the polar plot and the system becomes unstable.

2 As $\omega \to +\infty$, the magnitude of the angle contributed by the transport lag increases indefinitely. This yields a spiraling curve as $|\mathbf{G}(j\omega)\mathbf{H}(j\omega)| \to 0$.

A transport lag therefore tends to make a system less stable.

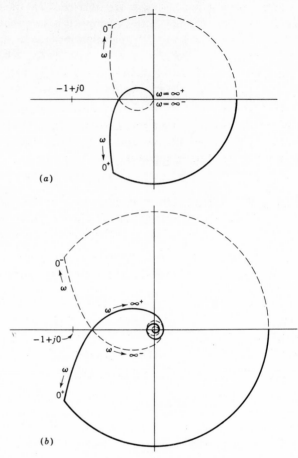

FIGURE 8-37
Complete polar plots of: (*a*) Eq. (8-71); (*b*) Eq. (8-72).

8-16 NYQUIST'S STABILITY CRITERION APPLIED TO THE INVERSE POLAR PLOTS

The criterion applied to the inverse plots is the same as that for the direct plots except for one minor modification. As previously stipulated, for a system to be stable, none of the roots of the characteristic equation $B(s) = 1 + G(s)H(s) = 0$ can lie in the right-half s plane or on the $j\omega$ axis. Dividing this equation by $G(s)H(s)$ yields

$$B'(s) = \frac{1}{G(s)H(s)} + 1 = 0 \qquad (8\text{-}73)$$

In terms of $G = N_1/D_1$ and $H = N_2/D_2$, the zeros of $B'(s)$ are the roots of the characteristic equation

$$D_1 D_2 + N_1 N_2 = 0 \qquad (8\text{-}74)$$

which is seen to be the same as the zeros of $B(s)$. Equation (8-73) can be expressed as

$$B'(s) = \frac{(s - Z_1)(s - Z_2) \cdots (s - Z_n)}{(s - z_1)(s - z_2) \cdots (s - z_w)} \qquad (8\text{-}75)$$

where $Z_1, Z_2, \ldots, Z_n$ are the zeros of the functions $B(s)$ and $B'(s)$, and $z_1, z_1, \ldots, z_w$ are the poles of $B'(s)$, which are the same as the zeros of $G(s)H(s)$.

The mathematical basis for Nyquist's stability criterion as applied to the inverse plots is the same as that stipulated in Sec. 8-13 for the direct plots. When $B'(s)$ takes on the values on the contour Q which encircles the entire right-half of the s plane, the resulting equation is

$$Z_R = P'_R - N' \qquad (8\text{-}76)$$

where N' = number of net rotations of $B'(s)$

P'_R = number of zeros of $G(s)H(s)$ in right-half s plane

Z_R = number of roots of characteristic equation in right-half plane

The plot of $B'(s) = 1/G(s)H(s) + 1$ is most easily obtained by first plotting the known function $1/G(s)H(s)$. Then the origin is moved to the $-1 + j0$ point to obtain $B'(s)$.

Analysis of Path Q

Replacing s by $j\omega$ in $[G(s)H(s)]^{-1}$ and letting ω take on all values from $-\infty$ to $+\infty$ gives the portion of the plot corresponding to that portion of the closed contour Q on the imaginary axis of Fig. 8-27. For the case of $\omega = 0$ the value is a constant. Thus, as the point O on the closed contour Q passes through the point $0 + j0$, the corresponding portion of the $B'(s)$ plot crosses the real axis.

To complete the plot of $[G(s)H(s)]^{-1}$ the effect of moving point O on the semicircle of infinite radius on the contour, which ensures the inclusion of all poles and zeros in the right-half plane, must be investigated. This semicircle is represented by

$$s = re^{j\theta} \qquad (8\text{-}77)$$

where $r \rightarrow \infty$ and $\pi/2 \geq \theta \geq -\pi/2$. The general open-loop transfer function is

$$G(s)H(s) = \frac{K_m(s - z_1) \cdots (s - z_w)}{s^m(s - p_1)(s - p_2) \cdots (s - p_u)} \qquad (8\text{-}78)$$

The function $[G(s)H(s)]^{-1}$ for this portion of the contour Q has the value

$$\lim_{s \rightarrow \infty} \frac{1}{G(s)H(s)} = \frac{s^m s^u}{K_m s^w} = \frac{r^{(m+u-w)}}{K_m} e^{j(m+u-w)\theta} \qquad (8\text{-}79)$$

From this equation it is easily seen that, as the point O, for $r = \infty$, goes from $\pi/2$ to $-\pi/2$ on the path Q in the s plane, the plot of $1/G(s)H(s)$ makes $m + u - w$ semicircles of infinite radius about the origin.

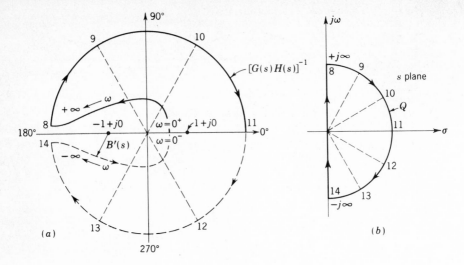

FIGURE 8-38
Complete contour of Eq. (8-80) and its correlation to the Q contour in the s plane.

8-17 EXAMPLES OF NYQUIST'S CRITERION USING THE INVERSE POLAR PLOTS

EXAMPLE 1 *Type 0* The inverse polar plot of

$$[G(s)H(s)]^{-1} = \frac{(1 + T_1 s)(1 + T_2 s)}{K_0} \qquad (8\text{-}80)$$

is illustrated in Fig. 8-38. For this example

$$\lim_{s \to \infty} [G(s)H(s)]^{-1} = \frac{T_1 T_2}{K_0} r^2 e^{j2\theta} \qquad (8\text{-}81)$$

By referring to Fig. 8-38 and using Eq. (8-81), the complete polar plot of $[G(s)H(s)]^{-1}$ shown in Fig. 8-38a is obtained. By moving the origin to the $-1 + j0$ point, the curve becomes $B'(s)$. When Nyquist's criterion is applied to this plot, the net encirclement of the origin is zero. From Eq. (8-80) the value $P'_R = 0$; therefore,

$$Z_R = P'_R - N' = 0 - 0 = 0 \qquad (8\text{-}82)$$

In other words, the system is stable. No matter how much the gain K_0 is increased, this particular system is never unstable.

EXAMPLE 2 *Type 1* The inverse polar plot of

$$[G(s)H(s)]^{-1} = \frac{s(1 + T_1 s)(1 + T_2 s)}{K_1} \qquad (8\text{-}83)$$

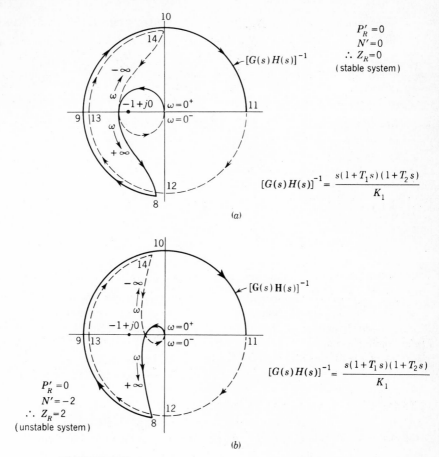

$$[G(s)H(s)]^{-1} = \frac{s(1+T_1 s)(1+T_2 s)}{K_1}$$

(a)

$$[G(s)H(s)]^{-1} = \frac{s(1+T_1 s)(1+T_2 s)}{K_1}$$

$P'_R = 0$
$N' = -2$
$\therefore Z_R = 2$
(unstable system)

(b)

FIGURE 8-39
A completed inverse transfer function for Eq. (8-83) indicating its correlation to
Fig. 8-38b: (a) stable; (b) unstable.

given in Fig. 8-39a shows that the system is stable. When the gain is increased, the
plot drawn in Fig. 8-39b shows that the system is unstable. The stability of this system
therefore depends on the gain.

8-18   DEFINITIONS OF PHASE MARGIN AND GAIN MARGIN AND THEIR RELATION TO STABILITY[8]

The stability and approximate degree of stability can be determined from the log
magnitude and phase diagram. The stability characteristic is specified in terms of the
following quantities:

Gain crossover   This is the point on the plot of the transfer function at which
the magnitude is unity [Lm $\mathbf{G}(j\omega)$ = 0 dB]. The frequency at gain crossover
is called the phase-margin frequency ω_ϕ.

Phase margin   This is $180°$ plus the negative trigonometrically considered angle of the transfer function at the gain-crossover point. It is designated as the angle γ, which can be expressed as $\gamma = 180° + \phi$, where ϕ is negative.

Phase crossover   This is the point on the plot of the transfer function at which the phase angle is $-180°$. The frequency at which phase crossover occurs is called the gain-margin frequency ω_c.

Gain margin   The gain margin is the factor a by which the gain must be changed in order to produce instability. Expressed in terms of the transfer function at the frequency ω_c, it is

$$|\mathbf{G}(j\omega_c)|a = 1 \qquad (8\text{-}84)$$

On the polar plot of $\mathbf{G}(j\omega)$ the value at ω_c is

$$|\mathbf{G}(j\omega_c)| = \frac{1}{a} \qquad (8\text{-}85)$$

In terms of the log magnitude, in decibels, this is

$$\text{Lm } a = -\text{Lm } \mathbf{G}(j\omega_c) \qquad (8\text{-}86)$$

which identifies the gain margin on the log magnitude diagram.

These quantities are illustrated in Fig. 8-40 on both the log and the polar curves. Note the algebraic sign associated with these two quantities as marked on the curves. Figure 8-40*a*, *b*, and *c* represents a stable system, and Fig. 8-40*d*, *e*, and *f* represents an unstable system.

The phase margin is the amount of phase shift at the frequency ω_ϕ that would just produce instability. For minimum-phase networks, the phase margin must be positive for a stable system, whereas a negative phase margin means that the system is unstable.

It can be shown that the phase margin is related to the effective damping ratio ζ of the system. This is discussed qualitatively in the next chapter. Satisfactory response is usually obtained with a phase margin of 45 to $60°$. As an individual gains experience and develops his own particular technique, the value of γ to be used for a particular system becomes more evident. This guideline for system performance applies only to those systems where behavior is equivalent to that of a second-order system. The gain margin must be positive when expressed in decibels (greater than unity as a numeric) for a stable system. A negative gain margin means that the system is unstable.

The damping ratio ζ of the system is also related to the gain margin. However, the phase margin gives a better estimate of damping ratio, and therefore of the transient overshoot of the system, than the gain margin.

Further information about the speed of response of the system can be obtained from the log magnitude–angle diagram, which defines the maximum value of the control ratio and the frequency at which this maximum occurs.

8-19   STABILITY CHARACTERISTICS OF THE LOG MAGNITUDE AND PHASE DIAGRAM

The earlier sections of this chapter discuss how the asymptotes of the log magnitude curve are related to each factor of the transfer function. For example, factors of the

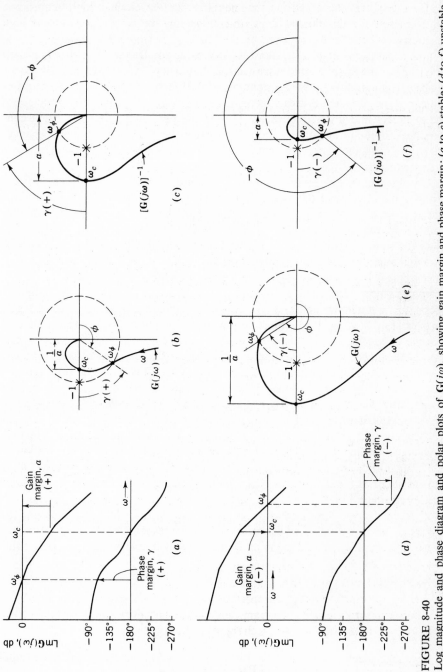

FIGURE 8-40

Log magnitude and phase diagram and polar plots of $\mathbf{G}(j\omega)$, showing gain margin and phase margin: (a to c) stable; (d to f) unstable.

form $(1 + j\omega)^{-1}$ have a negative slope equal to -20 dB/decade at high frequencies. Also, the angle for this factor varies from $0°$ at low frequencies to $-90°$ at high frequencies.

The total phase angle of a transfer function at any frequency is closely related to the slope of the log magnitude curve at that frequency. A slope of -20 dB/decade is related to an angle of $-90°$; a slope of -40 dB/decade is related to an angle of $-180°$; a slope of -60 dB/decade is related to an angle of $-270°$; etc. Changes of slope at higher and lower frequencies, around the particular frequency being considered, contribute to the total angle at that frequency. The farther away the changes of slope are from the particular frequency, the less they contribute to the total angle at that frequency.

By observing the asymptotes of the log magnitude curve, it is possible to estimate the approximate value of the angle. With reference to the example of Sec. 8-7, the angle at $\omega = 4$ is now investigated. The slope of the curve at $\omega = 4$ is -20 dB/decade; therefore the angle is near $-90°$. The slope changes at $\omega = 2$ and $\omega = 8$, and so the slopes beyond these frequencies contribute to the total angle at $\omega = 4$. The actual angle, as read from the graph of Fig. 8-7, is $-122°$. The farther away the corner frequencies occur, the closer the angle is to $-90°$.

As has been seen, the stability of a system requires that the phase margin be positive for a minimum-phase system. For this to be true, the angle at the gain crossover [Lm $\mathbf{G}(j\omega) = 0$ dB] must be greater than $-180°$. This places a limit on the slope of the log magnitude curve at the gain crossover. *The slope at the gain crossover should be more positive than -40 dB/decade if the adjacent corner frequencies are not close.* A slope of -20 dB/decade is preferable. This is derived from the consideration of a theorem by Bode. However, it can be seen qualitatively from the association of the slope of the log magnitude curve to the value of the phase. This guide should be used to assist in system design.

The log magnitude and phase diagram reveals some pertinent information, just as the polar plots do. For example, the gain can be adjusted (this raises or lowers the log magnitude curve) to produce a phase margin in the desirable range of 45 to $60°$. The shape of the low-frequency portion of the curve determines system type and therefore the degree of steady-state accuracy. The system type and the gain determine the error coefficients and therefore the steady-state error. The phase-margin frequency ω_ϕ gives a qualitative indication of the speed of response of a system. However, this is only a qualitative relationship. A more detailed analysis of this relationship is made in the next chapter.

8-20   STABILITY FROM THE NICHOLS PLOT (LOG MAGNITUDE–ANGLE DIAGRAM)

The log magnitude–angle diagram is drawn by picking for each frequency the values of log magnitude and angle from the log magnitude and phase diagram. The resultant curve has frequency as a parameter. The curve for the example of Sec. 8-7, sketched in Fig. 8-41, shows a positive gain margin and phase margin; therefore this represents a stable system. Changing the gain raises or lowers the curve without changing the

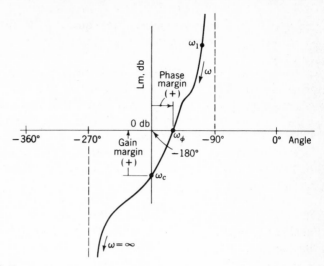

FIGURE 8-41
Log magnitude–angle diagram obtained from the curves of Figs. 8-6 and 8-7.

angle characteristics. Increasing the gain raises the curve, thereby decreasing the gain margin and phase margin, with the result that the stability is decreased. Increasing the gain so that the curve has a positive log magnitude at $-180°$ results in a negative gain and phase margin; therefore an unstable system results. Decreasing the gain lowers the curve and increases stability. However, a large gain is desired to reduce steady-state errors, as shown in Chap. 6.

The log magnitude–angle diagram for $G(s)H(s)$ can be drawn for all values of s on the contour Q of Fig. 8-30a. For minimum-phase systems the resultant curve is a closed contour. Nyquist's criterion can be applied to this contour by determining the number of points (having the values 0 dB and odd multiples of 180°) enclosed by the curve of $G(s)H(s)$. This number is the value of N which is used in the equation $N = Z_R$ to determine the value of Z_R. As an example, consider a control system whose transfer function is given by

$$G(s) = \frac{K_1}{s(1 + Ts)} \qquad (8\text{-}87)$$

Its log magnitude–angle diagram, for the contour Q, is shown in Fig. 8-42. From this figure it is seen that the value of N is zero and the system is stable. For a non-minimum-phase system the log magnitude–angle contour does not close; thus it is difficult to determine the value of N. For these cases the polar plot is easier to use to determine stability.

For minimum-phase systems it is not necessary to obtain the complete log magnitude–angle contour to determine stability. Only that portion of the contour is drawn representing $\mathbf{G}(j\omega)$ for the range of values $0^+ < \omega < \infty$. The stability is then determined from the position of the curve of $\mathbf{G}(j\omega)$ relative to the (0-dB, $-180°$) point. In other words, the curve is traced in the direction of increasing frequency.

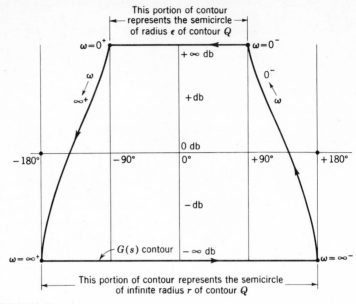

FIGURE 8-42
The log magnitude–angle contour for the minimum-phase system of Eq. (8-87).

The system is stable if the (0-dB, $-180°$) point is to the right of the curve. This is a simplified rule of thumb which is based on Nyquist's stability criterion.

A conditionally stable system (as defined in Sec. 8-14, Example 4) is one in which the curve crosses the $-180°$ axis at more than one point. Figure 8-43 shows the transfer-function plot for such a system with two stable and two unstable regions. The gain determines whether the system is stable or unstable. Additional stability information can be obtained from the log magnitude–angle diagram. This is shown in the next chapter.

8-21   SUMMARY

In this chapter different types of frequency-response plots are introduced. From the log magnitude and phase-angle diagrams the polar plot can be obtained with ease and rapidity. All these plots indicate the type of system under consideration and the necessary adjustments that must be made to improve its response. How these adjustments are made is discussed in the following chapters.

The methods presented for obtaining the frequency-response plots have stressed graphical techniques. For greater accuracy the curves can be determined analytically. A digital-computer program can be used to calculate these data.[4]

The methods described in this chapter for obtaining frequency-response plots are based upon the condition that the transfer function of a given system is known. These plots can also be obtained from experimental data, which do not require

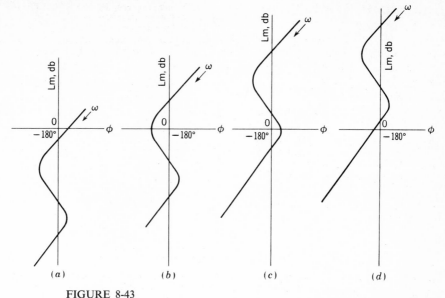

FIGURE 8-43
Log magnitude–angle diagram for a conditionally stable system: (*a* and *c*) stable; (*b* and *d*) unstable.

the analytical expression for the transfer function of a system to be known. These experimental data can be used to synthesize an analytical expression for the transfer function using the log magnitude plot.

In this chapter it is shown that the polar plot of the transfer function $G(s)H(s)$ or its inverse, in conjunction with Nyquist's stability criterion, gives a rapid means of determining whether a system is stable or unstable. The same information is obtained from the log magnitude and phase-angle diagrams and the log magnitude–angle diagram; the phase margin and gain margin are also used as a means of indicating stability. In the next chapter it is shown that other key information related to the time domain is readily obtained from any of these plots. Thus the designer can determine whether the given system is satisfactory.

Another useful application of the Nyquist criterion is the analysis of systems having the characteristic of dead time. The Nyquist criterion has one disadvantage in that poles or zeros on the $j\omega$ axis require special treatment. This treatment is not expounded since, in general, most problems are free of poles or zeros on the $j\omega$ axis.

REFERENCES

1   Maccoll, L. A.: "Fundamental Theory of Servomechanisms," Van Nostrand, Princeton, N.J., 1945.
2   James, H. M., N. B. Nichols, and R. S. Phillips: "Theory of Servomechanisms," McGraw-Hill, New York, 1947.

3   Nyquist, H.: Regeneration Theory, *Bell Syst. Tech. J.*, vol. 11, pp. 126–147, 1932.

4   Frequency Response Program (FREQR), School of Engineering, Air Force Institute of Technology, Wright-Patterson Air Force Base, Ohio, 1974.

5   Bruns, R. A., and R. M. Saunders: "Analysis of Feedback Control Systems," chap. 14, McGraw-Hill, New York, 1955.

6   Balabanian, N., and W. R. LePage: What Is a Minimum-Phase Network?, *Trans. AIEE*, vol. 74, pt. II, pp. 785–788, January 1956.

7   Bode, H. W.: "Network Analysis and Feedback Amplifier Design," chap. 8, Van Nostrand, Princeton, N.J., 1945.

8   Chestnut, H., and R. W. Mayer: "Servomechanisms and Regulating System Design," 2d ed., vol. 1, Wiley, New York, 1959.

9   Sanathanan, C. K., and H. Tsukui: Synthesis of Transfer Function from Frequency Response Data, *Intern. J. Systems Science*, vol. 5, no. 1, pp. 41–54, Jan. 1974.

10   Brown, G. S., and D. P. Campbell: "Principles of Servomechanisms," Wiley, New York, 1948.

CLOSED-LOOP PERFORMANCE BASED ON THE FREQUENCY RESPONSE

9-1 INTRODUCTION

Chapter 8 is devoted to plotting the open-loop transfer function $G(j\omega)H(j\omega)$. The three types of plots that are useful are the direct polar plot (Nyquist plot), the inverse polar plot, and the log magnitude-angle plot (Nichols plot). Also included in Chap. 8 is the determination of closed-loop stability in terms of the open-loop transfer function by use of the Nyquist stability criterion, which is illustrated in terms of all three types of plot. The result of the stability study places bounds on the permitted range of values of gain.

This chapter develops a correlation between the frequency and the time responses of a system, leading to a method of gain setting in order to achieve a specified closed-loop frequency response.[1] The closed-loop frequency response is obtained as a function of the open-loop frequency response.

9-2 DIRECT POLAR PLOT

Consider a simple control system with unity feedback, as shown in Fig. 9-1. The following equations describe the performance of this system with a sinusoidal input $R(j\omega)$:

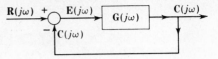

FIGURE 9-1
Block diagram of a simple control
system.

$$R(j\omega) - C(j\omega) = E(j\omega) \qquad (9\text{-}1)$$

$$\frac{C(j\omega)}{E(j\omega)} = G(j\omega) = |G(j\omega)|e^{j\phi} \qquad (9\text{-}2)$$

$$\frac{C(j\omega)}{R(j\omega)} = \frac{G(j\omega)}{1 + G(j\omega)} = M(j\omega) = M(\omega)e^{j\alpha(\omega)} \qquad (9\text{-}3)$$

$$\frac{E(j\omega)}{R(j\omega)} = \frac{1}{1 + G(j\omega)} \qquad (9\text{-}4)$$

In Fig. 9-2 is shown the polar plot of $G(j\omega)$ for this control system. As pointed out in Chap. 8, the directed line segment drawn from the $-1 + j0$ point to any point on the $G(j\omega)H(j\omega)$ curve represents the quantity $B(j\omega) = 1 + G(j\omega)H(j\omega)$. For the system under consideration $H(j\omega)$ is unity; therefore

$$B(j\omega) = |B(j\omega)|e^{j\lambda(\omega)} = 1 + G(j\omega) \qquad (9\text{-}5)$$

The frequency control ratio $C(j\omega)/R(j\omega)$ is therefore equal to the ratio of $A(j\omega)$ to $B(j\omega)$. In other words,

$$\frac{C(j\omega)}{R(j\omega)} = \frac{A(j\omega)}{B(j\omega)} = \frac{|A(j\omega)|e^{j\phi(\omega)}}{|B(j\omega)|e^{j\lambda(\omega)}} = \frac{G(j\omega)}{1 + G(j\omega)} \qquad (9\text{-}6)$$

$$\frac{C(j\omega)}{R(j\omega)} = \frac{|A(j\omega)|}{|B(j\omega)|} e^{j(\phi - \lambda)} = M(\omega)e^{j\alpha} \qquad (9\text{-}7)$$

Note that the angle $\alpha(\omega)$ can be determined directly from the construction shown in Fig. 9-2. Since the magnitude of the angle $\phi(\omega)$ is greater than the magnitude of the angle $\lambda(\omega)$, the value of the angle $\alpha(\omega)$ is negative. Remember that counterclockwise rotation is taken as positive.

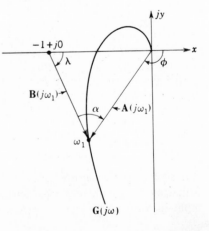

FIGURE 9-2
Polar plot of the $G(j\omega)$ for the control
system of Fig. 9-1.

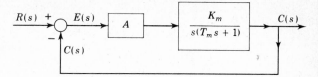

FIGURE 9-3
A position-control system.

Combining Eqs. (9-4) and (9-5) gives

$$\frac{E(j\omega)}{R(j\omega)} = \frac{1}{1 + G(j\omega)} = \frac{1}{|B(j\omega)|e^{j\lambda}} \tag{9-8}$$

From Eq. (9-8) it is seen that the greater the distance from the $-1 + j0$ point to a point on the $G(j\omega)$ locus, for a given frequency, the smaller the steady-state sinusoidal error for a stated sinusoidal input. Thus the usefulness and importance of the polar plot of $G(j\omega)$ have been enhanced.

9-3 DETERMINATION OF M_m AND ω_m FOR A SIMPLE SECOND-ORDER SYSTEM

The frequency at which the maximum value of $|C(j\omega)/R(j\omega)|$ occurs is referred to as the *resonant frequency* ω_m. The maximum value is labeled M_m. These two quantities are figures of merit of a system. The methods of compensation to improve system performance, using polar and log plots as shown in later chapters, are based upon a knowledge of these two factors.

Only for a *simple second-order system* can a direct and simple relationship be obtained for M_m and ω_m in terms of the system parameters. Consider the position control system of Fig. 9-3, which utilizes the servomotor described by Eq. (2-148). The forward and closed-loop transfer functions are, respectively,

$$\frac{C(s)}{E(s)} = \frac{K_1}{s(T_m s + 1)} \tag{9-9}$$

$$\frac{C(s)}{R(s)} = \frac{1}{(T_m/K_1)s^2 + (1/K_1)s + 1} = \frac{1}{s^2/\omega_n^2 + (2\zeta/\omega_n)s + 1} \tag{9-10}$$

where $K_1 = AK_m$. The damping ratio and the undamped natural frequency for this system, as determined from Eq. (9-10), are $\zeta = 1/(2\sqrt{K_1 T_m})$ and $\omega_n = \sqrt{K_1/T_m}$. The control ratio as a function of frequency is

$$\frac{C(j\omega)}{R(j\omega)} = \frac{1}{(1 - \omega^2/\omega_n^2) + j2\zeta(\omega/\omega_n)} = M(\omega)e^{j\alpha(\omega)} \tag{9-11}$$

For a particular value of ω_n, plots of $M(\omega)$ and $\alpha(\omega)$ vs. ω can be obtained for different values of ζ. For these plots to be applicable to all simple second-order systems with different values of ω_n, they are plotted vs. ω/ω_n, as shown in Fig. 9-4a and b.

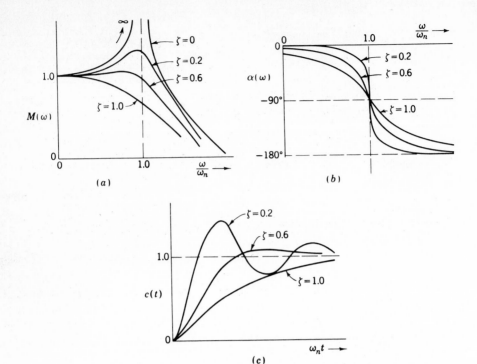

FIGURE 9-4

Plots of M and α vs. ω/ω_n for a simple second-order system, with the corresponding time plots for a step input.

For a unit step input the inverse Laplace transform of $C(s)$ gives the time response, for $\zeta < 1$, as

$$c(t) = 1 - \frac{1}{\sqrt{1 - \zeta^2}} e^{-\zeta\omega_n t} \sin\left(\omega_n\sqrt{1 - \zeta^2}\, t + \cos^{-1}\zeta\right) \qquad (9\text{-}12)$$

This time solution is derived in Sec. 3-9. The plot of $c(t)$ for several values of damping ratio ζ is drawn in Fig. 9-4c.

Next, consider the magnitude M^2, as derived from Eq. (9-11):

$$M^2 = \frac{1}{(1 - \omega^2/\omega_n^2)^2 + 4\zeta^2(\omega^2/\omega_n^2)} \qquad (9\text{-}13)$$

To find the maximum value of M and the frequency at which it occurs, Eq. (9-13) is differentiated with respect to frequency and set equal to zero:

$$\frac{dM^2}{d\omega} = -\frac{-4(1 - \omega^2/\omega_n^2)(\omega/\omega_n^2) + 8\zeta^2(\omega/\omega_n^2)}{[(1 - \omega^2/\omega_n^2)^2 + 4\zeta^2(\omega^2/\omega_n^2)]^2} = 0 \qquad (9\text{-}14)$$

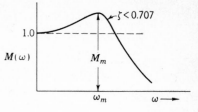

FIGURE 9-5
A closed-loop frequency-response curve
indicating M_m and ω_m.

The frequency ω_m at which the value M exhibits a peak (see Fig. 9-5), as found from
Eq. (9-14), is

$$\omega_m = \omega_n\sqrt{1 - 2\zeta^2} \qquad (9\text{-}15)$$

This value of frequency is substituted into Eq. (9-13) to yield

$$M_m = \frac{1}{2\zeta\sqrt{1 - \zeta^2}} \qquad (9\text{-}16)$$

From these equations it is seen that the curve of M vs. ω has a peak value, other than
at $\omega = 0$, only for $\zeta < 0.707$. Figure 9-6 shows a plot of M_m vs. ζ for a simple second-
order system. For values of $\zeta < 0.4$ it is seen that M_m increases very rapidly in
magnitude; the transient oscillatory response is therefore excessively large and might
damage the physical equipment.

In Secs. 3-5 and 7-11 it is determined that the damped natural frequency ω_d
for the transient of the simple second-order system is

$$\omega_d = \omega_n\sqrt{1 - \zeta^2} \qquad (9\text{-}17)$$

Also, in Sec. 3-9 the expression for the peak value M_p for a unit step input to this
simple system is determined. It is repeated here:

$$M_p = \frac{c_p}{r} = 1 + \exp\left(-\frac{\zeta\pi}{\sqrt{1 - \zeta^2}}\right) \qquad (9\text{-}18)$$

Therefore, for this simple second-order system the following conclusions are obtained
in correlating the frequency and time responses:

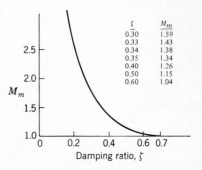

ζ	M_m
0.30	1.59
0.33	1.43
0.34	1.38
0.35	1.34
0.40	1.26
0.50	1.15
0.60	1.04

FIGURE 9-6
A plot of M_m vs. ζ for a simple second-
order system.

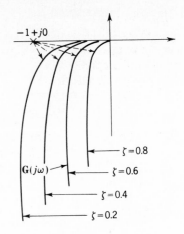

FIGURE 9-7

Polar plots of $G(j\omega)$ for different damping ratios for the system shown in Fig. 9-3.

1 Inspection of Eq. (9-15) reveals that ω_m is a function of both ω_n and ζ. For a given ζ, the larger the value of ω_m, the larger ω_n, and the faster the transient time of response for this system.

2 Inspection of Eqs. (9-16) and (9-18) shows that both M_m and M_p are functions of ζ. The smaller ζ becomes, the larger in value M_m and M_p become. Thus, it is concluded that the larger the value of M_m, the larger the value of M_p. For values of $\zeta < 0.4$ the correspondence between M_m and M_p is only qualitative for this simple case. In other words, for $\zeta = 0$ the time domain yields $M_p = 2$, whereas the frequency domain yields $M_m = \infty$. In practice, systems with $\zeta < 0.4$ are not utilized. When $\zeta > 0.4$, there is a close correspondence between M_m and M_p. As an example, for ζ equal to 0.6, $M_m = 1.04$ and $M_p = 1.09$.

3 In Fig. 9-7 are shown polar plots for different damping ratios for the simple second-order system. Note that the shorter the distance between the $-1 + j0$ point and a particular $G(j\omega)$ plot, as indicated by the dashed lines, the smaller the damping ratio. Thus M_m is larger and consequently M_p is also larger.

As a result of these characteristics, a designer can obtain a good approximation of the time response of a simple second-order system by knowing only the M_m and ω_m of its frequency response.

The procedure used above of setting the derivative of $C(j\omega)/R(j\omega)$ with respect to ω equal to zero works very well with a simple system. But as $C(j\omega)/R(j\omega)$ becomes more complicated, the differentiation and solution become tedious. This analytical procedure can be simplified, but a graphic procedure is generally used, as shown in the following sections.[2]

9-4   CORRELATION OF SINUSOIDAL AND TIME RESPONSES[3]

Although the correlation in the preceding section is for a simple second-order system, it has been found by experience that M_m is also a function of the *effective* ζ and ω_n for higher-order systems. The effective ζ and ω_n of a higher-order system (see Sec. 7-11) is dependent upon the ζ and ω_n of each second-order term and the values of the real roots in the characteristic equation of $C(s)/R(s)$. Thus, in order to alter the M_m, the location of some of the roots must be changed. Which ones should be altered depends on which are dominant in the time domain.

From the analysis for a simple second-order system, whenever the frequency response has the shape shown in Fig. 9-5, the following correlation exists between the frequency and time responses for systems of any order:

1 The larger ω_m is made, the faster the time of response for the system.
2 The value of M_m gives a good approximation of M_p within the acceptable range of the *effective* damping ratio $0.4 < \zeta < 0.707$. In terms of M_m, the acceptable range is $1 < M_m < 1.4$.
3 The closer the $\mathbf{G}(j\omega)$ curve comes to the $-1 + j0$ point, the larger the value of M_m.

To these three items can be added one more factor that is developed in Chap. 6: the larger K_p, K_v, or K_a is made, the greater the steady-state accuracy for a step, a ramp, and a parabolic input, respectively. In terms of the polar plot, the farther the point $\mathbf{G}(j\omega)]_{\omega=0} = K_0$ for a Type 0 system is from the origin, the more accurate is the steady-state time response for a step input. For a Type 1 system, the farther the low-frequency asymptote (as $\omega \to 0$) is from the imaginary axis, the more accurate is the steady-state time response for a ramp input.

It must be remembered that all the factors mentioned above are merely *guideposts* in the *frequency domain* to assist the designer in obtaining an *approximate* idea of the time response of a particular system. This serves as a "stop-and-go signal" with respect to whether one is headed in the right direction in achieving the desired time response. If the desired performance specifications are not satisfactorily met, compensation techniques (see later chapters) must be used. After compensation the exact time response can be obtained by taking the inverse Laplace transform of $C(s)$, if desired. It should be obvious by now that the approximate approach saves much valuable time. Exceptions to the above analysis do occur for higher-order systems. The digital computer is a valuable tool for obtaining the exact closed-loop frequency and time responses.[4,5] Thus a direct correlation between them can easily be made.

9-5 CONSTANT $M(\omega)$ AND $\alpha(\omega)$ CONTOURS OF $C(j\omega)/R(j\omega)$ ON THE COMPLEX PLANE (DIRECT PLOT)

The open-loop transfer function and its polar plot for a given feedback control system have provided the following information, so far:

1 The stability or instability of the system
2 If the system is stable, the degree of its stability
3 The system type
4 The degree of steady-state accuracy
5 A graphical method of determining $\mathbf{C}(j\omega)/\mathbf{R}(j\omega)$

All these items permit a qualitative idea about the system's time response. The value of the polar plot would be greatly enhanced if the values of M_m and ω_m could be readily obtained from the plot. The importance of knowing these two values for a given system is stressed in the previous sections.

The contours of constant values of M drawn in the complex plane yield a rapid means of determining the values of M_m and ω_m. In conjunction with the contours of constant values of $\alpha(\omega)$, also drawn in the complex plane, the plot of $C(j\omega)/R(j\omega)$ can be obtained more rapidly than by the graphical method indicated in Sec. 9-2. The M and α contours are developed only for unity-feedback systems. At the end of this section it is shown how these contours can be applied to a non-unity-feedback system.

Equation of a Circle

The equation of a circle with its center at the origin of a plane is $x^2 + y^2 = r^2$. A circle with its center on the x axis but displaced by a distance a from the origin has an equation of the form

$$(x - a)^2 + y^2 = r^2 \qquad (9\text{-}19)$$

When the center of the circle has its center at the point (a,b), the equation of the circle is

$$(x - a)^2 + (y - b)^2 = r^2 \qquad (9\text{-}20)$$

It should be realized that the x and y axes have identical scales. For a negative value of a the circle is displaced to the left of the origin. These equations of a circle in the xy plane are utilized later in this section to express contours of constant M and α.

$M(\omega)$ Contours

Figure 9-2 is the polar plot of a forward transfer function $G(j\omega)$ of a *unity-feedback system*. From Eq. (9-3), the magnitude M of the control ratio is

$$M(\omega) = \frac{|A(j\omega)|}{|B(j\omega)|} = \frac{|G(j\omega)|}{|1 + G(j\omega)|} \qquad (9\text{-}21)$$

The question at hand is: How many points are there in the complex plane for which the ratios of the magnitudes of the phasors $A(j\omega)$ and $B(j\omega)$ have the same value of $M(\omega)$? For example, referring to Fig. 9-2, for the frequency $\omega = \omega_1$, M has a value of M_a. It is desirable to determine all the other points in the complex plane for which

$$\frac{|A(j\omega)|}{|B(j\omega)|} = M_a$$

To derive the constant M locus, express the transfer function in the rectangular coordinates. That is,

$$G(j\omega) = x + jy \qquad (9\text{-}22)$$

Substituting this equation into Eq. (9-21) gives

$$M = \frac{|x + jy|}{|1 + x + jy|} = \left[\frac{x^2 + y^2}{(1 + x)^2 + y^2} \right]^{1/2}$$

or

$$M^2 = \frac{x^2 + y^2}{(1 + x)^2 + y^2}$$

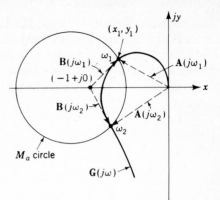

FIGURE 9-8
A plot of the transfer function $G(j\omega)$
and the M_a circle.

Rearranging the terms of this equation yields

$$\left(x + \frac{M^2}{M^2 - 1}\right)^2 + y^2 = \frac{M^2}{(M^2 - 1)^2} \qquad (9\text{-}23)$$

By comparison with Eq. (9-19) it is seen that Eq. (9-23) is the equation of a circle with its center on the real axis with M as a parameter. The center is located at

$$x_0 = -\frac{M^2}{M^2 - 1} \qquad (9\text{-}24)$$

$$y_0 = 0 \qquad (9\text{-}25)$$

and the radius is

$$r_0 = \left|\frac{M}{M^2 - 1}\right| \qquad (9\text{-}26)$$

Inserting a given value of $M = M_a$ into Eq. (9-23) results in a circle in the complex plane having a radius r_0 and its center at (x_0, y_0). This circle is called a constant M contour for $M = M_a$.

The ratio of magnitudes of the phasors $A(j\omega)$ and $B(j\omega)$ drawn to any point on the M_a circle has the same value. As an example, refer to Fig. 9-8, in which the M_a circle and the $G(j\omega)$ function are plotted. Since the circle intersects the $G(j\omega)$ plot at the two frequencies ω_1 and ω_2, there are two frequencies for which

$$\frac{|A(j\omega_1)|}{|B(j\omega_1)|} = \frac{|A(j\omega_2)|}{|B(j\omega_2)|} = M_a$$

In other words, a given point (x_1, y_1) is simultaneously a point on a particular transfer function $G(j\omega)$ and a point on the M circle passing through it.

The plot of Fig. 9-8 is redrawn in Fig. 9-9 with two more M circles added. In this figure the circle $M = M_b$ is just tangent to the $G(j\omega)$ plot. There is only one point (x_3, y_3) for $G(j\omega_3)$ in the complex plane for which the ratio $|A(j\omega)/B(j\omega)|$ is equal to M_b. Also, the M_c circle does not intersect and is not tangent to the $G(j\omega)$ plot. This indicates that there are no points in the plane that can simultaneously satisfy Eqs. (9-21) and (9-23) for $M = M_c$.

Figure 9-10 shows a family of circles in the complex plane for different values of M. In this figure it is noticed that the larger the value M, the smaller its corresponding M circle. Thus, for the example shown in Fig. 9-9 the ratio $C(j\omega)/R(j\omega)$, for a unity-feedback control system, has a maximum value of M equal to $M_m = M_b$. A further inspection of Fig. 9-10 and Eq. (9-24) reveals the following:

1 For $M \to \infty$, which represents a condition of oscillation ($\zeta \to 0$), the center of the M circle $x_0 \to -1 + j0$ and the radius $r_0 \to 0$. This agrees with the statement made previously that as the $G(j\omega)$ plot comes closer to the $-1 + j0$ point, the system's effective ζ becomes smaller and the degree of its stability becomes less.

2 For $M(\omega) = 1$, which represents the condition where $C(j\omega) = R(j\omega)$, $r_0 \to \infty$ and the M contour becomes a straight line perpendicular to the real axis at $x = -\frac{1}{2}$.

3 For $M \to 0$, the center of the M circle $x_0 \to 0$ and the radius $r_0 \to 0$.

4 For $M > 1$ the centers of the circles lie to the left of $x = -1 + j0$, and for $M < 1$ the centers of the circles lie to the right of $x = 0$. All centers are on the real axis.

$\alpha(\omega)$ Contours

The $\alpha(\omega)$ contours, representing constant values of phase angle $\alpha(\omega)$ for $C(j\omega)/R(j\omega)$, can also be determined in the same manner as the M contours. Substituting Eq. (9-22) into Eq. (9-3) yields

$$M(\omega)e^{j\alpha(\omega)} = \frac{x + jy}{(1 + x) + jy} = \frac{A(j\omega)}{B(j\omega)} \qquad (9\text{-}27)$$

The question at hand now is: How many points are there in the complex plane for which the ratio of the phasors $A(j\omega)$ and $B(j\omega)$ yields the same angle α? To answer this question, express the angle α obtained from Eq. (9-27) as follows:

$$\alpha = \tan^{-1}\frac{y}{x} - \tan^{-1}\frac{y}{1 + x}$$

$$= \tan^{-1}\frac{y/x - y/(1 + x)}{1 + (y/x)[y/(1 + x)]} = \tan^{-1}\frac{y}{x^2 + x + y^2}$$

$$\tan \alpha = \frac{y}{x^2 + x + y^2} = N \qquad (9\text{-}28)$$

For a constant value of the angle α, $\tan \alpha$, as expressed by the letter N, is also constant. Rearranging Eq. (9-28) results in

$$(x + \tfrac{1}{2})^2 + \left(y - \frac{1}{2N}\right)^2 = \frac{1}{4}\frac{N^2 + 1}{N^2} \qquad (9\text{-}29)$$

By comparing with Eq. (9-20) it is seen that Eq. (9-29) is the equation of a circle with N as a parameter. It has its center at

$$x_q = -\tfrac{1}{2} \qquad (9\text{-}30)$$

$$y_q = \frac{1}{2N} \qquad (9\text{-}31)$$

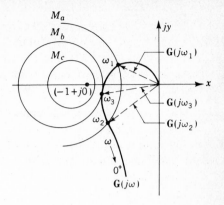

FIGURE 9-9
M contours and a $\mathbf{G}(j\omega)$ plot.

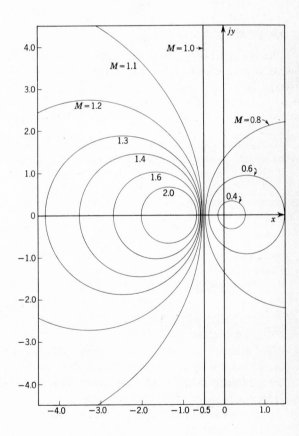

FIGURE 9-10
Constant M contours.

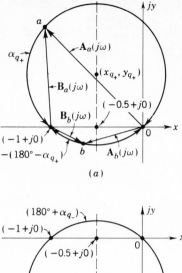

(a)

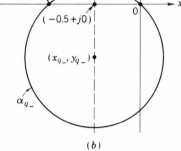

(b)

FIGURE 9-11
Arcs of constant α.

and has a radius of

$$r_q = \frac{1}{2}\left(\frac{N^2 + 1}{N^2}\right)^{1/2} \qquad (9\text{-}32)$$

Inserting a given value of $N_q = \tan \alpha_q$ into Eq. (9-29) results in a circle (see Fig. 9-11) of radius r_q with its center at (x_q, y_q).

The tangent of angles in the first and third quadrant is positive. Therefore the y_q coordinate given by Eq. (9-31) is the same for an angle in the first quadrant and for the negative of its supplement, which is in the third quadrant. As a result, the constant α contour is only an arc of the circle. In other words, the $\alpha = -310°$ and $-130°$ arcs are part of the same circle, as shown in Fig. 9-11a. Similarly, angles α in the second and fourth quadrants have the same value y_q if they are negative supplements of each other. The constant α contours for these angles are shown in Fig. 9-11b.

For all points on the α_q arc the ratio of the complex quantities $A(j\omega)$ and $B(j\omega)$ yields the same phase angle α_q, that is,

$$\underline{/A(j\omega)} - \underline{/B(j\omega)} = \alpha_q$$

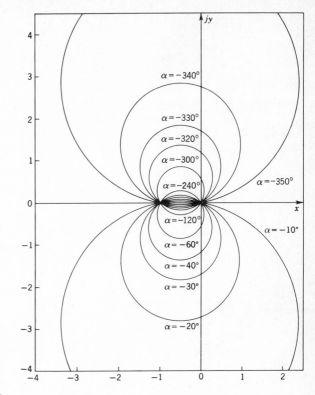

FIGURE 9-12
Constant α contours.

To avoid ambiguity in determining the angle α, a good procedure is to start at zero frequency, where α = 0, and to proceed to higher frequencies, knowing that the angle is a continuous function.

For different values of α there results a family of circles in the complex plane with centers on the line represented by $(-\frac{1}{2}, y)$, as illustrated in Fig. 9-12.

Figure 9-13 shows a plot of $\mathbf{G}(j\omega)$ with constant M and α contours superimposed. From this figure, Table 9-1 gives the magnitude M and the angle α of the control ratio $\mathbf{C}(j\omega)/\mathbf{R}(j\omega)$ for a number of values of frequency. While a graphical interpretation of M and α as related to $\mathbf{G}(j\omega)$ is shown in Fig. 9-13, it should be realized that these data are easily obtainable by use of a digital-computer program.[4] The control-ratio characteristics are plotted in Fig. 9-14.

Tangents to the M Circles

The line drawn through the origin of the complex plane tangent to a given M circle plays an important part in the gain setting of the $\mathbf{G}(j\omega)$ function. By referring to

308

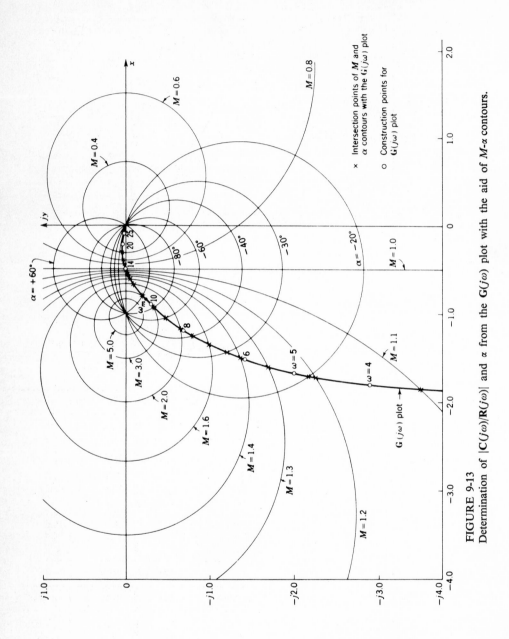

FIGURE 9-13
Determination of $|C(j\omega)/R(j\omega)|$ and α from the $G(j\omega)$ plot with the aid of M-α contours.

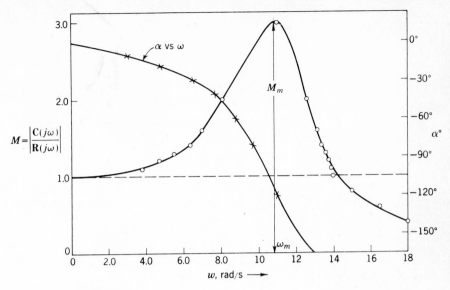

$$M = \left| \frac{C(j\omega)}{R(j\omega)} \right|$$

FIGURE 9-14
Resultant $|C(j\omega)/R(j\omega)|$ and α vs. ω, obtained from Fig. 9-13.

Table 9-1   VALUES OBTAINED FROM
FIG. 9-13 TO PLOT THE
CURVES IN FIG. 9-14

Values for $|C(j\omega)/R(j\omega)|$ plot			
ω, rad/s	M	ω, rad/s	M
3.8	1.1	13.4	1.4
4.7	1.2	13.6	1.3
5.5	1.3	13.8	1.2
6.2	1.4	13.9	1.1
7.0	1.6	14.0	1.0
8.1	2.0	15.0	0.8
11.0	3.0	16.5	0.6
12.6	2.0	18.0	0.4
13.2	1.6		

Values for α plot			
ω, rad/s	α, deg	ω, rad/s	α, deg
3	-10	8.8	-60
4.8	-20	9.7	-80
6.5	-30	11.3	-120
7.7	-40	13.0	-180

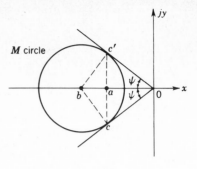

FIGURE 9-15
Determination of sin ψ.

Fig. 9-15 and remembering that $bc = r_0$ is the radius and $ob = x_0$ is the distance to the center of the particular M circle, then

$$\sin \psi = \frac{bc}{ob} = \frac{M/(M^2 - 1)}{M^2/(M^2 - 1)} = \frac{1}{M} \qquad (9\text{-}33)$$

This relationship is utilized in the section on gain adjustment.

Utilized along with the tangent is the fact that the point a in Fig. 9-15 is the $-1 + j0$ point. This is proved as follows:

$$(oc)^2 = (ob)^2 - (bc)^2$$

$$ac = oc \sin \psi$$

and

$$(oa)^2 = (oc)^2 - (ac)^2$$

Combining these equations yields

$$(oa)^2 = \frac{M^2 - 1}{M^2} [(ob)^2 - (bc)^2] \qquad (9\text{-}34)$$

The values of the distances to the center ob and of the radius bc of the M circle are then substituted into Eq. (9-34), which results in

$$oa = 1 \qquad (9\text{-}35)$$

The values necessary for constructing M and α contours, for typical values of M and α, are given in Tables 9-2 and 9-3, respectively.

Nonunity Feedback Control System

To apply the M and α contours to a nonunity feedback system, the system is first represented by an equivalent block diagram. The control ratio can be put in the form

$$\frac{C(j\omega)}{R(j\omega)} = \frac{G(j\omega)}{1 + G(j\omega)H(j\omega)} = \frac{1}{H(j\omega)} \frac{G(j\omega)H(j\omega)}{1 + G(j\omega)H(j\omega)} \qquad (9\text{-}36)$$

or

$$\frac{C(j\omega)}{R(j\omega)} = \frac{1}{H(j\omega)} \frac{G_0(j\omega)}{1 + G_0(j\omega)} = \frac{1}{H(j\omega)} \frac{C_0(j\omega)}{R_0(j\omega)} \qquad (9\text{-}37)$$

where

$$G_0(j\omega) = G(j\omega)H(j\omega) \qquad (9\text{-}38)$$

The M and α contours can be applied to $G_0(j\omega)$ to obtain $C_0(j\omega)/R_0(j\omega)$. Multiplying $C_0(j\omega)/R_0(j\omega)$ by $1/H(j\omega)$ gives $C(j\omega)/R(j\omega)$.

Table 9-2 VALUES FOR CONSTRUCTING M CIRCLES

M	Center $x_0 = -\dfrac{M^2}{M^2 - 1}$	Radius $r_0 = \dfrac{M}{M^2 - 1}$	Angle, deg, $\psi = \sin^{-1}\dfrac{1}{M}$
0.5	0.333	0.67	—
0.7	0.960	1.37	—
0.9	4.26	4.74	—
1.0	∞	∞	90
1.05	−10.74	10.24	72.3
1.1	−5.76	5.24	65.4
1.15	−4.1	3.57	60.3
1.2	−3.27	2.73	56.4
1.25	−2.78	2.22	53.2
1.3	−2.45	1.88	50.3
1.35	−2.215	1.64	47.7
1.4	−2.04	1.46	45.6
1.5	−1.8	1.20	41.8
1.6	−1.64	1.03	38.7
1.7	−1.53	0.90	36.0
1.8	−1.45	0.80	33.7
1.9	−1.38	0.729	31.7
2.0	−1.33	0.67	30.0

Table 9-3 VALUES FOR CONSTRUCTING α CONTOURS

$\alpha \pm 180°m,$ deg	N	Radius $r_q = \dfrac{1}{2N}\sqrt{N^2+1}$	Center $y_q = \dfrac{1}{2N}$
−90	−∞	0.500	0
−80	−5.67	0.528	−0.0882
−70	−2.75	0.531	−0.182
−60	−1.73	0.577	−0.289
−50	−1.19	0.656	−0.420
−40	−0.838	0.775	−0.596
−30	−0.577	1.000	−0.866
−20	−0.364	1.460	−1.370
−10	−0.176	2.88	−2.84
0	0.0	∞	∞
10	0.176	2.88	2.84
30	0.577	1.000	0.866
50	1.19	0.656	0.42
70	2.75	0.531	0.182
90	∞	0.5	0

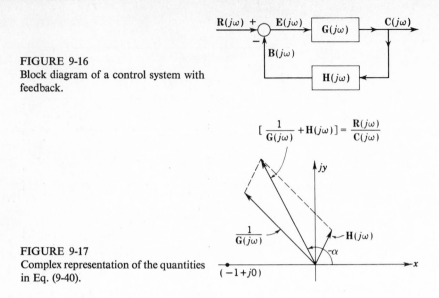

FIGURE 9-16
Block diagram of a control system with feedback.

FIGURE 9-17
Complex representation of the quantities in Eq. (9-40).

9-6 CONTOURS IN THE INVERSE POLAR PLANE

The use of direct polar plots for feedback control systems that utilize nonunity feedback is rather tedious when compared with the use of inverse plots. Also, the effect of the $H(j\omega)$ term is more evident in the inverse plot. Even for unity-feedback systems the value of required gain can be determined just as rapidly as on the direct plot, as described in Sec. 9-7.

Inverse Polar Plot

Figure 9-16 illustrates a control system with a feedback network $H(j\omega)$, whose performance is described by the following equations:

$$\frac{C(j\omega)}{R(j\omega)} = \frac{G(j\omega)}{1 + G(j\omega)H(j\omega)} = Me^{j\alpha} \qquad (9\text{-}39)$$

$$\frac{R(j\omega)}{C(j\omega)} = \frac{1}{G(j\omega)} + H(j\omega) = \frac{1}{M}e^{-j\alpha} \qquad (9\text{-}40)$$

Note that Eq. (9-40) is composed of two complex quantities that can be readily plotted in the complex plane, as shown in Fig. 9-17. Thus, by plotting the complex quantities $1/G(j\omega)$ and $H(j\omega)$, the $R(j\omega)/C(j\omega)$ term can be calculated graphically. It is now seen that the effect on $R(j\omega)/C(j\omega)$ of changing the $H(j\omega)$ term is more evident than it would be with direct plots.

Constant $1/M$ and α Contours (Unity Feedback)

Constant $1/M$ and α contours for the inverse plots are developed first for unity-feedback systems. Figure 9-18 illustrates the quantities in Eq. (9-40) for a unity-

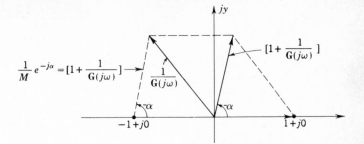

FIGURE 9-18
Representation of the quantities in Eq. (9-40) with $H(j\omega) = 1$.

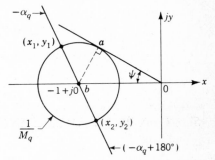

FIGURE 9-19
Contours for a particular magnitude $|R(j\omega)/C(j\omega)|$.

feedback system, with $H(j\omega) = 1$. Note that because of the geometry of construction the directed line segment drawn from the $-1 + j0$ point is also the quantity $1 + 1/G(j\omega)$ or $(1/M)e^{-j\alpha}$. Thus the inverse plot yields very readily the form and location of the contours of constant $1/M$ and α. In other words, in the inverse plane:

1 Contours of constant values of M are circles whose centers are at the $-1 + j0$ point, and the radii are equal to $1/M$.
2 Contours of constant values of $-\alpha$ are radial lines passing through the $-1 + j0$ point.

Figure 9-19 indicates the contours for a particular magnitude of $R(j\omega)/C(j\omega)$ and a tangent drawn to the M circle from the origin. In Fig. 9-19

$$\sin \psi = \left|\frac{ab}{ob}\right| = \frac{1/M}{1} = \frac{1}{M}$$

which is the same as for the contours in the direct plane. The directed line segment from $-1 + j0$ to the point (x_1, y_1) has a value of $1/M_q$ and an angle of $-\alpha_q$. If the polar plot of $1/G(j\omega)$ passes through the (x_1, y_1) point, then $R(j\omega_1)/C(j\omega_1)$ has these values. The directed line segment from $-1 + j0$ to the point (x_2, y_2) has a magnitude of $1/M_q$ and an angle of $-\alpha_q - 180°$. If the polar plot of $1/G(j\omega)$ passes through the (x_2, y_2) point, then $|R(j\omega_2)/C(j\omega_2)|$ has these values. Figure 9-20 illustrates families of typical M and α contours.

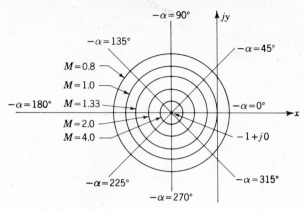

FIGURE 9.20
Typical M and α contours on the inverse polar plane for unity feedback.

The values of $\mathbf{C}(j\omega)/\mathbf{R}(j\omega)$ for different values of frequency can be determined in the manner given in the example in Sec. 9-5. By superimposing the M and α contours on the $1/\mathbf{G}(j\omega)$ plot, the points of intersection provide data for plotting M and α vs. ω.

Non-Unity-Feedback Control System

The case of nonunity feedback is handled more simply on the inverse plot than on the direct plot since

$$\frac{\mathbf{R}(j\omega)}{\mathbf{C}(j\omega)} = \frac{1 + \mathbf{G}(j\omega)\mathbf{H}(j\omega)}{\mathbf{G}(j\omega)} = \frac{1}{\mathbf{G}(j\omega)} + \mathbf{H}(j\omega) \qquad (9\text{-}41)$$

The curves $1/\mathbf{G}(j\omega)$ and $\mathbf{H}(j\omega)$ can be drawn separately and then added, as shown in Fig. 9-17. From the origin the quantity $\mathbf{R}(j\omega)/\mathbf{C}(j\omega)$ is measured directly in both magnitude and angle. The $1/M$ circles and α lines are the same as those for the unity-feedback case but are now drawn from the origin.

EXAMPLE

$$G(s) = \frac{K_0}{(1 + T_1 s)(1 + T_2 s)} \qquad (9\text{-}42)$$

$$H(s) = K_t s \qquad (9\text{-}43)$$

The curves $1/\mathbf{G}(j\omega)$ and $\mathbf{H}(j\omega)$ are plotted in Fig. 9-21. Their sum, $1/\mathbf{G}(j\omega) + \mathbf{H}(j\omega) = \mathbf{R}(j\omega)/\mathbf{C}(j\omega)$, is also shown in this figure.

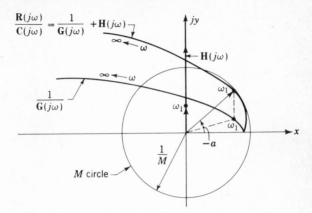

$$\frac{R(j\omega)}{C(j\omega)} = \frac{1}{G(j\omega)} + H(j\omega)$$

FIGURE 9-21
An example of the correlation of M and α contours for nonunity feedback.

9-7 GAIN ADJUSTMENT FOR A DESIRED M_m OF A UNITY-FEEDBACK SYSTEM: DIRECT POLAR PLOT

The root-locus method provides a means of determining the locus of all roots, with gain as the independent variable. Once the locus is determined, a particular value of gain is selected that meets the given set of desired response specifications as closely as possible. Note that setting the gain is the first and easiest step in adjusting the system for the optimum performance. This is also true with respect to polar plots. If satisfactory response cannot be achieved by gain adjustment alone, compensation techniques must be utilized, as described in later chapters.

Figure 9-22a shows $\mathbf{G}_x(j\omega)$ with its respective M_m circle in the complex plane. Since

$$\mathbf{G}_x(j\omega) = x + jy = K_x\mathbf{G}_x'(j\omega) = K_x(x' + jy') \qquad (9\text{-}44)$$

then

$$x' + jy' = \frac{x}{K_x} + j\frac{y}{K_x}$$

where $\mathbf{G}_x'(j\omega) = \mathbf{G}_x(j\omega)/K_x$ is defined as the frequency-sensitive portion of $\mathbf{G}_x(j\omega)$ with unity gain. Note that changing the gain merely changes the amplitude and not the angle of the locus of points of $\mathbf{G}_x(j\omega)$. Thus, if in Fig. 9-22a a change of scale is made by dividing the x, y coordinates by K_x so that the new coordinates are x', y', the following are true:

1 The $\mathbf{G}_x(j\omega)$ plot becomes the $\mathbf{G}_x'(j\omega)$ plot.
2 The M_m circle becomes *a circle* which is simultaneously tangent to $\mathbf{G}_x'(j\omega)$ and the line representing $\sin \psi = 1/M_m$.
3 The $-1 + j0$ point becomes the $-1/K_x + j0$ point.
4 The radius r_0 becomes $r_0' = r_0/K_x$.

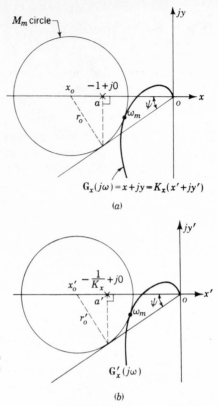

FIGURE 9-22
(a) A $G_x(j\omega)$ plot with the respective M_m circle. (b) A $G'_x(j\omega)$ plot with circle drawn tangent to it and to the line representing the angle $\sin \psi = 1/M_m$.

In other words, if $G'_x(j\omega)$ is drawn on a separate graph sheet from that of $G_x(j\omega)$ (see Fig. 9-22b), by superimposing the two graphs so that the axes coincide, the circles and the $G_x(j\omega)$ and $G'_x(j\omega)$ plots also coincide. Referring to Fig. 9-22a and b, note that $oa = -1$ and $oa' = -1/K_x$.

As a consequence, it is possible to determine the required gain to achieve a desired M_m for a given system by the following graphical procedure.

Step 1 If the original system has a transfer function

$$G_x(j\omega) = \frac{K_x(1 + j\omega T_1)(1 + j\omega T_2) \cdots}{(j\omega)^m(1 + j\omega T_a)(1 + j\omega T_b)(1 + j\omega T_c) \cdots} \qquad (9\text{-}45)$$

with an original gain K_x, only the frequency-sensitive portion $G'_x(j\omega)$ is plotted.

Step 2 Draw the line representing the angle $\psi = \sin^{-1}(1/M_m)$.

Step 3   By trial and error, find a circle whose center lies on the negative real axis and is simultaneously tangent to the $\mathbf{G}'_x(j\omega)$ plot and the line representing the angle ψ.

Step 4   Having found this circle, draw a line from the point of tangency on the ψ-angle line perpendicular to the real axis. Label the point where this line intersects the axis as a'.

Step 5   For this circle to be an M circle representing M_m, the point a' must be the $-1 + j0$ point. Thus, the x', y' coordinates must be multiplied by a gain factor K_m in order to convert this plot into a plot of $\mathbf{G}(j\omega)$. From the graphical construction the value of K_m is

$$K_m = \frac{1}{oa'}$$

Step 6   Thus the original gain must be changed by a factor

$$A = \frac{K_m}{K_x}$$

Note that if $\mathbf{G}_x(j\omega)$ is already plotted, this plot can become $\mathbf{G}'_x(j\omega)$ by merely changing the scale. That is, divide the x, y coordinates by K_x. However, it is possible to work directly with the plot of the function $\mathbf{G}_x(j\omega)$ which includes a gain K_x. Following the procedure outlined above results in the determination of the *additional* gain required to produce the specified M_m; that is, the additional gain is

$$A = \frac{K_m}{K_x} = \frac{1}{oa''}$$

EXAMPLE   It is desired that the system which has the transfer function

$$\mathbf{G}_x(j\omega) = \frac{1.47}{j\omega(1 + j0.25\omega)(1 + j0.1\omega)}$$

have an $M_m = 1.3$. The problem is to determine the actual gain K_1 needed and the amount by which the original gain K_x must be changed to obtain this M_m. The procedure previously outlined is applied to this problem. $\mathbf{G}'_x(j\omega)$ is plotted in Fig. 9-23 and results in

$$K_1 = \frac{1}{oa'} \approx \frac{1}{0.34} = 2.94 \text{ s}^{-1}$$

The additional gain required is

$$A = \frac{K_1}{K_x} = \frac{2.94}{1.47} = 2.0$$

In other words, the original gain must be doubled to obtain an M_m of 1.3 for $\mathbf{C}(j\omega)/\mathbf{R}(j\omega)$.

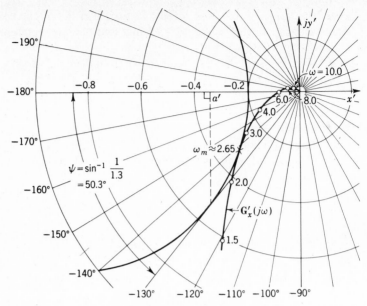

FIGURE 9-23
Gain adjustment utilizing the procedure outlined in Sec. 9-7 for

$$G'_x(j\omega) = \frac{1}{j\omega(1 + j0.25\omega)(1 + j0.1\omega)}$$

9-8   GAIN ADJUSTMENT FOR A DESIRED M_m OF A UNITY-FEEDBACK SYSTEM: INVERSE POLAR PLOT

The approach is similar to that for the direct plots in establishing the procedure for gain setting for the inverse plot. From Fig. 9-24

$$\frac{1}{G_x(j\omega)} = \frac{1}{K_x G'_x(j\omega)} = G_x^{-1}(j\omega) = x + jy \qquad (9\text{-}46a)$$

Then
$$\frac{K_x}{G_x(j\omega)} = \frac{1}{G'_x(j\omega)} = K_x x + jK_x y = x' + jy' \qquad (9\text{-}46b)$$

Thus, if a change of scale is made in Fig. 9-24 by multiplying the x, y coordinates by K_x so that the new coordinates are x', y', the following are true:

1   The $G_x^{-1}(j\omega)$ plot becomes the $[G'_x(j\omega)]^{-1}$ plot.
2   The $1/M_m$ circle becomes *a circle* which is simultaneously tangent to $[G'_x(j\omega)]^{-1}$ and the line drawn at the angle $\sin \psi = 1/M_m$.
3   The $-1 + j0$ point becomes the $-K_x + j0$ point.
4   The radius r_0 becomes $r'_0 = K_x r_0$.

In the same manner as with the direct plot, if $[G'_x(j\omega)]^{-1}$ is drawn on a separate sheet from that of $G_x^{-1}(j\omega)$ (see Fig. 9-25), and the two graphs are superimposed so

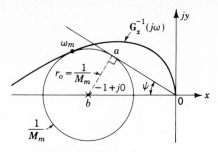

FIGURE 9-24
A $\mathbf{G}_x^{-1}(j\omega)$ plot and the $1/M_m$ circle.

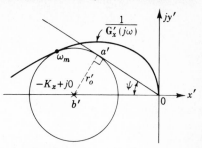

FIGURE 9-25
A $[\mathbf{G}_x'(j\omega)]^{-1}$ plot with circle drawn tangent to it and to the line at the angle $\sin\psi = 1/M_m$.

that the axes coincide, the circles and the $\mathbf{G}_x^{-1}(j\omega)$ and $[\mathbf{G}_x'(j\omega)]^{-1}$ plots also coincide. Referring to Figs. 9-24 and 9-25, note that $ob = -1$ and that $ob' = -K_x$.

To determine the gain necessary for a given M_m from the inverse plots, the following graphical procedure can be used.

Step 1   If the original system has an inverse transfer function of

$$\mathbf{G}_x^{-1}(j\omega) = \frac{(j\omega)^m(1 + j\omega T_a)(1 + j\omega T_b)(1 + j\omega T_c)\cdots}{K_x(1 + j\omega T_1)(1 + j\omega T_2)\cdots} \qquad (9\text{-}47)$$

with an original gain K_x, only the frequency-sensitive portion $[\mathbf{G}_x'(j\omega)]^{-1}$ is plotted.

Step 2   Draw the line at the angle $\psi = \sin^{-1}(1/M_m)$.

Step 3   By trial and error, find a circle whose center lies on the negative real axis and is simultaneously tangent to both the $[\mathbf{G}_x'(j\omega)]^{-1}$ plot and the line representing the angle ψ.

Step 4   For this circle to be a $1/M_m$ circle representing M_m, the point b' must be the $-1 + j0$ point. Thus, the x', y' coordinates must be divided by the gain constant K_m in order to convert this plot into a plot of $\mathbf{G}^{-1}(j\omega)$. From the graphical construction the value of K_m is

$$K_m = ob'$$

Step 5   Thus the original gain must be changed by a factor

$$A = \frac{K_m}{K_x}$$

Just as for the direct plot, if $\mathbf{G}_x^{-1}(j\omega)$ is already plotted, this plot can become $[\mathbf{G}_x'(j\omega)]^{-1}$ by merely changing the scale, i.e., multiplying the x, y coordinates by K_x. However, one can use the $\mathbf{G}_x^{-1}(j\omega)$ plot for the determination of the *additional* gain required for the desired M_m. The value of the additional gain is equal to ob''.

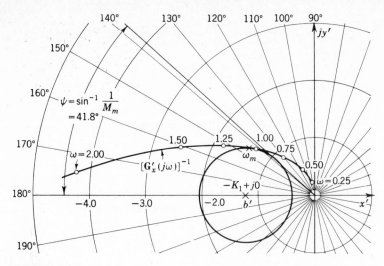

FIGURE 9-26
Gain adjustment utilizing the procedure outlined in Sec. 9-8 for $[G_x'(j\omega)]^{-1} = j\omega(1 + j0.8\omega)(1 + j0.25\omega)$.

EXAMPLE Assume that a system having the inverse transfer function

$$\mathbf{G}_x^{-1}(j\omega) = \frac{j\omega(1 + j0.8\omega)(1 + j0.25\omega)}{0.5}$$

must have an $M_m = 1.5$. In Fig. 9-26 the above procedure is utilized in order to determine the required gain K_1. Thus

$$K_1 = 1.25 \ \mathrm{s}^{-1}$$

Then the amount by which the original gain K_x must be changed is

$$A = \frac{K_1}{K_x} = \frac{1.25}{0.5} = 2.5$$

Therefore, increasing the original gain by a factor of 2.5 gives the desired $M_m = 1.5$ for $\mathbf{C}(j\omega)/\mathbf{R}(j\omega)$.

9-9 CONSTANT M AND α CURVES ON THE LOG MAGNITUDE–ANGLE DIAGRAM (NICHOLS CHART)[6]

As derived earlier in this chapter, the constant M curves on the direct and inverse polar plots are circles. The next necessary step is to transform these curves to the log magnitude–angle diagram. This is done more easily by starting from the inverse polar plot since all the M circles have the same center. This requires a change of sign of the log magnitude and angle obtained, since the transformation is from the inverse transfer function on the polar plot to the direct transfer function on the log magnitude plot.

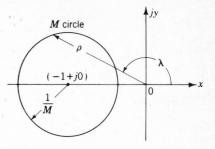

FIGURE 9-27
Constant M circle on the inverse polar plot.

A constant M circle is shown in Fig. 9-27 on the inverse polar plot. The magnitude ρ and angle λ drawn to any point on this circle are shown.

The equation for this M circle is

$$y^2 + (1 + x)^2 = \frac{1}{M^2} \qquad (9\text{-}48)$$

where

$$x = \rho \cos \lambda \qquad y = \rho \sin \lambda \qquad (9\text{-}49)$$

Combining these equations produces

$$\rho^2 M^2 + 2\rho M^2 \cos \lambda + M^2 - 1 = 0 \qquad (9\text{-}50)$$

Solving for ρ and λ yields

$$\rho = -\cos \lambda \pm \left(\cos^2 \lambda - \frac{M^2 - 1}{M^2} \right)^{1/2} \qquad (9\text{-}51)$$

$$\lambda = \cos^{-1} \frac{1 - M^2 - \rho^2 M^2}{2\rho M^2} \qquad (9\text{-}52)$$

These equations are derived from the inverse polar plot $1/\mathbf{G}(j\omega)$. Since the log magnitude–angle diagram is drawn for the transfer function $\mathbf{G}(j\omega)$ and not for its reciprocal, a change in the equations must be made by substituting

$$r = \frac{1}{\rho} \qquad \phi = -\lambda$$

Since Lm $r = -$Lm ρ, it is only necessary to change the sign of Lm ρ. For any value of M a series of values of angle λ can be inserted in Eq. (9-51) to solve for ρ. This magnitude must be changed to decibels. Alternatively, for any value of M a series of values of ρ can be inserted in Eq. (9-52) to solve for the corresponding angle λ. Therefore the constant M curve on the log magnitude–angle diagram can be plotted by using either of these two equations.

In a similar fashion the constant α curves can be drawn on the log magnitude–angle diagram. The α curves on the inverse polar plot are semi-infinite straight lines terminating on the -1 point and are given by

$$\tan \alpha + x \tan \alpha + y = 0 \qquad (9\text{-}53)$$

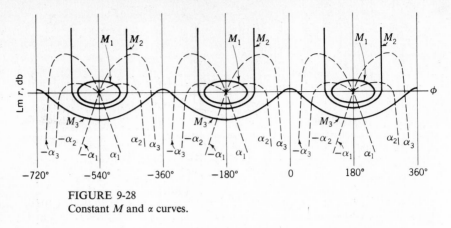

FIGURE 9-28
Constant M and α curves.

Combining with Eq. (9-49) produces

$$\tan \alpha + \rho \cos \lambda \tan \alpha + \rho \sin \lambda = 0 \qquad (9\text{-}54)$$

$$\tan \alpha = \frac{-y}{1 + x} = \frac{-\rho \sin \lambda}{1 + \rho \cos \lambda} \qquad (9\text{-}55)$$

For constant values of α a series of values of λ can be inserted in this equation to solve for ρ. The constant-α semi-infinite curves can then be plotted on the log magnitude–angle diagram.

Constant M and α curves, as shown in Fig. 9-28, repeat for every 360°. Also, there is symmetry at every 180° interval. An expanded 300° section of the constant M and α graph is shown in Fig. 9-29. Note that the $M = 1$ (0-dB) curve is asymptotic to $\phi = -90$ and $-270°$ and that the curves for $M < \frac{1}{2}$ (−6 dB) are always negative. $M = \infty$ is the point at 0 dB, −180°, and the curves for $M > 1$ are closed curves inside the limits $\phi = -90°$ and $\phi = -270°$.

9-10 ADJUSTMENT OF GAIN BY USE OF THE LOG MAGNITUDE–ANGLE DIAGRAM

The log magnitude–angle diagram for

$$G(j\omega) = \frac{2.04(1 + j2\omega/3)}{j\omega(1 + j\omega)(1 + j0.2\omega)(1 + j0.2\omega/3)} \qquad (9\text{-}56)$$

is drawn as the solid curve in Fig. 9-30 on graph paper which has the constant M and α contours. (The log magnitude and phase-angle diagram must be drawn first to get the data for this curve.) It is convenient to use the same log magnitude and angle scales on both the log magnitude and phase diagram and the log magnitude–angle diagram. A pair of dividers can then be used to move the transfer-function locus to the log magnitude–angle diagram. The $M = 1.12$ (1-dB) curve is tangent to the curve at

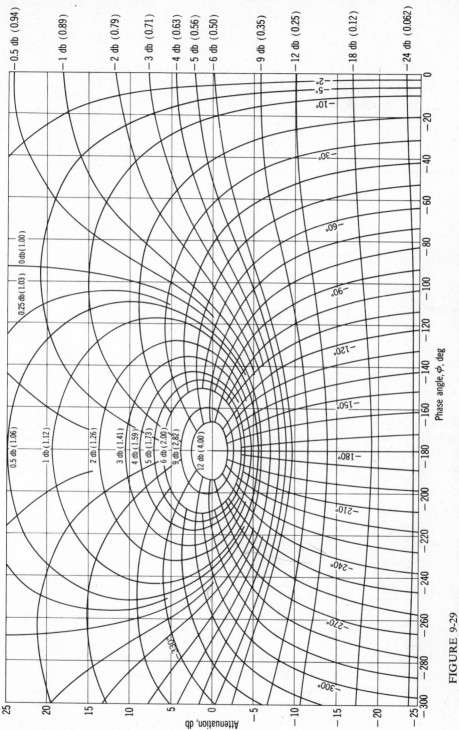

FIGURE 9-29
Constant M and α curves in the log magnitude–angle plane.

324

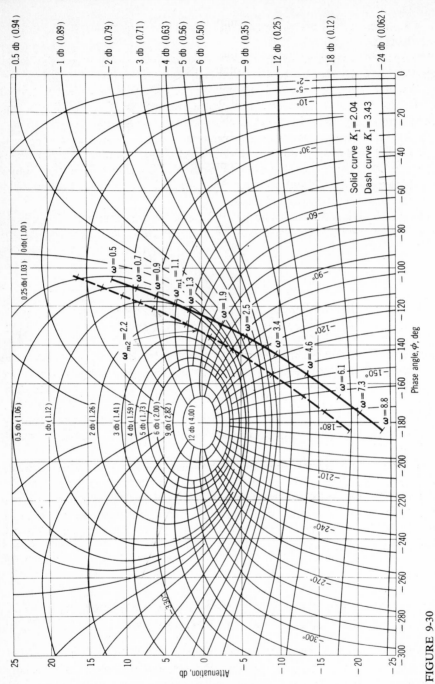

FIGURE 9-30

Log magnitude–angle diagram for Eq. (9-56).

$\omega_{m1} = 1.1$. These values are the maximum value of the control ratio M_m and the resonant frequency ω_m.

It is specified that the system be adjusted to produce an $M_m = 1.26$ (2 dB) by changing the gain. The dashed curve is obtained by raising the transfer-function curve until it is tangent to the $M_m = 1.26$ (2-dB) curve. The resonant frequency is now equal to $\omega_{m2} = 2.2$. The curve has been raised by the amount Lm $A = 4.5$ dB, meaning that an additional gain of $A = 1.68$ must be put into the system.

In practice it is easier to make a template of the desired M_m curve. Then the template can be moved up or down until it is tangent to the transfer-function curve. The amount that the template is moved represents the gain change required. Graph paper with the constant M and α curves superimposed is available commercially,† which simplifies system analysis and design.

After the gain has been adjusted for the desired M_m, there are several methods described in the literature for obtaining the approximate values of the closed-loop poles.[7]

9-11 CLOSED-LOOP FREQUENCY RESPONSE FROM THE LOG MAGNITUDE–ANGLE DIAGRAM

When the gain has been adjusted for the desired M_m, the closed-loop frequency response can be found either graphically or by use of the digital computer. For any frequency point on the transfer-function curve the values of M and α can be read from the graph. For example, with M_m made equal to 1.26 in Fig. 9-30, the value of M at $\omega = 3.4$ is -0.5 dB and α is $-110°$. The closed-loop frequency response, both log magnitude and angle, obtained from Fig. 9-30 is plotted in Fig. 9-31 for both values of gain.

The feedback system which has now been adjusted may be a portion of a larger system. Analysis and design of the larger system can proceed in a similar fashion.

The closed-loop response with the resultant gain adjustment may have too low a resonant frequency for the desired performance. The next design step would be to compensate the system to improve the frequency response. Compensation is studied in detail in Chaps. 10, 11, and 12.

9-12 CORRELATION OF POLE-ZERO DIAGRAM WITH FREQUENCY AND TIME RESPONSES

Whenever the closed-loop control ratio $\mathbf{M}(j\omega)$ has the characteristic form shown in Fig. 9-5, the system may be approximated as a second-order system This usually implies that the poles, other than the dominant complex pair, are either far to the left of the dominant complex poles or are close to zeros. When these conditions are not

---

† "Nichols Chart," Boonshaft and Fuchs, Inc., Hatboro, Pa. 19040.

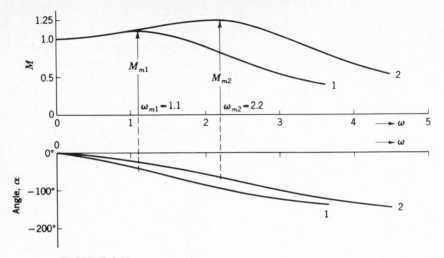

FIGURE 9-31
Control ratio vs. frequency obtained from the log magnitude–angle diagram of
Fig. 9-30.

satisfied, the frequency response may have other shapes. This can be illustrated by
considering the following three control ratios:

$$\frac{C(s)}{R(s)} = \frac{1}{s^2 + s + 1} \tag{9-57}$$

$$\frac{C(s)}{R(s)} = \frac{0.313(s + 0.8)}{(s + 0.25)(s^2 + 0.3s + 1)} \tag{9-58}$$

$$\frac{C(s)}{R(s)} = \frac{4}{(s^2 + s + 1)(s^2 + 0.4s + 4)} \tag{9-59}$$

The pole-zero diagram, the frequency response, and the time response to a step input
for each of these equations are shown in Fig. 9-32.

From Fig. 9-32a, which represents Eq. (9-57), the following characteristics are
noted:

1   The control ratio has only two complex poles, which are dominant, and no
    zeros.
2   The frequency-response curve has the following characteristics:
    a   A single peak, $M_m = 1.157$.
    b $1 < M < M_m$ in the frequency range $0 < \omega < 1$.
3   The time response has the typical waveform described in Chap. 3. That is,
    the first maximum of $c(t)$ due to the oscillatory term is greater than $c(t)_{ss}$ and
    the $c(t)$ response after this maximum oscillates around the value of $c(t)_{ss}$.

From Fig. 9-32b, for Eq. (9-58), the following characteristics are noted:

1   The control ratio has two complex poles and one real pole, all dominant, and
    one real zero.

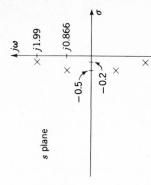

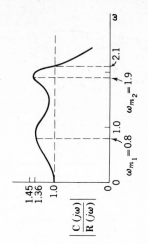

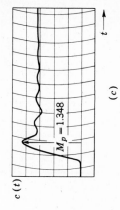

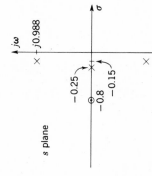

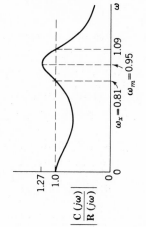

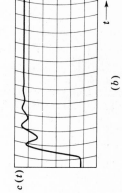

(a)

(b)

(c)

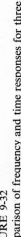

FIGURE 9-32

Comparison of frequency and time responses for three pole-zero patterns.

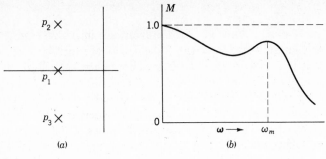

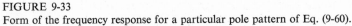

(a)   (b)

FIGURE 9-33
Form of the frequency response for a particular pole pattern of Eq. (9-60).

2 The frequency-response curve has the following characteristics:
  a   A single peak, $M_m = 1.27$.
  b $M < 1$ in the frequency range $0 < \omega < \omega_x$.
  c   The peak M_m occurs at $\omega_m = 0.95 > \omega_x$.
3 The time response does not have the conventional waveform. That is, the first maximum of $c(t)$ due to the oscillatory term is less than $c(t)_{ss}$.

From Fig. 9-32c, for Eq. (9-59), the following characteristics are noted:

1 The control ratio has four complex poles, all dominant, and no zeros.
2 The frequency-response curve has the following characteristics:
  a   There are two peaks, $M_{m1} = 1.36$ and $M_{m2} = 1.45$.
  b $1 < M < 1.45$ in the frequency range of $0 < \omega < 2.1$.
  c   The time response does not have the simple second-order waveform. That is, the first maximum of $c(t)$ in the oscillation is greater than $c(t)_{ss}$, and the oscillatory portion of $c(t)$ does not oscillate about a value of $c(t)_{ss}$.

Another example is the system represented by

$$M(s) = \frac{K}{(s - p_1)(s^2 + 2\zeta\omega_n s + \omega_n^2)} \qquad (9\text{-}60)$$

When the real pole p_1 and real part of the complex poles are equal, the frequency response is as shown in Fig. 9-33b. The magnitude at ω_m is less than unity. The corresponding time response to a step input is monotonic; i.e., there is no overshoot. This may be considered as a critically damped response.

The examples discussed in this section show that the time-response waveform is closely related to the frequency response of the system. In other words, the time-response waveform of the system can be predicted from the shape of the frequency-response plot. Thus, as illustrated in this section, the frequency-response plot may be utilized as a guide in determining (or predicting) time-response characteristics.

9-13 SUMMARY

In summary, this chapter is devoted to indicating the correlation between the frequency and time responses. The figures of merit M_m and ω_m are established as guideposts for evaluating a system's performance. The addition of a pole to an open-loop
transfer function produces a clockwise shift of the direct polar plot, which results in a
larger value of M_m. The time response also suffers because ω_m becomes smaller. The
reverse is true if a zero is added to the open-loop transfer function. This agrees with
the analysis of the root locus, which shows that the addition of a pole or zero results
in a less stable or more stable system, respectively. Thus the qualitative correlation
between the root locus and the frequency response is enhanced. The M and α contours are developed as a graphical aid in obtaining the closed-loop frequency response
and in adjusting the gain to obtain a desired M_m. The methods described for setting
the gain for a desired M_m are based on the fact that generally the desired values of
M_m are greater than 1. The procedure for gain adjustment may not yield a satisfactory
value of ω_m. In this case the system must be compensated in order to increase ω_m
without changing the value of M_m. Compensation procedures are covered in the
following chapters.

The procedure used in the frequency-response method is summarized as follows.

Step 1 Derive the open-loop transfer function $G(s)H(s)$ of the system.

Step 2 Put the transfer function into the form $\mathbf{G}(j\omega)\mathbf{H}(j\omega)$.

Step 3 Arrange the various factors of the transfer function so that they are in the
complex form $j\omega$, $1 + j\omega T$, and $1 + aj\omega + b(j\omega)^2$.

Step 4 Plot the log magnitude and phase angle diagram for $\mathbf{G}(j\omega)\mathbf{H}(j\omega)$. The
graphical methods of Chap. 8 may be used, but a digital computer can provide more
extensive data.[4]

Step 5 Transfer the data from the plots in step 4 to any of the following: (*a*) log
magnitude–angle diagram, (*b*) inverse polar plot, or (*c*) direct polar plot.

Step 6 Apply the Nyquist stability criterion and adjust the gain for the desired
degree of stability M_m of the system. Then check the correlation to the time response
for a step input signal. This correlation reveals some qualitative information about the
time response.

Step 7 If the qualitative response does not meet the desired specifications, determine
the shape that the plot must have to meet these specifications.

Step 8 Synthesize the compensator that must be inserted into the system, if other
than just gain adjustment is required, to make the necessary modification on the
original plot. This procedure is developed in the following chapters.

The digital computer is frequently available and is used extensively in system
design. A standard computer program can be used to obtain the frequency response

of both the open-loop and the closed-loop transfer functions.[4] The computer techniques are especially useful for systems containing frequency-sensitive feedback. The exact time response is then obtained to provide a correlation between the frequency and time responses.

REFERENCES

1　Brown, G. S., and D. P. Campbell: "Principles of Servomechanisms," chaps. 6 and 8, Wiley, New York, 1948.

2　Higgins, T. J., and C. M. Siegel: Determination of the Maximum Modulus, or the Specified Gain, of a Servomechanism by Complex Variable Differentiation, *Trans. AIEE*, vol. 72, pt. II, p. 467, January 1954.

3　Chu, Y.: Correlation between Frequency and Transient Responses of Feedback Control Systems, *Trans. AIEE*, vol. 72, pt. II, pp. 81–92, May 1953.

4　Frequency Response Program (FREQR), School of Engineering, Air Force Institute of Technology, Wright-Patterson Air Force Base, Ohio, 1974.

5　Heaviside Partial Fraction Expansion and Time Response Program (PARTL), School of Engineering, Air Force Institute of Technology, Wright-Patterson Air Force Base, Ohio, 1974.

6　James, H. M., N. B. Nichols, and R. S. Phillips: "Theory of Servomechanisms," chap. 4, McGraw-Hill, New York, 1947.

7　Chen, K.: A Quick Method for Estimating Closed-Loop Poles of Control Systems, *Trans. AIEE*, vol. 76, pt. II, pp. 80–87, May 1957.

ROOT-LOCUS COMPENSATION

10-1 INTRODUCTION[1]

The preceding chapters have dealt with basic feedback control systems comprising the minimum amount of hardware in conjunction with the controlled element. Refinements are not made until the designer has analyzed the performance of the basic system. This chapter presents additional topics which enhance the design of a control system. Graphical means, utilizing a pole-zero diagram, are presented for calculating the figures of merit of the system and the effect of additional significant nondominant poles. Typical pole-zero patterns are employed to demonstrate the correlation between the pole-zero diagram and the frequency and time responses. As a result of this analysis, equipment may be added to the control system to achieve the desired time response. The next few chapters are devoted to achieving the necessary refinements.

Introducing additional equipment into a system to reshape its root locus in order to improve system performance is called *compensation* or stabilization. When the system is compensated, it is stable, has a satisfactory transient response, and has a large enough gain to ensure that the steady-state error does not exceed the specified maximum. Compensation devices may consist of electric networks or mechanical equipment containing levers, springs, dashpots, etc. The compensator (also called a filter) may be placed in cascade with the forward transfer function (cascade or series compensation), as shown in Fig. 10-1*a*, or in the feedback path (feedback or parallel

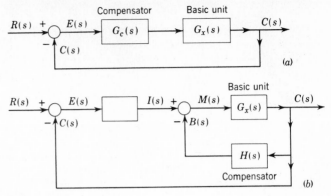

FIGURE 10-1
Block diagrams showing the location of compensators: (*a*) cascade; (*b*) feedback.

compensation), as shown in Fig. 10-1*b*. The selection of the location for inserting the compensator depends largely on the control system, the necessary physical modifications, and the results desired. The cascade compensator is inserted at the low-energy point in the forward path so that power dissipation is very small. This also requires that the input impedance be high. Isolation amplifiers may be necessary to avoid loading of or by the compensating network. The networks used for compensation are generally called lag, lead, and lag-lead compensators. The examples used for each of these compensators are not the only ones available but are intended primarily to show the methods of applying compensation.

10-2   TRANSIENT RESPONSE: DOMINANT COMPLEX POLES[2]

The root-locus plot permits selection of the best poles for the control ratio. The criteria for determining which poles are best must come from the specifications of system performance and from practical considerations. For example, the presence of backlash, dead zone, and coulomb friction can produce a steady-state error with a step input, even though the linear portion of the system is Type 1 or higher. The response obtained from the pole-zero locations of the control ratio does not show the presence of a steady-state error because the nonlinearities have been neglected. The effect of some nonlinearities can be minimized by making the system underdamped.†
Therefore, the system gain is adjusted so that there is a dominant pair of complex poles and the response to a step input has the form shown in Fig. 10-2*a*. In this case the transient contributions from the other poles must be small. From Eq. (4-68), the necessary conditions for the time response to be dominated by one pair of complex poles require the pole-zero pattern of Fig. 10-2*b* and have the following characteristics.

† For a detailed study of the effect of nonlinear characteristics on system performance, the reader is referred to the literature.[3,4]

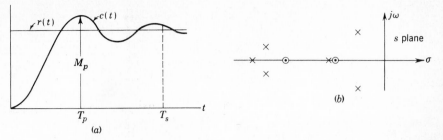

FIGURE 10-2

(a) Transient response to a step input with dominant complex poles. (b) Pole-zero pattern of C/R for the desired response.

1 The other poles must be far to the left of the dominant poles, so that the transients due to these other poles are small in amplitude and die out rapidly.

2 Any other pole which is not far to the left of the dominant complex poles must be near a zero so that the magnitude of the transient term due to that pole is small.

With the system designed so that the response to a unit step input has the underdamped form shown in Fig. 10-2a, the following figures of merit (described in Secs. 3-9 and 3-10) are used to judge its performance:

1 M_p, peak overshoot, is the amplitude of the first overshoot.

2 T_p, peak time, is the time to reach the peak overshoot.

3 T_s, settling time, is the time for the response to first reach and thereafter remain within 2 percent of the final value.

4 N is the number of oscillations in the response up to the settling time.

Consider the non-unity-feedback system shown in Fig. 10-3, using

$$G(s) = \frac{N_1}{D_1} = \frac{K_G \prod (s - z_g)}{\prod (s - p_g)} \tag{10-1}$$

$$H(s) = \frac{N_2}{D_2} = \frac{K_H \prod (s - z_i)}{\prod (s - p_i)} \tag{10-2}$$

$$G(s)H(s) = \frac{N_1 N_2}{D_1 D_2} = \frac{K_G K_H \prod_{h=1}^{w} (s - z_h)}{\prod_{c=1}^{n} (s - p_c)} \tag{10-3}$$

The product $K_G K_H = K$ is defined as the loop sensitivity. For $G(s)H(s)$ the degree of the numerator is w and the degree of the denominator is n. These symbols are used throughout Chap. 6. The control ratio is

$$\frac{C(s)}{R(s)} = \frac{P(s)}{Q(s)} = \frac{N_1 D_2}{D_1 D_2 + N_1 N_2} = \frac{K_G \prod_{m=1}^{w'} (s - z_m)}{\prod_{k=1}^{n} (s - p_k)} \tag{10-4}$$

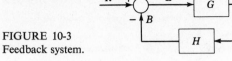

FIGURE 10-3
Feedback system.

Note that the constant K_G in Eq. (10-4) is not the same as the loop sensitivity unless the system has unity feedback. The degree n of the denominator of $C(s)/R(s)$ is the same as for $G(s)H(s)$, regardless of whether the system has unity or nonunity feedback. The degree w' of the numerator of $C(s)/R(s)$ is equal to the sum of the degrees of N_1 and D_2. For a unity feedback system the degree of the numerator is $w' = w$.

For a unit step input the output of Fig. 10-3 is

$$C(s) = \frac{P(s)}{sQ(s)} = \frac{K_G \prod\limits_{m=1}^{w'} (s - z_m)}{s \prod\limits_{k=1}^{n} (s - p_k)}$$

$$= \frac{A_0}{s} + \frac{A_1}{s - p_1} + \cdots + \frac{A_k}{s - p_k} + \cdots + \frac{A_n}{s - p_n} \qquad (10\text{-}5)$$

The pole $s = 0$ which comes from $R(s)$ must be included as a pole of $C(s)$, as shown in Eq. (10-5). The values of the coefficients can be obtained graphically from the pole-zero diagram by the method described in Sec. 4-9 and Eq. (4-68) or by means of a computer program.[5] Assume that the system represented by Eq. (10-5) has a dominant complex pole $p_1 = \sigma + j\omega_d$. The complete time solution is

$$c(t) = \frac{P(0)}{Q(0)} + 2 \left| \frac{K_G \prod\limits_{m=1}^{w'} (p_1 - z_m)}{p_1 \prod\limits_{k=2}^{n} (p_1 - p_k)} \right| e^{\sigma t} \cos \left[\omega_d t + \underline{/P(p_1)} - \underline{/p_1} - \underline{/Q'(p_1)} \right]$$

$$+ \sum_{k=3}^{n} \left[\frac{P(p_k)}{p_k Q'(p_k)} \right] e^{p_k t} \qquad (10\text{-}6)$$

where
$$Q'(p_k) = \frac{dQ(s)}{ds} \bigg]_{s = p_k} = \frac{Q(s)}{s - p_k} \bigg]_{s = p_k}$$

By assuming the pole-zero pattern of Fig. 10-2, the last term of Eq. (10-6) may be neglected. The time response is therefore approximated by

$$c(t) \approx \frac{P(0)}{Q(0)} + 2 \left| \frac{K_G \prod\limits_{m=1}^{w'} (p_1 - z_m)}{p_1 \prod\limits_{k=2}^{n} (p_1 - p_k)} \right| e^{\sigma t} \cos \left[\omega_d t + \underline{/P(p_1)} - \underline{/p_1} - \underline{/Q'(p_1)} \right]$$

$$(10\text{-}7)$$

Note that although the transient terms due to the other poles have been neglected, the effect of those poles on the amplitude and phase angle of the dominant transient has not been neglected.

The peak time T_p is obtained by setting the derivative with respect to time of Eq. (10-7) equal to zero. This gives

$$T_p = \frac{1}{\omega_d}\left[\frac{\pi}{2} - \underline{/P(p_1)} + \underline{/Q'(p_1)}\right] \qquad (10\text{-}8)$$

which may be stated as follows:

$$T_p = \frac{1}{\omega_d}\left\{\frac{\pi}{2} - \left[\text{sum of angles from zeros of } \frac{C(s)}{R(s)} \text{ to dominant pole } p_1\right]\right.$$

$$\left. + \left[\begin{array}{l}\text{sum of angles from all other poles of } \frac{C(s)}{R(s)} \text{ to} \\ \text{dominant pole } p_1\text{, including conjugate pole}\end{array}\right]\right\} \qquad (10\text{-}9)$$

From a physical consideration of Eq. (10-7) it can be seen that the phase angle $\phi = \underline{/P(p_1)} - \underline{/p_1} - \underline{/Q'(p_1)}$ cannot have a value greater than 2π. This same limitation must also be applied to the angles of Eq. (10-9), i.e., the value T_p must occur within one "cycle" of the transient. Inserting this value of T_p into Eq. (10-7) gives the peak overshoot M_p. By using the value $\cos(\pi/2 - \underline{/p_1}) = \omega_d/\omega_n$, the value M_p can be expressed as

$$M_p = \frac{P(0)}{Q(0)} + \frac{2\omega_d}{\omega_n^2}\left|\frac{K_G \displaystyle\prod_{m=1}^{w'}(p_1 - z_m)}{\displaystyle\prod_{k=2}^{n}(p_1 - p_k)}\right|e^{\sigma T_p} \qquad (10\text{-}10)$$

The first term in Eq. (10-10) represents the final value, and the second term represents the overshoot M_o. For a unity-feedback system which is Type 1 or higher, the equation for M_o can be put in an alternative form. For such a system there is a zero steady-state error when the input is a step function. Therefore the first term on the right side of Eq. (10-7) is equal to unity. The same expression can also be obtained by applying the final-value theorem to Eq. (10-5). This equality is used to solve for K_G. Note that under these conditions $K_G = K$ and $w' = w$:

$$K_G = \frac{\displaystyle\prod_{k=1}^{n}(-p_k)}{\displaystyle\prod_{m=1}^{w}(-z_m)} \qquad (10\text{-}11)$$

The value M_o can therefore be expressed as

$$M_o = \frac{2\omega_d}{\omega_n^2}\left|\frac{\displaystyle\prod_{k=1}^{n}(-p_k)\displaystyle\prod_{m=1}^{w}(p_1 - z_m)}{\displaystyle\prod_{m=1}^{w}(-z_m)\displaystyle\prod_{k=2}^{n}(p_1 - p_k)}\right|e^{\sigma T_p} \qquad (10\text{-}12)$$

Some terms inside the brackets in Eq. (10-12) cancel the terms in front so that M_o can be expressed in words as

$$M_o = \left[\frac{\begin{array}{c}\text{product of distances} \\ \text{from all poles of} \\ C(s)/R(s) \text{ to origin,} \\ \text{excluding distances} \\ \text{of two dominant} \\ \text{poles from origin}\end{array}}{\begin{array}{c}\text{product of distances} \\ \text{from all other poles} \\ \text{of } C(s)/R(s) \text{ to} \\ \text{dominant pole } p_1, \\ \text{excluding distance} \\ \text{between dominant} \\ \text{poles}\end{array}} \cdot \frac{\begin{array}{c}\text{product of distances} \\ \text{from all zeros of} \\ C(s)/R(s) \text{ to dominant} \\ \text{pole } p_1\end{array}}{\begin{array}{c}\text{product of distances} \\ \text{from all zeros of} \\ C(s)/R(s) \text{ to origin}\end{array}} \right] e^{\sigma T_p} \qquad (10\text{-}13)$$

If there are no finite zeros of $C(s)/R(s)$, the factors in M_o involving zeros become unity. Equation (10-13) is valid only for a unity-feedback system that is Type 1 or higher. It may be sufficiently accurate for a Type 0 system that has a large value for K_0. The value of M_o can be calculated either from the right-hand term of Eq. (10-10) or from Eq. (10-13), whichever is more convenient. The effect on M_o of other poles, which cannot be neglected, is discussed in the next section.

The values of T_s and N can be obtained approximately from the dominant roots. Section 3-9 shows that T_s is four time constants for 2 percent error:

$$T_s = \frac{4}{|\sigma|} = \frac{4}{\zeta \omega_n} \qquad (10\text{-}14)$$

$$N = \frac{\text{settling time}}{\text{period}} = \frac{T_s}{2\pi/\omega_d} = \frac{2\omega_d}{\pi|\sigma|} = \frac{2\sqrt{1-\zeta^2}}{\pi} \frac{1}{\zeta} \qquad (10\text{-}15)$$

An examination of Eq. (10-8) or (10-9) reveals that zeros of the control ratio cause a decrease in the peak time T_p, whereas poles increase the peak time. Peak time can also be decreased by shifting zeros to the right or poles (other than the dominant poles) to the left. Equation (10-13) shows that the larger the value of T_p, the smaller the value of M_o because $e^{\sigma T_p}$ decreases. M_o can also be decreased by reducing the ratios $p_k/(p_k - p_1)$ and $(z_m - p_1)/z_m$. But this can have an adverse effect on T_p. The conditions on pole-zero locations that lead to a small M_o may therefore produce a large T_p. Conversely, the conditions that lead to a small T_p may be obtained at the expense of a large M_o. Thus a compromise is required in the peak time and peak overshoot that are attainable. Some improvements can be obtained by introducing additional poles and zeros into the system and locating them appropriately. This is covered in the following sections on compensation.

The approximate equations, given in this section, for the response of a system to a step-function input are based on the fundamental premise that there is a pair of dominant complex poles and that the effect of other poles is small. By using a computer program[5] to calculate $c(t)$ it is not necessary to use the approximations in this

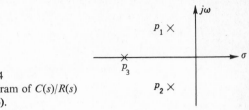

FIGURE 10-4
Pole-zero diagram of $C(s)/R(s)$
for Eq. (10-16).

section. In that case the values of M_p, t_p, and t_s can be obtained from the computer listing for $c(t)$.

10-3 ADDITIONAL SIGNIFICANT POLES[6]

When there are two dominant complex poles, the approximations developed in Sec. 10-2 give accurate results. However, there are cases where an additional pole of $C(s)/R(s)$ is significant. Figure 10-4 shows a pole-zero diagram which contains dominant complex poles and an additional real pole p_3. The control ratio is given by

$$\frac{C(s)}{R(s)} = \frac{K}{(s^2 + 2\zeta\omega_n s + \omega_n^2)(s - p_3)} \qquad (10\text{-}16)$$

With a unit step input the time response is $= (s + p)(s + p)(\quad)$

$$c(t) = 1 + 2|A_1|e^{-\zeta\omega_n t} \sin(\omega_n \sqrt{1 - \zeta^2}\, t + \phi) + A_3 e^{p_3 t} \qquad (10\text{-}17)$$

The transient term due to the real pole p_3 has the form $A_3 e^{p_3 t}$, where A_3 *is always negative.* Thus the overshoot M_p is reduced, and settling time t_s may be increased or decreased. This is the typical effect of an additional real pole. The magnitude A_3 depends on the location of p_3 relative to the complex poles. The further to the left the pole p_3 is located, the smaller the magnitude of A_3, therefore the smaller its effect on the total response. A pole which is 6 times as far to the left as the complex poles has negligible effect on the time response. The typical time response is shown in Fig. 10-5a. As the pole p_3 moves to the right, the magnitude of A_3 increases and the overshoot becomes smaller. As p_3 approaches but is still to the left of the complex poles, the first maximum in the time response is less than the final value. Overshoot of the final value can occur at the second or a later maximum. Figure 10-5b shows such a time response.

When p_3 is located at the real-axis projection of the complex poles, the response is monotonic; i.e., there is no overshoot. This represents the critically damped situation, as shown in Fig. 10-5c.

When p_3 is located to the right of the complex poles, it is the dominant pole and the response is overdamped. The complex poles contribute a "ripple" to the time response, as shown in Fig. 10-5d.

When the real pole p_3 is to the left of the complex poles, the peak time T_p is approximately given by Eq. (10-9). Although the effect of the *real* pole is to increase the peak time, this change is small if the real pole is fairly far to the left. A first-order

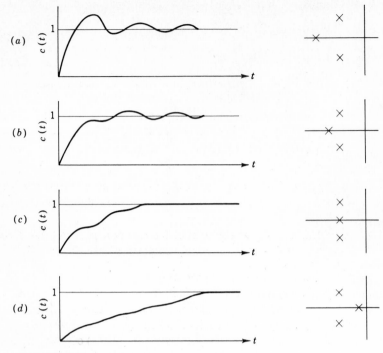

FIGURE 10-5
Typical time response as a function of the real-pole location.

correction can be made to the peak overshoot M_o given by Eq. (10-13) by adding the value of $A_3 e^{p_3 T_p}$ to the peak overshoot due to the complex poles. The effect of the real pole on the actual settling time t_s can be estimated by calculating $A_3 e^{p_3 T_s}$ and comparing it with the size and sign of the underdamped transient at time T_s obtained from Eq. (10-14). If both are negative at T_s, then the true settling time t_s is increased. If they have opposite signs, then the value $T_s = 4T$ based on the complex roots is a good approximation for t_s.

The presence of a real zero in addition to the real pole further modifies the transient response. Figure 10-6 shows a possible pole-zero diagram for which the control ratio is

$$\frac{C(s)}{R(s)} = \frac{K(s - z)}{(s^2 + 2\zeta\omega_n s + \omega_n^2)(s - p_3)} \qquad (10\text{-}18)$$

The complete time response to a unit step-function input still has the form given in Eq. (10-17). However, the sign of A_3 depends on the relative locations of the real pole and the real zero. A_3 is negative if the zero is to the left of p_3, and it is positive if the zero is to the right of p_3. Also, the magnitude of A_3 is proportional to the distance from p_3 to z [see Eq. (4-68)]. Therefore, if the zero is close to the pole, A_3 is small and the contribution of this transient term is correspondingly small. Compared with the response for Eq. (10-16), as shown in Fig. 10-5:

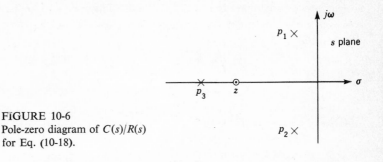

FIGURE 10-6
Pole-zero diagram of $C(s)/R(s)$
for Eq. (10-18).

1   If the zero z is to the left of the real pole p_3, the response is qualitatively the same as that for a system with only complex poles but the peak overshoot is smaller.

2   If the zero z is to the right of the real pole p_3, the peak overshoot is greater than that for a system with only complex poles.

When the real pole p_3 and the real zero z are close together, the change in the time response is small compared with that of a system which has only complex poles. The value of peak time T_p obtained from Eq. (10-9) can therefore be considered essentially correct. Actually T_p decreases if the zero is to the right of the real pole, and vice versa. A first-order correction to M_o from Eq. (10-13) can be obtained by adding the contribution of $A_3 e^{p_3 t}$ at the time T_p. The analysis of the effect on t_s is similar to that described above for the case of just an additional real pole.

Many control systems can be approximated by one having the following characteristics: (1) two complex poles; (2) two complex poles and one real pole; and (3) two complex poles, one real pole, and one real zero. For case 1 the relations T_p, M_o, T_s, and N developed in Sec. 10-2 give an accurate representation of the time response. For case 2 these approximate values can be corrected if the real pole is far enough to the left. Then the contribution of the additional transient term is small, and the total response remains essentially the sum of a constant and an underdamped sinusoid. For case 3 the approximate values can be corrected provided the zero is near the real pole so that the amplitude of the additional transient term is small. More exact calculation of the figures of merit can be obtained by plotting the exact response as a function of time. This is conveniently accomplished by simulating the system on a computer.[5]

10-4   RESHAPING THE ROOT LOCUS[7]

The root-locus plots described in Chap. 7 show the relationship between the gain of the system and the time response. Depending on the specifications established for the system, the gain that best achieves the desired performance is selected. The performance specifications may be based on the desired damping ratio, undamped natural frequency, time constant, or steady-state error. The root locus may show that the desired performance cannot be achieved just by adjustment of the gain. In fact, the system may be unstable for all values of gain. The control-systems engineer must then investigate the methods for reshaping the root locus to meet the performance specifications.

The purpose of reshaping the root locus generally falls into one of the following categories:

1 A given system is stable and its transient response is satisfactory, but its steady-state error is too large. Thus, the gain must be increased to reduce the steady-state error (see Chap. 6). This must be accomplished without appreciably reducing the system stability.

2 A given system is stable, but its transient response is unsatisfactory. Thus, the root locus must be reshaped so that it is moved farther to the left, away from the imaginary axis.

3 A given system is stable, but both its transient response and its steady-state response are unsatisfactory. Thus, the locus must be moved to the left and the gain must be increased.

4 A given system is unstable for all values of gain. Thus, the root locus must be reshaped so that part of each branch falls in the left-half s plane, thereby making the system stable.

Compensation of a system by the introduction of poles and zeros is used to improve the operating performance. However, each additional compensator pole increases the number of roots of the closed-loop characteristic equation. To make a comparison of the time responses, the contribution of each root must be taken into account. If an underdamped response of the form shown in Fig. 10-1 is desired, the system gain must be adjusted so that there is a pair of dominant complex poles. This requires that any other pole be far to the left or near a zero so that its transient has a small amplitude and therefore has a small effect on the total time response. The required pole-zero diagram is shown in Fig. 10-2. The approximate values of peak overshoot M_o, peak time T_p, settling time T_s, and the number of oscillations up to settling time N can be obtained from the pole-zero pattern as described in Sec. 10-2. The effect of compensator poles and zeros on these quantities can therefore be evaluated fairly rapidly.

10-5   IDEAL INTEGRAL CASCADE COMPENSATION

When the transient response of a feedback control system is considered satisfactory but the steady-state error is too large, it is possible to eliminate the error by increasing the system type. This must be accomplished *without appreciably changing the dominant roots of the characteristic equation.* The system type can be increased by operating on the actuating signal to produce one that is proportional to both the magnitude and the integral of this signal. This is shown in Fig. 10-7, where

$$E_1(s) = \left(1 + \frac{K_i}{s}\right) E(s) \qquad (10\text{-}19)$$

$$G_c(s) = \frac{E_1(s)}{E(s)} = \frac{s + K_i}{s} \qquad (10\text{-}20)$$

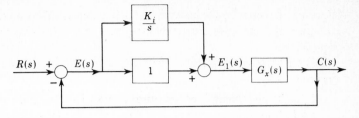

FIGURE 10-7
Ideal integral plus proportional control.

In this system the quantity $e_1(t)$ continues to increase as long as an error $e(t)$ is present. Eventually $e_1(t)$ becomes large enough to produce an output $c(t)$ equal to the input $r(t)$. The error $e(t)$ is then equal to zero. The constant K_i (generally very small) and the overall gain of the system must be selected to produce satisfactory roots of the characteristic equation. The resulting system performance has been improved. Since the system type has been increased, the corresponding error coefficient is equal to infinity. Provided that the new roots of the characteristic equation can be satisfactorily located, the transient response is still acceptable.

The locations of the pole and zero of this ideal integral plus proportional compensator are shown in Fig. 10-8. The pole alone would move the root locus to the right, thereby slowing down the time response. The zero must be near the origin in order to minimize the increase in response time of the complete system. The main limitation on the use of ideal integral plus proportional control is the equipment required to obtain the integral signal. This generally requires an amplifier in a positive-feedback system. Mechanically the integral signal can be obtained by an integrating gyroscope, which is used in aerospace vehicles where the improved performance justifies the cost. Frequently, however, a passive electric network consisting of resistors and capacitors sufficiently approximates the proportional plus integral action.

10-6 CASCADE LAG COMPENSATION USING PASSIVE ELEMENTS

Figure 10-9 shows a network which approximates a proportional plus integral output and is used as a lag compensator. Putting an amplifier of gain A in series with this network yields the transfer function

$$G_c(s) = A\frac{1 + Ts}{1 + \alpha Ts} = \frac{A}{\alpha}\frac{s + 1/T}{s + 1/\alpha T} \qquad (10\text{-}21)$$

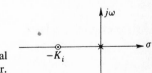

FIGURE 10-8
Location of the pole and zero of an ideal integral plus proportional compensator.

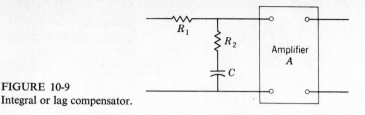

FIGURE 10-9
Integral or lag compensator.

where $\alpha = (R_1 + R_2)/R_2 > 1$ and $T = R_2C$. The pole $s = -1/\alpha T$ is therefore to the right of the zero $s = -1/T$. The locations of the pole and zero of $G_c(s)$ on the s plane can be made close to those of the ideal compensator. Besides furnishing the necessary gain, the amplifier also acts as an isolating unit to prevent any loading effects between the compensator and the original system.

Assume that the original forward transfer function is

$$G_x(s) = \frac{K \prod_{h=1}^{w} (s - z_h)}{\prod_{c=1}^{n} (s - p_c)} \qquad (10\text{-}22)$$

For the original system the loop sensitivity K for the selected root s is

$$K = \frac{\prod_{c=1}^{n} |s - p_c|}{\prod_{h=1}^{w} |s - z_h|} \qquad (10\text{-}23)$$

With the addition of the lag compensator in cascade, the new forward transfer function is

$$G(s) = G_c(s)G_x(s) = \frac{AK}{\alpha} \frac{\left(s + \dfrac{1}{T}\right) \prod_{h=1}^{w} (s - z_h)}{\left(s + \dfrac{1}{\alpha T}\right) \prod_{c=1}^{n} (s - p_c)} \qquad (10\text{-}24)$$

When the desired roots of the characteristic equation are located on the root locus, the magnitude of the loop sensitivity for the new root s' becomes

$$K' = \frac{AK}{\alpha} = \frac{\left|s' + \dfrac{1}{\alpha T}\right| \prod_{c=1}^{n} |s' - p_c|}{\left|s' + \dfrac{1}{T}\right| \prod_{h=1}^{w} |s' - z_h|} \qquad (10\text{-}25)$$

As an example, a lag compensator is applied to a Type 0 system. To improve the steady-state accuracy, the error coefficient must be increased. The value of K_0

before and after the addition of the compensator is calculated by using the definition $K_0 = \lim_{s \to 0} G(s)$ from Eqs. (10-22) and (10-24), respectively,

$$K_0 = \frac{\prod_{h=1}^{w} (-z_h)}{\prod_{c=1}^{n} (-p_c)} K \qquad (10\text{-}26)$$

$$K_0' = \frac{\prod_{h=1}^{w} (-z_h)}{\prod_{c=1}^{n} (-p_c)} \alpha K' \qquad (10\text{-}27)$$

The following procedure is used to design the passive lag cascade compensator. First, the pole $s = -1/\alpha T$ and the zero $s = -1/T$ of the compensator are placed very close together. This means that most of the original root locus remains practically unchanged. If the angle contributed by the compensator *at the original closed-loop dominant root* is less than 5°, the new locus is displaced only slightly. This 5° figure is only a guide and should not be applied arbitrarily. The new closed-loop pole s' is therefore essentially unchanged from the uncompensated value. This satisfies the restriction that the transient response must not change appreciably. As a result, the values $s' + 1/\alpha T$ and $s' + 1/T$ are almost equal, and the values K and K' in Eqs. (10-23) and (10-25) are approximately equal. The values K_0 and K_0' in Eqs. (10-26) and (10-27) now differ only by the factor α so that $K_0' \approx \alpha K_0$. The gain required to produce the new root s' therefore increases approximately by the factor α, which is the ratio of the compensator zero and pole. Summarizing, the necessary conditions on the compensator are that (1) the pole and zero must be close together and (2) the ratio α of the zero and pole must approximately equal the desired increase in gain. These requirements can be achieved by placing the compensator pole and zero very close to the origin. The size of α is limited by the physical parameters required in the network. A value $\alpha = 10$ is often used.

Although the statements above are based on a Type 0 system, the same conditions apply equally well for a Type 1 or higher system.

Example of Lag Compensation Applied to a Type 1 System

A control system with unity feedback has a forward transfer function

$$G_x(s) = \frac{K_1}{s(1 + s)(1 + 0.2s)} = \frac{K}{s(s + 1)(s + 5)} \qquad (10\text{-}28)$$

which yields the root locus shown in Fig. 10-10. For the basic control system a damping ratio $\zeta = 0.45$ yields the following pertinent data:

Dominant roots:

$$s_{1,2} = -0.404 \pm j0.804$$

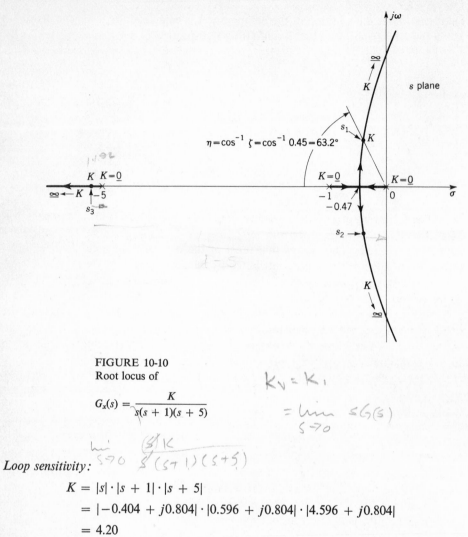

FIGURE 10-10
Root locus of

$$G_x(s) = \frac{K}{s(s + 1)(s + 5)}$$

Loop sensitivity:

$$K = |s| \cdot |s + 1| \cdot |s + 5|$$
$$= |-0.404 + j0.804| \cdot |0.596 + j0.804| \cdot |4.596 + j0.804|$$
$$= 4.20$$

Ramp error coefficient:

$$K_1 = \lim_{s \to 0} sG(s)$$

$$= \frac{K}{5} = \frac{4.20}{5} = 0.84 \text{ s}^{-1}$$

Undamped natural frequency:

$$\omega_n = 0.89 \text{ rad/s}$$

Third root:

$$s_3 = -5.192$$

The values of peak time T_p, peak overshoot M_o, and settling time T_s are obtained from Eqs. (10-9), (10-13), and (10-14) as $T_p = 4.11$ s, $M_o = 0.203$, and $T_s = 9.9$ s. To increase the gain, a lag compensator is put in cascade with the forward

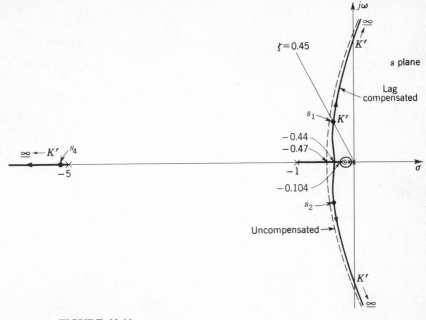

FIGURE 10-11
Root locus of

$$G(s) = \frac{K'(s + 0.05)}{s(s + 1)(s + 5)(s + 0.005)}$$

transfer function. With the criteria discussed in the preceding section and $\alpha = 10$, the compensator pole is located at $s = -0.005$ and the zero at $s = -0.05$. This selection is based on the requirements of a practical network. For the network of Fig. 10-9, the values $R_1 = 18\ \text{M}\Omega$, $R_2 = 2\ \text{M}\Omega$, and $C = 10\ \mu\text{F}$ produce the pole and zero required. The angle of the compensator at the original dominant roots is about $2.6°$ and is acceptable. Since the compensator pole and zero are very close together, they make only a small change in the new root locus in the vicinity of the original roots. The compensator transfer function is

$$G_c(s) = \frac{A}{\alpha}\frac{s + 1/T}{s + 1/\alpha T} = \frac{A}{10}\frac{s + 0.05}{s + 0.005} \qquad (10\text{-}29)$$

Figure 10-11 (not to scale) shows the new root locus for $G(s) = G_x(s)G_c(s)$ as solid lines and the original locus as dashed lines. Note that for the damping ratio $\zeta = 0.45$ the new locus and the original locus are close together. The new dominant roots are $s_{1,2} = -0.384 \pm j0.764$; thus the roots are essentially unchanged.

For the compensated system adjusted to a damping ratio 0.45 the following results are obtained:

Dominant roots:

$$s_{1,2} = -0.384 \pm j0.764$$

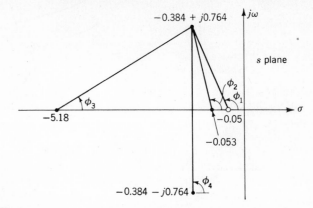

FIGURE 10-12
Angles used to evaluate T_p.

Loop sensitivity:

$$K' = \frac{|s| \cdot |s + 1| \cdot |s + 5| \cdot |s + 0.005|}{|s + 0.05|} = 4.019$$

Ramp error coefficient:

$$K_1' = \frac{K'\alpha}{5} = \frac{(4.019)(10)}{5} = 8.04 \text{ s}^{-1}$$

Increase in gain:

$$A = \frac{K_1'}{K_1} = \frac{8.04}{0.84} = 9.57$$

Undamped natural frequency:

$$\omega_n = 0.87 \text{ rad/s}$$

Other roots:

$$s_3 = -5.18 \quad \text{and} \quad s_4 = -0.053$$

The value of T_p for the compensated system is determined by use of Eq. (10-9), with the values shown in Fig. 10-12:

$$T_p = \frac{1}{\omega_d}\left[\frac{\pi}{2} - \phi_1 + (\phi_2 + \phi_3 + \phi_4)\right]$$

$$= \frac{1}{0.764}[90° - 113.6° + (113.4° + 9.2° + 90°)]\frac{\pi \text{ rad}}{180°}$$

$$= 4.32 \text{ s}$$

The value of M_o for the compensated system is determined by use of Eq. (10-13), with the values shown in Fig. 10-13;

$$M_o = \frac{l_1 l_2}{l_3 l_4}\frac{(l)_1}{(l)_2} e^{\sigma T_p}$$

$$= \frac{(5.18)(0.053)}{(4.76)(0.832)}\frac{.0834}{0.05} e^{(-0.384)(4.32)}$$

$$= 1.1563 e^{-1.659} = 0.22$$

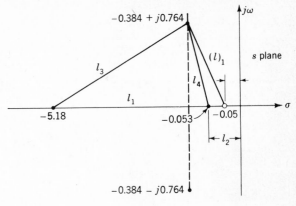

FIGURE 10-13
Lengths used to evaluate M_o.

A comparison of the uncompensated system and the system with integral compensation shows that the ramp error coefficient K_1' has increased by a factor of 9.57 but the undamped natural frequency has been decreased slightly from 0.89 to 0.87 rad/s. This means that the steady-state error with a ramp input has decreased but the settling time has been increased. Provided that the increased settling time and peak overshoot are acceptable, the system has been improved. Although the complete root locus is drawn in Fig. 10-11, in practice this is not necessary. Only the general shape is sketched and the particular needed points accurately determined.

10-7   IDEAL DERIVATIVE CASCADE COMPENSATION

When the transient response of a feedback system must be improved, it is necessary to reshape the root locus so that it is moved farther to the left of the imaginary axis. Section 7-3 showed that introducing an additional zero in the forward transfer function produces this effect. A zero can be produced by operating on the actuating signal to produce one that is proportional to both the magnitude and the derivative (rate of change) of this signal. This is shown in Fig. 10-14, where

$$E_1(s) = (1 + K_d s)E(s) \qquad (10\text{-}30)$$

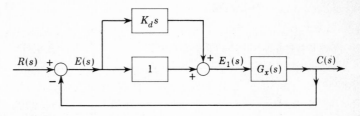

FIGURE 10-14
Ideal derivative plus proportional control.

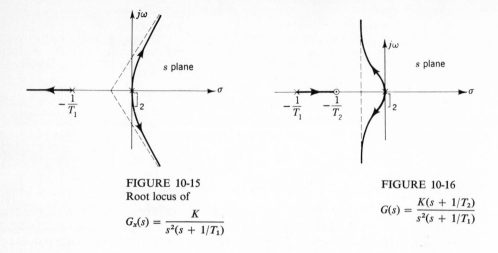

FIGURE 10-15
Root locus of

$$G_x(s) = \frac{K}{s^2(s + 1/T_1)}$$

FIGURE 10-16

$$G(s) = \frac{K(s + 1/T_2)}{s^2(s + 1/T_1)}$$

Physically the effect can be described as introducing anticipation into the system. The system reacts not only to the magnitude of the error but also to its rate of change. If the error is changing rapidly, then $e_1(t)$ is large and the system responds faster. The net result is to speed up the response of the system. On the root locus the introduction of a zero has the effect of shifting the curves to the left, as shown in the following example.

EXAMPLE   The root locus for a Type 2 system which is unstable for all values of loop sensitivity K is shown in Fig. 10-15. Adding a zero to the system by means of the compensator $G_c(s)$ (an ideal derivative plus proportional compensator) has the effect of moving the locus to the left, as shown in Fig. 10-16. The addition of a zero $s = -1/T_2$ between the origin and the pole at $-1/T_1$ stabilizes the system for all positive values of gain.

This example has shown how the introduction of a zero in the forward transfer function by a derivative plus proportional compensator has modified and improved the feedback control system. However, the synthesis of such derivative action requires the use of an active network. In addition, the derivative action amplifies any spurious signal or noise that may be present in the actuating signal. This noise amplification may saturate electronic amplifiers so that the system does not operate properly. Compensators used to synthesize derivative action involve approximations, so that the proportional plus derivative action is frequently obtained by a simple passive network, as shown in the next section.

10-8   LEAD COMPENSATION USING PASSIVE ELEMENTS

Figure 10-17 shows a derivative or lead compensator made up of electrical elements. To this network an amplifier of gain A is added in series. The transfer function is

$$G_c(s) = A\alpha \frac{1 + Ts}{1 + \alpha Ts} = A \frac{s + 1/T}{s + 1/\alpha T} = A \frac{s - z_c}{s - p_c} \qquad (10\text{-}31)$$

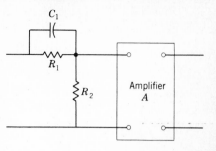

FIGURE 10-17
Derivative or lead compensator.

where $\alpha = R_2/(R_1 + R_2) < 1$ and $T = R_1 C$. This lead compensator introduces a zero at $s = -1/T$ and a pole at $s = -1/\alpha T$. By making α sufficiently small, the location of the pole is far to the left and has small effect on the important part of the root locus. Near the zero the net angle of the compensator is due predominantly to the zero. Figure 10-18 shows the location of the lead-compensator pole and zero and the angles contributed by each at a point s_1. The best location of the zero must be determined by trial and error. It is found that the gain of the compensated system is often increased. The maximum increases in gain and in the real part $(\zeta\omega_n)$ of the dominant root of the characteristic equation do not coincide. The compensator zero location must then be determined, by making several trials, for the desired optimum performance.

It can be shown that the loop sensitivity is proportional to the ratio of $|s + 1/\alpha T|$ to $|s + 1/T|$. Therefore, as α decreases, the loop sensitivity increases. The minimum value of α is limited by the size of the parameters needed in the network to obtain the minimum input impedance required. Note also from Eq. (10-31) that a small α requires a large value of additional gain A from the amplifier. The value $\alpha = 0.1$ is a common choice.

Example of Lead Compensation Applied to a Type 1 System

The same type of system used in Sec. 10-6 is now used with lead compensation. The locations of the pole and zero of the compensator are first selected by trial. At the conclusion of this section some rules are given to show the best location.

The forward transfer function of the original system is

$$G_x(s) = \frac{K_1}{s(1 + s)(1 + 0.2s)} = \frac{K}{s(s + 1)(s + 5)} \qquad (10\text{-}32)$$

FIGURE 10-18
Location of pole and zero of a lead compensator.

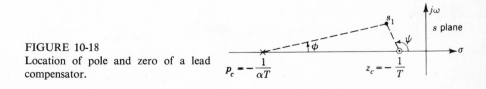

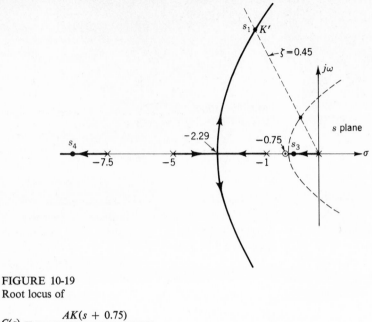

FIGURE 10-19
Root locus of

$$G(s) = \frac{AK(s + 0.75)}{s(s + 1)(s + 5)(s + 7.5)}$$

The transfer function of the compensator, using a value $\alpha = 0.1$, is

$$G_c(s) = 0.1A \frac{1 + Ts}{1 + 0.1Ts} = A \frac{s + 1/T}{s + 1/0.1T} \qquad (10\text{-}33)$$

The new forward transfer function is

$$G(s) = G_x(s)G_c(s) = \frac{AK(s + 1/T)}{s(s + 1)(s + 5)(s + 1/0.1T)} \qquad (10\text{-}34)$$

Three selections for the position of the zero $s = -1/T$ are made. A comparison of the results shows the relative merits of each. The three locations of the zero are $s = -0.75$, $s = -1.00$, and $s = -1.50$. Figures 10-19 to 10-21 show the resultant loci for these three cases. In each figure the dashed lines are the locus of the original uncompensated system.

The loop sensitivity is evaluated in each case from the expression

$$K' = AK = \frac{|s| \cdot |s + 1| \cdot |s + 5| \cdot |s + 1/\alpha T|}{|s + 1/T|} \qquad (10\text{-}35)$$

and the ramp error coefficient is evaluated from the equation

$$K_1' = \frac{K'\alpha}{5} \qquad (10\text{-}36)$$

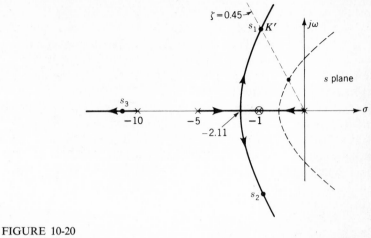

FIGURE 10-20
Root locus of

$$G(s) = \frac{AK(s + 1)}{s(s + 1)(s + 5)(s + 10)}$$

The results, given in Table 10-1, show that a large increase in ω_n has been obtained by adding the lead compensators. Since the value of ζ is held constant, the system has a much faster response time. A comparison of the individual lead compensators shows the following.

A zero of the compensator to the right of the pole at $s = -1$ results in a root

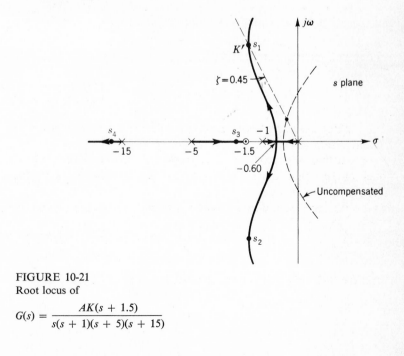

FIGURE 10-21
Root locus of

$$G(s) = \frac{AK(s + 1.5)}{s(s + 1)(s + 5)(s + 15)}$$

on the negative real axis close to the origin (see Fig. 10-19). From Eq. (4-68) it can be seen that the coefficient associated with this transient term is negative. Therefore it has the desirable effect of decreasing the peak overshoot and the undesirable characteristic of having a long time constant. If the size of this transient term is small (the magnitude is proportional to the distance from the real root to the zero), the effect on the settling time is small.

The decrease in M_o can be determined by calculating the contribution of the transient term due to the real root, as discussed in Sec. 10-3. At the time $t = T_s$ the sign of the underdamped transient can be determined. If it is negative, the actual settling time has increased. If it is positive, the settling time may be decreased. In this example both transients are negative at T_s; therefore the actual settling time is increased. The exact settling time can be obtained from a complete solution of the time response, which is shown in Fig. 10-22. The correct settling time is 2.8 s.

The best transient response due to the complex roots occurs when the zero of the compensator cancels the pole $s = -1$ of the original transfer function (see Fig. 10-20). There is no real root near the imaginary axis in this case. The actual setting time is $t_s = 2.46$ s. The order of the characteristic equation is unchanged.

The largest gain occurs when the zero of the compensator is located at the left of the pole $s = -1$ of the original transfer function (see Fig. 10-21). The transient due to the root $s = -1.76$ is positive. Therefore it increases the peak overshoot of the response. The value of the transient due to the real root can be added to the peak value of the underdamped sinusoid to obtain the correct value of the peak overshoot. The real root may affect the settling time. The change of T_s can be estimated as outlined in case 1. In this case the real root decreased the peak time and increased the settling time. From Fig. 10-22 the actual values are $t_p = 1.03$ s and $t_s = 2.55$ s.

From the root-locus plots it is evident that increasing the gain increases the peak overshoot and the settling time. The response with the three lead compensators is shown in Fig. 10-22. The effect of the additional real root is to change the peak overshoot and the settling time. The designer must make a choice between the largest gain attainable, the smallest peak overshoot, and the shortest settling time.

Table 10-1 RESULTS OF LEAD COMPENSATION

Compensator	Dominant complex roots	ω_n	K_1	Additional roots	t_s, s
Uncompensated	$-0.404 + j0.804$	0.89	0.83	-5.19	9.6
$\dfrac{s + 0.75}{s + 7.5}$	$-1.522 + j3.022$	3.38	2.05	$-0.689; -9.77$	2.80
$\dfrac{s + 1.0}{s + 10}$	$-1.588 + j3.151$	3.52	2.94	-11.82	2.46
$\dfrac{s + 1.50}{s + 15.0}$	$-1.531 + j3.040$	3.40	4.40	$-1.76; -16.18$	2.55

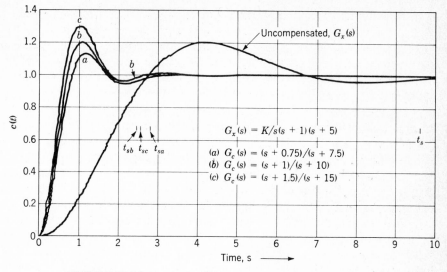

FIGURE 10-22
Time responses with a step input for the uncompensated and compensated cases of Table 10-1.

Based on this example, the following guidelines are used to apply cascade lead compensators to a Type 1 or higher system:

1 If the zero $s = -1/T$ of the lead compensator is superimposed and cancels the largest real pole (excluding the pole at zero) of the original transfer function, a good improvement in the transient response is obtained.

2 If a larger gain is desired than that obtained by guideline 1, several trials should be made with the zero of the compensator moved to the left or right. The location of the zero that results in the desired gain and roots is selected.

For a Type 0 system it will often be found that a better time response and a larger gain can be obtained by placing the compensator zero so that it cancels or is close to the second largest real pole of the original transfer function.

10-9 GENERAL LEAD-COMPENSATOR DESIGN

Additional methods are available for the design of an appropriate lead compensator G_c which is placed in cascade with a basic transfer function G_x, as shown in Fig. 10-1. Figure 10-23 shows the original root locus of a control system. For a specified damping ratio ζ the dominant root of the uncompensated system is s_1. Also shown is s_2, which is the desired root of the system's characteristic equation. Selection of s_2 as a desired root is based on the performance required for the system. The design problem is to select a lead compensator that results in s_2 being a root. The first step is to find the sum of the angles at the point s_2 due to the poles and zeros of the original system.

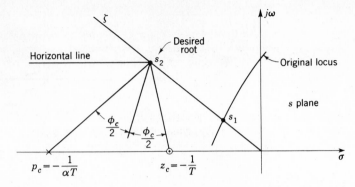

FIGURE 10-23
Graphical construction for locating the pole and zero of a simple lead compensator.

This angle is $180° + \phi$. For s_2 to be on the new root locus, it is necessary for the lead compensator to contribute an angle $\phi_c = -\phi$ at this point. The total angle at s_2 is then $180°$, and it is a point on the new root locus. A simple lead compensator represented by Eq. (10-31), with its zero to the right of its pole, can be used to provide the angle ϕ_c at the point s_2.

Method 1 Actually, there are many possible locations of the compensator pole and zero that will produce the necessary angle ϕ_c at the point s_2. Cancellation of poles of the original open-loop transfer function by means of compensator zeros may simplify the root locus and thereby reduce the complexity of the problem. The compensator zero is simply placed over a real pole. Then the compensator pole is placed further to the left at a location which makes s_2 a point on the new root locus. The pole to be canceled depends on the system type. For a Type 1 system the largest real pole (excluding the pole at zero) should be canceled. For a Type 0 system the second largest pole should be canceled.

Method 2 The following construction is based on obtaining the maximum value of α, which is the ratio of the compensator zero and pole. The steps in the location of the lead-compensator pole and zero (see Fig. 10-23) are:

1 Locate the desired root s_2. Draw a line from this root to the origin and a horizontal line to the left of s_2.
2 Bisect the angle between the two lines drawn in step 1.
3 Measure the angle $\phi_c/2$ on either side of the bisector drawn in step 2.
4 The intersections of these lines with the real axis locate the compensator pole p_c and zero z_c.

The construction outlined above is shown in Fig. 10-23. It is left for the reader to show that this construction results in the largest possible value of α. Since α is less than unity and appears as the gain of the compensator [see Eq. (10-31)], the largest value of α requires the smallest additional gain A.

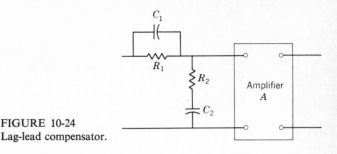

FIGURE 10-24
Lag-lead compensator.

Method 3 The procedures outlined above give the desired location of a root. However, no conditions have been placed on the desired system gain. A minimum value of gain may be specified to restrict the maximum steady-state error. Changing the locations of the compensator pole and zero can produce a range of values for the gain while maintaining the desired root s_2. A specific procedure for achieving both the desired root s_2 and the system gain K_m is given in Refs. 8 and 9.

10-10 LAG-LEAD CASCADE COMPENSATION

The preceding sections have shown that (1) the insertion of an integral or lag compensator in cascade results in a large increase in gain and a small reduction in the undamped natural frequency and (2) that the insertion of a derivative or lead compensator in cascade results in a small increase in gain and a large increase in the undamped natural frequency. By inserting both the lag and the lead compensators in cascade with the original transfer function, the advantages of both can be realized simultaneously; i.e., a large increase in gain and a large increase in the undamped natural frequency can be obtained. Instead of using two separate networks, it is possible to use one network that acts as both a lead and a lag compensator. Such a network, shown in Fig. 10-24 with an amplifier added in cascade, is called a lag-lead compensator. The transfer function of this compensator (see Sec. 8-10) is

$$G_c(s) = A\,\frac{(1 + T_1 s)(1 + T_2 s)}{(1 + \alpha T_1 s)[1 + (T_2/\alpha)s]} = A\,\frac{(s + 1/T_1)(s + 1/T_2)}{(s + 1/\alpha T_1)(s + \alpha/T_2)} \qquad (10\text{-}37)$$

where $\alpha > 1$ and $T_1 > T_2$.

The fraction $(1 + T_1 s)/(1 + \alpha T_1 s)$ represents the lag compensator, and the fraction $(1 + T_2 s)/[1 + (T_2/\alpha)s]$ represents the lead compensator. The values T_1, T_2, and α are selected to achieve the desired improvement in system performance. The specific values of the compensator components are obtained from the relationships $T_1 = R_1 C_1$, $T_2 = R_2 C_2$, $\alpha T_1 + T_2/\alpha = R_1 C_1 + R_2 C_2 + R_1 C_2$. It may also be desirable to specify a minimum value of the input impedance over the passband frequency range.

The procedure for applying the lag-lead compensator is a combination of the procedures for the individual units.

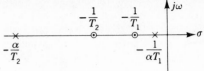

FIGURE 10-25
Location of the poles and zeros of the
lag-lead compensator.

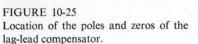

1 For the integral or lag component:

a The zero $s = -1/T_1$ and the pole $s = -1/\alpha T_1$ are selected close together, with α set to a large value such as $\alpha = 10$.

b The pole and zero are located to the left of and close to the origin. This
results in an increased gain.

2 For the derivative or lead component the zero $s = -1/T_2$ is superimposed
on a pole of the original system. This results in moving the root locus to
the left and therefore increases the undamped natural frequency.

The relative positions of the poles and zeros of the lag-lead compensators are
shown in Fig. 10-25.

Example of Lag-Lead Compensation Applied to a Type 1 System

The system described in Secs. 10-6 and 10-8 is now used with a lag-lead compensator.
The forward transfer function is

$$G_x(s) = \frac{K_1}{s(1 + s)(1 + 0.2s)} = \frac{K}{s(s + 1)(s + 5)} \qquad (10\text{-}38)$$

The lag-lead compensator used is shown in Fig. 10-24, and the transfer function
$G_c(s)$ is given by Eq. (10-37). The new complete forward transfer function is

$$G(s) = G_x(s)G_c(s) = \frac{AK(s + 1/T_1)(s + 1/T_2)}{s(s + 1)(s + 5)(s + 1/\alpha T_1)(s + \alpha/T_2)} \qquad (10\text{-}39)$$

The poles and zeros of the compensator are selected in accordance with the principles
outlined previously in this section. They coincide with the values used for the integral
and derivative compensators when applied individually in Secs. 10-6 and 10-8. With
an $\alpha = 10$, the poles and zeros are

$$\frac{1}{T_1} = 0.05 \qquad \frac{1}{T_2} = 1 \qquad \frac{1}{\alpha T_1} = 0.005 \qquad \frac{\alpha}{T_2} = 10$$

The forward transfer function becomes

$$G(s) = G_x(s)G_c(s) = \frac{K'(s + 0.05)}{s(s + 0.005)(s + 5)(s + 10)} \qquad (10\text{-}40)$$

The root locus (not to scale) for this system is shown in Fig. 10-26.

For a damping ratio $\zeta = 0.45$ the following pertinent data are obtained for the
compensated system:

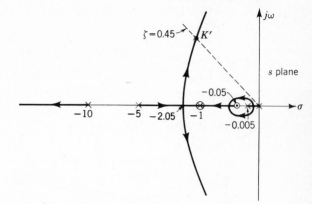

FIGURE 10-26
Root locus of

$$G(s) = \frac{K'(s + 1)(s + 0.05)}{s(s + 1)(s + 0.005)(s + 5)(s + 10)}$$

Dominant roots:

$$s_{1,2} = -1.571 \pm j3.120$$

Loop sensitivity (evaluated at $s = s_1$):

$$K' = \frac{|s| \cdot |s + 0.005| \cdot |s + 5| \cdot |s + 10|}{|s + 0.05|} = 146.4$$

Ramp error coefficient:

$$K_1' = \frac{K'(0.05)}{(0.005)(5)(10)} = 29.3 \text{ s}^{-1}$$

Undamped natural frequency:

$$\omega_n = 3.40 \text{ rad/s}$$

The additional roots are $s_3 = -11.81$ and $s_4 = -0.051$. These results, when compared with those of the uncompensated system, show a large increase in both the gain and the undamped natural frequency. They indicate that the advantages of the lag-lead compensator are equivalent to the combined improvement obtained by the lag and the lead compensators. Note the practicality in the choice of $\alpha = 10$, which is based on experience.

10-11   COMPARISON OF CASCADE COMPENSATORS

Table 10-2 shows a comparison of the system response obtained when a lag, a lead, and a lag-lead compensator are added in cascade with the same basic feedback system. These results are for the examples of Secs. 10-6, 10-8, and 10-10. The locations of the poles of the closed-loop system roughly satisfy the conditions of Sec. 10-2. Therefore the approximate equations given in Sec. 10-2 are used to calculate M_o, T_p, and T_s.

These results are typical of the changes introduced by the use of cascade compensators, which can be summarized as follows:

1 Lag compensator:

    a Results in a large increase in gain K_m (by a factor almost equal to α), which means a much smaller steady-state error.

    b Decreases ω_n and therefore has the disadvantage of producing a small increase in the settling time.

2 Lead compensator:

    a Results in a moderate increase in gain K_m, thereby improving steady-state accuracy.

    b Results in a large increase in ω_n and therefore reduces the settling time considerably.

    c The transfer function of the lead compensator, using the passive network of Fig. 10-17, contains the gain α, which is less than unity. Therefore the additional gain A, which must be added, is larger than the increase in K_m for the system.

Table 10-2   COMPARISON OF PERFORMANCE OF SEVERAL CASCADE COMPENSATORS FOR THE SYSTEM OF EQ. (10-28) WITH $\zeta = 0.45$

Compensator	Dominant complex roots	Un-damped natural fre-quency ω_n	Other roots	Ramp error coeffi-cient K_1	T_p, s	M_o	T_s, s	Addi-tional gain required A
Uncompensated	$-0.404 + j0.804$	0.89	-5.192	0.84	4.11	0.203	9.9	—
Lag: $\dfrac{s + 0.05}{s + 0.005}$	$-0.384 + j0.764$	0.87	-0.053 -5.18	8.04	4.32	0.22	10.4	9.57
Lead: $\dfrac{s + 1.5}{s + 15}$	$-1.531 + j3.040$	3.40	-1.76 -16.18	4.40	1.12	0.24	2.61*	52.38
Lead: $\dfrac{s + 1}{s + 10}$	$-1.588 + j3.151$	3.52	-11.82	2.94	1.09	0.19	2.52*	35.0
Lag-lead: $\dfrac{(s + 0.05)(s + 1)}{(s + 0.005)(s + 10)}$	$-1.571 + j3.120$	3.40	-0.051 -11.81	29.3	1.1	0.199	2.54	34.9

* Values of t_s are given in Table 10-1.

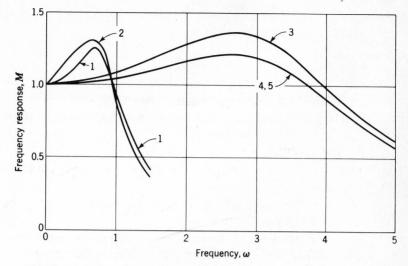

FIGURE 10-27
Frequency response for systems of Table 10-2: (1) original basic system; (2) lag-compensated; (3) lead-compensated, zero at -1.5; (4) lead-compensated, zero at -1.0; (5) lag-lead-compensated.

3 Lag-lead compensator (essentially combines the desirable characteristics of the lag and the lead compensators):

a Results in a large increase in gain K_m, which improves the steady-state response.

b Results in a large increase in ω_n, which improves the transient-response settling time.

Figure 10-27 shows the frequency response for the original and compensated systems corresponding to Table 10-2. There is a close qualitative correlation between the time-response characteristics listed in Table 10-2 and the corresponding frequency-response curves shown in Fig. 10-27. The peak value M_p is directly proportional to the maximum value M_m, and the settling time T_s is a direct function of the passband frequency. Curves 3, 4, and 5 have a much wider passband than curves 1 and 2; therefore these systems have a faster settling time.

For systems other than the one used as an example, these simple compensators may not produce the desired changes. However, the basic principles described here are applicable to any system for which the open-loop poles closest to the imaginary axis are on the negative real axis. For these systems the conventional compensators may be used, and good results are obtained.

When the complex open-loop poles are dominant, the conventional compensators produce only small improvements. A pair of complex poles near the imaginary axis often occurs in aircraft and missile transfer functions. Effective compensation would remove the open-loop dominant complex poles and replace them with poles

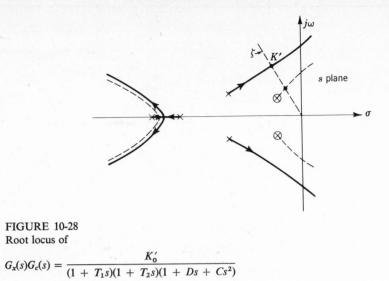

FIGURE 10-28
Root locus of

$$G_x(s)G_c(s) = \frac{K_0'}{(1 + T_1s)(1 + T_2s)(1 + Ds + Cs^2)}$$

located farther to the left. This means that the compensator should have a transfer function of the form

$$G_c(s) = \frac{1 + Bs + As^2}{1 + Ds + Cs^2} \qquad (10\text{-}41)$$

With exact cancellations, the zeros of $G_c(s)$ would coincide with the complex poles of the original transfer function. Consider the feedback system which has the forward transfer function

$$G(s) = \frac{K_0}{(1 + T_1s)(1 + T_2s)(1 + Bs + As^2)} \qquad T_1 < T_2 \qquad (10\text{-}42)$$

The composite forward transfer function with the compensator would be

$$G(s) = G_c(s)G_x(s) = \frac{K_0'}{(1 + T_1s)(1 + T_2s)(1 + Ds + Cs^2)} \qquad (10\text{-}43)$$

and the root locus is shown in Fig. 10-28. The solid lines are the new locus, and the dashed lines are the original locus. The new poles introduced by the compensator can be real instead of complex. In this case the root locus is shown in Fig. 10-29. In both cases the transient response has been improved. The problem remaining to be solved is the synthesis of a network with the desired poles and zeros. The reader is referred to books that cover network synthesis.[10-12]

10-12 INTRODUCTION TO FEEDBACK COMPENSATION

System performance may be improved by using either cascade or feedback compensation. The factors that must be considered when making that decision are listed in Sec. 6-10. The effectiveness of cascade compensation has been illustrated in the

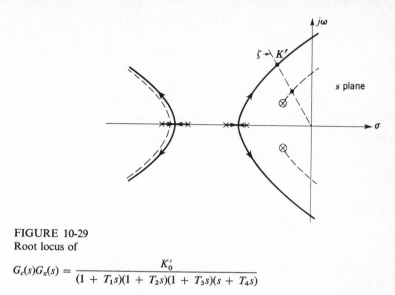

FIGURE 10-29
Root locus of

$$G_c(s)G_x(s) = \frac{K_0'}{(1 + T_1 s)(1 + T_2 s)(1 + T_3 s)(s + T_4 s)}$$

preceding sections of this chapter. The use of feedback compensation is demonstrated in the remainder of this chapter. The location of a feedback compensator depends on the complexity of the basic system, the accessibility of insertion points, the form of the signal being fed back, the signal with which it is being compared, and the desired improvement. Figure 10-30 shows various forms of feedback-compensated systems. For each case there is a unity-feedback loop (major loop) besides the compensation loop (minor loop) in order to maintain a direct correspondence between the output and input.

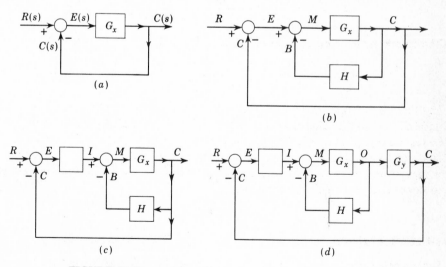

FIGURE 10-30
Various forms of feedback systems: (a) uncompensated system; (b to d) feedback-compensated systems.

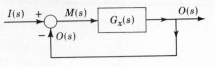

FIGURE 10-31
Transforming non-Type 0 control
element to Type 0.

In applying feedback compensation, it is often necessary to maintain the type of the basic system. Care must therefore be exercised in determining the number of differentiating terms which are permitted in the numerator of $H(s)$. Table 10-3 presents the basic forms that $H(s)$ can have so that the system type remains the same after the minor feedback-compensation loop is added. $H(s)$ may have additional factors in the numerator and denominator. It can be shown by evaluating $G(s)$ that the degree of the factor s in the numerator of $H(s)$ must be equal to or higher than the type of the forward transfer function $G_x(s)$ shown in Fig. 10-30.

By the use of a direct or unity-feedback loop any non-Type 0 control element is converted into a Type 0 element. This effect is advantageous when going from a low-energy-level input to a high-energy-level output for a given control element and also when the forms of the input and output energy are different. In Fig. 10-31 the non-Type 0 $G_x(s)$ function may represent either a single control element or a group of control elements that form part of a control system and whose transfer function is of the general form

$$G_x(s) = \frac{K_m(1 + T_1 s) \cdots}{s^m(1 + T_a s)(1 + T_b s) \cdots} \qquad (10\text{-}44)$$

where $m \neq 0$. The overall ratio of output to input is given by

$$\frac{O(s)}{I(s)} = \frac{G_x(s)}{1 + G_x(s)} = G(s) \qquad (10\text{-}45)$$

or

$$G(s) = \frac{K_m(1 + T_1 s) \cdots}{s^m(1 + T_a s)(1 + T_b s) \cdots + K_m(1 + T_1 s) \cdots} \qquad (10\text{-}46)$$

Equation (10-46) represents an equivalent forward transfer function between $O(s)$ and $I(s)$ of Fig. 10-31 and has the form of a Type 0 control element. Thus the transformation of type has been effected. Therefore, for a step input, $O_{ss}(t) = \lim_{s \to 0} I(s)G(s) = 1$. In other words, for this equivalent single element an input of constant

Table 10-3   BASIC FORMS OF $H(s)$ TO
BE USED TO PREVENT
CHANGE OF SYSTEM
TYPE

Basic form of $H(s)$	Type with which it can be used
K	0
Ks	0 and 1
Ks^2	0, 1, and 2
Ks^3	0, 1, 2, and 3

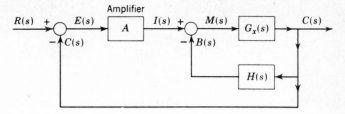

FIGURE 10-32
Block diagram for feedback compensation.

value produces an output of constant value which satisfies the definition of a Type 0 element. As a consequence, a given input signal tends to produce an output signal of equal magnitude but at a higher energy level and/or with a different form of energy. The polar plots of $\mathbf{G}_x(j\omega)$ and $\mathbf{O}(j\omega)/\mathbf{I}(j\omega)$ would show that for any given frequency the phase shift is less with feedback, thus indicating a more stable system (a smaller value of M_m).

10-13 FEEDBACK COMPENSATION: DESIGN PROCEDURES

Techniques are now described for applying feedback compensation to the root-locus method of system design.[13] The system to be investigated is shown in the block diagram of Fig. 10-32. A minor feedback loop has been formed around the original transfer function $G_x(s)$. The minor-loop feedback transfer function $H(s)$ consists of elements and networks described in the following pages. Either the gain portion of $G_x(s)$ or the cascade amplifier A is adjusted to fix the system characteristics. For more complex systems one can develop a design approach based on the knowledge obtained for this simple system.

The method of attack is similar to that used for series compensation. The characteristic equation of the complete system may be obtained and the root locus plotted, by use of the *partition*[1] method, as a function of a gain that appears in the system. A more general approach is first to adjust the roots of the characteristic equation of the inner loop. These roots are then the poles of the forward transfer function and are used to draw the root locus for the overall system. Both methods, using various types of feedback compensators, are used in the succeeding sections. The techniques developed with series compensation should be kept in mind. Basically, the addition of poles and zeros changes the root locus. To improve the system time response the locus must be moved to the left, away from the imaginary axis.

10-14 SIMPLIFIED RATE FEEDBACK COMPENSATION

The feedback system shown in Fig. 10-32 provides the opportunity for varying three functions, $G_x(s)$, $H(s)$, and A. The larger the number of variables inserted into a system, the better the opportunity for achieving a specified performance. However, the design problem becomes much more complex. In order to develop a feel for the

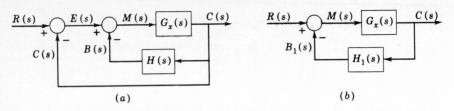

FIGURE 10-33
(a) Minor-loop compensation.   (b) Simplified block diagram, $H_1(s) = 1 + H(s)$.

changes that can be introduced by feedback compensation, a simplification is first made by letting $A = 1$. The system of Fig. 10-32 then reduces to that shown in Fig. 10-33a. It can be further simplified as shown in Fig. 10-33b.

The first and simplest case of feedback compensation to be considered is the use of a transducer which has the transfer function $H(s) = K_t s$. The original forward transfer function is

$$G_x(s) = \frac{K}{s(s + 1)(s + 5)} \qquad (10\text{-}47)$$

The control ratio is

$$\frac{C(s)}{R(s)} = \frac{G_x(s)}{1 + G_x(s)H_1(s)} = \frac{G_x(s)}{1 + G_x(s)[1 + H(s)]} \qquad (10\text{-}48)$$

The characteristic equation is

$$1 + G_x(s)[1 + H(s)] = 1 + \frac{K(1 + K_t s)}{s(s + 1)(s + 5)} = 0 \qquad (10\text{-}49)$$

Equation (10-49) shows that the transducer plus the unity feedback have introduced the ideal proportional plus derivative control that can also be achieved by a cascade compensator. The results should therefore be qualitatively the same.

The characteristic equation is partitioned to yield the general format $F(s) = -1$ which is required for obtaining a root locus. Partitioning Eq. (10-49) yields

$$\frac{KK_t(s + 1/K_t)}{s(s + 1)(s + 5)} = -1 \qquad (10\text{-}50)$$

It is seen from Eq. (10-50) that introduction of rate feedback results in the introduction of an open-loop zero. It therefore has the effect of moving the locus to the left and improving the time response. *This is not the same as introducing a cascade zero because no zero appears in the rationalized equation of the control ratio:*

$$\frac{C(s)}{R(s)} = \frac{K}{s(s + 1)(s + 5) + KK_t(s + 1/K_t)} \qquad (10\text{-}51)$$

The rate constant must be chosen before the root locus can be drawn for Eq. (10-50).

One must be careful in selecting the value of K_t. For example, if $K_t = 1$ is used, Eq. (10-50) reduces to

$$\frac{K}{s(s + 5)} = -1 \qquad (10\text{-}52)$$

and one can get the impression that $C(s)/R(s)$ has only two poles. *This is not the case,* as can be seen from Eq. (10-51). Letting $K_t = 1$ provides the common factor $s + 1$ in the characteristic equation so that it becomes

$$(s + 1)[s(s + 5) + K] = 0 \qquad (10\text{-}53)$$

One closed-loop pole has been made equal to $s = -1$, and the other two poles are obtained from the root locus of Eq. (10-52). If complex roots are selected from the root locus of Eq. (10-52), it is easily shown that the pole $s = -1$ is dominant and the response of the closed-loop system is overdamped (see Fig. 10-5*d*).

The proper design procedure is to select a number of trial locations for the zero $s = -1/K_t$, to tabulate or plot the system characteristics that result, and then to select the best combination. As an example, with $\zeta = 0.45$ as the criterion for the dominant closed-loop poles, a comparison of the original system and the rate-compensated system with $K_t = 0.4$ is given in Table 10-4. The gain K_1 must be obtained from the forward transfer function:

$$K_1 = \lim_{s \to 0} sG(s) = \lim_{s \to 0} \frac{sG_x(s)}{1 + G_x(s)H(s)} = \lim_{s \to 0} \frac{sK}{s[(s + 1)(s + 5) + KK_t]}$$

$$= \frac{K}{5 + KK_t} \qquad (10\text{-}54)$$

An analysis of the results of Table 10-4 shows that the effects of rate feedback are similar to those of a cascade lead compensator but the gain K_1 is not so large (see Table 10-2, Sec. 10-11). The control ratio of the system with rate feedback is

$$\frac{C(s)}{R(s)} = \frac{20}{(s + 1 - j2)(s + 1 + j2)(s + 4)} \qquad (10\text{-}55)$$

Table 10-4 COMPARISON OF PERFORMANCES

System	Dominant complex poles	Other roots	Ramp error coefficient K_1	T_p, s	M_o	T_s, s
Original	$-0.40 \pm j0.8$	-5.19	0.84	4.11	0.203	9.9
Rate feedback	$-1.0 \pm j2.0$	-4.0	1.54	2.83	0.17	4.0

10-15 RATE FEEDBACK

The more general case of feedback compensation shown in Fig. 10-32, with the amplifier gain A not equal to unity, is considered in this section. Two methods of attacking this problem are shown. With the addition of the inner feedback loop, the forward transfer function of the complete system is

$$G(s) = \frac{AG_x(s)}{1 + G_x(s)H(s)} = \frac{AK}{s(s + 1)(s + 5) + KK_t s} \qquad (10\text{-}56)$$

Method 1 The first method involves the characteristic equation of the complete system, which is

$$1 + G(s) = 1 + \frac{AK}{s(s + 1)(s + 5) + KK_t s} = 0 \qquad (10\text{-}57)$$

By clearing fractions and rewriting, the characteristic equation is

$$s(s + 1)(s + 5) + KK_t \left(s + \frac{A}{K_t} \right) = 0 \qquad (10\text{-}58)$$

This equation can be partitioned and put in the form that permits drawing a root locus. This consists in dividing the equation by $s(s + 1)(s + 5)$ to obtain the form

$$\frac{KK_t(s + A/K_t)}{s(s + 1)(s + 5)} = -1 \qquad (10\text{-}59)$$

Since the numerator and denominator are factored, the roots can be obtained by plotting a root locus. The angle condition of $(1 + 2h)180°$ must be satisfied for the root locus of Eq. (10-59). It should be kept in mind, however, that Eq. (10-59) is *not the forward transfer function of this system*; it is *just an equation whose roots are the poles of the control ratio*. The three roots of this characteristic equation vary as K, K_t, and A are varied. Obviously, it is necessary to fix two of these quantities and to determine the roots as a function of the third. A sketch of the root locus as a function of K is shown in Fig. 10-34 for arbitrary values of A and K_t. It should be noted that the rate feedback has resulted in an equivalent open-loop zero, so that the system is stable for all values of gain for the value of A/K_t chosen. This is typical of the stabilizing effect which can be achieved by derivative feedback and corresponds to the effect achieved with an ideal derivative plus proportional compensator in cascade.

The angles of the asymptotes of two of the branches of Eq. (10-59) are $\pm 90°$. For the selection A/K_t, shown in Fig. 10-34, these asymptotes are in the left half of the s plane, and the system is stable for any value of K. However, if $A/K_t > 5$, the asymptotes are in the right half of the s plane and the system can become unstable if K is large enough. For the best transient response the root locus should be moved as far to the left as possible. This is determined by the location of the zero $s = -A/K_t$. To avoid having a dominant real root the value of A/K_t is made greater than unity. The selection of these two parameters also determines the ramp error coefficient, since

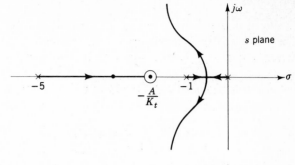

FIGURE 10-34
Root locus for Eq. (10-59).

Eq. (10-56) gives the value $K_1' = AK/(5 + KK_t)$. The designer must select K_t to meet the system specifications.

The values $A = 2.0$ and $K_t = 0.8$ result in a zero at $s = -2.5$, the same as the example in Sec. 10-14. For $\zeta = 0.45$, the control ratio is given by Eq. (10-55) and the performance is the same as that listed in Table 10-4. The use of the additional amplifier A does not change the performance; instead, it permits greater flexibility in the allocation of gains. For example, K can be decreased and both A and K_t increased by a corresponding amount with no change in performance.

Method 2 A more general method is to start with the inner loop, adjusting the gain to select the poles. Then successive loops are adjusted, setting the poles for each loop to desired values. For the system being studied, the transfer function of the inner loop is

$$\frac{C(s)}{I(s)} = \frac{K}{s(s + 1)(s + 5) + KK_t s} = \frac{K}{s[(s + 1)(s + 5) + KK_t]}$$

$$= \frac{K}{s(s - p_1)(s - p_2)} \tag{10-60}$$

The equation for which the root locus is to be plotted to find the roots p_1 and p_2 is

$$(s + 1)(s + 5) + KK_t = 0$$

which can be rewritten as

$$\frac{KK_t}{(s + 1)(s + 5)} = -1 \tag{10-61}$$

The locus is plotted in Fig. 10-35, where several values of the inner-loop damping ratio ζ_i for the roots are shown. Selection of a ζ_i permits evaluation of the product KK_t by use of Eq. (10-61). Upon assignment of a value to one of these quantities, the other can then be determined. The poles of the inner loop are the points on the locus that are determined by the ζ_i selected. They also are the poles of the open-loop

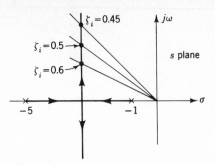

FIGURE 10-35
Poles of the inner loop for several values
of ζ_i.

transfer function for the overall system as represented in Eq. (10-60). It is therefore
necessary that the damping ratio ζ_i of the inner loop have a higher value than the
damping ratio of the overall system. The forward transfer function of the complete
system becomes

$$G(s) = \frac{C(s)}{E(s)} = \frac{AK}{s(s^2 + 2\zeta_i\omega_{n,i}s + \omega_{n,i}^2)} \qquad (10\text{-}62)$$

The root locus of the complete system is shown in Fig. 10-36 for several values of the
inner-loop damping ratio ζ_i. These loci are plotted as a function of AK.

The control ratio has the form

$$\frac{C(s)}{R(s)} = \frac{AK}{s(s^2 + 2\zeta_i\omega_{n,i}s + \omega_{n,i}^2) + AK}$$

$$= \frac{AK}{(s + \alpha)(s^2 + 2\zeta\omega_n s + \omega_n^2)}$$

$$= \frac{AK}{(s + \alpha)(s + \zeta\omega_n - j\omega_n\sqrt{1 - \zeta^2})(s + \zeta\omega_n + j\omega_n\sqrt{1 - \zeta^2})} \qquad (10\text{-}63)$$

In this problem the damping ratio ζ of the complex poles of the control ratio is used
as the basis for design. Both the undamped natural frequency ω_n and the ramp error

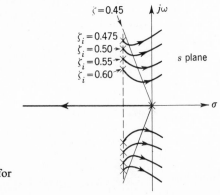

FIGURE 10-36
Root locus of complete system for
several values of ζ_i.

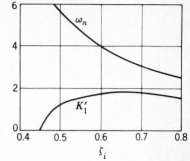

FIGURE 10-37
Variation of K_1' and ω_n with ζ_i.

coefficient $K_1' = AK/(5 + KK_t)$ depend on the selection of the inner-loop damping ratio ζ_i. Several trials are necessary to show how these quantities vary with the selection of ζ_i. This information is shown graphically in Fig. 10-37. The system performance specifications must be used to determine suitable values of K_1' and ω_n. Once the values of K_1' and ω_n are selected, the values of A, K, and K_t can be determined.

10-16 FEEDBACK OF SECOND DERIVATIVE OF OUTPUT

To further improve the system performance, a signal proportional to the second derivative of the output may be fed back. For a position system an accelerometer will generate such a signal. An approximation to this desired signal may be generated by modifying the output of a rate transducer with a high-pass filter. An example of this technique for a position-control system is shown in Fig. 10-38.

The transfer function for the inner feedback is

$$H(s) = \frac{B(s)}{C(s)} = \frac{K_t s^2}{s + 1/RC} \qquad (10\text{-}64)$$

Method 1 is used to show the application of this compensator. The characteristic equation of the complete system is

$$1 + G(s) = 1 + \frac{AK(s + 1/RC)}{s(s + 1)(s + 5)(s + 1/RC) + KK_t s^2} \qquad (10\text{-}65)$$

By clearing fractions, this equation becomes

$$s(s + 1)(s + 5)\left(s + \frac{1}{RC}\right) + KK_t\left(s^2 + \frac{A}{K_t}s + \frac{A}{K_t RC}\right) = 0 \qquad (10\text{-}66)$$

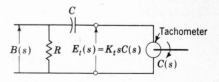

FIGURE 10-38
Tachometer and RC feedback network.

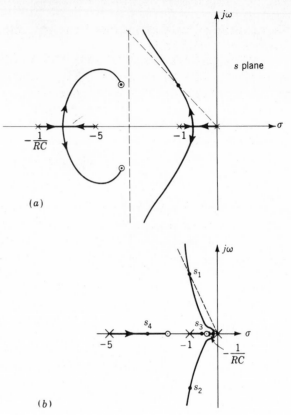

FIGURE 10-39
Root locus of Eq. (10-67).

The root locus can be drawn as the gain K varies from zero to infinity from the partitioned equation

$$\frac{KK_t(s^2 + As/K_t + A/K_tRC)}{s(s + 1)(s + 5)(s + 1/RC)} = -1 \qquad (10\text{-}67)$$

Since A/K_t and $1/RC$ must be selected before the locus can be drawn, it is well to look at the expression for K_1' to determine the limits to be placed on these values:

$$K_1' = \frac{AK}{5} \qquad (10\text{-}68)$$

It appears that K_1' is independent of RC; therefore the value of RC is chosen to produce the most desirable root locus. There are two zeros which are the factors of $s^2 + (A/K_t)s + A/K_tRC$, and they may be either real or complex. The zeros are located by $s^2 + As/K_t + A/K_tRC = 0$:

$$s_{a,b} = -\frac{A}{2K_t} \pm \sqrt{\left(\frac{A}{2K_t}\right)^2 - \frac{A}{K_tRC}} \qquad (10\text{-}69)$$

Two possible sketches of the root locus are shown in Fig. 10-39 for an arbitrary selection of A, K_t, and RC.

The values of $A = 1$, $K_t = 0.344$, $1/RC = 0.262$ give zeros at $s_a = -0.29$ and $s_b = -2.62$. Note that the zero s_a and the pole $s = -1/RC$ have the properties of a cascade lag compensator. Therefore an increase in gain can be expected. The zero s_b results in an improved time response. By specifying a damping ratio $\zeta = 0.35$ for the dominant roots, the result is $K = 53.3$, $K_1' = 10.6$, and $\omega_n = 3.79$, and the overall control ratio is

$$\frac{C(s)}{R(s)} = \frac{53.3(s + 0.262)}{(s + 1.32 + j3.55)(s + 1.32 - j3.55)(s + 3.34)(s + 0.291)} \qquad (10\text{-}70)$$

Note that a passive lead network in the feedback loop acts like a passive lag network in cascade and has resulted in a large increase in K_1'. Several trials may be necessary to determine the best selection of the closed-loop poles and ramp error coefficient. A $\zeta = 0.35$ is chosen instead of $\zeta = 0.45$ to obtain a larger value for K_1'. Since the third pole $s = -3.34$ reduces the peak overshoot, a smaller value of ζ is justifiable.

The use of an RC network in the feedback path may not always produce a desired increase in gain. Also, the proper location of the zeros and the corresponding values of A, K_t, and RC in Eq. (10-69) may require much work. Therefore, a possible alternative is to use only rate feedback to improve the time response and a cascade lag compensator to increase the gain. The corresponding design steps are essentially independent.

10-17 RESULTS OF FEEDBACK COMPENSATION

The preceding sections have shown the application of feedback compensation by the root-locus method. Basically, it has been shown that rate feedback is similar to ideal derivative control in cascade. The addition of a high-pass RC filter in the feedback path permits a large increase in the error coefficient. For good results the feedback compensator $H(s)$ should have a zero, $s = 0$, of higher order than the type of the original transfer function $G_x(s)$. The method of adjusting the poles of the inner loop and then adjusting successively larger loops is applicable to all multiloop systems. The results of feedback compensation applied in the preceding sections are summarized in Table 10-5. Results of cascade compensation are also listed for comparison. This tabulation shows that comparable results are obtained by cascade and feedback compensation.

10-18 SUMMARY

This chapter has shown the applications of cascade and feedback compensators. The procedures for applying the compensators are given so that they can be used with any system. They involve a certain amount of trial and error and the exercise of judgment based on past experience. The improvements produced by each compensator are shown by application to a specific system. In cases where the conventional

Table 10-5 COMPARISON OF CASCADE AND
FEEDBACK COMPENSATION USING
THE ROOT LOCUS

System	K_1	ω_n	ζ
Uncompensated	0.84	0.89	0.45
Rate feedback	1.8	3.92	0.45
Second derivative of output	10.6	3.79	0.35
Cascade lag compensator	8.04	0.87	0.45
Cascade lead compensator	2.94	3.52	0.45
Cascade lag-lead compensator	29.3	3.40	0.45

compensators do not produce the desired results, other methods of applying compensators are available in the literature.

It is important to emphasize that the conventional compensators cannot be applied indiscriminately. For example, the necessary improvements may not fit the four categories listed in Sec. 10-5, or the use of the conventional compensator may cause a deterioration of other performance characteristics. In such cases the designer must use his ingenuity in locating the compensator poles and zeros to achieve the desired results. The use of several compensating networks in cascade and/or in feedback may be required to achieve the desired root locations.

Realization of compensator pole-zero locations by means of passive networks requires an investigation of network synthesis techniques. Detail design procedures are available for specialized networks which produce two poles and two zeros.[10-12]

It may occur to the reader that the methods of compensation in this and the following chapters can become laborious because there is some trial and error involved. The use of computers[5,7] to solve design problems of this type minimizes the work involved. Using a computer means that an acceptable design of the compensator can be determined in a shorter time, but a computer is not always available. If one is available, it is still necessary to obtain by an analytical or graphical method a few values of the solution to ensure that the problem has been properly set up and that the computer program is operating correctly.

REFERENCES

1 D'Azzo, J. J., and C. H. Houpis: "Feedback Control System Analysis and Synthesis," 2d ed., McGraw-Hill, New York, 1966.
2 Chu, Y.: Synthesis of Feedback Control System by Phase-Angle Loci, *Trans. AIEE*, vol. 71, pt. II, pp. 330–339, November 1952.
3 Gibson, J. E.: "Nonlinear Automatic Control," McGraw-Hill, New York, 1963.
4 Thaler, G. J., and M. P. Pastel: "Analysis and Design of Nonlinear Feedback Control Systems," McGraw-Hill, New York, 1960.
5 Heaviside Partial Fraction Expansion and Time Response Digital Computer Program (PARTL), School of Engineering, Air Force Institute of Technology, Wright-Patterson Air Force Base, Ohio, 1974.

6   Elgerd, O. I., and W. C. Stephens: Effect of Closed-Loop Transfer Function Pole and Zero Locations on the Transient Response of Linear Control Systems, *Trans. AIEE*, vol. 78, pt. II, pp. 121–127, May 1959.

7   "User's Manual for a Digital Computer Routine to Calculate the Root Locus (ROOTL)," School of Engineering, Air Force Institute of Technology, Wright-Patterson Air Force Base, Ohio, 1974.

8   Ross, E. R., T. C. Warren, and G. J. Thaler: Design of Servo Compensation Based on the Root Locus Approach, *Trans. AIEE*, vol. 79, pt. II, pp. 272–277, September 1960.

9   Pena, L. Q.: Designing Servocompensators Graphically, *Control Eng.*, vol. 82, pp. 79–81, January 1964.

10   Lazear, T. J., and A. B. Rosenstein: On Pole-Zero Synthesis and the General Twin-T, *Trans. IEEE*, vol. 83, pt. II, pp. 389–393, November 1964.

11   Kuo, F. F.: "Network Analysis and Synthesis," 2d ed., Wiley, New York, 1966.

12   Mitra, S. K.: "Analysis and Synthesis of Linear Active Networks," Wiley, New York, 1969.

13   Truxal, J. G.: "Automatic Feedback Control System Synthesis," McGraw-Hill, New York, 1955.

11

CASCADE AND FEEDBACK COMPENSATION: FREQUENCY-RESPONSE PLOTS

11-1 INTRODUCTION

General methods have been developed for improving the frequency response of a feedback control system when the analysis is performed in the frequency domain. In the preceding chapter the determining factors for the selection of a lag, lead, or lag-lead network and their corresponding effects on the time response are presented. The choice of the poles and zeros of a compensator by the use of the root-locus method is often readily determinable for any system. The designer can place the poles and zeros of the proposed compensator on the s plane and determine the poles of the closed-loop system, which in turn permits evaluation of the closed-loop time response. Similarly, he can determine the new values of M_m, ω_m, and K_m by the frequency-response method, using either the polar plot or the log plot. However, the closed-loop poles are not explicitly determined. Furthermore, the correlation between the frequency-response parameters M_m and ω_m and the time response is only qualitative, as discussed in Secs. 9-3, 9-4, and 9-12. The presence of real roots near the complex dominant roots further changes the correlation between the frequency-response parameters and the time response, as discussed in Sec. 9-12.

This chapter is concerned with the changes that can be made in the frequency-response characteristics by use of the three types of cascade compensators (lag, lead, and lag-lead) and by feedback compensation. They are only representative of the compensators that can be used. A study is made of the effect of these compensators on the overall system. Design procedures to obtain improvement of the system performance are described. With practical experience behind him, a designer can extend or modify these design procedures, which are based on the presence of one pair of complex dominant roots, to those systems where there are dominant roots besides the main complex pair.[1] Both the log plots and the polar plots are utilized in this chapter in applying the frequency-response compensation criteria to show that either type of plot can be used; the choice depends on individual preference.

The performance of a closed-loop system can be described in terms of M_m, ω_m, and the system error coefficient. The value of M_m essentially describes the damping ratio and therefore the amount of overshoot in the transient response. If M_m is fixed, ω_m determines the undamped natural frequency ω_n, which in turn determines the response time of the system. The system error coefficient is important because it determines the steady-state error with the appropriate standard input. The design procedure is usually based on selecting a value for M_m and using the methods described in Chap. 9 to find the corresponding values of ω_m and the required gain K_m. Once this is accomplished, if the desired performance specifications are not met, compensating devices must be used. These devices alter the shape of the frequency-response plot to try to meet the performance specifications. Also, for those systems that are unstable for all values of gain it is mandatory that a stabilizing or compensating network be inserted in the system. The compensator may be placed in cascade or in a minor feedback loop.

The reasons for reshaping the frequency-response plot generally fall into the following categories:

1 A given system is stable, and its M_m and ω_m (and therefore the transient response) are satisfactory, but its steady-state error is too large. The gain must therefore be increased to reduce the steady-state error (see Chap. 6) without appreciably altering the values of M_m and ω_m. It is shown later that in this case the high-frequency portion of the frequency-response plot is satisfactory but the low-frequency end is not.

2 A given system is stable but its transient settling time is unsatisfactory; i.e., the M_m is set to a satisfactory value, but the ω_m is too low. The gain may be satisfactory, or a small increase may be desirable. The high-frequency portion of the frequency-response plot must be altered in order to increase the value of ω_m.

3 A given system is stable and has a desired M_m, but both its transient response and its steady-state response are unsatisfactory. Therefore the values of both ω_m and K_m must be increased. The portion of the frequency plot in the vicinity of the $-1 + j0$ or $(-180°, 0\text{-dB})$ point must be altered to yield the desired ω_m, and the low-frequency end must be changed to obtain the increase in gain desired.

4 A given system is unstable for all values of gain. The frequency-response plot must be altered in the vicinity of the $-1 + j0$ or $(-180°, 0\text{-dB})$ point to produce a stable system with a desired M_m and ω_m.

Thus the objective of compensation is to reshape, by means of a compensator, the frequency-response plot of the basic system to try to achieve the performance specifications. Examples which demonstrate this objective are presented in the follow-ing sections.

11-2   SELECTION OF A CASCADE COMPENSATOR

Consider the following:

$G_x(j\omega)$ = basic feedback control system forward transfer function
$G(j\omega)$ = forward transfer function which results in required
    stability and steady-state accuracy

Dividing $G(j\omega)$ by $G_x(j\omega)$ gives the necessary performance requirements of the compensator:

$$G_c(j\omega) = \frac{G(j\omega)}{G_x(j\omega)} \qquad (11.1)$$

The physical network for the compensator described by Eq. (11-1) can be synthesized by the techniques of Foster, Cauer, Brune, and Guillemin. In the pre-ceding and present chapters the compensators are limited to relatively simple RC networks. Based upon the design criteria presented, the parameters of these simple networks can be evaluated in a rather direct manner. The simple networks quite often provide the desired improvement in the control system's performance. When they do not, more complex networks must be synthesized.[2,3] Physically realizable networks are limited in their performance over the whole frequency spectrum; this makes it difficult to obtain the transfer-function characteristic given by Eq. (11-1). Actually this is not a serious limitation, since the desired performance specifications are interpreted in terms of a comparatively small bandwidth of the frequency spec-trum. A compensator can be constructed to have the desired magnitude and angle characteristics over this bandwidth.

The characteristics of the lag and lead compensators and their effects upon a given polar plot are shown in Fig. 11-1a and b. The cascade lag compensator results in a clockwise rotation of $G(j\omega)$ and can produce a large increase in gain K_m with a small decrease in the resonant frequency ω_m. The cascade lead compensator results in a counterclockwise rotation of $G(j\omega)$ and can produce a large increase in ω_m with a small increase in K_m. Where the characteristics of both the lag network and the lead network are necessary to achieve the desired specifications, a lag-lead compensator can be used. The transfer function and polar plot of this lag-lead compensator are shown in Fig. 11-1c. The corresponding log plots for each of these compensators are included with the design procedures later in the chapter.

The choice of a compensator depends on the characteristics of the given (or basic) feedback control system and the desired specifications. Since these networks are to be inserted in cascade, they must be of the (1) proportional plus integral and (2) proportional plus derivative types. The plot of the lag compensator shown in Fig. 11-1a approximates the ideal integral plus proportional characteristics shown in Fig. 11-2a at high frequencies. Also, the plot of the lead compensator shown in Fig. 11-1b approximates the ideal derivative plus proportional characteristics shown

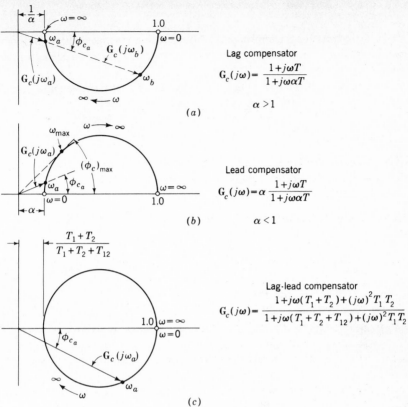

Lag compensator

$$G_c(j\omega) = \frac{1+j\omega T}{1+j\omega\alpha T}$$

$$\alpha > 1$$

(a)

Lead compensator

$$G_c(j\omega) = \alpha\frac{1+j\omega T}{1+j\omega\alpha T}$$

$$\alpha < 1$$

(b)

Lag-lead compensator

$$G_c(j\omega) = \frac{1+j\omega(T_1+T_2)+(j\omega)^2 T_1 T_2}{1+j\omega(T_1+T_2+T_{12})+(j\omega)^2 T_1 T_2}$$

(c)

FIGURE 11-1
Polar plots: (a) lag compensator; (b) lead compensator; (c) lag-lead compensator.

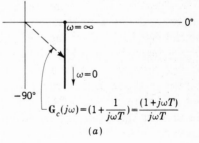

$$G_c(j\omega) = \left(1+\frac{1}{j\omega T}\right) = \frac{(1+j\omega T)}{j\omega T}$$

(a)

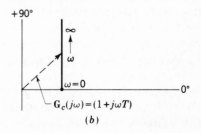

$$G_c(j\omega) = (1+j\omega T)$$

(b)

FIGURE 11-2
(a) Ideal integral plus proportional compensator; (b) ideal derivative plus proportional compensator.

in Fig. 11-2b at low frequencies. These are the frequency ranges where the proportional plus integral and proportional plus derivative controls are needed to provide the necessary compensation.

The equations of $G_c(j\omega)$ in Fig. 11-1, which represent the lag and the lead compensators used in this chapter, can be expressed as

$$G_c(j\omega) = |G_c(j\omega)|\underline{/\phi_c} \qquad (11\text{-}2)$$

Solving for the angle ϕ_c yields

$$\phi_c = \tan^{-1}\omega T - \tan^{-1}\omega\alpha T = \tan^{-1}\frac{\omega T - \omega\alpha T}{1 + \omega^2\alpha T^2} \qquad (11\text{-}3)$$

or

$$T^2 + \frac{\alpha - 1}{\omega\alpha \tan \phi_c}T + \frac{1}{\omega^2\alpha} = 0 \qquad (11\text{-}4)$$

where $\alpha > 1$ for the lag compensator and $\alpha < 1$ for the lead compensator. It is shown in later sections that in the design procedure a frequency ω is specified at which a particular value of ϕ is desired. For selected values of α, ϕ_c, and ω it is possible to determine, by use of Eq. (11-4), the value T required for the compensator. For a lag compensator a small value of ϕ_c is required, whereas for a lead compensator a large value of ϕ_c is required. For a lead compensator the value ω_{max} at which the maximum phase shift occurs, for a given T and α, can be determined by setting the derivative of ϕ_c with respect to the frequency ω equal to zero. Therefore, from Eq. (11-3) the maximum angle occurs at the frequency.

$$\omega_{max} = \frac{1}{T\sqrt{\alpha}} \qquad (11\text{-}5)$$

Inserting this value of frequency into Eq. (11-3) gives the maximum phase shift

$$(\phi_c)_{max} = \sin^{-1}\frac{1 - \alpha}{1 + \alpha} \qquad (11\text{-}6)$$

Typical values for a lead compensator are

α	0	0.1	0.2	0.3	0.4	0.5
$(\phi_c)_{max}$	90°	54.9°	41.8°	32.6°	25.4°	19.5°

A knowledge of the maximum phase shift is useful in the application of lead compensators.

The design procedures developed in the following sections utilize the characteristics of compensators discussed in this section.

11-3 CASCADE LAG COMPENSATOR

When the M_m and ω_m (that is, the transient response) of a feedback control system are considered satisfactory but the steady-state error is too large, it is necessary to increase the gain without appreciably altering that portion of the given log magnitude–

angle plot in the vicinity of the $(-180°, 0\text{-dB})$ point (category 1). This can be done by introducing an integral or lag compensator in cascade with the original forward transfer function. Shown in Fig. 10-9 is a representative lag network having the transfer function

$$G_c(s) = A \frac{1 + Ts}{1 + \alpha Ts} \qquad (11\text{-}7)$$

where $\alpha > 1$. As an example of its application, consider a basic Type 0 system. For an improvement in the steady-state accuracy the system step error coefficient K_p must be increased. Its values before and after the addition of the compensator are

$$K_p = \lim_{s \to 0} G_x(s) = K_0 \qquad (11\text{-}8)$$

$$K_p' = \lim_{s \to 0} G_c(s)G_x(s) = AK_0 = K_0' \qquad (11\text{-}9)$$

The compensator amplifier gain A must have a value which gives the desired increase in the value of the step error coefficient. Thus

$$A = \frac{K_0'}{K_0} \qquad (11\text{-}10)$$

This increase in gain must be achieved while maintaining the same M_m and without appreciably changing the value of ω_m. However, the lag compensator has a negative angle which moves the original log magnitude–angle plot to the left, or closer to the $(-180°, 0\text{-dB})$ point. This has a destabilizing effect and reduces ω_m. To limit this decrease in ω_m, the lag compensator is designed so that it introduces a small angle, generally no more than $-5°$, at the original resonant frequency ω_{m1}. The value of $-5°$ at $\omega = \omega_{m1}$ is an empirical value, determined from practical experience in applying this type of network. A slight variation in method, which gives equally good results, is to select the lag compensator so that the magnitude of its angle is $5°$ or less at the original phase-margin frequency $\omega_{\phi 1}$.

The overall characteristics of lag compensators can best be visualized from the log plot. The log magnitude and phase-angle equations for the compensator of Eq. (11-7) are

$$\text{Lm } G_c'(j\omega) = \text{Lm } (1 + j\omega T) - \text{Lm } (1 + j\omega \alpha T) \qquad (11\text{-}11)$$

$$\underline{/G_c(j\omega)} = \underline{/1 + j\omega T} - \underline{/1 + j\omega \alpha T} \qquad (11\text{-}12)$$

For various values of α there is a family of curves for the log magnitude and phase-angle diagram, as shown in Fig. 11-3. Engineers who use this compensator repeatedly may find it convenient to make templates of these curves. From the shape of these curves it is seen that an attenuation equal to Lm α has been introduced above $\omega T = 1$. Assume that a system has the original forward transfer function $G_x(j\omega)$ and that the gain has been adjusted for the desired phase margin. The log magnitude and phase-angle diagram of $G_x(j\omega)$ is sketched in Fig. 11-4. The addition of the lag compensator reduces the log magnitude by Lm α for the frequencies above $\omega = 1/T$. The value of T of the compensator is selected so that the attenuation Lm α occurs at the original phase-margin frequency $\omega_{\phi 1}$. Notice that the phase margin at $\omega_{\phi 1}$ has been reduced.

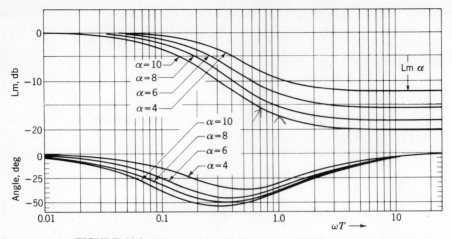

FIGURE 11-3
Log magnitude and phase-angle diagram for lag compensator

$$\mathbf{G}'_c(j\omega) = \frac{1 + j\omega T}{1 + j\omega\alpha T}$$

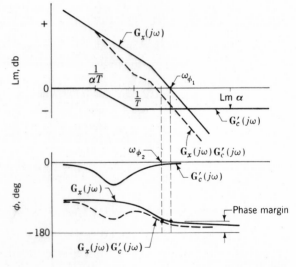

FIGURE 11-4
Original and lag-compensated curves.

To maintain the specified phase margin, the phase-margin frequency has been reduced to $\omega_{\phi 2}$. It is now possible to increase the gain of $\mathbf{G}_x(j\omega)\mathbf{G}'_c(j\omega)$ by selecting the value A so that the Lm curve will have the value 0 dB at the frequency $\omega_{\phi 2}$.

The effects of the lag compensator can now be analyzed. The reduction in log magnitude at the higher frequencies due to the compensator is desirable. This permits

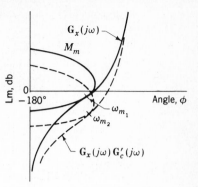

FIGURE 11-5
Original and lag-compensated curves.

an increase in the gain to maintain the desired phase margin. Unfortunately, the negative angle of the compensator lowers the angle curve. To maintain the desired phase margin, the phase-margin frequency is reduced. To keep this reduction small, the value of $\omega_{cf} = 1/T$ should be made as small as possible.

The log magnitude–angle diagram permits gain adjustment for the desired M_m. The curves for $\mathbf{G}_x(j\omega)$ and $\mathbf{G}_x(j\omega)\mathbf{G}_c'(j\omega)$ are shown in Fig. 11-5. The increase in gain to obtain the same M_m is the amount that the curve of $\mathbf{G}_x(j\omega)\mathbf{G}_c'(j\omega)$ must be raised in order to be tangent to the M_m curve. Because of the negative angle introduced by the lag compensator, the new resonant frequency ω_{m2} is smaller than the original resonant frequency ω_{m1}. Provided that the decrease in ω_m is small and the increase in gain is sufficient to meet the specifications, the system can be considered satisfactorily compensated.

In summary, the lag compensator is basically a low-pass filter; the low frequencies are passed and the higher frequencies are attenuated. The attenuation characteristic of the lag compensator is useful and permits an increase in the gain. The negative-phase-shift characteristic is detrimental to system performance but must be tolerated. Because the predominant and useful feature of the compensator is attenuation, a more appropriate name for it is *high-frequency attenuation compensator*.

11-4   EXAMPLE: CASCADE LAG COMPENSATION

To give a basis for comparison of all methods of compensation, the examples of Chap. 10 are used in this chapter. For the unity-feedback Type 1 system of Sec. 10-6 the forward transfer function is

$$G_x(s) = \frac{K_1}{s(1 + s)(1 + 0.2s)} \quad (11\text{-}13)$$

By the methods of Secs. 9-7 and 9-10 the gain required for this system to have an M_m of 1.26 (2 dB) is $K_1 = 0.87$ s^{-1}, and the resulting resonant frequency is $\omega_{m1} = 0.72$ rad/s. With these values the transient response is considered to be satisfactory, but the steady-state error is too large. Putting a lag compensator in cascade with a basic system (falling into category 1) permits an increase in gain and therefore reduces the steady-state error.

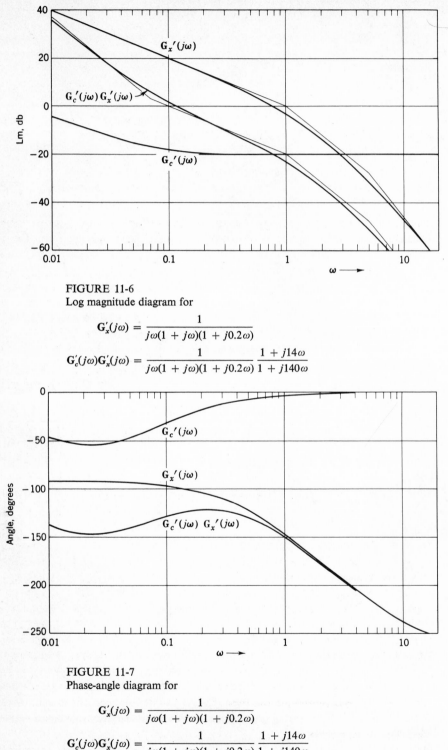

FIGURE 11-6
Log magnitude diagram for

$$G_x'(j\omega) = \frac{1}{j\omega(1 + j\omega)(1 + j0.2\omega)}$$

$$G_c'(j\omega)G_x'(j\omega) = \frac{1}{j\omega(1 + j\omega)(1 + j0.2\omega)}\frac{1 + j14\omega}{1 + j140\omega}$$

FIGURE 11-7
Phase-angle diagram for

$$G_x'(j\omega) = \frac{1}{j\omega(1 + j\omega)(1 + j0.2\omega)}$$

$$G_c'(j\omega)G_x'(j\omega) = \frac{1}{j\omega(1 + j\omega)(1 + j0.2\omega)}\frac{1 + j14\omega}{1 + j140\omega}$$

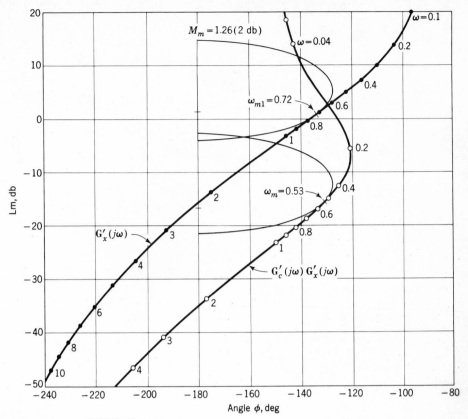

FIGURE 11-8

Log magnitude–angle diagram: original and lag-compensated system.

The cascade compensator to be used has the transfer function given by Eq. (11-7). Selection of an $\alpha = 10$ means that an increase in gain of almost 10 is desired. Based on the criterion discussed in the last section (a phase shift of $-5°$ at the original ω_m), the compensator time constant can be determined. Note that in Fig. 11-3, $\phi_c = -5°$ occurs for values of ωT approximately equal to 0.01 and 10. These two points correspond, respectively, to points ω_b and ω_a in Fig. 11-1a. The choice $\omega T = 10 = \omega_{m1}T$ (or $\omega_a = \omega_{m1}$) is made since this provides the maximum attenuation at ω_{m1}. This ensures a maximum possible gain increase in $G(j\omega)$ to achieve the desired M_m. Thus for $\omega_{m1} = 0.72$ the value $T \approx 13.9$ is obtained. Note that the value of T can also be obtained from Eq. (11-4). For ease of calculation, the value of the time constant is rounded off to 14.0, which does not appreciably affect the results.

In Figs. 11-6 and 11-7 the log magnitude and phase-angle diagrams for the compensator are shown, together with the curves for $G'_x(j\omega)$. The sum of the curves is also shown and represents the function $G'_x(j\omega)G'_c(j\omega)$. The new curves of $G'_x(j\omega) \cdot G'_c(j\omega)$ from Figs. 11-6 and 11-7 are used to draw the log magnitude-angle diagram of Fig. 11-8. By using this curve, the gain is adjusted to obtain an $M_m = 1.26$. The results are $K'_1 = 6.68$ and $\omega_m = 0.53$. Thus an improvement in K_1 has been obtained, but a

Rm 502
Pace

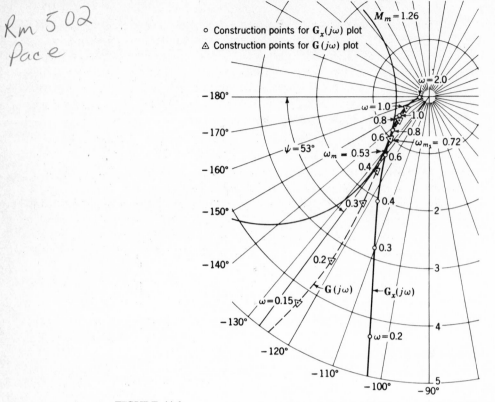

FIGURE 11-9
Polar plots of a Type 1 control system, with and without lag compensation:

$$G_x(j\omega) = \frac{0.87}{j\omega(1 + j\omega)(1 + j0.2\omega)}$$

$$G(j\omega) = G_x(j\omega)G_c(j\omega) = \frac{60.68(1 + j14\omega)}{j\omega(1 + j\omega)(1 + j0.2\omega)(1 + j140\omega)}$$

penalty of a smaller ω_m results. An increase in gain of approximately $A = 7.7$ has resulted from the use of a lag compensator with $\alpha = 10$, which is typical of the gain increase permitted by a given α. In actual practice this additional gain may be obtained from an amplifier present in the basic system if it is not already operating at maximum gain. Note that the value of gain is somewhat less than the chosen value of α. Provided that the decrease in ω_m and the increase in gain are acceptable, the system has been suitably compensated.

To further show the effects of compensation, the polar plots representing the basic and compensated systems are shown in Fig. 11-9. Note that there is essentially no rotation of the high-frequency portion of $G(j\omega)$ compared with $G_x(j\omega)$. However, in the low-frequency region there is a clockwise rotation of $G(j\omega)$ due to the lag compensator.

Table 11-1 presents a comprehensive comparison between the root-locus method and the frequency-plot method of applying a lag compensator to a basic control system. Note the similarity of results obtained by these methods. The selection of the method depends upon individual preference. The compensators developed in the root-locus and frequency-plot methods are different, but the difference in the results is small. To show this explicitly, Table 11-1 includes the case of the compensator developed by the root-locus method and applied to the frequency-response method. In the frequency-response method, either the log or the polar plot may be used.

11-5 LEAD COMPENSATOR

Figure 10-17 shows a lead compensator made up of passive elements which has the transfer function

$$\mathbf{G}_c'(j\omega) = \alpha\, \frac{1 + j\omega T}{1 + j\omega \alpha T} \qquad \alpha < 1 \qquad (11\text{-}14a)$$

This equation is marked with a prime since it does not contain the gain A. The log magnitude and phase-angle equations for this compensator are

$$\text{Lm } \mathbf{G}_c'(j\omega) = \text{Lm } \alpha + \text{Lm } (1 + j\omega T) - \text{Lm } (1 + j\omega \alpha T) \qquad (11\text{-}14b)$$

$$\underline{/\mathbf{G}_c'(j\omega)} = \underline{/1 + j\omega T} - \underline{/1 + j\omega \alpha T} \qquad (11\text{-}14c)$$

A family of curves for various values of α is shown in Fig. 11-10. It may be convenient to make templates of these log magnitude and phase-angle curves for the lead compensator. It is seen from the shape of the log magnitude curves that an attenuation equal to Lm α has been introduced at frequencies below $\omega T = 1$. Thus the lead network is basically a high-pass filter: the high frequencies are passed, and the low frequencies are attenuated. Also, an appreciable lead angle is introduced in the frequency range from $\omega = 1/T$ to $\omega = 1/\alpha T$ ($\omega T = 1$ to $\omega T = 1/\alpha$). Because of

Table 11-1 COMPARISON BETWEEN ROOT-LOCUS AND FREQUENCY METHODS OF APPLYING A CASCADE LAG COMPENSATOR TO A BASIC CONTROL SYSTEM

Method	$G_c'(s)$	ω_{m1}	ω_m	ω_{n1}	ω_n	K_1	K_1'	Damping ratio (held constant)	Maximum peak overshoot
Root locus	$\dfrac{1 + 20s}{1 + 200s}$	—	0.68	0.89	0.87	0.84	8.04	ζ of dominant roots = 0.45	$M_p = 1.22$ $M_m = 1.30$
Frequency plot	$\dfrac{1 + 14s}{1 + 140s}$	0.72	0.53	—	—	0.87	6.68	Effective $\zeta = 0.45$	$M_m = 1.26$
	$\dfrac{1 + 20s}{1 + 200s}$		0.59				7.35		

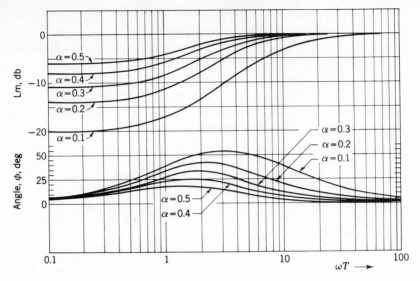

FIGURE 11-10
Log magnitude and phase-angle diagram for lead compensator

$$G'_c(j\omega) = \alpha \frac{1 + j\omega T}{1 + j\omega \alpha T}$$

its angle characteristic a lead network can be used to increase the bandwidth for a system that falls in category 2.

Application of the lead compensator can be based on adjusting the phase margin and the phase-margin frequency. Assume that a system has the original forward transfer function $G_x(j\omega)$ and that the gain has been adjusted for the desired phase margin or for the desired M_m. The log magnitude and phase-angle diagram of $G_x(j\omega)$ is sketched in Fig. 11-11. The purpose of the lead compensator is to increase the phase-margin frequency and therefore to increase ω_m. The lead compensator introduces a positive angle over a relatively narrow bandwidth. By properly selecting the value of T, the phase-margin frequency can be increased from $\omega_{\phi 1}$ to $\omega_{\phi 2}$. Selection of T can be accomplished by physically placing the angle curve for the compensator on the same graph with the angle curve of the original system. The location of the compensator angle curve must be such as to produce the specified phase margin at the highest possible frequency; this location determines the value of the time constant T. The gain of $G_x(j\omega)G'_c(j\omega)$ must be increased for the log magnitude curve to have the value 0 dB at the frequency $\omega_{\phi 2}$. For a given α, an analysis of the log magnitude curve of Fig. 11-10 shows that the farther to the right the compensator curves are placed; i.e., the smaller T is made, the larger the new gain of the system.

It is also possible to use the criterion derived in the previous chapter for the selection of T. This criterion is to select T, for Type 1 or higher systems, equal to or slightly smaller than the largest time constant of the original forward transfer function.

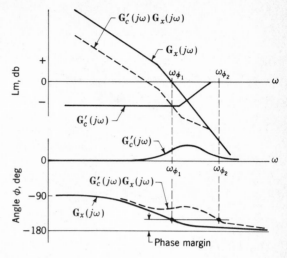

FIGURE 11-11
Original and lead-compensated curves.

This is the procedure used in the example of the next section. For a Type 0 system the compensator time constant T is made equal to or slightly smaller than the second largest time constant of the original system. Several locations of the compensator curves should be tested and the best results selected.

More accurate application of the lead compensator is based on adjusting the gain to obtain a desired M_m by use of the log magnitude–angle diagram. Figure 11-12 shows the original curve $\mathbf{G}_x(j\omega)$ and the new curve $\mathbf{G}_x(j\omega)\mathbf{G}_c'(j\omega)$. The increase in gain to obtain the same M_m is the amount that the curve $\mathbf{G}_x(j\omega)\mathbf{G}_c'(j\omega)$ must be raised to be tangent to the M_m curve. Because of the positive angle introduced by the lead compensator, the new resonant frequency ω_{m2} is larger than the original resonant frequency ω_{m1}. The increase in gain is not as large as that obtained by use of the lag compensator.

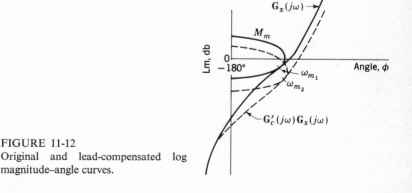

FIGURE 11-12
Original and lead-compensated log magnitude–angle curves.

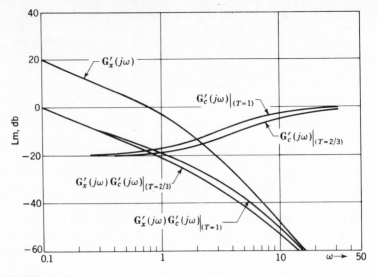

FIGURE 11-13
Log magnitude diagram of

$$G'_x(j\omega) = \frac{1}{j\omega(1 + j\omega)(1 + j0.2\omega)}$$

$$G'_x(j\omega)G'_c(j\omega) = \frac{1}{j\omega(1 + j\omega)(1 + j0.2\omega)} \frac{0.1(1 + j\omega T)}{1 + j0.1\omega T}$$

11-6   EXAMPLE: CASCADE LEAD COMPENSATION

For the unity-feedback Type 1 system of Sec. 10-6 the forward transfer function is

$$G_x(s) = \frac{K_1}{s(1 + s)(1 + 0.2s)} \qquad (11\text{-}15)$$

From Sec. 11-4 for $M_m = 1.26$ it is found that $K_1 = 0.87$ s^{-1} and $\omega_{m1} = 0.72$ rad/s. In this example the value of ω_m is considered to be too low. From previous considerations, putting a lead compensator in cascade with the basic system increases the value of ω_m and therefore improves the time response of the system. This situation falls in category 2.

The choice of values of α and T of the compensator must be such that its angle ϕ_c, at frequency ω_{m1}, adds a sizable positive phase shift to the overall forward transfer function. The nominal value of $\alpha = 0.1$ is often utilized; thus only the determination of the value of T remains. The values of T selected in Sec. 10-8 give satisfactory results. Therefore these values of T are used in this example so that a comparison can be made between the root-locus and frequency-response techniques. This example is worked with only two of the three values of T used in Sec. 10-8. The results are used to establish the design criteria for lead compensators.

In Figs. 11-13 and 11-14 are drawn the log magnitude and phase-angle diagrams for the basic system $G'_x(j\omega)$, the two compensators $G'_c(j\omega)$, and the two composite

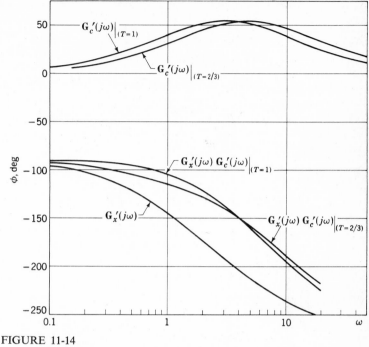

FIGURE 11-14
Phase-angle diagram for Fig. 11-13.

transfer functions $\mathbf{G}_x'(j\omega)\mathbf{G}_c'(j\omega)$. With the curves of $\mathbf{G}_x'(j\omega)\mathbf{G}_c'(j\omega)$ from Figs. 11-13 and 11-14, the log magnitude–angle diagrams are drawn in Fig. 11-15. The gain is again adjusted to obtain an $M_m = 1.26$ (2 dB).

The corresponding polar plots of the original system $\mathbf{G}_x(j\omega)$ and the compensated systems $\mathbf{G}(j\omega) = \mathbf{G}_x(j\omega)\mathbf{G}_c(j\omega)$ are shown in Fig. 11-16.

Table 11-2 presents a comprehensive comparison between the root-locus method and the frequency-response method of applying a lead compensator to a basic control system. The results show that a lead compensator increases the resonant frequency ω_m. However, a range of values of ω_m is possible, depending on the compensator time constant used. The larger the value of ω_m, the smaller the value of K_1. The characteristics selected must be based on the system specifications. The increase in gain is not as large as that obtained with a lag compensator. Note in the table the similarity of results obtained by the root-locus and frequency-response methods. Because of this similarity, the design rules of Sec. 10-8 can be interpreted in terms of the frequency-response method as follows.

Rule 1   The time constant T in the numerator of $\mathbf{G}_c(j\omega)$ given by Eq. (11-14) should be set equal to the value of the largest time constant in the denominator of the original transfer function for a Type 1 or higher system. This usually results in the largest increase in ω_m (best time response), and there is an increase in system gain K_m.

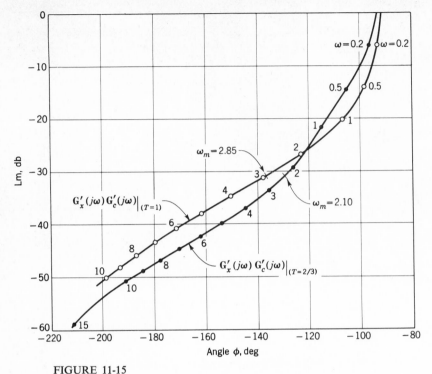

FIGURE 11-15
Log magnitude–angle diagrams with lead compensators from Figs. 11-13 and 11-14.

Rule 2 The time constant T in the numerator of $G_c(j\omega)$ should be set to a value slightly smaller than the largest time constant in the denominator of the original transfer function for a Type 1 or higher system. This results in achieving the largest gain increase for the system with an appreciable increase in ω_m.

If the maximum improvement in the time response is desired with whatever gain increase is obtainable, Rule 1 is applied to the design of the lead compensator. Where maximum gain increase and a good improvement in the time of response are desired, Rule 2 is applicable.

For a Type 0 system, Rules 1 and 2 are modified so that the lead-compensator time constant is selected either equal to or slightly smaller than the second largest time constant of the original system.

Remember that the lag and lead networks are designed for the same basic system; thus from Tables 11-1 and 11-2 it is seen that both result in a gain increase. The distinction between the two types of compensation is that the lag network gives the largest increase in gain (with the best improvement in steady-state accuracy) at the expense of increasing the response time, whereas the lead network gives an appreciable improvement in the time of response and a small improvement in steady-state accuracy. The particular problem at hand dictates the type of compensation to be used.

Table 11-2   COMPARISON BETWEEN ROOT-LOCUS AND FREQUENCY-RESPONSE METHODS OF APPLYING A LEAD COMPENSATOR TO A BASIC CONTROL SYSTEM

Method	$G'_c(s)$	ω_{m1}	ω_m	ω_{n1}	ω_n	K_1	K'_1	Damping ratio (held constant)	Maximum peak overshoot
Root locus	$0.1\dfrac{1+s}{1+0.1s}$	—	2.68	0.89	3.52	0.84	2.94	ζ of dominant roots = 0.45	$M_p = 1.19$ $M_m = 1.21$
	$0.1\dfrac{1+0.667s}{1+0.0667s}$		2.70		3.40		4.40		$M_p = 1.24$ $M_m = 1.38$
Frequency plot	$0.1\dfrac{1+s}{1+0.1s}$	0.72	2.85	—	—	0.87	3.13	Effective $\zeta = 0.45$	$M_m = 1.26$
	$0.1\dfrac{1+0.667s}{1+0.0667s}$		2.10				3.50		

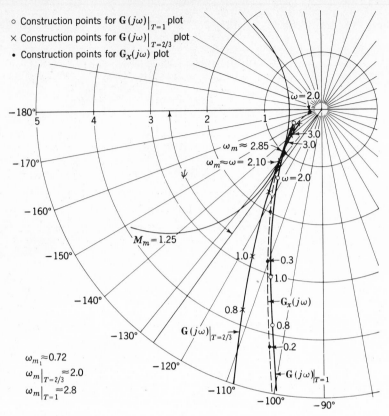

FIGURE 11-16
Polar plots of Type 1 control system with and without lead compensation:

$$G_x(j\omega) = \frac{0.87}{j\omega(1 + j\omega)(1 + j0.2\omega)}$$

$$G(j\omega)]_{T=2/3} = \frac{3.50(1 + j2\omega/3)}{j\omega(1 + j\omega)(1 + j0.2\omega)(1 + j0.2\omega/3)}$$

$$G(j\omega)]_{T=1} = \frac{3.13}{j\omega(1 + j0.2\omega)(1 + j0.2\omega)}$$

11-7 LAG-LEAD COMPENSATOR

The previous sections have demonstrated that the introduction of a lag compensator results in an increase in the gain and a lead compensator results in an increase in the resonant frequency ω_m. These changes in the time response reduce the steady-state error and improve the response time, respectively. If both changes are required, both a lag and a lead compensator must be utilized. This improvement can be accomplished by inserting the individual lag and lead networks in cascade. However, it is more economical in equipment to use a new network that has both the lag and the

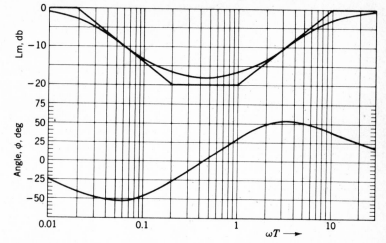

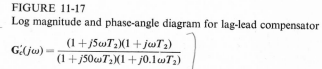

FIGURE 11-17
Log magnitude and phase-angle diagram for lag-lead compensator

$$G_c'(j\omega) = \frac{(1+j5\omega T_2)(1+j\omega T_2)}{(1+j50\omega T_2)(1+j0.1\omega T_2)}$$

lead characteristics. Such a network, shown in Fig. 10-24, is called a lag-lead compensator. Its transfer function is

$$G_c(j\omega) = \frac{A(1+j\omega T_1)(1+j\omega T_2)}{(1+j\omega\alpha T_1)(1+j\omega T_2/\alpha)} \quad (11\text{-}16)$$

where $\alpha > 1$ and $T_1 > T_2$. The first half of this transfer function produces the lag effect, and the second half produces the lead effect. The log magnitude and phase-angle diagram for a representative lag-lead compensator is shown in Fig. 11-17, in which the selection is made $T_1 = 5T_2$. A simple design method is to make T_2 equal to a time constant in the denominator of the original system. An alternative procedure is to locate the compensator curves on the log plots of the original system to produce the highest possible phase-margin frequency. Figure 11-18 shows a sketch of the log magnitude and phase-angle diagram for the original forward transfer function $G_x(j\omega)$, the compensator $G_c'(j\omega)$, and the combination $G_x(j\omega)G_c'(j\omega)$. The phase-margin frequency has been increased from $\omega_{\phi1}$ to $\omega_{\phi2}$, and the additional gain is labeled Lm A. The new value of M_m can be obtained from the log magnitude–angle diagram in the usual manner.

Another method is simply the combination of the design procedures given for the use of the lag and lead networks, respectively. Basically, the log magnitude and phase-angle curves of the lag compensator can be located as described in Sec. 11-3. The lag-compensator curves are given by Fig. 11-3 and are located so that the negative angle introduced at either the original phase-margin frequency $\omega_{\phi1}$ or the original resonant frequency ω_{m1} is small—of the order of $-5°$. This permits determination of

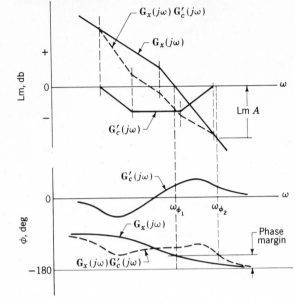

FIGURE 11-18
Original and lag-lead compensated curves.

the time constant T_1. The log magnitude and phase-angle curves of the lead compensator can be located as described in Sec. 11-5. The lead-compensator curves are given by Fig. 11-10 except that the log magnitude curve should be raised to have a value of 0 dB at the low frequencies. This is necessary because the lag-lead compensator transfer function does not contain α as a factor. Either the time constant T_2 of the compensator can be made equal to the largest time constant of the original system (for a Type 1 or higher system), or the angle curve is located to produce the specified phase margin at the highest possible frequency. In the latter case, the location chosen for the angle curve permits determination of the time constant T_2. To utilize the lag-lead network of Eq. (11-16) the value of α used for the lag compensation must be the reciprocal of the α used for the lead compensation. The example given in the next section utilizes this second method since it is more flexible.

11-8   EXAMPLE: CASCADE LAG-LEAD COMPENSATION

For the basic system treated in Secs. 11-4 and 11-6, the forward transfer function is

$$\mathbf{G}_x(j\omega) = \frac{K_1}{j\omega(1 + j\omega)(1 + j0.2\omega)} \qquad (11\text{-}17)$$

For an $M_m = 1.26$ the result is $K_1 = 0.87$ and $\omega_{m1} = 0.72$ rad/s. According to the second method described above, the lag-lead compensator is the combination of

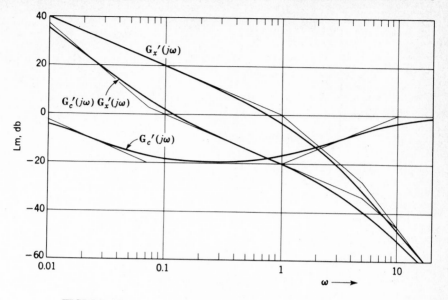

FIGURE 11-19
Log magnitude diagrams of Eqs. (11-17) to (11-19).

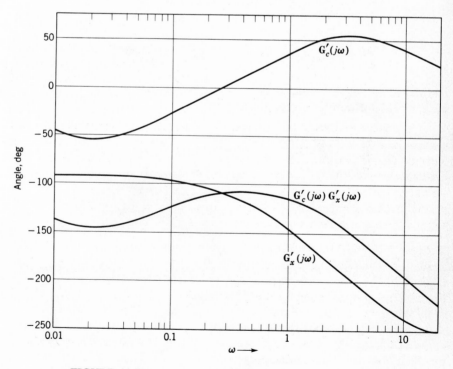

FIGURE 11-20
Phase-angle diagrams of Eqs. (11-17) to (11-19).

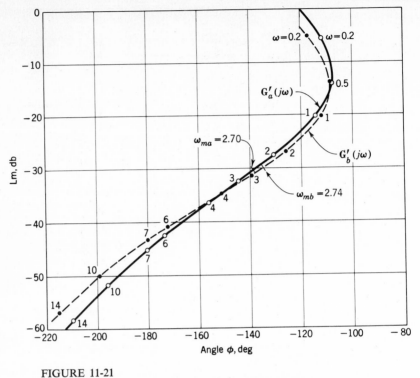

FIGURE 11-21
Log magnitude–angle diagram of $G'(j\omega)$ from Eq. (11-19).

the individual lag and lead compensators designed in Secs. 11-4 and 11-6. Thus the compensator transfer function is

$$\mathbf{G}_c(j\omega) = A\,\frac{(1 + j14\omega)(1 + j\omega)}{(1 + j140\omega)(1 + j0.1\omega)} \qquad \text{for } \alpha = 10 \qquad (11\text{-}18)$$

The new forward transfer function is

$$\mathbf{G}(j\omega) = \mathbf{G}_x(j\omega)\mathbf{G}_c(j\omega) = \frac{AK_1(1 + j14\omega)}{j\omega(1 + j0.2\omega)(1 + j140\omega)(1 + j0.1\omega)} \qquad (11\text{-}19)$$

In Figs. 11-19 and 11-20 are drawn the log magnitude and phase-angle diagrams of the basic system $\mathbf{G}_x'(j\omega)$, the compensator $\mathbf{G}_c'(j\omega)$, and the composite transfer function $\mathbf{G}_x'(j\omega)\mathbf{G}_c'(j\omega)$. From these curves, the log magnitude–angle diagram is drawn in Fig. 11-21. Upon adjusting for an $M_m = 1.26$ the results are $K_1' = 29.9$ and $\omega_{ma} = 2.70$. When these values are compared with those of Tables 11-1 and 11-2, it is seen that lag-lead compensation results in a value of K_1 approximately equal to the product of the K_1's of the lag- and lead-compensated systems. Also, the ω_m is only slightly less than that obtained with the lead-compensated system. Thus a larger increase in gain is achieved than with lag compensation, and the value of ω_m is almost as large as the value obtained by using the lead-compensated system. One may therefore be inclined to use a lag-lead compensator exclusively.

Table 11-3   COMPARISON BETWEEN ROOT-LOCUS AND FREQUENCY-RESPONSE METHODS OF APPLYING A LAG-LEAD COMPENSATOR TO A BASIC CONTROL SYSTEM

Method	$G_c(s)$	ω_{m1}	ω_m	ω_{n1}	ω_n	K_1	K_1'	Damping ratio (held constant)	Maximum peak overshoot
Root locus	$\dfrac{(1+s)(1+20s)}{(1+0.1s)(1+200s)}$	—	2.63	0.89	3.40	0.84	29.3	ζ of dominant roots $= 0.45$	$M_p = 1.20$ $M_m = 1.23$
Frequency plot	$\dfrac{(1+s)(1+20s)}{(1+0.1s)(1+200s)}$	0.72	2.74	—	—	0.87	30.4	Effective $\zeta = 0.45$	$M_m = 1.26$
	$\dfrac{(1+s)(1+14s)}{(1+0.1s)(1+140s)}$		2.70				29.9		

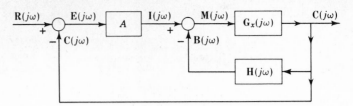

FIGURE 11-22
Block diagram for feedback compensation.

The resulting error coefficient (gain) due to the insertion of a lag-lead compensator is approximately equal to the product of the following:

1 The original error coefficient K_m of the uncompensated system
2 The increase in K_m due to the insertion of a lag compensator
3 The increase in K_m due to the insertion of a lead compensator

Table 11-3 presents a comparison of the root-locus and log-plot methods of applying a lag-lead compensator to a basic control system. The compensator designed from the log plots is different from the one designed by the root-locus method. To show that the difference in performance is small, Table 11-3 includes both compensators applied to the frequency plots. Both plots are shown in Fig. 11-21. The plot using Eq. (10-40) yields $\omega_{mb} = 2.74$ as shown in this figure.

11-9   FEEDBACK COMPENSATION USING LOG PLOTS[1]

The previous sections have demonstrated the procedures for applying cascade compensation. Those methods can be applied in a straightforward manner. When feedback compensation is applied, additional parameters must be determined. Besides the gain constant of the basic forward transfer function, the gain constant and frequency characteristics of the minor feedback must also be selected. Therefore, additional steps must be introduced in the design procedure. The general effects of feedback compensation are demonstrated by application to the specific system which is used throughout this text. The use of a minor loop for feedback compensation is shown in Fig. 11-22. The transfer function $G_x(j\omega)$ represents the basic system, $H(j\omega)$ is the feedback compensator forming a minor loop around $G_x(j\omega)$, and A is an amplifier which is used to adjust the overall performance. The forward transfer function of this system is

$$G_1(j\omega) = \frac{C(j\omega)}{I(j\omega)} = \frac{G_x(j\omega)}{1 + G_x(j\omega)H(j\omega)} \qquad (11\text{-}20)$$

In order to apply feedback compensation new techniques must be developed. This is done by first using some approximations and the straight-line log magnitude curves and then developing an exact procedure. Consider the case when

$$|G_x(j\omega)H(j\omega)| \ll 1$$

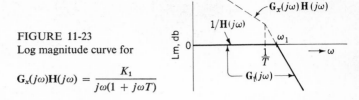

FIGURE 11-23
Log magnitude curve for

$$\mathbf{G}_x(j\omega)\mathbf{H}(j\omega) = \frac{K_1}{j\omega(1 + j\omega T)}$$

The forward transfer function can be approximated by

$$\mathbf{G}_1(j\omega) \approx \mathbf{G}_x(j\omega) \qquad \text{for } |\mathbf{G}_x(j\omega)\mathbf{H}(j\omega)| \ll 1 \qquad (11\text{-}21)$$

The next condition is $|\mathbf{G}_x(j\omega)\mathbf{H}(j\omega)| \gg 1$. Then the forward transfer function can be approximated by

$$\mathbf{G}_1(j\omega) \approx \frac{1}{\mathbf{H}(j\omega)} \qquad \text{for } |\mathbf{G}_x(j\omega)\mathbf{H}(j\omega)| \gg 1 \qquad (11\text{-}22)$$

Still undefined is the condition when $|\mathbf{G}_x(j\omega)\mathbf{H}(j\omega)| \approx 1$, in which case neither Eq. (11-21) nor Eq. (11-22) is applicable. In the approximate procedure this condition is neglected, and Eqs. (11-21) and (11-22) are used when $|\mathbf{G}_x(j\omega)\mathbf{H}(j\omega)| < 1$ and $|\mathbf{G}_x(j\omega)\mathbf{H}(j\omega)| > 1$, respectively. This approximation allows investigation of the qualitative results to be obtained. After these results are found to be satisfactory, the refinements for an exact solution are introduced.

An example illustrates the use of these approximations. Assume that $\mathbf{G}_x(j\omega)$ represents a motor having inertia and damping. The transfer function derived in Sec. 2-14 can be represented by

$$\mathbf{G}_x(j\omega) = \frac{K_m}{j\omega(1 + j\omega T_m)} \qquad (11\text{-}23)$$

Let the feedback $\mathbf{H}(j\omega) = 1/\underline{0^\circ}$. This problem is sufficiently simple to be solved exactly algebraically. However, use is made of the log magnitude curve and the approximate conditions. In Fig. 11-23 is sketched the log magnitude curve for $\mathbf{G}_x(j\omega)\mathbf{H}(j\omega)$.

From Fig. 11-23 it is seen that $|\mathbf{G}_x(j\omega)\mathbf{H}(j\omega)| > 1$ for all frequencies below ω_1. With the approximation of Eq. (11-22), $\mathbf{G}_1(j\omega)$ can be represented by $1/\mathbf{H}(j\omega)$ for frequencies below ω_1. Also, $|\mathbf{G}_x(j\omega)\mathbf{H}(j\omega)| < 1$ for all frequencies above ω_1. Therefore $\mathbf{G}_1(j\omega)$ can be represented, as shown, by the line of zero slope and 0 dB for frequencies up to ω_1 and the line of -12-dB slope above ω_1. The equation of $\mathbf{G}_1(j\omega)$ therefore has a quadratic in the denominator with $\omega_n = \omega_1$:

$$\mathbf{G}_1(j\omega) = \frac{1}{1 + 2\zeta j(\omega/\omega_1) + [j(\omega/\omega_1)]^2} \qquad (11\text{-}24)$$

Of course $\mathbf{G}(j\omega)$ can be obtained algebraically for this simple case from Eq. (11-20), with the result given as

$$\mathbf{G}_1(j\omega) = \frac{1}{1 + (1/K_m)j\omega + (j\omega)^2 T/K_m} \qquad (11\text{-}25)$$

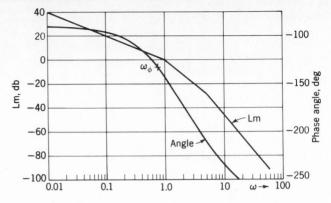

FIGURE 11-24
Log magnitude–phase-angle diagram of

$$G'_x(j\omega) = \frac{1}{j\omega(1 + j\omega)(1 + j0.2\omega)}$$

where $\omega_1 = \omega_n = (K_m/T)^{1/2}$ and $\zeta = 1/[2(K_mT)^{1/2}]$. Note that the approximate result is basically correct but some detailed information, in this case the value of ζ, is missing. The approximate angle curve can be drawn to correspond to the log magnitude curve or to Eq. (11-24).

11-10   EXAMPLE: FEEDBACK COMPENSATION (LOG PLOTS)

The system of Fig. 11-22 is investigated with the value of $G_x(j\omega)$ given by

$$G_x(j\omega) = \frac{K_x}{j\omega(1 + j\omega)(1 + j0.2\omega)} \qquad (11\text{-}26)$$

The system having this transfer function has been used throughout this text for the various methods of compensation. The log magnitude and phase-angle diagram using the straight-line log magnitude curve is drawn in Fig. 11-24. A phase margin of 45° occurs at the frequency $\omega_{\phi 1} = 0.8$ with a gain of -2 dB. The object in applying compensation is to increase both the phase-margin frequency and the gain. In order to select a feedback compensator $H(j\omega)$, consider the following facts:

1 The system type should be maintained. In accordance with the conditions outlined in Sec. 10-12 and the results given in Sec. 10-17 for the root-locus method, it is known that good improvement is achieved by using a feedback compensator $H(j\omega)$ which has a zero, $s = 0$, of order equal to (or preferably higher than) the type of the original transfer function $G_x(j\omega)$.

2 It is shown in Sec. 11-9 that the new forward transfer function can be approximated by

$$G_1(j\omega) \approx \begin{cases} G_x(j\omega) & \text{for } |G_x(j\omega)H(j\omega)| < 1 \\ \dfrac{1}{H(j\omega)} & \text{for } |G_x(j\omega)H(j\omega)| > 1 \end{cases}$$

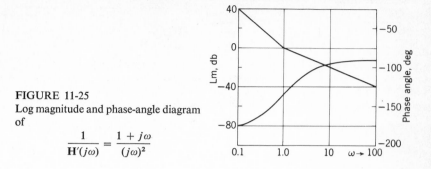

FIGURE 11-25
Log magnitude and phase-angle diagram
of

$$\frac{1}{H'(j\omega)} = \frac{1 + j\omega}{(j\omega)^2}$$

Now apply this information to a consideration of the proper $H(j\omega)$ to use. Consider the possibility of replacing a portion of the curves of $G_x(j\omega)$ for a range of frequencies by the curves of $1/H(j\omega)$ with the intent of increasing the values of ω_ϕ and the gain. This requires that the value of $H(j\omega)$ be such that $|G_x(j\omega)H(j\omega| > 1$ for that range of frequencies. A feedback unit using a tachometer and an RC derivative network is considered. The circuit is shown in Fig. 10-38, and the transfer function is

$$H(j\omega) = \frac{K_t T(j\omega)^2}{1 + j\omega T} \qquad (11\text{-}27)$$

The log magnitude and phase-angle diagram for $1/H'(j\omega)$ is shown in Fig. 11-25 with a value of $T = 1$. These curves shift to the left or to the right according as T is increased or decreased. The angle curve $1/H(j\omega)$ shows that the phase margin can be obtained as a function of the compensator time constant T. This shows promise for increasing the phase margin, provided that the magnitude $|G_x(j\omega)H(j\omega)|$ can be made greater than unity over the correct range of frequencies. Also, the larger $K_t T$ is made, the smaller the magnitude of $1/H(j\omega)$ at the phase-margin frequency. This permits a large value of A and therefore a large error coefficient.

The selection of T and K_t is based on a trial-and-error procedure. The object is to produce a section of the $G_1(j\omega)$ log magnitude curve with a slope of $-20\text{dB}/$ decade and with a smaller magnitude than (i.e., it is below) the $G_x(j\omega)$ log magnitude curve. This new section of the $G_1(j\omega)$ curve must occur at a higher frequency than the original phase-margin frequency. Section 8-19 describes the desirability of a slope of -20 dB/decade to produce a large phase margin. To achieve compensation, the log magnitude curve $1/H(j\omega)$ is placed over the $G_x(j\omega)$ curve so that there are one or two points of intersection. There is no reason to restrict the value of K_x. One such arrangement is shown in Fig. 11-26, with $K_x = 2$, $T = 1$, $K_t T = 25$, and intersections between the two curves occur at $\omega_1 = 0.021$ and $\omega_2 = 16$. The log magnitude plot of $G_x(j\omega)H(j\omega)$ is also drawn in Fig. 11-26 and shows that

$$|G_x(j\omega)H(j\omega)| \begin{cases} < 1 & \text{for } \omega_2 < \omega < \omega_1 \\ > 1 & \text{for } \omega_2 > \omega > \omega_1 \end{cases}$$

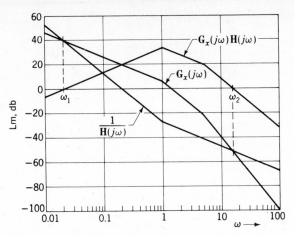

FIGURE 11-26
Log magnitude plots of $1/H(j\omega)$, $G_x(j\omega)$, and $G_x(j\omega)H(j\omega)$

Therefore $G_1(j\omega)$ can be represented approximately by

$$G_1(j\omega) \approx \begin{cases} G_x(j\omega) & \text{for } \omega_2 < \omega < \omega_1 \\ \dfrac{1}{H(j\omega)} & \text{for } \omega_2 > \omega > \omega_1 \end{cases}$$

The composite curve can therefore be approximately represented by

$$G_1(j\omega) = \frac{2(1 + j\omega)}{j\omega(1 + j47.5\omega)(1 + j0.0625\omega)^2} \qquad (11\text{-}28)$$

The log magnitude and phase-angle diagram for Eq. (11-28) is shown in Fig. 11-27. The new value of phase-margin frequency is $\omega_\phi = 4.5$, and the gain is 41 dB. This shows a considerable improvement of the system performance.

An exact curve of $G_1(j\omega)$ can be obtained analytically. The functions $G_x(j\omega)$ and $H(j\omega)$ were found to be

$$G_x(s) = \frac{2}{s(1 + s)(1 + 0.2s)} \qquad H(s) = \frac{25s^2}{1 + s}$$

The function $G_1(j\omega)$ is obtained as follows:

$$G_1(j\omega) = \frac{G_x(j\omega)}{1 + G_x(j\omega)H(j\omega)}$$

$$= \frac{2(1 + j\omega)}{j\omega(1 + j\omega)^2(1 + j0.2\omega) + 50(j\omega)^2}$$

$$= \frac{2(1 + j\omega)}{j\omega[0.2(j\omega)^3 + 1.4(j\omega)^2 + 52.2(j\omega) + 1]}$$

$$= \frac{2(1 + j\omega)}{j\omega(1 + j52\omega)[0.00385(j\omega)^2 + 0.027(j\omega) + 1]} \qquad (11\text{-}29)$$

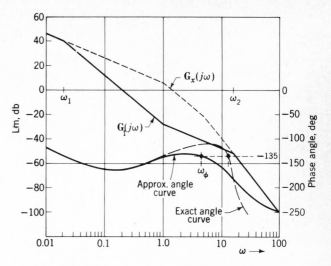

FIGURE 11-27
Log magnitude and phase-angle diagrams of

$$G_1(j\omega) = \frac{2(1 + j\omega)}{j\omega(1 + j47.5\omega)(1 + j0.0625\omega)^2}$$

A comparison of Eqs. (11-28) and (11-29) shows that there is little difference between the approximate and the exact equations. The main difference is that the approximate equation has the term $(1 + j0.0625\omega)^2$, which assumes a damping ratio $\zeta = 1$ for this quadratic factor. The actual damping ratio is $\zeta = 0.217$ in the correct quadratic factor $[0.00385(j\omega)^2 + 0.027(j\omega) + 1]$. The corner frequency is the same for both cases, $\omega_n = 16$. As a result, the exact angle curve, given in Fig. 11-27, shows a higher phase-margin frequency $(\omega = 13)$ than that obtained for the approximate curve $(\omega = 4.5)$. If the basis of design is the value of M_m, the exact curve should be used. The data for the exact curve are readily available from a computer.

The adjustment of the amplifier gain A is based on either the phase margin or M_m. An approximate value for A is the gain necessary to raise the curve of Lm $G_1(j\omega)$ so that it crosses the 0-dB line at ω_ϕ. A more precise value for A is obtained by plotting $G_1(j\omega)$ on the log magnitude–angle diagram and adjusting the system for a desired value of M_m. The exact Lm $G_1(j\omega)$ curve should be used for best results.

The shape of the Nichols plot for $G_1(j\omega)$ is shown in Fig. 11-28. When adjusting the gain for a desired M_m the objectives are to achieve a large resonant frequency ω_m and a large gain constant K_1. The tangency of the M_m curve at $\omega_c = 15.3$ requires a gain $A = 95.5$ and yields $K_1 = 191$. However, the rapid change in phase angle in the vicinity of ω_c can produce unexpected results. The M vs. ω characteristic curve c, shown in Fig. 11-29, has two peaks. This is not the form desired and must be rejected. This example serves to call attention to the fact that one point, the peak M value, is not sufficient to determine the suitability of the system response.

Tangency at $\omega_b = 1.2$ in Fig. 11-28 requires $A = 57.5$ and yields $K_1 = 115$. Curve b in Fig. 11-29 has the form desired and is acceptable. Compared with the

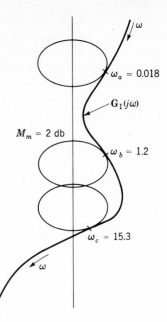

FIGURE 11-28
Nichols plot of $G_1(j\omega)$.

original system, the minor-loop feedback compensation has produced a modest increase in ω_m from 0.72 to 1.2. There is a very large increase in K_1 from 0.88 to 115.

Tangency at ω_a is not considered, as it would represent a degradation of the uncompensated performance.

This example has demonstrated qualitatively that feedback compensation can be used to produce a section of the log magnitude curve of the forward transfer function $G(j\omega)$ with a slope of -20 dB/decade. This section of the curve can be placed in the desired frequency range, with the result that the new phase-margin

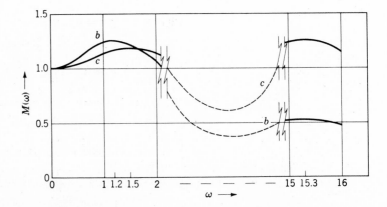

FIGURE 11-29
Closed-loop response.

frequency is larger than that of the original system. However, to obtain the correct quantitative results the exact curves should be used instead of the straight-line approximations. Also, the use of the log magnitude–angle diagram as the last step in the design process, to adjust the gain for a specified M_m, gives a better indication of performance than the phase margin does.

The reader may wish to try other values of K_t and T in the feedback compensator in order to modify the system performance. Also, it is worth considering whether the RC network should be used at all as part of the feedback compensator. The results using the root-locus method (see Secs. 10-14 to 10-17) show that rate feedback alone does improve system performance. The use of second-derivative feedback produces only qualified or limited improvements.

11-11   SUMMARY

A basic unity-feedback control system has been used throughout this and other chapters. It is therefore possible to compare the results obtained by adding each of the compensators to the system. The results obtained are consolidated in Table 11-4.

Table 11-4   SUMMARY OF CASCADE AND FEEDBACK COMPENSATION OF A BASIC CONTROL SYSTEM USING THE FREQUENCY-RESPONSE METHOD

Compensator	ω_m	K_1	Effective ζ	M_m	Additional gain required
Uncompensated	0.72	0.87	0.45	1.26	—
Lag					
$\dfrac{1 + 14s}{1 + 140s}$	0.53	6.68	0.45	1.26	7.7
$\dfrac{1 + 20s}{1 + 200s}$	0.59	7.35			8.5
Lead					
$0.1\,\dfrac{1 + s}{1 + 0.1s}$	2.85	3.13	0.45	1.26	36.0
$0.1\,\dfrac{1 + 2s/3}{1 + 0.2s/3}$	2.10	3.50			40.2
Lag-lead					
$\dfrac{(1 + s)(1 + 14s)}{(1 + 0.1s)(1 + 140s)}$	2.70	29.9	0.45	1.26	34.4
$\dfrac{(1 + s)(1 + 20s)}{(1 + 0.1s)(1 + 200s)}$	2.74	30.4			35.0
Feedback					
$H(s) = \dfrac{25s^2}{1 + s}$	1.20	115	0.45	1.26	57.5

The results achieved by using the frequency response are comparable with those obtained by using the root locus (see Chap. 10). The design methods which have been demonstrated use some trial-and-error techniques. A digital computer can expedite the work.[4,5]

The following properties can be attributed to each compensator:

Cascade lag compensator Results in an increase in the gain K_m and a small reduction in ω_m. This reduces steady-state error but increases transient settling time.

Cascade lead compensator Results in an increase in ω_m, thus reducing the settling time. There may be a small increase in gain.

Cascade lag-lead compensator Results in both an increase in gain K_m and in resonant frequency ω_m. This combines the improvements of the lag and the lead compensators.

Feedback compensator Results in an increase in both the gain K_m and resonant frequency ω_m. The basic improvement in ω_m is achieved by derivative feedback. This may be modified by adding a filter such as an *RC* network.

REFERENCES

1 Chestnut, H., and R. W. Mayer: "Servomechanisms and Regulating System Design," 2d ed., vol. 1, chaps. 10 and 12, Wiley, New York, 1959.
2 Kuo, F. F.: "Network Analysis and Synthesis," 2d ed., Wiley, New York, 1966.
3 Mitra, S. K.: "Analysis and Synthesis of Linear Active Networks," Wiley, New York, 1969.
4 Frequency Response Program (FREQR), School of Engineering, Air Force Institute of Technology, Wright-Patterson Air Force Base, Ohio, 1974.
5 Heaviside Partial Fraction Expansion and Time Response Program (PARTL), School of Engineering, Air Force Institute of Technology, Wright-Patterson Air Force Base, Ohio, 1974.

12

SYSTEM DESIGN THROUGH CLOSED-LOOP POLE SPECIFICATION (STATE-VARIABLE FEEDBACK)

12-1 INTRODUCTION

Chapters 10 and 11 dealt with the improvement of system performance by conventional design procedures, based on improving the closed-loop time response by designing cascade and/or feedback compensators from an analysis of the open-loop transfer function. The design methods presented in this chapter are based upon achieving a desired or specified overall transfer function; i.e., they are pole-zero placement methods. Section 12-2 presents the Guillemin-Truxal method of designing a cascade compensator to yield this desired transfer function. The Guillemin-Truxal compensator and those discussed in Chaps. 10 and 11 utilize a forward and/or a feedback input signal to actuate them. With the advent of modern control theory the concept of utilizing all the system states to provide the desired improvement in system performance has been introduced. This concept requires that all states be accessible in a physical system. For most systems this requirement is not met; i.e., some of the states are inaccessible. A technique for handling systems with inaccessible states is also presented in this chapter. Thus this chapter illustrates how conventional control-theory design concepts are applied to a system utilizing state-variable feedback. The state-variable-feedback design method presented in this chapter is based upon achieving a desired control-ratio transfer function for a single-input single-output system.

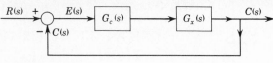

FIGURE 12-1
A cascade-compensated control system.

12-2 GUILLEMIN-TRUXAL DESIGN PROCEDURE[1]

In contrast to designing the cascade compensator $G_c(s)$ based upon the analysis of the open-loop transfer function, the Guillemin-Truxal method is based upon designing $G_c(s)$ to yield a specified or desired control ratio $C(s)/R(s)$. The control ratio for the unity-feedback cascade-compensated control system of Fig. 12-1 is

$$M(s) = \frac{C(s)}{R(s)} = \frac{N(s)}{D(s)} = \frac{G_c(s)G_x(s)}{1 + G_c(s)G_x(s)} \qquad (12\text{-}1)$$

where $N(s)$ and $D(s)$ represent the specified zeros and poles, respectively, of the desired control ratio. Solving this equation for $G_c(s)$ yields

$$G_c(s) = \frac{M(s)}{[1 - M(s)]G_x(s)} = \frac{N(s)}{[D(s) - N(s)]G_x(s)} \qquad (12\text{-}2)$$

In other words, when $N(s)$ and $D(s)$ have been specified, the required transfer function for $G_c(s)$ is obtained. Network-synthesis techniques are then utilized to synthesize a network having the required transfer function of Eq. (12-2). Usually a passive network containing only resistors and capacitors is desired.

EXAMPLE The desired control ratio of a unity-feedback control system is

$$\frac{C(s)}{R(s)} = \frac{210(s + 1.5)}{(s + 1.75)(s + 16)(s + 1.5 \pm j3)} = \frac{N(s)}{D(s)} \qquad (12\text{-}3)$$

The numerator constant has been selected to yield zero steady-state error with a step input. The basic plant is

$$G_x(s) = \frac{4}{s(s + 1)(s + 5)} \qquad (12\text{-}4)$$

The required cascade compensator, determined by substituting Eqs. (12-3) and (12-4) into Eq. (12-2), is

$$G_c(s) = \frac{52.5s(s + 1)(s + 1.5)(s + 5)}{s^4 + 20.75s^3 + 92.6s^2 + 73.69s}$$

$$= \frac{52.5s(s + 1)(s + 1.5)(s + 5)}{s(s + 1.02)(s + 4.88)(s + 14.86)} \approx \frac{52.5(s + 1.5)}{s + 14.86} \qquad (12\text{-}5)$$

The simplification is made in $G_c(s)$ in order to yield a simple, physically realizable, minimum-phase network with an $\alpha = 0.101$. Based on a root-locus analysis of

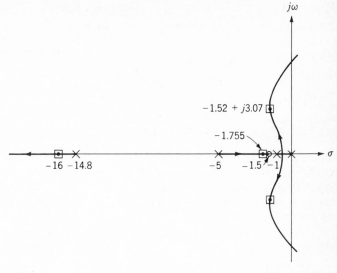

FIGURE 12-2
Root-locus plot.

$G_c(s)G_x(s)$, the $(s + 1)/(s + 1.02)$ and the $(s + 5)/(s + 4.88)$ terms of $G_c(s)$ have a negligible effect on the location of the desired dominant poles of $C(s)/R(s)$. With the simplified $G_c(s)$ the root-locus plot is shown in Fig. 12-2, and the control ratio achieved is

$$\frac{C(s)}{R(s)} = \frac{210(s + 1.5)}{(s + 1.755)(s + 16.0)(s + 1.52 \pm j3.07)}$$

Since this equation is very close to that of Eq. (12-3), it is satisfactory. Note that the root-locus plot of $G_c(s)G_x(s) = -1$ has asymptotes of 180 and $\pm 60°$; therefore the system becomes unstable for high gain.

The Guillemin-Truxal design method, as illustrated by this example, involves three steps:

1 Specifying the desired zeros, poles, and numerator constant of the desired closed-loop function $C(s)/R(s)$
2 Solving for the required cascade compensator transfer function $G_c(s)$ from Eq. (12-2)
3 Synthesizing a physically realizable network, preferably a passive unit

The first step involves the determination of $N(s)$ and $D(s)$ from the given desired control-system performance specification. This determination is based upon a knowledge of the transient and frequency response of the basic system. The synthesis of the desired $C(s)/R(s)$ function can be a difficult task. When the deficiencies of the basic system are known, it is possible to synthesize a more desirable $C(s)/R(s)$. For practical cascade compensators the order of the closed-loop system must be equal to or greater than the order of the basic system. This must include the poles and zeros introduced by the cascade compensator. In order to ensure that the simplest $G_c(s)$ function is achieved and can be realized by RC networks, all the poles of $G_c(s)G_x(s)$ must lie on the negative real axis. The practical aspects of system synthesis impose the limitation that the poles of $G_x(s)$ lie in the left-half s plane. Failure to exactly cancel poles in the right-half s plane would result in an unstable closed-loop system. The state-variable-feedback method, described in the remainder of this chapter, is a compensation method that achieves the desired $C(s)/R(s)$ without this limitation.

In the Guillemin-Truxal example above the desired control ratio $Y(s)/R(s)$ is specified in Eq. (12-3). This specification is also required for the state-variable design method presented in the remainder of this chapter. In general, the designer must select the proper control ratio to meet the system specifications. This requires that the poles and zeros be located in order to achieve the desired time response. This is called the *pole-zero placement technique*.

For the simple second-order system the three time-response parameters M_p, t_p, and T_s are described by Eqs. (3-60), (3-61), and (3-64). The specification of any two of these quantities can be used to locate the dominant poles of $Y(s)/R(s)$. A third real pole may be located to assist in satisfying the specifications. The remaining poles of an all-pole plant can be specified to be nondominant.

For a plant having w zeros, w of the remaining poles must be located "very close" to these w zeros. The remaining $n - 2 - w$ poles are located in the nondominant region of the s plane. Another possible approach is to include a third real root s_3 to improve the time response so that all three parameters are acceptable. Another method of selecting the location of the poles and zeros is given in Sec. 14-6.

12-3 STATE-VARIABLE FEEDBACK[2]

The simple open-loop position-control system of Fig. 12-3 is used to illustrate the effects of state-variable feedback. The dynamic equations for this basic system are

$$e_a = Ae \quad A > 0 \qquad (12\text{-}6)$$

$$e_a - e_m = (R_m + L_m D)i_m \qquad (12\text{-}7)$$

$$e_m = K_b \omega_m \qquad (12\text{-}8)$$

$$T = K_T i_m = JD\omega_m + B\omega_m \qquad (12\text{-}9)$$

$$\omega_m = D\theta_m \qquad (12\text{-}10)$$

In order to obtain the maximum number of accessible (measurable) states, the *physical variables are utilized as state variables*. The two variables i_m and ω_m and the desired output quantity θ_m are identified as the three state variables: $x_1 = \theta_m = y$, $x_2 = \omega_m = \dot{x}_1$, and $x_3 = i_m$, and the input is $u = e$. Figure 12-3 shows that all three state variables are accessible for measurement. The sensors are selected to produce voltages which are proportional to the state variables.

Figure 12-4a is the block diagram of a closed-loop position-control system utilizing the system of Fig. 12-3. Each of the states is fed back through amplifiers

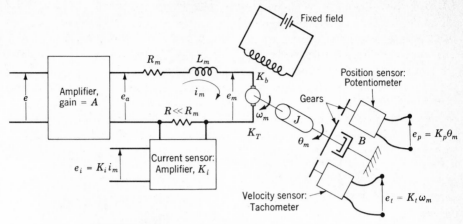

FIGURE 12-3
Open-loop position-control system.

whose gain values k_1, k_2, and k_3 are called *feedback coefficients*. These feedback coefficients include the sensor constants and any additional gain required by the system design to yield the required values of k_1, k_2, and k_3. The summation of the three feedback quantities, where **k** is the *feedback vector*, is

$$k_1 X_1 + k_2 X_2 + k_3 X_3 = \mathbf{k}^T \mathbf{X} \qquad (12\text{-}11)$$

Using block-diagram manipulation techniques, Fig. 12-4a is simplified to the block diagram of Fig. 12-4b. A further reduction is achieved as shown in Fig. 12-4c. Note that the forward transfer function $G(s) = X_1(s)/U(s)$ shown in this figure is identical to that obtained from Fig. 12-4a with zero-state feedback coefficients. Thus, state-variable feedback is *equivalent* to inserting a feedback compensator $H_{eq}(s)$ in the feedback loop of a conventional feedback control system, as shown in Fig. 12-4d. The reduction of Fig. 12-4a to that of Fig. 12-4c or Fig. 12-4d is referred to as the *H-equivalent reduction*. The system-control ratio† is therefore

$$\frac{Y(s)}{R(s)} = \frac{G(s)}{1 + G(s)H_{eq}(s)} \qquad (12\text{-}12)$$

For Fig. 12-4c the following equations are obtained:

$$H_{eq}(s) = \frac{k_3 s^2 + (2k_2 + k_3)s + 2k_1}{2} \qquad (12\text{-}13)$$

$$G(s) = \frac{200A}{s(s^2 + 6s + 25)} \qquad (12\text{-}14)$$

$$G(s)H_{eq}(s) = \frac{100A[k_3 s^2 + (2k_2 + k_3)s + 2k_1]}{s(s^2 + 6s + 25)} \qquad (12\text{-}15)$$

$$\frac{Y(s)}{R(s)} = \frac{200A}{s^3 + (6 + 100k_3 A)s^2 + (25 + 200k_2 A + 100k_3 A)s + 200k_1 A} \qquad (12\text{-}16)$$

† See also Prob. 12-3.

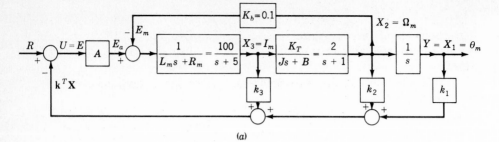

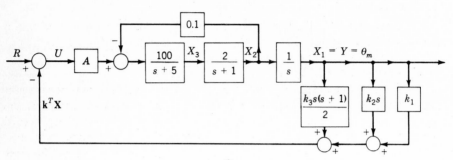

(a)

(b)

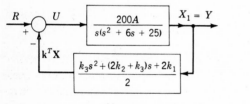

(c)

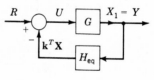

(d)

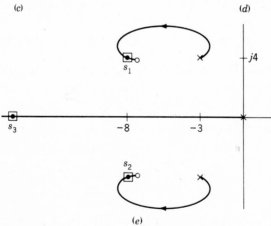

(e)

FIGURE 12-4
A position-control system with state feedback and its H_{eq} representation.

The equivalent feedback transfer function $H_{eq}(s)$ must be designed so that the system's desired figures of merit are satisfied. The following observations are noted from Eqs. (12-13) to (12-16) for this system in which $G(s)$ has no zeros; i.e., it is an all-pole plant.

1. In order to achieve zero steady-state error for a step input, $R(s) = R_0/s$, Eq. (12-16) must yield

$$y(t)_{ss} = \lim_{s \to 0} s Y(s) = \frac{R_0}{k_1} = R_0$$

This requires that $k_1 = 1$.

2. The numerator of $H_{eq}(s)$ is of second order; i.e., the denominator of $G(s)$ is of degree $n = 3$ and the numerator of $H_{eq}(s)$ is of degree $(n - 1) = 2$.

3. The poles of $G(s)H_{eq}(s)$ are the poles of $G(s)$.

4. The zeros of $G(s)H_{eq}(s)$ are the zeros of $H_{eq}(s)$.

5. The use of state feedback produces additional zeros in the open-loop transfer function $G(s)H_{eq}(s)$ without adding poles. This is in contrast to cascade compensation using passive networks. The zeros of $H_{eq}(s)$ are to be located to yield the desired response. It is known that the addition of zeros to the open-loop transfer function moves the root locus to the left; i.e., they make the system more stable and improve the time-response characteristics.

6. Since the root-locus plot of $G(s)H_{eq}(s) = -1$ has one asymptote, $\gamma = -180°$ for $k_3 > 0$, it is possible to pick the locations of the zeros of $H_{eq}(s)$ to ensure a completely stable system for all positive values of gain. Figure 12-4e shows a possible selection of the zeros and the resulting root locus. Note that when the conventional cascade compensators of Chap. 10 are utilized, the angles of the asymptotes yield branches, in general, that go into the right-half s plane. The maximum gain is therefore limited in order to maintain system stability. This restriction is removed with state feedback.

7. The desired figures of merit, for an underdamped response, are satisfied by selecting the desired locations of the poles of $Y(s)/R(s)$ of Eq. (12-16) at $s = -a \pm jb$ and $s = -c$. For these specified pole locations the known characteristic equation has the form.

$$(s + a \pm jb)(s + c) = s^3 + d_2 s^2 + d_1 s + d_0 = 0 \qquad (12\text{-}17)$$

Equating the denominator of Eq. (12-16) to Eq. (12-17) yields

$$d_2 = 6 + 100k_3 A \qquad (12\text{-}18)$$

$$d_1 = 25 + 200k_2 A + 100k_3 A \qquad (12\text{-}19)$$

$$d_0 = 200k_1 A = 200A \qquad (12\text{-}20)$$

These three equations with three unknowns can be solved for the required values of A, k_2, and k_3.

It should be noted that if the basic system $G(s)$ had any zeros, they become poles of $H_{eq}(s)$. Thus, the zeros of $G(s)$ cancel the poles of $H_{eq}(s)$ in $G(s)H_{eq}(s)$. Also, the zeros of $G(s)$ become the zeros of the control ratio. The reader can verify this property by replacing $2/(s + 1)$ in Fig. 12-4a by $2(s + \alpha)/(s + 1)$ and performing an H-equivalent reduction. This property is also illustrated in the examples of Sec. 12-9.

12-4 GENERAL PROPERTIES OF STATE FEEDBACK (USING PHASE VARIABLES)

In order to obtain general state-variable-feedback properties that are applicable to any $G(s)$ transfer function, consider the minimum-phase transfer function

$$G(s) = \frac{Y(s)}{U(s)} = \frac{K_G(s^w + c_{w-1}s^{w-1} + \cdots + c_1 s + c_0)}{s^n + a_{n-1}s^{n-1} + \cdots + a_1 s + a_0} \quad (12\text{-}21)$$

where $K_G > 0$, $w < n$, and the c_i elements are all positive quantities. The system characteristics and properties are most readily obtained by representing this system in terms of phase variables (see Secs. 3-15 and 5-7, with $b_w = 1$ and u replaced by $K_G u$):

$$\dot{x} = \begin{bmatrix} 0 & 1 & 0 & \cdots & 0 \\ 0 & 0 & 1 & \cdots & 0 \\ \cdots\cdots\cdots\cdots\cdots\cdots\cdots & 1 \\ -a_0 & -a_1 & -a_2 & \cdots & -a_{n-1} \end{bmatrix} x + \begin{bmatrix} 0 \\ 0 \\ \cdot \\ K_G \end{bmatrix} u \quad (12\text{-}22)$$

$$y = \begin{bmatrix} c_0 & c_1 & \cdots & c_{w-1} & 1 & 0 & \cdots & 0 \end{bmatrix} x = c^T x \quad (12\text{-}23)$$

From Fig. 12-4d and Eq. (12-23)

$$H_{eq}(s) = \frac{k^T X(s)}{Y(s)} = \frac{k^T X(s)}{c^T X(s)} = \frac{k_1 X_1(s) + k_2 X_2(s) + \cdots + k_n X_n(s)}{c_0 X_1(s) + c_1 X_2(s) + \cdots + X_{w+1}} \quad (12\text{-}24)$$

When the phase-variable relationship $X_j(s) = s^{j-1} X_1(s)$ is utilized, Eq. (12-24) becomes

$$H_{eq}(s) = \frac{k_n s^{n-1} + k_{n-1}s^{n-2} + \cdots + k_2 s + k_1}{s^w + c_{w-1}s^{w-1} + \cdots + c_1 s + c_0} \quad (12\text{-}25)$$

Thus, from Eqs. (12-21) and (12-25)

$$G(s)H_{eq}(s) = \frac{K_G(k_n s^{n-1} + k_{n-1}s^{n-2} + \cdots + k_2 s + k_1)}{s^n + a_{n-1}s^{n-1} + \cdots + a_1 s + a_0} \quad (12\text{-}26a)$$

An alternate form of this equation is

$$G(s)H_{eq}(s) = \frac{K_G k_n(s^{n-1} + \cdots + \alpha_1 s + \alpha_0)}{s^n + a_{n-1}s^{n-1} + \cdots + a_1 s + a_0} \quad (12\text{-}26b)$$

where the loop sensitivity is $K_G k_n = K$. Substituting Eqs. (12-21) and (12-26a) into Eq. (12-12) yields

$$\frac{Y(s)}{R(s)} = \frac{K_G(s^w + c_{w-1}s^{w-1} + \cdots + c_0)}{s^n + (a_{n-1} + K_G k_n)s^{n-1} + \cdots + (a_0 + K_G k_1)} \quad (12\text{-}27)$$

An analysis of Eqs. (12-21), (12-25), (12-26), and (12-27) yields the following conclusions.

The numerator of $H_{eq}(s)$ is a polynomial of degree $n - 1$ in s; that is, it has $n - 1$ zeros. The values of the coefficients of this polynomial, after k_n is factored from it, can be selected by the system designer to yield the desired system performance.

The numerator of $G(s)$, except for K_G, is equal to the denominator of $H_{eq}(s)$.

Therefore, the open-loop transfer function $G(s)H_{eq}(s)$ has the same n poles as $G(s)$ and has the $n - 1$ zeros of $H_{eq}(s)$. $G(s)H_{eq}(s)$ has $n - 1$ zeros, which are determined in the design process. (An exception to this occurs when physical variables are utilized and the transfer function $X_n(s)/U(s)$ contains a zero. In that case this zero does not appear as a pole of $H_{eq}(s)$, and the numerator and denominator of $G(s)H_{eq}(s)$ are both of degree n.)

The root-locus diagram, based on Eq. (12-26b) for $k_n > 0$, reveals that a single asymptote exists at $\gamma = -180°$, that $n - 1$ branches terminate on the $n - 1$ arbitrary zeros which can be located anywhere in the s plane, and that one branch terminates on the negative real axis at $s = -\infty$. System stability is ensured for high values of K_G if all the zeros of $H_{eq}(s)$ are located in the left-half s plane. For $k_n < 0$ a single asymptote exists at $\gamma = 0°$, thus the system becomes unstable for high values of K_G.

State-variable feedback provides the ability to select the locations of the zeros of $G(s)H_{eq}(s)$, as given by Eq. (12-26b), in order to locate closed-loop system poles where desired. From Eq. (12-27) it is noted that the result of state feedback through constant elements is to change the closed-loop pole positions and to leave the system zeros, if any, the same as those of $G(s)$. It may also be desirable to specify that one or more of the closed-loop poles cancel one or more unwanted zeros in $Y(s)/R(s)$.

In synthesizing the desired $Y(s)/R(s)$ the following limitation must be observed: the number of poles minus the number of zeros of $Y(s)/R(s)$ must equal or be less than the number of poles minus the number of zeros of $G(s)$.

A thorough steady-state error analysis of Eq. (12-27) is made in Sec. 12-5 for the standard step, ramp, and parabolic inputs. From the conclusions obtained in that section, some or all of the feedback gains k_1, k_2, and k_3 are specified.

Summarizing, the $H_{eq}(s)$ can be designed to yield the specified closed-loop transfer function in the manner described for the system of Fig. 12-4. A root-locus sketch of $G(s)H_{eq}(s) = \pm 1$ can be used as an aid in synthesizing the desired $Y(s)/R(s)$ function. From this sketch it is possible to determine the locations of the closed-loop poles, in relation to any system zeros, that will achieve the desired performance specifications. Note that although the zeros of $G(s)$ are *not* involved in obtaining the root locus for $G(s)H_{eq}(s)$, they do appear in $Y(s)/R(s)$ and thus affect the time function $y(t)$. The design procedure is summarized as follows.

Step 1 Draw the system state-variable block diagram.

Step 2 Assume that all state variables are accessible.

Step 3 Obtain $H_{eq}(s)$ and $Y(s)/R(s)$ in terms of the k_i's.

Step 4 Determine the value of k_1 required for zero steady-state error for a step input, $u_{-1}(t)$; that is, $y(t)_{ss} = \lim_{s \to 0} sY(s) = R_0 = 1$.

Step 5 Synthesize the desired $Y(s)/R(s)$ from the desired performance specifications (M_p, t_p, t_s, and steady-state error requirements). It may be desirable to specify that one or more of the closed-loop poles will cancel unwanted zeros of $Y(s)/R(s)$.

Step 6 Set the closed-loop transfer functions obtained in steps 3 and 5 equal to each other. Equate the coefficients of like powers of s of the denominator polynomials and solve for the required values of the k_i's. If phase variables are used in the design, a linear transformation must be made to convert the phase-variable feedback coefficients to the feedback coefficients required for the physical variables present in the control system.

Step 7 An alternate procedure to step 6 is to sketch the root locus of $G(s)H_{eq}(s) = \pm 1$ through the desired roots to provide a feel for the required zeros of $H_{eq}(s)$. Select the zeros to produce roots in the desired regions. Then determine the value of gain required to produce the dominant roots at the desired value of damping ratio. The remaining roots are then determined for this value of gain. If they are not satisfactory, a new selection is made for the zero locations. The procedure is repeated until satisfactory roots are obtained. The polynomial formed from the zeros is equated to the numerator of $H_{eq}(s)$ to determine the k_i's.

If some of the states are not accessible, suitable cascade or minor-loop compensators can be determined using the known values of the k_i's. This method is discussed in Sec. 12-10.

The procedure for designing *a state-variable feedback system* is illustrated by means of examples using physical variables in Secs. 12-7 and 12-9. Before the examples are considered, however, the steady-state error properties are investigated in Secs. 12-5 and 12-6.

12-5 STATE-VARIABLE FEEDBACK STEADY-STATE ERROR ANALYSIS[3]

As pointed out in Chap. 6, a steady-state error analysis of a *stable* control system is an important design consideration. The three standard inputs are utilized for a steady-state analysis of a state-variable feedback control system. The phase-variable representation is used in the following developments, in which the necessary values of the feedback coefficients are determined.

Step Input $r(t) = R_0 u_{-1}(t)$, $R(s) = R_0/s$

Solving Eq. (12-27), which is based upon the phase-variable representation, for $Y(s)$ and applying the final value theorem yields, for a step input,

$$y(t)_{ss} = \frac{K_G c_0 R_0}{a_0 + K_G k_1} \qquad (12\text{-}28)$$

The system error is defined by

$$e(t) = r(t) - y(t) \qquad (12\text{-}29)$$

For zero steady-state error, $e(t)_{ss} = 0$, the steady-state output must be equal to the input:

$$y(t)_{ss} = r(t) = R_0 = \frac{K_G c_0 R_0}{a_0 + K_G k_1} \qquad (12\text{-}30)$$

where $c_0 \neq 0$. This requires that $K_G c_0 = a_0 + K_G k_1$, or

$$k_1 = c_0 - \frac{a_0}{K_G} \qquad (12\text{-}31)$$

Thus, zero steady-state error with a step input can be achieved with state-variable feedback even if $G(s)$ is Type 0. For an all-pole plant $c_0 = 1$, and for a Type 1 or higher plant $a_0 = 0$. In that case the required value is $k_1 = 1$. *Therefore, the feedback coefficient k_1 is determined by the plant parameters once zero steady-state error for a step input is specified.* This specification is maintained for all state-variable-feedback control-system designs throughout the remainder of this text. From Eq. (12-27), the system characteristic equation is

$$s^n + (a_{n-1} + K_G k_n)s^{n-1} + \cdots + (a_2 + K_G k_3)s^2 + (a_1 + K_G k_2)s$$
$$+ (a_0 + K_G k_1) = 0 \qquad (12\text{-}32)$$

Note that the constant term of the polynomial is a positive quantity. This satisfies the requirement that all coefficients in Eq. (12-32) be positive in order that a stable response exist.

Ramp Input $\quad r(t) = R_1 u_{-2}(t) = R_1 t u_{-1}(t),\ R(s) = R_1/s^2$

To perform the analysis for ramp and parabolic inputs the error function $E(s)$ must first be determined. Solving for $Y(s)$ from Eq. (12-27) and substituting it into $E(s) = R(s) - Y(s)$ yields

$$E(s) = R(s)\ \frac{\begin{array}{c} s^n + (a_{n-1} + K_G k_n)s^{n-1} + \cdots + (a_0 + K_G k_1) \\ -\ K_G(s^w + c_{w-1}s^{w-1} + \cdots + c_1 s + c_0) \end{array}}{\begin{array}{c} s^n + (a_{n-1} + K_G k_n)s^{n-1} + \cdots \\ +\ (a_1 + K_G k_2)s + (a_0 + K_G k_1) \end{array}} \qquad (12\text{-}33)$$

Substituting Eq. (12-31), the requirement for zero steady-state error for a step input, into Eq. (12-33) produces a cancellation of the constant terms in the numerator.

$$E(s) = R(s)\ \frac{\begin{array}{c} s^n + (a_{n-1} + K_G k_n)s^{n-1} + \cdots + (a_1 + K_G k_2)s \\ -\ K_G(s^w + c_{w-1}s^{w-1} + \cdots + c_1 s) \end{array}}{s^n + (a_{n-1} + K_G k_n)s^{n-1} + \cdots + (a_1 + K_G k_2)s + K_G c_0} \qquad (12\text{-}34)$$

For a ramp input,

$$e(t)_{ss} = \lim_{s \to 0} sE(s) = \frac{a_1 + K_G k_2 - K_G c_1}{K_G c_0} R_1 \qquad (12\text{-}35)$$

In order to obtain $e(t)_{ss} = 0$ it is necessary that $a_1 + K_G k_2 - K_G c_1 = 0$. This is achieved when

$$k_2 = c_1 - \frac{a_1}{K_G} \qquad (12\text{-}36)$$

For an all-pole plant $c_0 = 1$ and $c_1 = 0$; thus $k_2 = -a_1/K_G$. In that case the first power of s in the system characteristic equation, Eq. (12-32), has a zero coefficient, and the system is unstable. *Therefore, zero steady-state error for a ramp input cannot be achieved by state-variable feedback for an all-pole stable system.* If $a_1 + K_G k_2 > 0$, in order to maintain a positive coefficient of s in Eq. (12-32), the system may be stable but a finite error exists. The error in an all-pole plant is

$$e(t)_{ss} = \left(k_2 + \frac{a_1}{K_G} \right) R_1 \qquad (12\text{-}37)$$

Assume that the value of k_2 given by Eq. (12-36) is specified to achieve zero steady-state error for a ramp input to a pole-zero system in which $c_1 > 0$. Substituting Eq. (12-36) into Eq. (12-32) results in the positive coefficient $K_G c_1$ for the first power of s. *Thus, a stable state-variable feedback system having zero steady-state error with both a step and a ramp input can be achieved only when the control ratio has at least one zero.* Note that this occurs even though $G(s)$ may be Type 0 or 1.

Parabolic Input $r(t) = R_2 u_{-3}(t) = (R_2 t^2/2) u_{-1}(t)$, $R(s) = R_2/s^3$

It is best to consider the all-pole and pole-zero plants separately for this type of input. For the all-pole plant, with Eq. (12-31) satisfied, Eq. (12-33) reduces to

$$E(s) = \frac{R_2}{s^3} \frac{s^n + (a_{n-1} + K_G k_n)s^{n-1} + \cdots + (a_1 + K_G k_2)s}{s^n + (a_{n-1} + K_G k_n)s^{n-1} + \cdots + (a_1 + K_G k_2)s + K_G c_0} \qquad (12\text{-}38)$$

Since $e(t)_{ss} = \lim_{s \to 0} sE(s) = \infty$ *it is concluded that an all-pole plant cannot follow a parabolic input.* This is true even if Eq. (12-31) is not satisfied.

Under the condition that $e(t)_{ss} = 0$ for both step and ramp inputs to a pole-zero stable system, Eqs. (12-31) and (12-36) are substituted into Eq. (12-34) to yield

$$E(s) = \frac{R_2}{s^3} \frac{\begin{array}{c} s^n + (a_{n-1} + K_G k_n)s^{n-1} + \cdots + (a_2 + K_G k_3)s^2 \\ - K_G(c_w s^w + c_{w-1}s^{w-1} + \cdots + c_2 s^2) \end{array}}{s^n + (a_{n-1} + K_G k_n)s^{n-1} + \cdots + K_G c_1 s + K_G c_0} \qquad (12\text{-}39)$$

Note that s^2 is the lowest power term in the numerator. Applying the final-value theorem to this equation yields

$$e(t)_{ss} = \lim_{s \to 0} sE(s) = \frac{a_2 + K_G k_3 - K_G c_2}{K_G c_0} R_2 \qquad (12\text{-}40)$$

In order to achieve $e(t)_{ss} = 0$ the requirement is that $a_2 + K_G k_3 - K_G c_2 = 0$. Thus

$$k_3 = c_2 - \frac{a_2}{K_G} \qquad (12\text{-}41)$$

With this value of k_3 the s^2 term in Eq. (12-32) has the positive coefficient $K_G c_2$ only when $w \geq 2$. *Thus, for a pole-zero system with two or more zeros, a stable system having zero steady-state error with a parabolic input can be achieved.*

In a system having only one zero ($w = 1$), the value $c_2 = 0$ and $k_3 = -a_2/K_G$ produces a zero coefficient for the s^2 term in the characteristic equation (12-32), resulting in an unstable system. *Therefore, zero steady-state error with a parabolic*

FIGURE 12-5
G-equivalent block diagram.

input cannot be achieved by a state-variable feedback pole-zero stable system with one zero. If $k_3 \neq -a_2/K_G$ and $a_2 + K_G k_3 > 0$ are satisfied to maintain a positive coefficient in Eq. (12-32), a finite error exists. The system can be stable, and the error is

$$e(t)_{ss} = \frac{R_2(a_2 + K_G k_3)}{K_G c_0} \qquad (12\text{-}42)$$

12-6 USE OF STEADY-STATE ERROR COEFFICIENTS

For the example represented by Fig. 12-4a the feedback coefficient blocks are manipulated to yield the *H*-equivalent block diagram in Fig. 12-4d. Similarly a *G-equivalent* block diagram, as shown in Fig. 12-5, can be obtained. The overall transfer functions for Figs. 12-4d and 12-5 are, respectively,

$$M(s) = \frac{Y(s)}{R(s)} = \frac{G(s)}{1 + G(s)H_{eq}(s)} \qquad (12\text{-}43)$$

$$M(s) = \frac{Y(s)}{R(s)} = \frac{G_{eq}(s)}{1 + G_{eq}(s)} \qquad (12\text{-}44)$$

Solving for $G_{eq}(s)$ from Eq. (12-44) yields

$$G_{eq}(s) = \frac{M(s)}{1 - M(s)} \qquad (12\text{-}45)$$

Utilizing the control ratio $M(s)$ from Eq. (12-27) in Eq. (12-45) yields

$$G_{eq}(s) = \frac{K_G(s^w + \cdots + c_1 s + c_0)}{s^n + (a_{n-1} + K_G k_n)s^{n-1} + \cdots + (a_1 + K_G k_2)s + (a_0 + K_G k_1) - K_G(s^w + \cdots + c_1 s + c_0)} \qquad (12\text{-}46)$$

This is a Type 0 transfer function and has the step error coefficient

$$K_p = \frac{K_G c_0}{a_0 + K_G k_1 - K_G c_0}$$

Satisfying the condition $k_1 = c_0 - a_0/K_G$, for $e(t)_{ss} = 0$ with a step input, Eq. (12-46) reduces to

$$G_{eq}(s) = \frac{K_G(s^w + \cdots + c_1 s + c_0)}{s^n + (a_{n-1} + K_G k_n)s^{n-1} + \cdots + (a_1 + K_G k_2)s - K_G(s^w + \cdots + c_1 s)} \qquad (12\text{-}47)$$

Since this is a Type 1 transfer function, the step error coefficient is $K_p = \infty$, and the ramp error coefficient is $K_v = K_G c_0/(a_1 + K_G k_2 - K_G c_1)$.

Satisfying the conditions for $e(t)_{ss} = 0$ for both step and ramp inputs, as given by Eqs. (12-31) and (12-36), Eq. (12-47) reduces to

$$G_{eq}(s) = \frac{K_G(s^w + \cdots + c_1 s + c_0)}{s^n + (a_{n-1} + K_G k_n)s^{n-1} + \cdots + (a_2 + K_G k_3)s^2 - K_G(s^w + \cdots + c_2 s^2)} \qquad (12\text{-}48)$$

Since this is a Type 2 transfer function, there is zero steady-state error with a ramp input. It follows a parabolic input with a steady-state error. The error coefficients are $K_p = \infty$, $K_v = \infty$, and $K_a = K_G c_0/(a_2 + K_G k_3 - K_G c_2)$.

For the case $w \geq 2$, when the conditions for $e(t)_{ss} = 0$ for all three standard inputs are satisfied [Eqs. (12-31), (12-36), and (12-41)], Eq. (12-48) reduces to

$$G_{eq}(s) = \frac{K_G(s^w + \cdots + c_1 s + c_0)}{s^n + (a_{n-1} + K_G k_n)s^{n-1} + \cdots + (a_3 + K_G k_4)s^3 - K_G(s^w + \cdots + c_3 s^3)} \qquad (12\text{-}49)$$

This is a Type 3 transfer function which has zero steady-state error with a parabolic input. The error coefficients are $K_p = K_v = K_a = \infty$. The results of the previous analysis are contained in Table 12-1. Table 12-2 summarizes the zero steady-state error requirements for a stable state-variable feedback control system. As Tables 12-1 and 12-2 indicate, the specifications on the type of input that the system must follow determine the number of zeros that must be included in $Y(s)/R(s)$. The number of zeros determines the ability of the system to follow a particular input with no steady-state error and to achieve a stable system.

With state-variable feedback any desired system type can be achieved for $G_{eq}(s)$, regardless of the type represented by $G(s)$. In order to achieve a Type m system the requirement, for $m > 0$, is that

$$a_{i-1} + K_G k_i - K_G c_{i-1} = 0 \qquad \text{for all } i = 1, 2, 3, \ldots, m$$

Thus, a system in which the plant is not Type m can be effectively made a Type m system by use of state feedback. This is achieved without adding pure integrators in cascade with the plant. *In order to achieve a Type m system it is seen from Table* 12-2

Table 12-1   STATE-VARIABLE FEEDBACK SYSTEM: STEADY-STATE ERROR COEFFICIENTS†

State-variable feedback system	Number of zeros required	K_p	K_v	K_a
All-pole plant	$w = 0$	∞	$\dfrac{K_G}{a_1 + K_G k_2}$	0
Pole-zero plant	$w = 1$	∞	∞	$\dfrac{K_G c_0}{a_2 + K_G k_3}$
	$w \geq 2$			∞

† For phase-variable representation.

that the control ratio must have a minimum of $m - 1$ zeros. This is an important characteristic of state-variable feedback systems.

The requirement for $e(t)_{ss} = 0$ for a ramp input can also be specified in terms of the locations of the zeros of the plant with respect to the desired locations of the poles of the control ratio.[1] Expanding

$$\frac{E(s)}{R(s)} = \frac{1}{1 + G_{eq}(s)} \qquad (12\text{-}50)$$

in a Maclaurin series in s yields

$$F(s) = \frac{E(s)}{R(s)} = C_0 + C_1 s + C_2 s^2 + \cdots + C_i s^i + \cdots \qquad (12\text{-}51)$$

where the coefficients $C_0, C_1, C_2, \ldots$ are called the *generalized error coefficients*[4] and are given by

$$C_i = \frac{1}{i!} \left[\frac{d^i F(s)}{ds^i} \right]_{s=0} \qquad (12\text{-}52)$$

From this equation, C_0 and C_1 are

$$C_0 = \frac{1}{1 + K_p} \qquad (12\text{-}53)$$

$$C_1 = \frac{1}{K_v} \qquad (12\text{-}54)$$

Note that C_0 and C_1 are expressed in terms of the system error coefficients defined in Chap. 6. From Eq. (12-47), in which the requirement of Eq. (12-31) is satisfied, $K_p = \infty$. Solving for $E(s)$ from Eq. (12-51) and substituting the result into $Y(s) = R(s) - E(s)$ yields

$$\frac{Y(s)}{R(s)} = 1 - C_0 - C_1 s - C_2 s^2 - \cdots \qquad (12\text{-}55)$$

Table 12-2 STABLE STATE-VARIABLE FEEDBACK SYSTEM: REQUIREMENTS FOR ZERO STEADY-STATE ERROR†

System	Number of zeros required	Input			$G_{eq}(s)$ type
		Step	Ramp	Parabola	
All-pole plant	$w = 0$	$k_1 = 1 - \dfrac{a_0}{K_G}$	‡	§	1
Pole-zero plant	$w = 1$	$k_1 = c_0 - \dfrac{a_0}{K_G}$	$k_2 = c_1 - \dfrac{a_1}{K_G}$	‡	2
	$w \geq 2$			$k_3 = c_2 - \dfrac{a_2}{K_G}$	3

† For phase-variable representation.
‡ A steady-state error exists.
§ Cannot follow.

The relation between K_v and the poles and zeros of $Y(s)/R(s)$ is readily determined if the control ratio is written in factored form:

$$\frac{Y(s)}{R(s)} = \frac{K_G(s - z_1)(s - z_2)\cdots(s - z_w)}{(s - p_1)(s - p_2)\cdots(s - p_n)} \qquad (12\text{-}56)$$

The derivative of $Y(s)/R(s)$ in Eq. (12-55) with respect to s, evaluated at $s = 0$, yields $-C_1$ which is equal to $-1/K_v$.

When it is noted that $[C(s)/R(s)]_{s=0} = 1$ for a Type 1 or higher system with unity feedback, the following equation can be written:

$$\frac{1}{K_v} = -\frac{\left[\dfrac{d}{ds}\left(\dfrac{Y}{R}\right)\right]_{s=0}}{\left[\dfrac{Y}{R}\right]_{s=0}} = -\left\{\frac{d}{ds}\ln\left[\frac{Y(s)}{R(s)}\right]\right\}_{s=0} \qquad (12\text{-}57)$$

Substitution of Eq. (12-56) into Eq. (12-57) yields

$$\frac{1}{K_v} = -\left(\frac{1}{s - z_1} + \cdots + \frac{1}{s - z_w} - \frac{1}{s - p_1} - \cdots - \frac{1}{s - p_n}\right)_{s=0}$$

$$= \sum_{j=1}^{n}\frac{1}{p_j} - \sum_{j=1}^{w}\frac{1}{z_j} \qquad (12\text{-}58)$$

Thus, the condition for $e(t)_{ss} = 0$ for a ramp input, which occurs for $K_v = \infty$, that is, $1/K_v = 0$, is satisfied when

$$\sum_{j=1}^{n}\frac{1}{p_j} = \sum_{j=1}^{w}\frac{1}{z_j} \qquad (12\text{-}59)$$

Therefore, in synthesizing the desired control ratio, its poles can be located not only to yield the desired transient response characteristics but also to yield the value $K_v = \infty$. In addition, it must contain a zero that is not canceled by a pole.

The values of k_1, k_2, and k_3 determined for $e(t)_{ss} = 0$ for the respective input from Eqs. (12-31), (12-36), and (12-41) are based upon the phase-variable representation of the system. These values may not be compatible with the desired roots of the characteristic equation. In that case it is necessary to leave at least one root unspecified so that these feedback gains may be used. If the resulting location of the unspecified root(s) is not satisfactory, other root locations may be tried in order to achieve a satisfactory performance. Once a system is designed which satisfies both the steady-state and the transient time responses, the feedback coefficients for the physical system can be calculated. This is accomplished by equating the corresponding coefficients of the characteristic equations representing respectively the physical- and the phase-variable systems. It is also possible to achieve the desired performance by working with the system represented entirely by physical variables, as illustrated by the examples in the following sections.

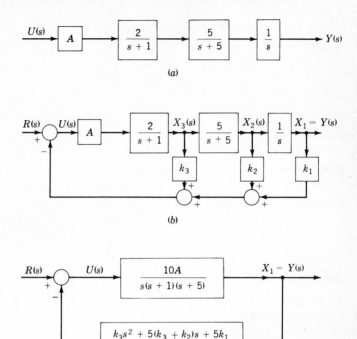

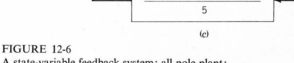

FIGURE 12-6
A state-variable feedback system; all-pole plant:
(a) plant; (b) state-variable feedback; (c) $H_{eq}(s)$ form.

12-7 STATE-VARIABLE FEEDBACK: ALL-POLE PLANT

The design procedure presented in Sec. 12-4 is now applied to an all-pole plant.

EXAMPLE   For the basic plant of Fig. 12-6a it is desired to use the state-variable feedback method in order to achieve an $M_p = 1.043$, $t_s = 5.65$ s, and zero steady-state error for a step input.

Steps 1 to 3   See Fig. 12-6b and c;

$$\frac{Y(s)}{R(s)} = \frac{10A}{s^3 + (6 + 2Ak_3)s^2 + [5 + 10A(k_3 + k_2)]s + 10Ak_1} \qquad (12\text{-}60)$$

Step 4   Inserting Eq. (12-60) into $y(t)_{ss} = \lim_{s \to 0} sY(s) = R_0$, for a step input, yields $k_1 = 1$.

Step 5 Assuming that the complex poles of Eq. (12-60) are truly dominant, from Eq. (3-61)

$$M_p = 1.043 = 1 + \exp\left(-\frac{\zeta\pi}{\sqrt{1 - \zeta^2}}\right)$$

or

$$\zeta \approx 0.7076$$

From $T_s = 4/|\sigma|$, $|\sigma| = 4/5.65 = 0.708$. For these values of ζ and $|\sigma|$, the desired closed-loop complex conjugate poles are $s_{1,2} = -0.708 \pm j0.704$. Since the denominator of $G(s)$ is of third order, the control ratio also has a third-order denominator, as shown in Eq. (12-60). To ensure the dominancy of the complex poles the third pole is arbitrarily taken as $s_3 = -100$. Therefore, the desired closed-loop transfer function is

$$\frac{Y(s)}{R(s)} = \frac{10A}{(s + 100)(s + 0.708 \pm j0.704)}$$

$$= \frac{10A}{s^3 + 101.4s^2 + 142.7s + 100} \tag{12-61}$$

Step 6 Equating corresponding terms in the denominators of Eqs. (12-60) and (12-61) and using $k_1 = 1$ yields

$$10Ak_1 = 100 \qquad A = 10$$
$$6 + 2Ak_3 = 101.4 \qquad k_3 = 4.77$$
$$5 + 100(k_3 + k_2) = 142.7 \qquad k_2 = -3.393$$

Inserting Eq. (12-61) into Eq. (12-45) to obtain the *G*-equivalent form (see Fig. 12-5) yields

$$G_{eq}(s) = \frac{100}{s(s^2 + 101.4s + 142.7)} \tag{12-62}$$

Thus, the ramp error coefficient is $K_1 = 0.701$.

An analysis of the root locus for the unity-feedback and the state-feedback systems, shown in Fig. 12-7a and b, respectively, reveals several points.

For large values of gain the unity-feedback system is unstable, with or without cascade compensation. For any variation in gain the corresponding change in the dominant roots changes the system response. Thus, the response is sensitive to changes of gain.

The state-variable feedback system is stable for all $A > 0$. Since the roots are very close to the zeros of $G(s)H_{eq}(S)$, any increase in gain from the value $A = 10$ produces negligible changes in the location of the roots. In other words, for high-forward-gain (hfg) operation, $A > 10$, the system's response is essentially unaffected by variations in A, provided the feedback coefficients remain fixed. The value of the hfg $K = 9.54A \geq K_{min}$, for which the state-variable feedback system's response is insensitive to gain variations, is obtained from the root-locus plot for $G(s)H_{eq}(s) = -1$. Thus, for the example represented by Fig. 12-7b, when $K = 9.54A \geq 95.4$, the change in the values of the dominant roots is small and there is little effect on the time response for a step input. In Chap. 14 another method is presented for determining

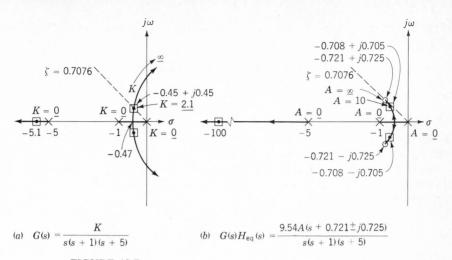

$$(a) \quad G(s) = \frac{K}{s(s+1)(s+5)} \qquad\qquad (b) \quad G(s)H_{eq}(s) = \frac{9.54A(s+0.721\pm j0.725)}{s(s+1)(s+5)}$$

FIGURE 12-7
Root-locus diagrams for (a) basic unity-feedback and (b) state-variable feedback control systems.

K_{min}. The insensitivity of the system response to the variation of the gain A is related to the system's sensitivity function, discussed in the next section. *This insensitivity is due to the fact that the two desired dominant roots are close to the two zeros of $H_{eq}(s)$ and the third root is nondominant. To achieve this insensitivity to gain variation there must be β dominant roots which are located close to β zeros of $H_{eq}(s)$. The remaining $n - \beta$ roots are nondominant.* Although in this example $\beta = n - 1 = 2$, in general $\beta \le n - 1$.

Both the conventional and state feedback systems can follow a step input with zero steady-state error. Also, they both follow a ramp input with a steady-state error. The state-variable feedback system cannot be designed to produce zero steady-state error for a ramp input because $G(s)$ does not contain a zero. In order to satisfy this condition a cascade compensator which adds a zero to $G(s)$ may be utilized in the state-variable feedback system.

In general, when selecting the poles and zeros of the $Y(s)/R(s)$ to satisfy the desired specifications, one pair of dominant complex-conjugate poles $s_{1,2}$ is often chosen. The other $n - 2$ poles are located far to the left of the dominant pair or close to zeros in order to ensure the dominancy of $s_{1,2}$. This approach may yield large values for K and the k_i's. It is also possible to have more than just a pair of dominant complex-conjugate poles to satisfy the desired specifications. For example, an additional pole of $Y(s)/R(s)$ can be located to the left of the dominant complex pair. When properly located, it can reduce the peak overshoot and may also reduce the settling time. Choosing $n - 1$ dominant roots yields lower values of K and k_i's. It is difficult with the techniques presented thus far to determine the required values of these $n - 1$ dominant roots. In Chap. 15 a method is presented that yields $n - 1$ dominant roots which may satisfy the desired specifications. *For state-variable feedback the value of A is always taken as positive.*

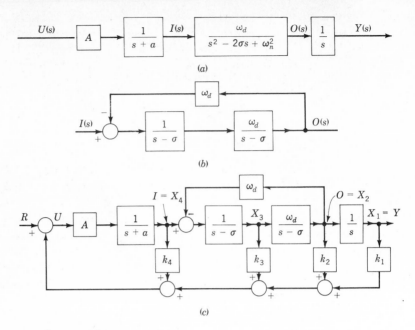

FIGURE 12-8
Complex-pole plant representation.

12-8 PLANT COMPLEX POLES

An example is used to show how to obtain the state-variable representation of a plant containing complex poles. Since the system shown in Fig. 12-8a is of the fourth order, four state variables must be identified. The transfer function $O(s)/I(s)$ involves two states. If the detailed block-diagram representation of $O(s)/I(s)$ contains two explicit states, it *should not* be combined into the quadratic form. By combining the blocks the identity of one of the states is lost. That situation is illustrated by the system of Fig. 12-3. As represented in Fig. 12-4a, the two states X_2 and X_3 are accessible, but in the combined mathematical form of Fig. 12-4c the physical variables are not identifiable. When one of the states is not initially identified, an equivalent model containing two states must be formed for the quadratic transfer function $O(s)/I(s)$ of Fig. 12-8a. This decomposition can be performed by the method of Sec. 5-10. The simulation diagram of Fig. 5-29 is shown in the block diagram of Fig. 12-8b. This decomposition is inserted into the complete control system utilizing state-variable feedback, as shown in Fig. 12-8c. If the state X_3 is inaccessible, then once the values of k_i are determined, the physical realization of the state feedback can be achieved by the block-diagram manipulation discussed in previous sections. An alternate approach is to represent the transfer function $O(s)/I(s)$ in state-variable form, as shown in Fig. 12-9. This representation is particularly attractive when X_3 is recognizable as a physical variable. If it is not, then the pick-off for k_3 can be moved to the right by block-diagram manipulation.

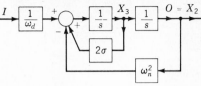

FIGURE 12-9
State-variable representation of
$O(s)/I(s)$ in Fig. 12-8a.

12-9 SENSITIVITY ANALYSIS[5,6]

In order to illustrate an advantage of state-variable feedback over the conventional method of achieving the desired system performance, a system sensitivity analysis is made for both approaches. The basic plant $G_x(s)$ of Fig. 12-6a is utilized for this comparison. The conventional control system is shown in Fig. 12-10, and the state-variable feedback system is shown in Fig. 12-6b. The analysis is based upon determining the value of ω_0 at $|\mathbf{M}(j\omega_o)| = M_0 = 0.707$ for the nominal system values of the conventional and state-variable feedback control systems. The system sensitivity function of Chap. 6, repeated here, is determined for each system and evaluated for the frequency range $0 \le \omega \le \omega_o$:

$$S_\delta^M(s) = \frac{\delta}{M} \frac{dM}{d\delta} \qquad (12\text{-}63)$$

The plots of $|\mathbf{S}_{p_1}{}^M(j\omega)|$, $|\mathbf{S}_{p_2}{}^M(j\omega)|$, and $|\mathbf{S}_A{}^M(j\omega)|$ vs. ω, shown in Fig. 12-11, are drawn for the nominal values of $p_1 = -1$, $p_2 = -5$, and $A = 10$ for the state-variable feedback system. For the conventional system the value of loop sensitivity equal to 2.1 corresponds to $\zeta = 0.7086$ on the root locus shown in Fig. 12-7a. The same damping ratio is used in the state-variable feedback system developed in the example in Sec. 12-7. Table 12-3 summarizes the values obtained.

For each system, at $\omega = \omega_o$, the value of $|\mathbf{M}(j\omega_o)|$ is determined for a 100 percent variation of the plant poles $p_1 = -1$, $p_2 = -5$, and of the forward gain A, respectively. The term $\%|\Delta M|$ is defined as

$$\%|\Delta M| = \frac{|\mathbf{M}(j\omega_o)| - |\mathbf{M}(j\omega_o)|_N}{|\mathbf{M}(j\omega_o)|_N} 100 \qquad (12\text{-}64)$$

where $|\mathbf{M}(j\omega_o)|_N$ is the value with nominal values of the plant parameters and $|(\mathbf{M}j\omega_o)|$ is the value with one parameter changed to the extreme value of its possible variation. The time-response data for systems 1 to 8 are given in Table 12-4 for a unit step input. Note that the response for the state-variable feedback system (5 to 8, except 7a) is essentially unaffected by parameter variations.

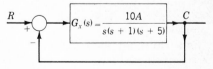

FIGURE 12-10
Conventional control system.

Table 12-3 SYSTEM SENSITIVITY ANALYSIS

Control system	$\lvert M(j\omega_o)\rvert$	$\%\lvert\Delta M(j\omega_o)\rvert$	$\lvert S_{p_1}{}^M(j\omega_o)\rvert$	$\lvert S_{p_2}{}^M(j\omega_o)\rvert$	$\lvert S_A{}^M(j\omega_o)\rvert$
Conventional system, $G(s)$					
1. Nominal ($A = 0.21$, $p_1 = -1, p_2 = -5$): $\dfrac{2.1}{s(s+1)(s+5)}$	0.707 ($\omega_o = 0.642$)	—	1.09	1.28	1.29
2. $p_1 = -2$: $\dfrac{2.1}{s(s+2)(s+5)}$	0.338	52.4	—	—	—
3. $p_2 = -10$: $\dfrac{2.1}{s(s+1)(s+10)}$	0.658	6.9	—	—	—
4. $A = 0.42$: $\dfrac{4.2}{s(s+1)(s+5)}$	1.230	74.0	—	—	—
State-variable feedback system, $G(s)H_{eq}(s)$ [M is given in Eq. (12-61)]					
5. Nominal: $\dfrac{95.4(s^2 + 1.443s + 1.0481)}{s(s+1)(s+5)}$	0.707 ($\omega_o = 1.0$)	—	0.036	0.05 for X_3 inaccessible —— 3.40 for X_3 accessible	0.051
6. $p_1 = -2$: $\dfrac{95.4(s^2 + 1.443s + 1.0481)}{s(s+2)(s+5)}$	0.683	3.39	—	—	—
7a. $p_2 = -10$, X_3 accessible: $\dfrac{95.4(s^2 + 6.443s + 1.0481)}{s(s+1)(s+10)}$	0.160	77.4	—	—	—
7b. $p_2 = -10$, X_3 inaccessible: $\dfrac{95.4(s^2 + 1.443s + 1.0481)}{s(s+1)(s+10)}$	0.682	3.54	—	—	—
8. $A = 20$: $\dfrac{190.8(s^2 + 1.443s + 1.0481)}{s(s+1)(s+5)}$	0.717	1.41	—	—	—

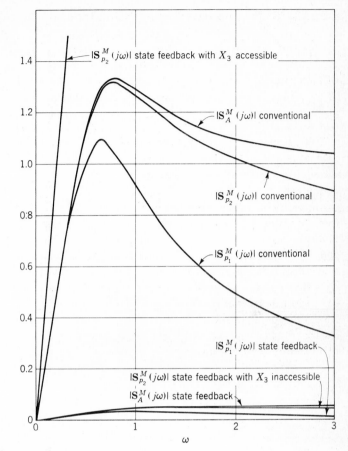

FIGURE 12-11
Sensitivity due to pole and gain variation.

As stated in Chap. 6, the $S_\delta{}^M$ function represents a measure of the change to be expected in the system performance for a change in a system parameter. This measure is borne out by comparing Tables 12-3 and 12-4. That is, in the frequency-domain analysis a value $|S_A{}^M(j\omega)| = 1.29$ at $\omega = \omega_o$ is indicated for the conventional system compared with 0.051 for the state-variable feedback system. Also, the percent change $\%|\Delta M(j\omega)|$ for a doubling of the forward gain is 74 percent for the conventional system and 1.41 percent for the state-variable feedback system. These are consistent with the changes in the peak overshoot in the time domain of $\%\Delta M_p(t) = 13.45$ and 0.192 percent, respectively. Thus, the magnitudes of $\%\Delta M(j\omega)$ and $\%\Delta M_p(t)$ are consistent with the relative magnitudes of $\%|S_A{}^M(j\omega_o)|$. Similar results are obtained for variations of the pole p_1.

An interesting result occurs when the pole $p_2 = -5$ is subject to variations. If the state X_3 (see Fig. 12-6) is observable and is fed back directly through k_3, the sensitivity $|S_{p_2}^M(j\omega)|$ is much larger than for the conventional system, having a value of 3.40 at $\omega_o = 1.0$. However, if the equivalent feedback is obtained from X_1 through the transfer function $H_3(s) = s(s + 5)k_3/5$, the sensitivity is considerably reduced to $|S_{p_2}^M(j1)| = 0.050$, compared with 1.28 for the conventional system. The components of the feedback unit may be selected so that the transfer function $H_3(s)$ is invariant. The time-response characteristic for state-variable feedback through $H_3(s)$ is essentially unchanged when p_2 doubles in value (see Table 12-4) compared with the conventional system.

In analyzing Fig. 12-11, the values of the sensitivities must be considered over the entire passband ($0 \le \omega \le \omega_o$). In this passband the state-variable feedback system has a much lower value of sensitivity than the conventional system. The state-variable feedback system performance is essentially unaffected by any variation in A, p_1, or p_2 (with restrictions) when feedback coefficients remain fixed, provided the forward gain satisfies the condition $K \ge K_{min}$ (see Probs. 12-4, 12-12, and 12-13). *The low-sensitivity state-variable feedback systems considered in this section satisfy the following conditions:* the system's characteristic equation must have (1) $\beta(<n)$ dominant roots which are located close to zeros of $H_{eq}(s)$ and (2) $n - \beta$ nondominant roots.

The analysis of this section reveals that for the state-variable feedback system any parameter variation between the control U and X_n state variable has minimal effect on system performance. In order to minimize the effect on system performance for a parameter variation between the X_n and X_1 states, the feedback signals should be moved to the right of the block containing the variable parameter, even if all the states are accessible. This yields a more invariant response to parameter variations than for the conventional system.

Table 12-4 TIME RESPONSE DATA

Control system	$M_p(t)$	t_p, s	t_s, s	$\%\Delta M_p(t) = \dfrac{\|M_p - M_{p_0}\|}{M_{p_0}} 100$
1	1.048	7.2	9.8	—
2	1.000	—	16.4	4.8
3	1.000	—	15.0	4.8
4	1.189	4.1	10.15	13.45
5	1.044	4.45	6	—
6	1.035	4.5	6^-	0.86
7a	1.000	—	24	4.4
7b	1.046	4.6	6.5	0.181
8	1.046	4.4	5.6^+	0.192

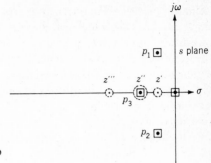

FIGURE 12-12
Pole-zero location of $Y(s)$ for a step input.

12-10 STATE-VARIABLE FEEDBACK: POLE-ZERO PLANT

Section 12-7 dealt with a system having an all-pole plant whose control ratio has no zeros. Locating the roots of the characteristic equation for such a system to yield the desired system response is relatively easy. (From the discussion in Secs. 12-2 and 12-3 the locations of the dominant and nondominant roots can be specified in order to achieve the desired time response.) This section deals with a system having a plant containing both poles and zeros. This results in the control ratio

$$\frac{Y(s)}{R(s)} = \frac{K_G(s - z_1)(s - z_2)\cdots(s - z_h)\cdots(s - z_w)}{(s - p_1)(s - p_2)\cdots(s - p_c)\cdots(s - p_n)} \qquad (12\text{-}65)$$

which has the same zeros as the plant. When open-loop zeros are present, the synthesis of the overall characteristic equation becomes more complex. Figure 12-12 represents the pole-zero diagram of $Y(s)$ for a third-order system ($n = 3$) with a step input. Consider three possible locations of the real zero when $Y(s)/R(s)$ has only one zero ($w = 1$): to the right of, canceling, and to the left of the real pole p_3. When the zero cancels the pole p_3, the value of M_p is determined only by the complex-conjugate poles, p_1 and p_2; when the zero is to the right of p_3, the value of M_p is larger; and when it is to the left, the value of M_p is smaller [see Eq. (4-68) and Sec. 10-8]. The amount of this increase or decrease is determined by how close the zero is to the pole p_3. Also, as noted in Table 12-2, having a zero in the control ratio permits the achievement of $e(t)_{ss} = 0$ for a ramp input. In synthesizing the desired control ratio, Eq. (12-59) can also be utilized to assist in achieving this condition while simultaneously satisfying the other desired figures of merit.

In synthesizing the desired control ratio, as expressed by Eq. (12-65), a system designer has a choice of one of the following approaches.

Method 1 Cancel some or all the zeros of the control ratio *by poles of the control ratio*. The number of zeros to be canceled α must be equal to or less than the excess of the number of poles n minus the number of desired dominant poles β. If necessary, cascade compensators of the form $G_c(s) = 1/(s - P_i)$ can be inserted into the system in order to increase the number of poles of $Y(s)/R(s)$ so that cancellation of zeros can be accomplished. The number of compensator poles required γ must increase the order of the system sufficiently to ensure that $\alpha \leq n + \gamma - \beta$. Theoretically it is possible to insert a compensator of this form just preceding the block that contains the unwanted zero to be canceled by the pole of this compensator. However, in many systems it is not physically possible to locate the compensator in this manner. Also, if P_i is set equal to the value of the unwanted zero z_h of the plant, this results in an unobservable state (see Chap. 6). In Chap. 14, the condition is specified that in order to achieve an optimal response the system must be controllable and observable. Thus, to achieve an optimal response and to cancel the unwanted zero, choose P_i such that $P_i \neq z_h$. It is still possible to achieve the desired $Y(s)/R(s)$ by using a cascade compensator with $P_i = z_h$; however, an optimal response, as defined in Secs. 14-7 to 14-12, is not achieved.

Method 2   Relocate some or all of the zeros by using cascade compensators of the form $G_c(s) = (s - z_i)/(s - P_i)$. The number γ of these compensators to be inserted equals the number of zeros α to be relocated. The values of z_i represent the desired location of the zeros. The values of P_i are selected to meet the condition that α *poles of the control ratio* cancel the α unwanted zeros.

Method 3   A combination of the above two approaches can be used, i.e., canceling some zeros by closed-loop poles and replacing some of them by the zeros of the cascade compensators $G_c(s)$.

Method 4   If the locations of the zeros are satisfactory, the design procedure is the same as for the all-pole case of the previous section. That is, the coefficients of the desired characteristic equation are equated to the corresponding coefficients of the numerator polynomial of $1 + G(s)H_{eq}(s)$.
    It may be possible to achieve low sensitivity in applying any of these approaches. To achieve this condition the following must be observed:

  1    $Y(s)/R(s)$ must have β dominant poles and *at least one nondominant pole*.
  2   The β dominant poles must be close to β zeros of $H_{eq}(s)$.

The latter can be achieved by a judicious choice of the compensator pole(s) and/or the nondominant poles.
    The cascade compensator utilized in approach 2 can be constructed in either of the forms shown in Fig. 12-13a or b. The input signal is shown as AU since, in general, this is the most accessible place to insert the compensator. The differential equations for x_{i+n} for the two forms are

Conventional: $$\dot{x}_{i+n} = P_i x_{i+n} - z_i Au + A\dot{u} \qquad (12\text{-}66)$$

Feedforward: $$\dot{x}'_{i+n} = P_i x'_{i+n} + (P_i - z_i)Au \qquad (12\text{-}67)$$

The presence of $\dot{u}$ in Eq. (12-66) does not permit describing the system with this compensator by the state equation $\dot{x} = Ax + bu$, and x_{i+n} in Fig. 12-13a is not a state variable. This situation can also occur in a system before a compensator is inserted. The design of such a system can be treated like that of a system utilizing the com-

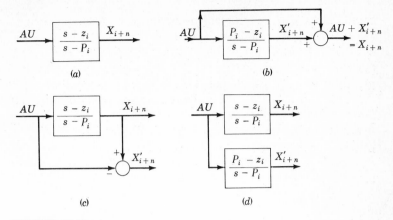

FIGURE 12-13
Cascade compensators,

$$G_c(s) = \frac{s - z_i}{s - P_i}$$

(a) conventional; (b) feedforward form; (c) generating the state variable X'_{i+n};
(d) mathematical model for (c).

pensator of Fig. 12-13a. The consequence of the $\dot{u}$ term in Eq. (12-66) is that the zero of $G_c(s)$ does not appear as a pole of $H_{eq}(s)$. Thus the root-locus characteristics discussed in Sec. 12-4 must be modified accordingly. There is no difficulty in expressing the system in state-variable form when the function of Fig. 12-13a is not the first frequency-sensitive term in the forward path. When the compensator of Fig. 12-13a is inserted at any place to the right of the G_1 block in the forward path shown in Fig. 12-14, the system can be described by a state equation. Thus, the root-locus characteristics discussed in Sec. 12-4 also apply for this case.

A system utilizing the feedforward compensator of Fig. 12-13b can be described by a state equation since $\dot{u}$ does not appear in Eq. (12-67). Thus, the poles of $H_{eq}(s)$ continue to be the zeros of $G(s)$. Therefore the root-locus characteristics discussed in Sec. 12-4 also apply for this case. This is a suitable method of constructing the compensator $G_c(s) = (s - z_i)/(s - P_i)$ since it produces the accessible state variable X'_{i+n}. The implementation of $G_c(s)$ consists of the lag term $K_c/(s - P_i)$, where $K_c = P_i - z_i$, with the unity-gain feedforward path.

An additional possibility is to generate a new state variable X'_{i+n} from the physical arrangement shown in Fig. 12-13c. This is equivalent to the form shown in

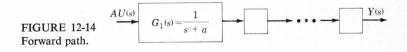

FIGURE 12-14
Forward path.

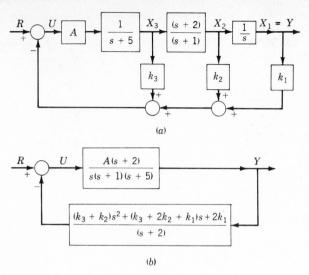

(a)

(b)

FIGURE 12-15
Closed-loop zero cancellation by state-variable feedback: (a) detailed block diagram; (b) H-equivalent reduction (where $K_G = A$).

Fig. 12-13d and is mathematically equivalent to Fig. 12-13b. Thus, X'_{i+n} is a valid state variable and may be used with a feedback coefficient k_{i+n}.

These approaches are now illustrated by examples having one zero in the plant transfer function. The procedure is essentially the same for systems having more than one zero. A typical design technique is illustrated in each example and can be modified as desired.

EXAMPLE 1   The control ratio for the system of Fig. 12-15 is

$$\frac{Y(s)}{R(s)} = \frac{A(s + 2)}{s^3 + [6 + (k_3 + k_2)A]s^2 + [5 + (k_3 + 2k_2 + k_1)A]s + 2k_1 A} \qquad (12\text{-}68)$$

The desired control ratio for this system is specified as

$$\left.\frac{Y(s)}{R(s)}\right|_{desired} = \frac{2}{s^2 + 2s + 2} \qquad (12\text{-}69)$$

If Eq. (12-68) is to reduce to Eq. (12-69), one pole of Eq. (12-68) must be at $s = -2$, in order to cancel the zero, and $A = 2$. Equation (12-69) is modified to reflect this requirement:

$$\left.\frac{Y(s)}{R(s)}\right|_{desired} = \frac{2(s + 2)}{(s^2 + 2s + 2)(s + 2)} = \frac{2(s + 2)}{s^3 + 4s^2 + 6s + 4} \qquad (12\text{-}70)$$

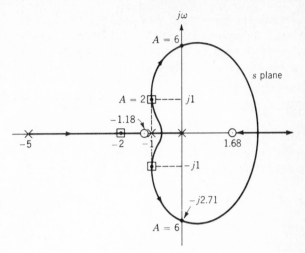

FIGURE 12-16
Root locus for Eq. (12-71).

Equating the corresponding coefficients of the denominators of Eqs. (12-68) and (12-70) yields $k_1 = 1$, $k_2 = \frac{1}{2}$, and $k_3 = -\frac{3}{2}$. Thus

$$G(s)H_{eq}(s) = \frac{-A(s^2 - s/2 - 2)}{s(s + 1)(s + 5)} = -1 \qquad (12\text{-}71)$$

Canceling the minus sign shows that the $0°$ root locus is required. The zeros are $s_1 = -1.18$ and $s_2 = 1.68$, and the root locus is shown in Fig. 12-16. Note that since Eq. (12-69) has no other poles in addition to the dominant poles, the conditions of Secs. 12-4 and 12-9 are not satisfied. As a result two branches of the root locus enter the right half of the s plane, yielding a system that can become unstable for high gain. Also, since the roots are not near zeros, the system is sensitive to parameter variations, as can be observed from the root locus in Fig. 12-16.

The response of this system to a unit step input has $M_p = 1.043$, $t_p = 3.1$ s, and $t_s = 4.2$ s. The system is highly sensitive to gain variation, as seen from the root locus of Fig. 12-16 and the sensitivity $|S_A{}^M(j\omega_o)| = 1.84$ at $\omega_o = 1.41$. The ramp error coefficient obtained from $G_{eq}(s)$ is $K_1 = 1.0$.

Example 1 illustrates that a very small value of $S_A{}^M(j\omega)$ and a system which is stable for all positive values of gain cannot be achieved when $Y(s)/R(s)$ does not have a nondominant pole. The synthesized function must have β dominant poles and $n + \gamma - \beta$ nondominant poles in order to satisfy these conditions. The next example illustrates this design concept.

EXAMPLE 2 For the third-order system of Fig. 12-15 it is desired to cancel the control ratio zero at $s = -2$ and improve the value of $|S_{K_G}{}^M(j\omega)|$ while effectively achieving the second-order response of Eq. (12-69). The time response is satisfied if $Y(s)/R(s)$ has three dominant poles $s_{1,2} = -1 \pm j1$ and $s_3 = -2$. In order to achieve insensitivity to gain changes, the transfer function $H_{eq}(s)$ must have three zeros. In accordance with Eq. (12-26a), the number of poles of $G(s)$ must be $n = 4$. Therefore, one pole must be added to $G(s)$ by using the cascade compensator $G_c(s) = 1/(s + a)$. Then $Y(s)/R(s)$ will have one nondominant pole p. The desired control ratio is

$$\left.\frac{Y(s)}{R(s)}\right|_{desired} = \frac{K_G(s + 2)}{(s^2 + 2s + 2)(s + 2)(s - p)} \qquad (12\text{-}72)$$

The selection of the values of K_G and p are interdependent since the condition of $e(t)_{ss} = 0$ for a step input is required. Thus

$$y(t)_{ss} = \lim_{s \to 0} sY(s) = \frac{-K_G}{2p} R_0 = R_0 \qquad (12\text{-}73)$$

or

$$K_G = -2p$$

As mentioned previously, the larger the value of the loop sensitivity for the root locus of $G(s)H_{eq}(s) = -1$, the closer the roots will be to the zeros of $H_{eq}(s)$. It is this "closeness" that provides a low value of $|S_{K_G}{}^M(j\omega)|$. Therefore, the value of K_G should be chosen as large as possible consistent with the capabilities of the system components. This large value of K_G may have to be limited to a value that does not accentuate the system noise to an unacceptable level. Therefore, for this example, assume that a value of $K_G = 100$ is satisfactory. This results in a value of $p = -50$. Inserting this value into Eq. (12-72) yields

$$\left.\frac{Y(s)}{R(s)}\right|_{desired} = \frac{100(s + 2)}{s^4 + 54s^3 + 206s^2 + 304s + 200} \qquad (12\text{-}74)$$

The detailed and H-equivalent block diagrams for the modified control system are shown in Fig. 12-17a and b, respectively,

In order to maintain a controllable and observable system the value of a can be chosen to be any value *other than* the value of the zero term of $G(s)$. Thus, assume $a = 1$. With this value of a and with $K_G = 100$, the control ratio of the system of Fig. 12-17b is

$$\frac{Y(s)}{R(s)} = \frac{100(s + 2)}{s^4 + (7 + 100k_4)s^3 + [11 + 100(6k_4 + k_3 + k_2)]s^2 + [5 + 100(5k_4 + k_3 + 2k_2 + k_1)]s + 200k_1} \qquad (12\text{-}75)$$

For Eq. (12-75) to yield $e(t)_{ss} = 0$ with a step input requires $k_1 = 1$. Equating the corresponding coefficients of the denominators of Eqs. (12-74) and (12-75) and solving for the feedback coefficients yields $k_2 = 0.51$, $k_3 = -1.38$, and $k_4 = 0.47$. Thus, from Fig. 12-17b:

$$G(s)H_{eq}(s) = \frac{0.47K_G(s^3 + 4.149s^2 + 6.362s + 4.255)}{s^4 + 7s^3 + 11s^2 + 5s} \qquad (12\text{-}76)$$

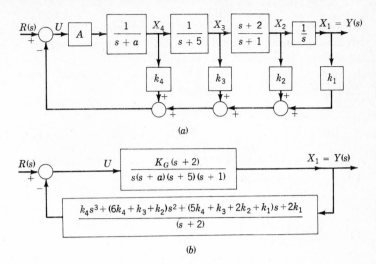

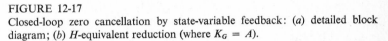

(b)

FIGURE 12-17
Closed-loop zero cancellation by state-variable feedback: (a) detailed block diagram; (b) H-equivalent reduction (where $K_G = A$).

The root locus shown in Fig. 12-18 reveals that the three desired dominant roots are close to the three zeros of $H_{eq}(s)$, the fourth root is nondominant, and the root locus lies entirely in the left-half s plane. Thus, the full benefits of state-variable feedback have been achieved. In other words, a completely stable system with low sensitivity to parameter variation has been realized. The figures of merit of this system are $M_p = 1.043$, $t_p = 3.2$ s, $t_s = 4.2$ s, $K_1 = 0.98$ s^{-1}, and $|S_{K_G}{}^M(j\omega_o)| = 0.064$.

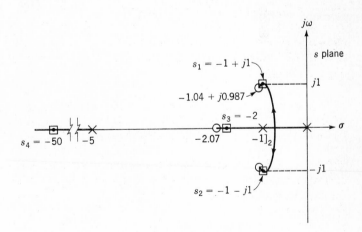

FIGURE 12-18
The root locus for Eq. (12-76).

EXAMPLE 3   A further improvement in the time response can be obtained by adding a pole and zero to the overall system function of Example 2. The control ratio to be achieved is

$$\frac{Y(s)}{R(s)} = \frac{K_G(s + 1.4)(s + 2)}{(s + 1)(s^2 + 2s + 2)(s + 2)(s - p)}$$

$$= \frac{K_G(s + 1.4)(s + 2)}{s^5 + (5 - p)s^4 + (10 - 5p)s^3 + (10 - 10p)s^2 \atop + (4 - 10p)s - 4p} \qquad (12\text{-}77)$$

Analysis of Eq. (12-77) shows the following features:

1 Since $Y(s)/R(s)$ has four dominant poles, the nondominant pole p yields the desired property of insensitivity to gain variation.
2 The pole-zero combination $(s + 1.4)/(s + 1)$ is added to reduce the peak overshoot in the transient response.
3 The factor $s + 2$ normally appears in the numerator because it is present in $G(s)$. Since it is not desired in $Y(s)/R(s)$, it is canceled by producing the same factor in the denominator.
4 In order to achieve zero steady-state error with a step input $r(t) = u_{-1}(t)$, the output obtained by applying the final-value theorem to $Y(s)$ is $y(t)_{ss} = K_G \times 1.4 \times 2/(-4p) = 1$. Therefore, $p = -0.7K_G$. Assuming the large value $K_G = 100$ to be satisfactory results in $p = -70$, which is nondominant, as required.

The required form of the modified system is shown in Fig. 12-19. The cascade compensator $1/(s + 1)$ is added to the system in order to achieve the degree of the denominator of Eq. (12-77). The cascade compensator $(s + 1.4)/(s + 3)$ is included to produce the desired numerator factor $s + 1.4$. The denominator is arbitrarily chosen as $s + 3$. This increases the degree of the denominator in order to permit canceling the factor $s + 2$ in $Y(s)/R(s)$.

The forward transfer function obtained from Fig. 12-19 is

$$G(s) = \frac{100(s + 1.4)(s + 2)}{s(s + 1)^2(s + 3)(s + 5)} = \frac{100(s^2 + 3.4s + 2.8)}{s^5 + 10s^4 + 32s^2 + 38s^2 + 15s} \qquad (12\text{-}78)$$

The value of $H_{eq}(s)$ is evaluated from Fig. 12-19 and is

$$H_{eq}(s) = \frac{\begin{array}{l}(k_5 + k_4)s^4 + (9k_5 + 7.4k_4 + k_3 + k_2)s^3 \\ + (23k_5 + 13.4k_4 + 2.4k_3 + 3.4k_2 + k_1)s^2 \\ + (15k_5 + 7k_4 + 1.4k_3 + 2.8k_2 + 3.4k_1)s + 2.8k_1\end{array}}{(s + 1.4)(s + 2)} \qquad (12\text{-}79)$$

The overall desired system function is obtained from Eq. (12-77), with the values $K_G = 100$ and $p = -70$, as follows:

$$\frac{Y(s)}{R(s)} = \frac{100(s + 1.4)(s + 2)}{s^5 + 75s^4 + 360s^3 + 710s^2 + 704s + 280} \qquad (12\text{-}80)$$

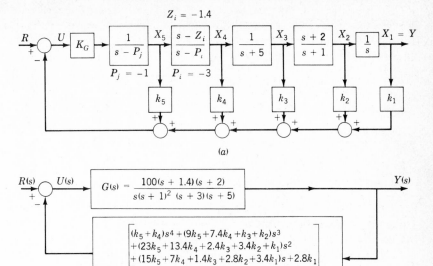

(a)

(b)

FIGURE 12-19
State-variable feedback diagram for Example 3.

The system function obtained by using $G(s)$ and $H_{eq}(s)$ is

$$\frac{Y(s)}{R(s)} = \frac{100(s + 1.4)(s + 2)}{\begin{aligned}s^5 &+ [10 + 100(k_5 + k_4)]s^4 \\ &+ [32 + 100(9k_5 + 7.4k_4 + k_3 + k_2)]s^3 \\ &+ [38 + 100(23k_5 + 13.4k_4 + 2.4k_3 + 3.4k_2 + k_1)]s^2 \\ &+ [15 + 100(15k_5 + 7k_4 + 1.4k_3 + 2.8k_2 + 3.4k_1)]s \\ &\qquad\qquad\qquad + 100(2.8k_1)\end{aligned}} \quad (12\text{-}81)$$

Equating the coefficients in the denominators of Eqs. (12-80) and (12-81) yields $280k_1 = 280$ or $k_1 = 1$.

$$\begin{aligned}
1{,}500k_5 + 700k_4 + 140k_3 + 280k_2 &= 349 \\
2{,}300k_5 + 1{,}340k_4 + 240k_3 + 340k_2 &= 572 \\
900k_5 + 740k_4 + 100k_3 + 100k_2 &= 328 \\
100k_5 + 100k_4 &= 65
\end{aligned} \qquad (12\text{-}82)$$

Solving these four simultaneous equations gives $k_2 = 1.0$, $k_3 = -2.44167$, $k_4 = 0.70528$, and $k_5 = -0.05528$.

In order to demonstrate features of this system the root locus is drawn in Fig. 12-20 for

$$\begin{aligned}
G(s)H_{eq}(s) &= \frac{100(0.65s^4 + 3.28s^3 + 6.72s^2 + 6.89s + 2.8)}{s(s + 1)^2(s + 3)(s + 5)} \\
&= \frac{65(s + 1.04532 \pm j1.05386)(s + 0.99972)(s + 1.95565)}{s(s + 1)^2(s + 3)(s + 5)}
\end{aligned} \qquad (12\text{-}83)$$

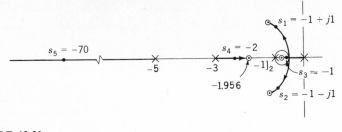

FIGURE 12-20
Root locus for Eq. (12-83).

This root locus shows that the desired zero of $Y(s)/R(s)$ at $s = -2$ has been canceled by the pole at $s = -2$. Because the root $s_4 = -2$ is near a zero in Fig. 12-20, its value is insensitive to increases in gain. This ensures the desired cancellation. The dominant roots are all near zeros in Fig. 12-20, thus assuring that the desired control ratio $Y(s)/R(s)$ as given by Eq. (12-77) has been achieved. The system is stable for all values of gain K_G. Also, state-variable feedback has produced a system in which $Y(s)/R(s)$ is unaffected by increases in K_G above 100. The time response $y(t)$ to a step input has a peak overshoot $M_p = 1.008$, a peak time $t_p = 3.92$, and a settling time $t_s = 2.94$. The ramp error coefficient evaluated for $G_{eq}(s)$ is $K_1 = 0.98$. These results show that the addition of the pole-zero combination $(s + 1.4)/(s + 1)$ to $Y(s)/R(s)$ has improved the performance. The sensitivity to gain variation at $\omega = 1.41$ is $|S_{K_G}^M(j1.41)| = 0.087$. Thus the system is insensitive to variation in gain.

The three examples in this section are intended to provide a firm foundation in the procedures for:

1   Synthesizing the desired system performance
2   Analyzing the root locus of $G(s)H_{eq}(s) = -1$ for the desired system performance and insensitivity to variations in gain
3   Determining the values of k_i to yield the desired system performance

These approaches are not intended to be all-inclusive, and variations may be developed.[2]

12-11   INACCESSIBLE STATES

In order to achieve the full benefit of a state-variable feedback system, all the states must be accessible. Although, in general, this requirement is not satisfied, the design procedures presented in this chapter are still valid. That is, the system's desired $Y(s)/R(s)$ is achieved by determining the value of **k** on the basis that all states are accessible. Then, for states that are not accessible, the corresponding k_i blocks are manipulated by block-diagram manipulation techniques to states that are accessible. As a result of these manipulations the full benefits of state-variable feedback may not

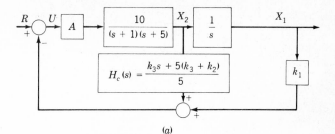

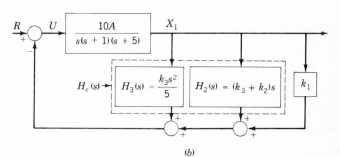

FIGURE 12-21
Minor-loop compensation (manipulation to the right) of the control system of
Fig. 12-6b.

be achieved (low sensitivity S_δ^M and a completely stable system), as described in Sec. 12-9. For the extreme case when the output $y = x_1$ is the only accessible state, the block-diagram manipulation to the G-equivalent reduces to the Guillemin-Truxal design. More sophisticated methods[7] for reconstructing (estimating) the values of the missing states do exist (Kalman filtering and Luenberger observed theory) but are beyond the scope of this text. They permit the accessible and the reconstructed (inaccessible) states to be all fed back through the feedback coefficients, thus eliminating the need for block-diagram manipulations and maintaining the full benefits of a state-variable feedback designed system. Even when all states are theoretically accessible, design and/or cost of instrumentation may make it prohibitive to measure some of the states.

For the control system of Fig. 12-6b assume that the X_3 state is inaccessible. Two possible block-diagram manipulations can be utilized: shift the k_3 block to the right (minor-loop compensation) or left (minor-loop or cascade compensation). Both approaches are now illustrated. Figure 12-21 represents the first. In Fig. 12-21a, $H_c(s)$ is a proportional plus derivative device. This unit *requires the use of a transducer or an active component*; i.e., if X_2 represents velocity, an *accelerometer* can be utilized to obtain the derivative action of $H_c(s)$. In the event that both the X_2 and X_3 states are inaccessible, the active minor-loop compensation of Fig. 12-21b can be used. This results in the requirement of first- and second-order derivative action in the minor loops. If X_1 represents position, an accelerometer and a tachometer can be used to realize $H_3(s)$ and $H_2(s)$, respectively. For the design requirement of the example of Sec. 12-4, the values of k_i have been determined. These values are used to design the

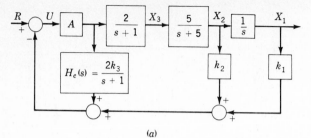

(a)

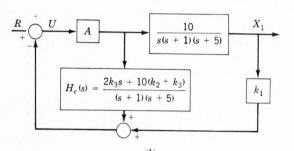

(b)

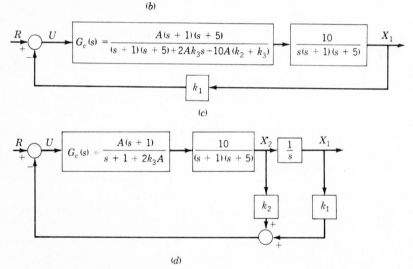

(c)

(d)

FIGURE 12-22
Manipulation to the left of the control system of Fig. 12-6b.

feedback compensator $H_c(s)$. From this example the following conclusion is evident: shifting k_i from the inaccessible state x_i to the right to the accessible state x_j requires $(i - j)$th-order derivative action in the minor-loop compensator $H_c(s)$, where $i > j$.

Figure 12-22 represents four possible cases of the manipulation to the left for the control system of Fig. 12-6: Fig. 12-22a and b results in the utilization of minor-loop compensators; Fig. 12-22c results in the utilization of a cascade compensator with all feedback other than k_1 eliminated; and Fig. 12-22d results in the utilization of both a cascade compensator and state feedback. The first and fourth cases assume that

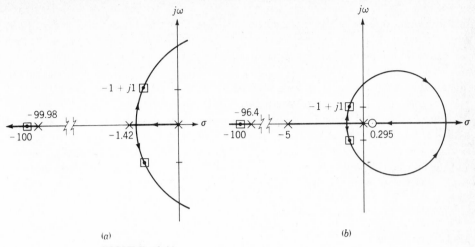

(a)   (b)

FIGURE 12-23
Root locus for: (a) Fig. 12-22c; (b) Fig. 12-22d.

only the X_3 state is inaccessible. The second and third cases assume that both the states X_2 and X_3 are inaccessible. The compensators in all four cases may be realized, utilizing network-synthesis techniques, by passive elements in conjunction with a cascade amplifier.

The compensators required for each case represented in Figs. 12-21 and 12-22, utilizing the values of $k_1 = 1$, $k_2 = -3.393$, $k_3 = 4.77$, and $A = 10$ determined in Sec. 12-6, are given in Table 12-5, which also gives the sensitivity function $|S_{K_G}^M(j\omega_o)|$ for each case at $\omega_o = 1.0$ and $M = 0.707$. As noted in Table 12-5, the low sensitivity of the output response to gain variations is maintained as long as feedback compensators $H_c(s)$ are utilized. In other words, the block-diagram manipulations that result in *only minor-loop* compensation, i.e., utilizing only *feedback* compensators $H_c(s)$, have the same $H_{eq}(s)$ as before the manipulations are performed. Therefore, the same stability and sensitivity characteristics are maintained for the manipulated as for the nonmanipulated systems. When the shifting is to the left, to yield a resulting cascade compensator $G_c(s)$, the low sensitivity to parameter variation is not achieved. This sensitivity characteristic is borne out by analyzing the root locus of Figs. 12-7b and 12-23. Figure 12-7b is the root locus representing the block-diagram manipulations utilizing only $H_c(s)$ networks. Figure 12-23 shows the root locus representing the block-diagram manipulations using cascade $G_c(s)$ networks. In Fig. 12-7b an increase in the gain does not produce any appreciable change in the closed-loop system response. In contrast, any gain variation in the systems represented by Fig. 12-23 greatly affects the closed-loop system response.

Note that the manipulated system of Fig. 12-22d, utilizing both cascade $G_c(s)$ and minor-loop $H_c(s) = k_2$ compensators, yields the same relative sensitivity as for the system of Fig. 12-22c, utilizing only cascade compensation. In general, it may be possible to achieve a relative sensitivity for a system utilizing both $G_c(s)$ and $H_c(s)$ compensators somewhere between that of a system utilizing only minor-loop compensation and one utilizing only cascade compensation. As pointed out in Sec. 12-9, in order to maintain low sensitivity for variations in parameters between the X_1 and X_n states, the feedback must be moved to the right.

The technique presented in this section for implementing a system having inaccessible states is rather simple and straightforward. It is very satisfactory for systems that are noise-free and time-invariant. For example, if in Fig. 12-6b the state X_3 is inaccessible and X_2 contains noise, it is best to shift the k_3 block to the left. This assumes that the amplifier output signal has minimal noise content. If all signals contain noise, the system performance is degraded from the desired performance. This technique is also satisfactory under parameter variations and/or if the poles and zeros of the transfer function $G_x(s)$ are not known exactly when operating under hfg.

12-12 SUMMARY

This chapter is devoted to designing a compensated control system based upon the specification of the desired closed-loop control ratio. The two methods that are utilized to yield this desired control ratio are the Guillemin-Truxal and the state-variable feedback methods. The root-locus properties and characterization, the steady-state error analysis, and the use of steady-state error coefficients, as they pertain to state-variable feedback systems (all pole and pole-zero plants) are discussed in detail. Also, the manner of handling inaccessible states and plant complex poles in state-variable feedback systems is covered. Parameter sensitivity is analyzed for both the unity-feedback and state-variable feedback systems. The use of computer programs can expedite the achievement of an acceptable design. These programs include the root locus,[8] time-response,[9] and state-variable feedback-coefficient determination.[10]

Table 12-5 COMPENSATORS AND SENSITIVITY FUNCTIONS FOR CONTROL SYSTEMS OF FIGS. 12-21 AND 12-22

Figure	Compensator	$\lvert S_A{}^M(j\omega_o)\rvert$
12-21a	$H_c(s) = 0.954s + 1.377$	0.0508
12-21b	$H_c(s) = 0.954s^2 + 1.377s$	0.0508
12-22a	$H_c(s) = \dfrac{9.54}{s + 1}$	0.0508
12-22b	$H_c(s) = \dfrac{9.54(s + 1.443)}{(s + 1)(s + 5)}$	0.0508
12-22c	$G_c(s) = \dfrac{10(s + 1)(s + 5)}{(s + 1.43)(s + 99.98)}$	0.51
12-22d	$G_c(s) = \dfrac{10(s + 1)}{s + 96.4}$	0.51

REFERENCES

1 Truxal, J. G.: "Automatic Feedback Control System Synthesis," McGraw-Hill, New York, 1955.

2 Schultz, D. G., and J. L. Melsa: "State Functions and Linear Control Systems," McGraw-Hill, New York, 1967.

3 Houpis, C. H.: The Relationship between the Conventional Control Theory Figures of Merit and the Performance Indices in Optimal Control Theory, Ph.D. dissertation, Department of Electrical Engineering, University of Wyoming, 1971.

4 James, H. M., N. B. Nichols, and R. S. Phillips: "Theory of Servomechanisms," McGraw-Hill, New York, 1947.

5 Sensitivity and Modal Response for Single-Loop and Multiloop Systems, *Flight Control Lab., ASD, AFSC, Tech. Doc. Rep.* ASD-TDR-62-812, Wright-Patterson Air Force Base, Ohio, January 1963.

6 Cruz, J. B., Jr.: "Feedback Systems," McGraw-Hill, New York, 1972.

7 Anderson, B. D. O., and J. B. Moore: "Linear Optimal Control," Prentice-Hall, Englewood Cliffs, N.J., 1971.

8 User's Manual for a Digital Computer Routine to Calculate the Root Locus (ROOTL), School of Engineering, Air Force Institute of Technology, Wright-Patterson Air Force Base, Ohio, 1974.

9 Heaviside Partial Fraction Expansion and Time Response Program (PARTL), School of Engineering, Air Force Institute of Technology, Wright-Patterson Air Force Base, Ohio, 1974.

10 State Variable Feedback (SVFB), School of Engineering, Air Force Institute of Technology, Wright-Patterson Air Force Base, Ohio, 1974.

13

LIAPUNOV'S SECOND METHOD

13-1 INTRODUCTION

The definition of stability for a linear time-invariant system is an easy concept to understand. The complete response to any input contains a particular solution which has the same form as the input and a complementary solution containing terms of the form $Ae^{\lambda_i t}$. When the eigenvalues λ_i have negative real parts, the transients associated with the complementary solution decay with time and the response is called stable. When the roots have positive real parts, the transients increase without bound and the system is called unstable. It is necessary to extend the concept of stability to nonlinear systems. Also, it is desired that stability be determined without explicitly solving for the eigenvalues. This is especially important for high-order systems. Liapunov attacked this problem through his *second, or direct, method.* The *sufficient but not necessary condition* of this method requires the determination of a scalar function V, called a *Liapunov function.* The function must approach an equilibrium point along a trajectory as time increases. A Liapunov function is readily determined for linear time-invariant systems; however, the determination is difficult for nonlinear and for time-variable systems. Despite these limitations, the Liapunov method provides a generalized approach to stability that is needed by the control engineer.

In order to develop Liapunov's second method properly, this chapter starts with an introduction to state-space trajectories. The kinds of singular points and the associated stability are categorized for both linear and nonlinear systems. The properties of quadratic functions and the definitions of definiteness are presented since the Liapunov function may be a quadratic function for linear time-invariant systems. Some useful definitions of stability are given. The Liapunov method is presented and illustrated by examples. Finally it is shown that the Liapunov function provides a measure of the system time response. This provides the basis for its use in Chap. 14 in defining optimum system response.

13-2   STATE-SPACE TRAJECTORIES[1-3]

The *state space* is defined as the n-dimensional space in which the components of the state vector represent its coordinate axes. The unforced response of a system having n state variables, released from any initial point $\mathbf{x}(t_0)$, traces a curve or trajectory in an n-dimensional state space. Time t is an implicit function along the trajectory. When the state variables are represented by phase variables, the state space may also be called a phase space. While it is impossible to visualize a space with more than three dimensions, the concept of a state space is nonetheless very useful in systems analysis. The behavior of second-order systems is conveniently viewed in the state space since it is two-dimensional and is in fact a state (or phase) plane. It is easy to obtain a graphical or geometrical representation and interpretation of second-order system behavior in the state plane. The concepts developed for the state plane can be extrapolated to the state space of higher-order systems. Examples of a number of second-order systems are presented in this section. These concepts are extended qualitatively to third-order systems. The symmetry that results in the state-plane trajectories is demonstrated for the cases when the **A** matrix is converted to normal (canonical) or modified normal (modified canonical) form. The second-order examples to be considered include the cases where the eigenvalues are (1) negative real and unequal (overdamped), (2) negative real and equal (critically damped), (3) imaginary (oscillatory), and (4) complex with negative real parts (underdamped). The system to be studied, with no forcing function, is represented by

$$\dot{\mathbf{x}} = \begin{bmatrix} a_{11} & a_{12} \\ a_{21} & a_{22} \end{bmatrix} \mathbf{x} \qquad (13\text{-}1)$$

Since there is no forcing function, the response results from energy stored in the system as represented by initial conditions.

EXAMPLE 1   *Overdamped response*   Consider the following example with the given initial conditions:

$$\dot{\mathbf{x}} = \begin{bmatrix} 0 & 3 \\ -1 & -4 \end{bmatrix} \mathbf{x} \qquad \mathbf{x}(0) = \begin{bmatrix} 0 \\ 2 \end{bmatrix} \qquad (13\text{-}2)$$

Since one state equation is $\dot{x}_1 = 3x_2$, x_2 is a phase variable. Therefore, the resulting state plane is called a *phase plane*.

The state transition matrix $\mathbf{\Phi}(t)$ is obtained by using the Sylvester expansion theorem or any other convenient method. The system response is overdamped, with

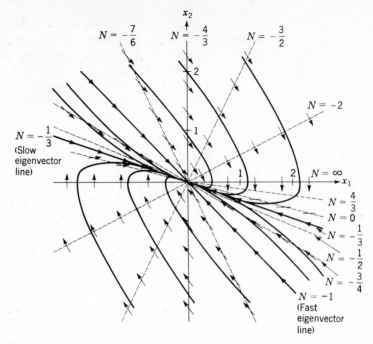

FIGURE 13-1
Phase-plane portrait for Eq. (13-2).

eigenvalues $\lambda_1 = -1$ and $\lambda_2 = -3$. For the given initial conditions the system response is

$$\mathbf{x}(t) = \mathbf{\Phi}(t)\mathbf{x}(0) = \begin{bmatrix} 3e^{-t} - 3e^{-3t} \\ -e^{-t} + 3e^{-3t} \end{bmatrix}$$

The trajectory represented by this equation can be plotted in the phase plane, x_2 versus x_1. A family of trajectories in the phase plane for a number of initial conditions is drawn in Fig. 13-1. Such a family of trajectories is called a *phase portrait*. The arrows show the direction of the states along the trajectories for increasing time. Motion of the point on a trajectory is in the clockwise direction about the origin since $\dot{x}_1 = 3x_2$. Thus, when x_2 is positive, x_1 must be increasing in value, and when x_2 is negative, x_1 must be decreasing in value. The trajectories always cross the x_1 axis in a perpendicular direction. The rate of change of x_1 is zero as the trajectories cross the x_1 axis.

One method of drawing the phase trajectory is to insert values of time t into the solution $\mathbf{x}(t)$ and to plot the results. For the example above the solutions are $x_1 = 3e^{-t} - 3e^{-3t}$ and $x_2 = -e^{-t} + 3e^{-3t}$. Another approach is to eliminate t from these equations to obtain an analytical expression for the trajectory. This can be done by first solving for e^{-t} and e^{-3t}.

$$e^{-t} = \frac{x_1 + x_2}{2}$$

$$e^{-3t} = \frac{x_1 + 3x_2}{6}$$

Raising e^{-t} to the third power and e^{-3t} to the first power and equating the results yields

$$\left(\frac{x_1 + x_2}{2}\right)^3 = \left(\frac{x_1 + 3x_2}{6}\right)^1$$

This equation is awkward to use and applies only for the specified initial conditions. Further, it is difficult to extend this procedure to nonlinear systems.

A practical graphical method for obtaining trajectories in the state plane is the *method of isoclines*. The previous example is used to illustrate this method. Taking the ratio of the state equations of this example yields

$$\frac{\dot{x}_2}{\dot{x}_1} = \frac{dx_2/dt}{dx_1/dt} = \frac{dx_2}{dx_1} = \frac{-x_1 - 4x_2}{3x_2} = N \qquad (13\text{-}3)$$

This equation represents the slope of the trajectory passing through any point in the state plane. The integration of this equation traces the trajectory starting at any initial condition. The method of isoclines is a graphical technique for performing this integration. When the slope of the trajectory in Eq. (13-3) is fixed at any specific value N, the result is the equation of a straight line represented by

$$x_2 = -\frac{1}{4 + 3N} x_1 = mx_1 \qquad (13\text{-}4)$$

This equation describes a family of curves which are called *isoclines*. They have the property that all trajectories crossing a particular isocline have the same slope at the crossing points. For linear time-invariant second-order systems the isoclines are all straight lines. A trajectory can be sketched by using the following procedure:

1 Draw a number of isoclines reasonably close together.
2 From any starting point A (see Fig. 13-2) draw two lines having slopes N_1 and N_2 and extending from the isocline for N_1 to the isocline for N_2. The trajectory must stay between these lines. Select point B midway in the segment of the isocline for N_2 between these lines.
3 Continue from point B using the procedure of step 2.
4 Connect the points with a smooth curve which crosses each isocline with the correct slope. An example is shown in Fig. 13-2.

The system represented by Example 1 is stable, as indicated by the negative eigenvalues. Since there is no forcing function, the response is due to the energy initially stored in the system. As this energy is dissipated, the response approaches equilibrium at $x_1 = x_2 = 0$. All the trajectories approach and terminate at this point. The slope of the trajectory going through this equilibrium point is therefore indeterminate, that is, $N = dx_2/dx_1 = 0/0$, representing a *singularity* point. This indeterminacy can be used to locate the equilibrium point by letting the slope of the trajectory in the state plane be $N = dx_2/dx_1 = 0/0$. Applying this condition to Eq. (13-3) yields the two equations $x_2 = 0$ and $-x_1 - 4x_2 = 0$. This confirms that the equilibrium point is $x_1 = x_2 = 0$. For the overdamped response this equilibrium point is called a *node*. The procedure can be extended to all systems of any order,

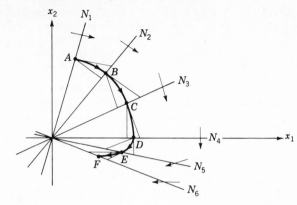

FIGURE 13-2
Construction of a trajectory using isoclines.

even when they are nonlinear. This is done by solving $\dot{\mathbf{x}} = \mathbf{0}$ for the equilibrium point(s).

It is interesting to note the symmetry that results in the phase portrait when the states are transformed to canonical form.[9] The methods for accomplishing this transformation are described in Sec. 5-9. The resulting state equations in canonical form for a second-order system are

$$\dot{z}_1 = \lambda_1 z_1 \qquad (13\text{-}5)$$

$$\dot{z}_2 = \lambda_2 z_2 \qquad (13\text{-}6)$$

When the eigenvalues $\lambda_1 \neq \lambda_2$, the ratio of these equations yields

$$\frac{dz_1}{dz_2} = \frac{\lambda_1 z_1}{\lambda_2 z_2} \qquad (13\text{-}7)$$

The variables are separable, and the resulting equation can be integrated to obtain the equation of the state-plane trajectory

$$z_2 = C z_1^{\lambda_2/\lambda_1} \qquad (13\text{-}8)$$

For a stable system with λ_1 and λ_2 both negative and real, where $\lambda_2 < \lambda_1 < 0$, the state-plane portraits are parabolic, as shown in Fig. 13-3a and b. All the trajectories are tangent to the z_1 axis. When returning to the $x_2 x_1$ plane, the trajectories are tangent to the line labeled z_1 in Fig. 13-3c and d. This line is the transformation of the z_1 axis into the $x_2 x_1$ plane. The slope of this line can be obtained from the **T** matrix, which transforms the **z** plane into the **x** plane. Using $\mathbf{T} = [v_{ij}]$ and $\mathbf{z} = \mathbf{T}^{-1}\mathbf{x}$ yields

$$z_1 = \frac{1}{\Delta_T}(v_{22}x_1 - v_{12}x_2) \qquad (13\text{-}9)$$

$$z_2 = \frac{1}{\Delta_T}(-v_{21}x_1 + v_{11}x_2) \qquad (13\text{-}10)$$

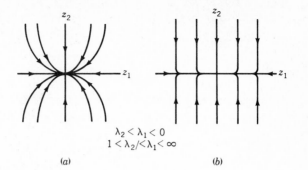

$$\lambda_2 < \lambda_1 < 0$$
$$1 < \lambda_2/<\lambda_1 < \infty$$

(a)   (b)

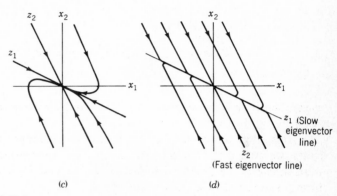

z_1 (Slow eigenvector line)

z_2 (Fast eigenvector line)

(c)   (d)

FIGURE 13-3

Stable node: (a) normal form: roots of similar magnitudes; (b) normal form: roots of widely different magnitudes; (c) x_2 vs. x_1 for roots with similar magnitudes; (d) x_2 vs. x_1 for roots with widely different magnitudes. (see Ref. 9).

The z_1 axis is defined by $z_2 = 0$. From Eq. (13-10) the slope of the z_1 axis after transformation to the $x_2 x_1$ plane is

$$m_1 = \frac{x_2}{x_1} = \frac{v_{21}}{v_{11}} \qquad (13\text{-}11)$$

Thus the slope is determined by that eigenvector which is the first column of the modal matrix $\mathbf{T}$. The ratio v_{21}/v_{11} can be evaluated by equating the elements of $\mathbf{AT} = \mathbf{T\Lambda}$ [see Eq. (5-76)], which yields

$$m_1 = \frac{v_{21}}{v_{11}} = \frac{\lambda_1 - a_{11}}{a_{12}} = \frac{a_{21}}{\lambda_1 - a_{22}} \qquad (13\text{-}12)$$

Since λ_1 yields the transient mode with the longest settling time, the line z_1 in Fig. 13-3c and d is labeled the "slow eigenvector line."

The trajectories in Fig. 13-3a and b are parallel to the z_2 axis at large distances from the origin. In the $x_2 x_1$ plane (see Fig. 13-3c and d) the trajectories retain this property except that they are parallel to the line labeled z_2. This line is the transformation of the z_2 axis into the x_2x_1 plane. The z_2 axis is defined by the condition $z_1 = 0$. Using Eq. (13-9), the slope of this axis transformed into the **x** plane is

$$m_2 = \frac{x_2}{x_1} = \frac{v_{22}}{v_{12}} \quad (13\text{-}13)$$

This slope is therefore determined by the eigenvector, the second column of **T**. From the coefficients of $\mathbf{AT} = \mathbf{T\Lambda}$ this slope is obtained as a function of λ_2:

$$m_2 = \frac{v_{22}}{v_{12}} = \frac{\lambda_2 - a_{11}}{a_{12}} = \frac{a_{21}}{\lambda_2 - a_{22}} \quad (13\text{-}14)$$

Since λ_2 yields the transient mode with the shortest settling time, the line z_2 in Fig. 13-3c and d is labeled the "fast eigenvector line."

For the system of Example 1 the slopes are $m_1 = -\frac{1}{3}$ and $m_2 = -1$. The lines going through the equilibrium point with these slopes are drawn in Fig. 13-1. The transient term $e^{\lambda_2 t}$ decays faster than the term $e^{\lambda_1 t}$. Therefore, near the origin the trajectories are tangent to the line determined by the slow eigenvalue, $m_1 = -\frac{1}{3}$.

When the real eigenvalues are equal and negative, the phase portrait is similar to that in Fig. 13-1 except that there is only one eigenvector line.

When any real eigenvalue is positive, the system response is unstable. For the second-order system with two positive eigenvalues the equilibrium point is an *unstable node*. In this case the trajectories depart from the equilibrium point and go to infinity.

EXAMPLE 2 *Underdamped response*   The system to be considered here is

$$\dot{\mathbf{x}} = \begin{bmatrix} 0 & 1 \\ -4 & -2 \end{bmatrix} \mathbf{x}$$

The characteristic equation is $\lambda^2 + 2\lambda + 4 = 0$, and the eigenvalues are $\lambda_{1,2} = -1 \pm j\sqrt{3}$, representing a damping ratio $\zeta = 0.5$. The equilibrium point is determined from the equation of the slope of the trajectory

$$\frac{dx_2}{dx_1} = \frac{-4x_1 - 2x_2}{x_2} = N \quad (13\text{-}15)$$

The singularity condition $N = 0/0$ determines the equilibrium point $x_1 = x_2 = 0$.

The isoclines are the straight lines represented by

$$x_2 = \frac{-4}{2 + N} x_1 \quad (13\text{-}16)$$

Since $x_2 = \dot{x}_1$, the state plane is a phase plane. Several phase-plane trajectories are shown in Fig. 13-4. The trajectories are modified logarithmic spirals.

The equilibrium point for an underdamped response is called a *focus*. For stable responses the motion of the point on the spiral trajectories is clockwise toward the focus.

When the state variables are transformed to canonical form, the equation

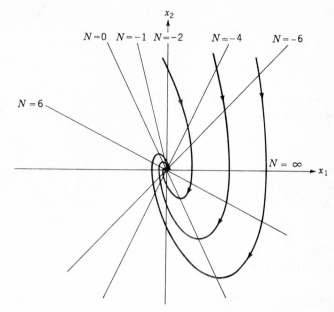

FIGURE 13-4
Underdamped stable phase-plane trajectories.

representing the trajectories is best represented in polar form, since the trajectories are known to be spiral. When the eigenvalues are complex conjugates, $\lambda_{1,2} = \sigma \pm j\omega_d$, a simplified form of the trajectory equation is obtained by assuming the canonical state variables are also complex conjugates:

$$z_1 = w_1 - jw_2 \qquad z_2 = w_1 + jw_2 \qquad (13\text{-}17)$$

Inserting these values into the canonical state equations $\dot{z}_1 = \lambda_1 z_1$ and $\dot{z}_2 = \lambda_2 z_2$ yields

$$\dot{w}_1 - j\dot{w}_2 = (\sigma + j\omega_d)(w_1 - jw_2)$$
$$\dot{w}_1 + j\dot{w}_2 = (\sigma - j\omega_d)(w_1 + jw_2) \qquad (13\text{-}18)$$

Separating the real and imaginary components of these equations produces

$$\dot{w}_1 = \sigma w_1 + \omega_d w_2 \qquad \dot{w}_2 = -\omega_d w_1 + \sigma w_2 \qquad (13\text{-}19)$$

These equations will be recognized as being in the modified normal form described in Sec. 5-10. The slope of the trajectories in the $w_1 w_2$ plane, obtained from Eq. (13-19), is

$$\frac{dw_2}{dw_1} = \frac{-\omega_d w_1 + \sigma w_2}{\sigma w_1 + \omega_d w_2} \qquad (13\text{-}20)$$

This equation is converted from the rectangular coordinates w_1 and w_2 to the polar coordinates r and ϕ by using $w_1 = r \cos \phi$ and $w_2 = r \sin \phi$. Then,

$$dw_1 = \cos \phi \, dr - r \sin \phi \, d\phi$$

and

$$dw_2 = \sin \phi \, dr + r \cos \phi \, d\phi$$

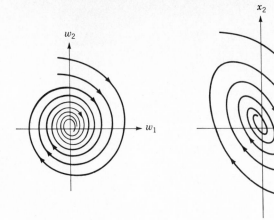

(a)   (b)

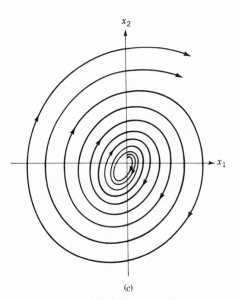

(c)

FIGURE 13-5
Underdamped trajectories: (a) stable, modified canonical form; (b) stable, general form; (c) unstable, general form.

After simplification the resulting equation representing the slope of the trajectory in the $w_1 w_2$ plane is

$$\frac{dr}{d\phi} = -\frac{\sigma}{\omega_d} r \qquad (13\text{-}21)$$

The variables can be separated, and integration of the equation yields

$$r = Ce^{-(\sigma/\omega_d)\phi} = Ce^{(\zeta/\sqrt{1-\zeta^2})\phi} \qquad (13\text{-}22)$$

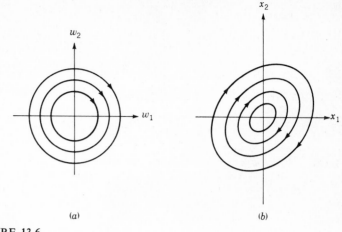

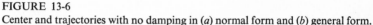

(a)   (b)

FIGURE 13-6
Center and trajectories with no damping in (a) normal form and (b) general form.

The trajectories represented by this equation are logarithmic spirals. Motion along the trajectories is clockwise for increasing time. For positive values of damping ratio, $1 > \zeta > 0$, the system response is stable, and the trajectories approach the equilibrium point (*stable focus*). Figure 13-5a and b shows the shapes of the trajectories in the **w** and **x** planes for a stable response. For negative damping ratio, $-1 < \zeta < 0$, the response is unstable, and the trajectories diverge from the equilibrium point (*unstable focus*). Figure 13-5c shows the trajectories for an unstable response in the **x** plane.

For the undamped case, $\zeta = 0$, the canonical and modified canonical coefficient matrices have the form

$$\Lambda = \begin{bmatrix} j\omega_d & 0 \\ 0 & -j\omega_d \end{bmatrix} \quad \text{and} \quad \Lambda_m = \begin{bmatrix} 0 & \omega_d \\ -\omega_d & 0 \end{bmatrix}$$

The trajectories, obtained from Eq. (13-22), are circles having a constant radius in the **w** plane. In the **x** plane they become ellipses. The equilibrium point is called a *center* or *vortex*. The portraits for the **w** and **x** planes are illustrated in Fig. 13-6. The principal axis of the ellipses in the **x** plane is a function of the **A** matrix. As an example, the matrix

$$\mathbf{A} = \begin{bmatrix} 1 & 1 \\ -2 & -1 \end{bmatrix}$$

has the eigenvalues $\lambda_{1,2} = \pm j1$ and the principal axis of the ellipses is shown in Fig. 13-6b.

EXAMPLE 3 *One positive and one negative real root* When the roots are real but of opposite sign, the equilibrium point is called a *saddle point* and represents an unstable equilibrium. The trajectories in canonical form are represented by Eq. (13-8), which may be rewritten as

$$z_2 = C z_1{}^m \qquad (13\text{-}23)$$

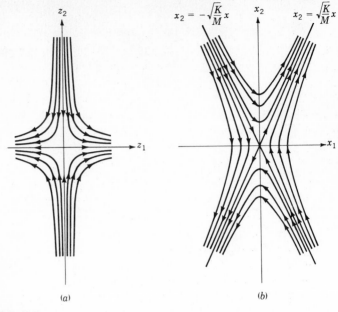

FIGURE 13-7
Saddle point and trajectories in (*a*) normal form; (*b*) general form.

where $m = \lambda_2/\lambda_1$. The trajectories are hyperbolas which are asymptotic to the z_1 and z_2 axes, as shown in Fig. 13-7*a*. Back in the **x** plane the trajectories are rotated and modified in shape as shown in Fig. 13-7*b*. The axes can be transformed from the **z** plane to the **x** plane. They are straight lines with the slopes given by Eqs. (13-11) to (13-14). These lines are called *separatrices* since they divide the state plane into separate regions.

The mechanical system containing mass, a "repulsive" spring force, and no damping is represented by the differential equation

$$M\ddot{x} - Kx = 0 \qquad (13\text{-}24)$$

This represents the motion of a pendulum in the neighborhood of the upper unstable equilibrium position. In terms of phase variables the state equations are

$$\begin{bmatrix} \dot{x}_1 \\ \dot{x}_2 \end{bmatrix} = \begin{bmatrix} 0 & 1 \\ \dfrac{K}{M} & 0 \end{bmatrix} \begin{bmatrix} x_1 \\ x_2 \end{bmatrix} \qquad (13\text{-}25)$$

The eigenvalues are $\lambda_{1,2} = \pm\sqrt{K/M}$. The slopes of the z_1 and z_2 axes transformed to the $x_1 x_2$ plane are obtained from Eqs. (13-12) and (13-14).

$$m_1 = \sqrt{\dfrac{K}{M}} = \lambda_1 \qquad m_2 = -\sqrt{\dfrac{K}{M}} = \lambda_2 \qquad (13\text{-}26)$$

The trajectories are shown in Fig. 13-7*b*.

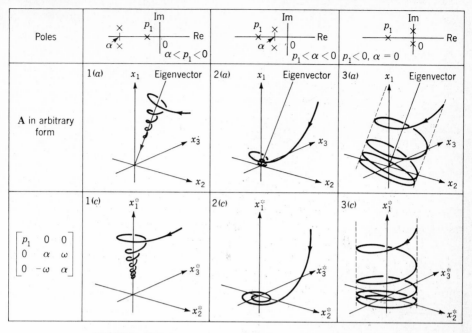

FIGURE 13-8
Trajectory patterns of third-order oscillatory systems. (*Reproduced by permission from Y. Takahashi, M. J. Rabins, and D. J. Anslander, "Control and Dynamic Systems," Addison-Wesley Publishing Company, Reading, Mass., 1970.*)

Third-order systems require a three-dimensional state space. At least one of the eigenvalues must be real. Figure 13-8 contains examples for third-order systems with a pair of complex-conjugate poles. It is observed that transforming the states to modified canonical form results in a symmetry of the trajectories with respect to the principal axes. The eigenvector for the real pole coincides with one of the axes.

The properties of trajectories in the state plane are developed in this section. The type of equilibrium points and trajectories can be extended conceptually to higher-order systems even when it is not possible to draw them in a higher-dimensional space. It is also possible to show the effect of nonlinearities on the shape of the trajectories. An example is presented in Sec. 13-3. An important consideration in the stability analysis of such systems is that the system response in the neighborhood of the singularities can be determined from the linearized equations. Use is made of this principle in later sections of this chapter.

13-3 LINEARIZATION (JACOBIAN MATRIX)[1,8]

A linear system with no forcing function (an autonomous system) has only one equilibrium point $\mathbf{x}_0$. All the examples in Sec. 13-2 have the equilibrium point at the origin $\mathbf{x}_0 = \mathbf{0}$. A nonlinear system, on the other hand, may have more than one

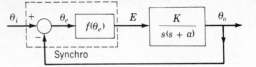

FIGURE 13-9
A nonlinear feedback control system.

equilibrium point. This is easily illustrated by considering the unity-feedback position-control system shown in Fig. 13-9. The feedback action is provided by synchros which generate the actuating signal $e = \sin(\theta_i - \theta_o)$. With no input, $\theta_i = 0$, the differential equation of the system is

$$\ddot{\theta}_o + a\dot{\theta}_o + K \sin \theta_o = 0 \qquad (13\text{-}27)$$

This is obviously a nonlinear differential equation. With the phase variables $x_1 = \theta_o$ and $x_2 = \dot{x}_1$, the corresponding state equations are

$$\dot{x}_1 = x_2 \qquad \dot{x}_2 = -K \sin x_1 - ax_2 \qquad (13\text{-}28)$$

The slope of the trajectories is obtained from

$$N = \frac{\dot{x}_2}{\dot{x}_1} = \frac{-K \sin x_1 - ax_2}{x_2} \qquad (13\text{-}29)$$

The equilibrium points exist at $\dot{x} = 0$, and so the singularities are $x_2 = 0$ and $x_1 = k\pi$, where k is an integer. The system therefore has multiple equilibrium points.

In a small neighborhood about the equilibrium points a nonlinear system behaves like a linear system. The states may be written as $x = x_0 + x^*$, where x_0 represents an equilibrium point and x^* represents the perturbation or *state deviation*. It is the variation of the states from the equilibrium conditions. In the general case the unforced nonlinear state equation is

$$\dot{x} = f(x) \qquad (13\text{-}30)$$

Each of the elements of $f(x)$ can be expanded in a Taylor series about one of the equilibrium points x_0. Assuming that x^* is restricted to a small neighborhood of the equilibrium point, the higher-order terms in the Taylor series may be neglected. Thus, the resulting linear variational state equation is

$$\dot{x}^* = \begin{bmatrix} \dfrac{\partial f_1}{\partial x_1} & \dfrac{\partial f_1}{\partial x_2} & \cdots & \dfrac{\partial f_1}{\partial x_n} \\[2mm] \dfrac{\partial f_2}{\partial x_1} & \dfrac{\partial f_2}{\partial x_2} & \cdots & \dfrac{\partial f_2}{\partial x_n} \\[1mm] \multicolumn{4}{c}{\cdots\cdots\cdots\cdots\cdots} \\[1mm] \dfrac{\partial f_n}{\partial x_1} & \dfrac{\partial f_n}{\partial x_2} & \cdots & \dfrac{\partial f_n}{\partial x_n} \end{bmatrix}_{x=x_0} x^* = J_x x^* \qquad (13\text{-}31)$$

where $J_x = \partial f / \partial x^T$ is called the *Jacobian matrix* and is evaluated at x_0.

For the system of Eq. (13-28) the motion about the equilibrium point $x_1 = x_2 = 0$ is represented by

$$\dot{x}^* = \begin{bmatrix} \dot{x}_1^* \\ \dot{x}_2^* \end{bmatrix} = \begin{bmatrix} 0 & 1 \\ -K & -a \end{bmatrix} \begin{bmatrix} x_1^* \\ x_2^* \end{bmatrix} = J_x x^* \qquad (13\text{-}32)$$

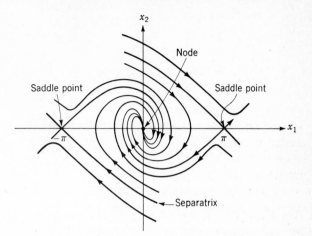

FIGURE 13-10
Phase portrait for Eq. (13-27).

For these linearized equations the eigenvalues are $\lambda_{1,2} = -a/2 \pm \sqrt{(a/2)^2 - K}$. The equilibrium point is stable and is either a node or a focus, depending upon the magnitudes of a and K.

For motion about the equilibrium point $x_1 = \pi$, $x_2 = 0$ the state equations are

$$\dot{\mathbf{x}}^* = \begin{bmatrix} 0 & 1 \\ K & -a \end{bmatrix} \mathbf{x}^* \qquad (13\text{-}33)$$

The eigenvalues of this $\mathbf{J}_x$ are $\lambda_{1,2} = -a/2 \pm \sqrt{(a/2)^2 + K}$. Thus, one eigenvalue is positive and the other is negative, and the equilibrium point represents an unstable saddle point. The motion around the saddle point is considered unstable since every point on all trajectories, except on the two separatrices, moves away from this equilibrium point. A phase-plane portrait for this system is obtained by the method of isoclines. From Eq. (13-29) the isocline equation is

$$x_2 = \frac{-K \sin x_1}{N + a} \qquad (13\text{-}34)$$

The phase portrait is shown in Fig. 13-10. Note that the linearized equations are applicable only in the neighborhood of the singular points. Thus, they describe stability *in the small*.

When an input $\mathbf{u}$ is present, the nonlinear state equation is

$$\dot{\mathbf{x}} = \mathbf{f}(\mathbf{x},\mathbf{u}) \qquad (13\text{-}35)$$

Its equilibrium point $\mathbf{x}_0$, when $\mathbf{u} = \mathbf{u}_0$ is a constant, is determined by letting $\dot{\mathbf{x}} = \mathbf{f}(\mathbf{x}_0,\mathbf{u}_0) = \mathbf{0}$. Then the linearized variational state equation describing the variation from the equilibrium point is obtained by using only the linear terms from

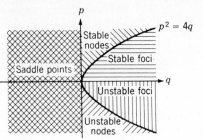

FIGURE 13-11
Type of equilibrium point for the characteristic equation $s^2 + ps + q = 0$.

the Taylor series expansion. For variations in $\mathbf{u}$, appearing as $\mathbf{u}^*$ in the input $\mathbf{u} = \mathbf{u}_0 + \mathbf{u}^*$, the linearized state equation is

$$
\dot{\mathbf{x}}^* =
\begin{bmatrix}
\dfrac{\partial f_1(\mathbf{x},\mathbf{u})}{\partial x_1} & \dfrac{\partial f_1(\mathbf{x},\mathbf{u})}{\partial x_2} & \cdots & \dfrac{\partial f_1(\mathbf{x},\mathbf{u})}{\partial x_n} \\[2mm]
\dfrac{\partial f_2(\mathbf{x},\mathbf{u})}{\partial x_1} & \dfrac{\partial f_2(\mathbf{x},\mathbf{u})}{\partial x_2} & \cdots & \dfrac{\partial f_2(\mathbf{x},\mathbf{u})}{\partial x_n} \\[2mm]
\cdots\cdots\cdots\cdots\cdots\cdots\cdots\cdots \\[2mm]
\dfrac{\partial f_n(\mathbf{x},\mathbf{u})}{\partial x_1} & \dfrac{\partial f_n(\mathbf{x},\mathbf{u})}{\partial x_2} & \cdots & \dfrac{\partial f_n(\mathbf{x},\mathbf{u})}{\partial x_n}
\end{bmatrix}_{\substack{\mathbf{x}=\mathbf{x}_0 \\ \mathbf{u}=\mathbf{u}_0}} \mathbf{x}^*
$$

$$
+
\begin{bmatrix}
\dfrac{\partial f_1(\mathbf{x},\mathbf{u})}{\partial u_1} & \cdots & \dfrac{\partial f_1(\mathbf{x},\mathbf{u})}{\partial u_r} \\[2mm]
\cdots\cdots\cdots\cdots\cdots\cdots \\[2mm]
\dfrac{\partial f_n(\mathbf{x},\mathbf{u})}{\partial u_1} & \cdots & \dfrac{\partial f_n(\mathbf{x},\mathbf{u})}{\partial u_r}
\end{bmatrix}_{\substack{\mathbf{x}=\mathbf{x}_0 \\ \mathbf{u}=\mathbf{u}_0}} \mathbf{u}^* = \mathbf{J}_x \mathbf{x}^* + \mathbf{J}_u \mathbf{u}^* \qquad (13\text{-}36)
$$

In analyzing the system performance in the vicinity of equilibrium, it is usually convenient to translate the origin of the state space to that point. This is done by inserting $\mathbf{x} = \mathbf{x}_0 + \mathbf{x}^*$ into the original equations. With the origin at $\mathbf{x}_0$, the variations from this point are described by $\mathbf{x}^*$. The characteristic equation describing the motion about the equilibrium point can be obtained from the linearized state equations. For a second-order system it has the form

$$
s^2 + ps + q = (s - \lambda_1)(s - \lambda_2) = 0 \qquad (13\text{-}37)
$$

The eigenvalues λ_1 and λ_2 determine whether the singular point is a node, a focus, or a saddle. For a node or focus the eigenvalues also determine whether the equilibrium point is stable or unstable. Figure 13-11 shows the type of equilibrium point, based on the coefficients p and q in Eq. (13-37).

Phase-plane portraits have been applied to provide a visual picture of the response of many linear and nonlinear systems. They are described extensively in the literature. The purpose in presenting them here is to provide some insight into trajectories and their relationships to the singular points. This understanding of phase-plane trajectories should provide the basis for a clearer understanding of the stability analysis for higher-order systems. The definitions of system stability and the techniques for evaluating stability are covered in later sections of this chapter.

13-4   QUADRATIC FORMS[4,5]

Some of the techniques used in determining stability of control systems and for optimizing their response utilize scalar functions expressed in *quadratic form*. The necessary background for expressing functions in quadratic form is developed in this section. Then some important properties of quadratic forms are presented.

Conjugate matrix   The elements of a matrix may be complex quantities. For example, a matrix **A** may have the elements $a_{ij} = \alpha_{ij} + j\beta_{ij}$. A *conjugate matrix* **B** has elements with the same real component and with imaginary components of the opposite sign, that is, $b_{ij} = \alpha_{ij} - j\beta_{ij}$. This conjugate property is expressed by

$$\mathbf{B} = \mathbf{A}^* \qquad (13\text{-}38)$$

Inner product   The inner product is also called a *scalar* (or *dot*) *product* since it yields a scalar function. The scalar product of vectors **x** and **y** is defined by

$$\langle \mathbf{x},\mathbf{y} \rangle = (\mathbf{x}^*)^T \mathbf{y} = \mathbf{y}^T \mathbf{x}^* = x_1^* y_1 + x_2^* y_2 + \cdots + x_n^* y_n = \sum_{i=1}^{n} x_i^* y_i \qquad (13\text{-}39)$$

When **x** and **y** are real vectors, the inner product becomes

$$\langle \mathbf{x},\mathbf{y} \rangle = \sum_{i=1}^{n} x_i y_i = x_1 y_1 + x_2 y_2 + \cdots + x_n y_n \qquad (13\text{-}40)$$

It should be noted that when **x** and **y** are complex, $\langle \mathbf{x},\mathbf{y} \rangle \neq \mathbf{x}^T \mathbf{y}^*$. However, when **x** and **y** are real,

$$\langle \mathbf{x},\mathbf{y} \rangle = \mathbf{x}^T \mathbf{y} = \mathbf{y}^T \mathbf{x} = \langle \mathbf{y},\mathbf{x} \rangle \qquad (13\text{-}41)$$

Bilinear form   A scalar homogeneous expression containing the product of the elements of vectors **x** and **y** is called a *bilinear form* in the variables x_i and y_i. When they are both of order n, the most general bilinear form in **x** and **y** is

$$\begin{aligned}
f(\mathbf{x},\mathbf{y}) = {} & a_{11}x_1 y_1 + a_{12}x_1 y_2 + \cdots + a_{1n}x_1 y_n \\
& + a_{21}x_2 y_1 + a_{22}x_2 y_2 + \cdots + a_{2n}x_2 y_n \\
& + \cdots\cdots\cdots\cdots\cdots\cdots\cdots\cdots\cdots \\
& + a_{n1}x_n y_1 + a_{n2}x_n y_2 + \cdots + a_{nn}x_n y_n \qquad (13\text{-}42)
\end{aligned}$$

This may be written more compactly as

$$\begin{aligned}
f(\mathbf{x},\mathbf{y}) &= \sum_{i=1}^{n} \sum_{j=1}^{n} a_{ij} x_i y_j \\
&= \begin{bmatrix} x_1 & x_2 & \cdots & x_n \end{bmatrix}
\begin{bmatrix}
a_{11} & a_{12} & \cdots & a_{1n} \\
a_{21} & a_{22} & \cdots & a_{2n} \\
\multicolumn{4}{c}{\cdots\cdots\cdots\cdots\cdots} \\
a_{n1} & a_{n2} & \cdots & a_{nn}
\end{bmatrix}
\begin{bmatrix} y_1 \\ y_2 \\ \vdots \\ y_n \end{bmatrix} \\
&= \mathbf{x}^T \mathbf{A} \mathbf{y} = \langle \mathbf{x}, \mathbf{A}\mathbf{y} \rangle \qquad (13\text{-}43)
\end{aligned}$$

The matrix **A** is called the coefficient matrix of the bilinear form, and the rank of **A** is called the rank of the bilinear form. A bilinear form is called *symmetric* if the matrix **A** is symmetric.

Quadratic form   A quadratic form V is a real homogeneous polynomial in the real variables $x_1, x_2, \cdots, x_n$ of the form

$$V = \sum_{i=1}^{n} \sum_{j=1}^{n} a_{ij} x_i x_j \qquad (13\text{-}44)$$

where all a_{ij} are real. This is the special case of Eq. (13-43) where $\mathbf{x} = \mathbf{y}$. The quadratic form V can be expressed as the inner product

$$V(\mathbf{x}) = \mathbf{x}^T \mathbf{A} \mathbf{x} = \langle \mathbf{x}, \mathbf{A}\mathbf{x} \rangle \qquad (13\text{-}45)$$

A homogeneous polynomial can always be expressed in terms of a symmetric matrix $\mathbf{A}$. In the expansion of Eq. (13-44) the cross-product terms $(i \neq j)$ all have the form $(a_{ij} + a_{ji})x_i x_j$. Choosing $a_{ij} = a_{ji}$ makes the matrix $\mathbf{A}$ for the quadratic form symmetric. This is illustrated in the following example.

EXAMPLE 1

$$V(\mathbf{x}) = x_1{}^2 - 4x_2{}^2 + 5x_3{}^2 + 6x_1 x_2 - 20 x_2 x_3$$

$$= \mathbf{x}^T \begin{bmatrix} 1 & 3 & 0 \\ 3 & -4 & -10 \\ 0 & -10 & 5 \end{bmatrix} \mathbf{x} = \mathbf{x}^T \mathbf{A} \mathbf{x} \qquad (13\text{-}46)$$

The rank of the matrix $\mathbf{A}$ is called the rank of the quadratic form. If the rank of $\mathbf{A}$ is $r < n$, the quadratic form is singular. If the rank of $\mathbf{A}$ is n, the quadratic form is nonsingular.

A nonsymmetric matrix can be converted into an equivalent symmetric matrix by replacing all sets of elements a_{ij} and a_{ji} by the average value $(a_{ij} + a_{ji})/2$. As illustrated in the following example, it is obvious that the principal diagonal is preserved.

EXAMPLE 2

$$\mathbf{A} = \begin{bmatrix} 1 & -1 & 2 \\ 5 & 3 & 7 \\ 0 & 1 & 2 \end{bmatrix} \qquad \mathbf{A}_{\text{sym}} = \begin{bmatrix} 1 & 2 & 1 \\ 2 & 3 & 4 \\ 1 & 4 & 2 \end{bmatrix}$$

Congruent transformation   It is often desirable to introduce a transformation so that Eq. (13-45) can be transformed to the quadratic form $V = \mathbf{y}^T \mathbf{B} \mathbf{y}$, in which the matrix $\mathbf{B}$ contains only diagonal elements. This can be accomplished by means of a congruent transformation. Two square matrices $\mathbf{A}$ and $\mathbf{B}$ of order n are called congruent if there exists a nonsingular transformation matrix $\mathbf{P}$ such that

$$\mathbf{B} = \mathbf{P}^T \mathbf{A} \mathbf{P} \qquad (13\text{-}47)$$

Then, with the linear transformation $\mathbf{x} = \mathbf{P}\mathbf{y}$, the quadratic form of Eq. (13-45) is transformed to

$$V = \mathbf{x}^T \mathbf{A} \mathbf{x} = \mathbf{y}^T (\mathbf{P}^T \mathbf{A} \mathbf{P}) \mathbf{y} = \mathbf{y}^T \mathbf{B} \mathbf{y} \qquad (13\text{-}48)$$

Canonical form By means of a congruent transformation the canonical matrix **C**, Eq. (13-49), can be obtained from a matrix **A** of order n and rank r, so that only $+1$, -1, and 0 are present along the principal diagonal:

$$\mathbf{C} = \mathbf{P}^T\mathbf{A}\mathbf{P} = \begin{bmatrix} \mathbf{I}_p & 0 & 0 \\ 0 & -\mathbf{I}_{r-p} & 0 \\ 0 & 0 & 0 \end{bmatrix} \quad (13\text{-}49)$$

The integer p is called the *index* of the matrix, and the integer $s = p - (r - p)$ is called the *signature* of the matrix. Congruent matrices are of the same order and have the same rank and index or the same rank and signature.

The pairs of elementary transformations that are permitted by the transforming matrix **P** are:

1 Interchanging the ith and jth rows and interchanging the ith and jth columns
2 Multiplying the ith row and column by a nonzero scalar k
3 Addition to the elements of the ith row of the corresponding elements of the jth row multiplied by a scalar k and addition to the ith column of the jth column multiplied by k

Premultiplying and postmultiplying the matrix **A** by elementary matrices performs these three operations. For example, for a third-order matrix **A**, premultiplying by the elementary matrices shown below results in operations on the rows.

$$\begin{bmatrix} 0 & 1 & 0 \\ 1 & 0 & 0 \\ 0 & 0 & 1 \end{bmatrix} \qquad \begin{bmatrix} 1 & 0 & 0 \\ 0 & 1 & 0 \\ 0 & 0 & k \end{bmatrix} \qquad \begin{bmatrix} 1 & 0 & 0 \\ 0 & 1 & k \\ 0 & 0 & 1 \end{bmatrix}$$

Interchanging rows 1 and 2   Multiplying row 3 by k Adding k times row 3 to row 2

Postmultiplying by similar elementary matrices performs the desired column operations.

EXAMPLE 3   The matrix **A** of Eq. (13-46) is transformed to diagonal form using the following procedure.

Step 1   Form the augmented matrix $[\mathbf{A} \mid \mathbf{I}]$, where the unit matrix **I** has the same order as **A**:

$$[\mathbf{AI}] = \begin{bmatrix} 1 & 3 & 0 & 1 & 0 & 0 \\ 3 & -4 & -10 & 0 & 1 & 0 \\ 0 & -10 & 5 & 0 & 0 & 1 \end{bmatrix}$$

Step 2   Subtract 3 times the first row from the second row. Then subtract 3 times the first column from the second column. The resulting matrix is

$$\begin{bmatrix} 1 & 0 & 0 & 1 & 0 & 0 \\ 0 & -13 & -10 & -3 & 1 & 0 \\ 0 & -10 & 5 & 0 & 0 & 1 \end{bmatrix}$$

Step 3   Add 2 times the third row to the second row; then add 2 times column 3 to column 2. This gives

$$\begin{bmatrix} 1 & 0 & 0 & 1 & 0 & 0 \\ 0 & -33 & 0 & -3 & 1 & 2 \\ 0 & 0 & 5 & 0 & 0 & 1 \end{bmatrix}$$

Step 4   Interchange rows 2 and 3; then interchange columns 2 and 3. Multiply row 2 and column 2 by $1/\sqrt{5}$. Multiply row 3 and column 3 by $1/\sqrt{33}$. The resulting matrix is

$$\begin{bmatrix} 1 & 0 & 0 & 1 & 0 & 0 \\ 0 & 1 & 0 & 0 & 0 & \dfrac{1}{\sqrt{5}} \\ 0 & 0 & -1 & -\dfrac{3}{\sqrt{33}} & \dfrac{1}{\sqrt{33}} & \dfrac{2}{\sqrt{33}} \end{bmatrix} = [\mathbf{C}\mathbf{P}^T] \qquad (13\text{-}50)$$

The resulting canonical matrix $\mathbf{C}$ has rank $r = 3$, index $p = 2$, and signature $s = 1$. The required transformation $\mathbf{P}^T$ is identified directly in the final matrix. The original matrix $\mathbf{A}$ and the resulting canonical matrix $\mathbf{C}$ are symmetric. However, the matrix $\mathbf{P}$ is not symmetric.

When the matrix $\mathbf{A}$ is transformed to the canonical form, the quadratic form of Eq. (13-46) becomes

$$V = \mathbf{y}^T \mathbf{C} \mathbf{y} = y_1^2 + y_2^2 + \cdots + y_p^2 - y_{p+1}^2 - \cdots - y_r^2 \qquad (13\text{-}51)$$

where p is the index and r is the rank of the quadratic form. The rank and index remain the same in a transformation to the canonical form.

Length of a vector   The length of a vector $\mathbf{x}$ is called the euclidean *norm* and is denoted by $\|\mathbf{x}\|$. It is defined as the square root of the inner product $\langle \mathbf{x}, \mathbf{x} \rangle$. For real vectors it is given by

$$\|\mathbf{x}\| = \sqrt{\langle \mathbf{x}, \mathbf{x} \rangle} = \sqrt{x_1^2 + x_2^2 + \cdots + x_n^2} \qquad (13\text{-}52)$$

A vector can be normalized so that its length is unity. In that case it is called a *unit vector*. The unit vector may be denoted by $\hat{\mathbf{x}}$ and is obtained by dividing each element of $\mathbf{x}$ by $\|\mathbf{x}\|$:

$$\hat{\mathbf{x}} = \frac{\mathbf{x}}{\|\mathbf{x}\|} \qquad (13\text{-}53)$$

Principal minors   A *principal minor* of a matrix $\mathbf{A}$ is obtained by deleting any row(s) and the same numbered column(s). The diagonal elements of a principal minor of $\mathbf{A}$ are therefore also diagonal elements of $\mathbf{A}$. The $|\mathbf{A}|$ is classified as a principal minor, with no rows and columns deleted. The number of principal minors can be determined from a Pascal triangle.[10]   The number of principal minors for a matrix $\mathbf{A}$ of order n is 1 for $n = 1$, 3 for $n = 2$, 7 for $n = 3$, 15 for $n = 4, \ldots$.

Leading principal minor There are n leading *principal minors*, formed by all the square arrays within **A** that contain a_{11}. The leading principal minors are given by

$$\Delta_1 = |a_{11}| \quad \Delta_{12} = \begin{vmatrix} a_{11} & a_{12} \\ a_{21} & a_{22} \end{vmatrix} \quad \Delta_{123} = \begin{vmatrix} a_{11} & a_{12} & a_{13} \\ a_{21} & a_{22} & a_{23} \\ a_{31} & a_{32} & a_{33} \end{vmatrix}, \dots, \Delta_{12 \cdots n} = |\mathbf{A}|$$

The subscripts of Δ are the rows and columns of **A** used to form the minor. The minor contains the common elements of these rows and columns.

Definiteness and semidefiniteness The sign definiteness for a scalar function of a vector, such as $V(\mathbf{x})$, is defined for a spherical region S about the origin described by $\|\mathbf{x}\| \leq K$ (a constant equal to the radius of S). The function of $V(\mathbf{x})$ and all $\partial V(\mathbf{x})/\partial x_i$, for $i = 1, \dots, n$, must be continuous within S. The definiteness of a quadratic form is determined by analyzing only the symmetric **A** matrix. If **A** is given as a non-symmetric matrix, it must first be converted to a symmetric matrix (see Example 2 of this section).

Positive definite (*PD*) A scalar function, such as the quadratic form $V(\mathbf{x}) = \langle \mathbf{x}, \mathbf{Ax} \rangle$, is called positive definite when $V(\mathbf{x}) = 0$ for $\mathbf{x} = 0$ and $V(\mathbf{x}) > 0$ for all other $\|\mathbf{x}\| \leq K$. The positive definite condition requires that $|\mathbf{A}| \neq 0$; thus the rank is equal to $r = n$. By reference to Eq. (13-49) it is also seen that the index p must be equal to the rank.

When a real quadratic form $V(\mathbf{x}) = \mathbf{x}^T \mathbf{Ax}$ is positive definite, then the matrix **A** is also called positive definite. The matrix **A** has the property that it is positive definite iff there exists a nonsingular matrix **H** such that $\mathbf{A} = \mathbf{H}^T \mathbf{H}$.

The matrix **A** can be reduced to diagonal or canonical form by means of a congruent transformation, as described earlier in this section. Another method of diagonalizing the **A** matrix is to determine a modal matrix **T** using the methods of Sec. 5-9. In that case the diagonalized matrix contains the characteristic values obtained from $|\lambda \mathbf{I} - \mathbf{A}| = 0$. In order for the matrix **A** to be positive definite, all the characteristic values must be positive.

An alternate method of determining positive definiteness is to calculate *all* the leading principal minors of the matrix **A**. If all the leading principal minors of **A** are positive, the real quadratic form is positive definite. This is sometimes called the Sylvester theorem. It should not be confused with the Sylvester expansion presented in Sec. 3-13. When the leading principal minors are all positive, then all the other principal minors are also positive.

Positive semidefinite (*PSD*) The quadratic form $V(\mathbf{x})$ is called positive semidefinite when $V(\mathbf{x}) = 0$ for $\mathbf{x} = 0$ and $V(\mathbf{x}) \geq 0$ for all $\|\mathbf{x}\| \leq K$. The function $V(\mathbf{x})$ is permitted to equal zero at points in S other than the origin, i.e., for some $\mathbf{x} \neq 0$, but it may not be negative. In this case $|\mathbf{A}| = 0$, the rank $r < n$, and the rank is equal to the index, that is, $r = p < n$. The positive semidefinite quadratic form can therefore be reduced to the form $y_1^2 + y_2^2 + \cdots + y_r^2$, where $r < n$. When $V(\mathbf{x})$ is positive semidefinite, then the matrix **A** is also called positive semidefinite. The matrix A is positive semidefinite iff all its characteristic values are ≥ 0.

In order to be positive semidefinite, *all* the principal minors[6] must be non-negative.

Negative definite (*ND*) **and negative semidefinite** (*NSD*)   The definitions of negative definite and negative semidefinite follow directly from the definitions above when the inequalities are applied to $-\mathbf{A}$. If a matrix $\mathbf{A}$ does not satisfy the conditions for positive definiteness or positive semidefiniteness, then $-\mathbf{A}$ is checked for these conditions. If $-\mathbf{A}$ satisfies the positive definite or positive semidefinite conditions, then $\mathbf{A}$ is said to be negative definite or negative semidefinite, respectively.

Indefinite   A scalar function $V(\mathbf{x})$ is *indefinite* if it assumes both positive and negative values within the region S described by $\|\mathbf{x}\| < K$.

If $\mathbf{A}$ is not positive definite, positive semidefinite, negative definite, or negative semidefinite, it is said to be indefinite. Note that when $\mathbf{A}$, in general form, has both positive and negative elements along the principal diagonal, it is indefinite. When $\mathbf{A}$ is transformed to a diagonal form, it is indefinite if some of the diagonal elements are positive and some are negative. This means that some of the eigenvalues are positive and some are negative.

EXAMPLE 4   A quadratic form is given in Example 1. The matrix $\mathbf{A}$ is transformed in Example 3 to the canonical form $\mathbf{C}$ represented in Eq. (13-50). Since some of the diagonal elements are positive and some are negative, the matrix is indefinite.

EXAMPLE 5   Use the principal minors to determine the definiteness of

$$\mathbf{A} = \begin{bmatrix} 1 & 1 & 1 \\ 1 & 1 & 1 \\ 1 & 1 & 0 \end{bmatrix}$$

The leading principal minors are evaluated to check for positive definiteness:

$$\Delta_1 = 1 \qquad \Delta_{12} = \begin{vmatrix} 1 & 1 \\ 1 & 1 \end{vmatrix} = 0 \qquad \Delta_{123} = \Delta = 0$$

The conditions for positive definiteness are not satisfied. Next the remaining principal minors are evaluated to check for positive semidefiniteness

$$\Delta_{13} = \begin{vmatrix} 1 & 1 \\ 1 & 0 \end{vmatrix} = -1 \qquad \Delta_{23} = \begin{vmatrix} 1 & 1 \\ 1 & 0 \end{vmatrix} = -1 \qquad \Delta_2 = 1 \qquad \Delta_3 = 0$$

Since the principal minors are not all nonnegative, the matrix $\mathbf{A}$ is not positive semidefinite. Similarly, $\mathbf{A}$ is not negative definite or negative semidefinite, therefore $\mathbf{A}$ is indefinite.

EXAMPLE 6   Transform $\mathbf{A}$ in Example 5 to the diagonal Jordan form $\mathbf{\Lambda}$ and check its definiteness. The characteristic equation is

$$|\lambda \mathbf{I} - \mathbf{A}| = \begin{vmatrix} \lambda - 1 & -1 & -1 \\ -1 & \lambda - 1 & -1 \\ -1 & -1 & \lambda \end{vmatrix} = \lambda^3 - 2\lambda^2 - 2\lambda = 0$$

The eigenvalues are

$$\lambda_1 = 0 \qquad \lambda_2 = 1 - \sqrt{3} = -0.732 \qquad \lambda_3 = 1 + \sqrt{3} = 2.732$$

Thus
$$\Lambda = \begin{bmatrix} 2.732 & 0 & 0 \\ 0 & -0.732 & 0 \\ 0 & 0 & 0 \end{bmatrix}$$

This confirms that the matrix is indefinite.

EXAMPLE 7 Determine the definiteness of

$$A = \begin{bmatrix} 2 & 1 & -1 \\ 1 & 2 & 0 \\ -1 & 0 & 2 \end{bmatrix}$$

The leading principal minors are

$$\Delta_1 = 2 \qquad \Delta_{12} = 3 \qquad \Delta = |A| = 4$$

Therefore **A** is positive definite. Note that the other principal minors are also positive. $\Delta_2 = 2, \Delta_3 = 2, \Delta_{13} = 3, \Delta_{23} = 4$.

13-5 STABILITY[7,8]

The output stability of a linear time-invariant system is expressed simply by the condition that the output response must be bounded when the input is bounded. It follows directly that this requires all eigenvalues to have negative real parts. This is a necessary and sufficient condition for stability. It tells one that the transient terms associated with those poles decrease with time and that the output response therefore approaches the particular solution which is determined by the input. An alternate way of expressing the necessary and sufficient condition for stability is that the integral of the system weighting function, or impulse response, must be finite. This is expressed by

$$\int_0^\infty |w(t)| \, dt < \infty \qquad (13\text{-}54)$$

A direct method of determining stability, when the characteristic equation is available, is to evaluate its roots. Forming the Routhian array or the Hurwitz determinants also permits the determination of stability without evaluating the roots.

The extension of stability evaluation to nonlinear systems is not so straightforward, since the concept of root location is no longer applicable. As a result, many different classes of stability have been described in the literature for such systems. In this chapter the intention is to present some new concepts for viewing system stability. Although this text is restricted primarily to linear time-invariant systems, these concepts can be extended to nonlinear and time-varying systems.

Stability in the sense of Liapunov Consider a region ε in the state space enclosing an equilibrium point x_0. This equilibrium point is stable provided that there is a region $\delta(\varepsilon)$, which is contained within ε, such that any trajectory starting in the region δ does

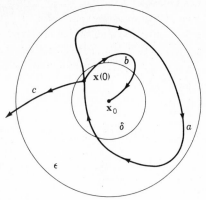

FIGURE 13-12
State-plane trajectories indicating (a)
Liapunov stability, (b) asymptotic sta-
bility, and (c) instability.

not leave the region ε. Note that with this definition it is not necessary for the trajectory to approach the equilibrium point. It is necessary only for the trajectory to stay within the region ε. This permits the existence of a continuous oscillation about the equilibrium point. The state-space trajectory for such an oscillation is a closed path called a *limit cycle*. The performance specifications for a control system must be used to determine whether or not a limit cycle can be permitted. The amplitude and frequency of the oscillation may influence whether it represents acceptable performance. When limit cycles are not acceptable, more stringent restraints must be chosen to exclude their possible existence.

Limit cycles are usually characterized as stable or unstable. If a limit cycle is stable, it means that trajectories in the state space on either side will approach the limit cycle. An unstable limit cycle is one in which the trajectories on either side diverge from the limit cycle. These trajectories may approach other limit cycles or equilibrium points.

Asymptotic stability An equilibrium point is *asymptotically stable* if, in addition to being stable in the sense of Liapunov, all trajectories approach the equilibrium point. This means that the variational solution $\mathbf{x}^*(t)$ approaches $\mathbf{0}$ as time t approaches infinity. This is the stability definition usually used in control-system design.

Figure 13-12 shows trajectories in the state plane illustrating the general principle of Liapunov stability, asymptotic stability, and instability. The trajectory a starting at the initial state point $\mathbf{x}(0)$ and remaining within the region ε meets the conditions for Liapunov stability. This trajectory is closed, indicating a continuous oscillation or limit cycle. Trajectory b terminates at the equilibrium point $\mathbf{x}_0$; thus it represents asymptotic stability. Trajectory c leaves the region ε and therefore indicates instability.

When the region δ includes the entire state space, the definitions of Liapunov and asymptotic stability are said to apply in a *global* sense. The stability or instability of a linear system is global because any initial state yields the same stability determination. A stable linear system is globally asymptotically stable.

In Sec. 13-4 the technique for linearizing nonlinear differential equations in the neighborhood of their singularities is presented. The validity of determining stability of the unperturbed solution near the singular points from the linearized equations was developed independently by Poincaré and Liapunov in 1892. Liapunov designated this as the *first method*. This stability determination is applicable only in a small region near the singularity and results in stability *in the small*. The next section considers Liapunov's *second method*, which is used to determine stability *in the large*. This larger region may include a finite portion or sometimes the whole region of the state space.

13-6   SECOND METHOD OF LIAPUNOV[7,8]

The second, or direct, method of Liapunov provides a means for determining the stability of a system without explicitly solving for the trajectories in the state space. This is in contrast to the first method of Liapunov, which requires the determination of the eigenvalues from the linearized equations about an equilibrium point. The second method is applicable for determining the behavior of higher-order systems which may be forced or unforced, linear or nonlinear, time-invariant or time-varying, and deterministic or stochastic. Solution of the differential equation is not required. The procedure requires the selection of a scalar function $V(\mathbf{x})$, which is tested for the conditions that indicate stability. When $V(\mathbf{x})$ successfully meets these conditions, it is called a *Liapunov function*. The principal difficulty in applying the method is in formulating a correct Liapunov function because the failure of one function to meet the stability conditions does not mean that a true Liapunov function does not exist. This difficulty is compounded by the fact that the Liapunov function is not unique. Nevertheless there is much interest in the second method. The further use of the Liapunov functions for evaluating quadratic performance criteria is covered in Chap. 14.

In order to show a simple example of a Liapunov function, consider the system of Sec. 13-3 represented by Eq. (13-27). The proposed function is the sum of the kinetic and potential stored energies, given by

$$V(\theta_o, \dot{\theta}_o) = \tfrac{1}{2}\dot{\theta}_o^2 + K(1 - \cos \theta_o) \qquad (13\text{-}55)$$

This function is positive for all values of θ_0 and $\dot{\theta}_0$, except at the origin where it is equal to zero. The rate of change of this energy function along any phase-plane trajectory is

$$\dot{V}(\theta_o, \dot{\theta}_o) = \dot{\theta}_o \ddot{\theta}_o + K(\sin \theta_o)\dot{\theta}_o = -a\dot{\theta}_o^2 \qquad (13\text{-}56)$$

The value of $\dot{V}$ is negative along any trajectory for all values of $\dot{\theta}_o$ except $\dot{\theta}_o = 0$. Note that the slope of $d\dot{\theta}_o/d\theta_o$ of the phase-plane trajectories is infinite along the line represented by $\dot{\theta}_o = 0$ except at the equilibrium points $\theta_o = n\pi$. Thus, the line $\dot{\theta}_o = 0$ does not represent an equilibrium, except at $\theta_o = n\pi$, and it is not a trajectory in the state plane. Since the energy stored in the system is continuously decreasing at all points except the equilibrium points, the equilibrium at the origin and at even multiples of $\theta_0 = n\pi$ is asymptotically stable. This $\dot{V}$ is NSD.

This example demonstrates that the total system energy may be used as the Liapunov function. When the equations of a large system are given in mathematical form, it is usually difficult to define the energy of the system. Thus, alternate Liapunov functions must be obtained. For any system of order n a positive, constant value of the proper positive definite Liapunov function $V(\mathbf{x})$ represents a closed surface, a

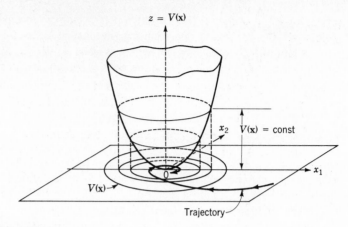

FIGURE 13-13
A positive definite function $V(x_1,x_2)$ and projections on the x_1x_2 plane.

hyperellipsoid, in the state space with its center at the origin. The entire state space is filled with such nonintersecting closed surfaces, each representing a different positive value of $V(\mathbf{x})$. When $\dot{V}(\mathbf{x})$ is negative for all points in the state space, it means that all trajectories cross the closed surfaces from the outside to the inside and eventually converge at the equilibrium point at the origin.

A qualitatively correct Liapunov function is shown in Fig. 13-13 for a second-order system. The function $V(x_1,x_2) = x_1{}^2 + x_2{}^2$ is positive definite and is represented by the paraboloid surface shown. The value $V(x_1,x_2) = k_i$ (a constant) is represented by the intersection of the surface $V(x_1,x_2)$ and the plane $z = k_i$. The projection of this intersection on the x_1x_2 plane is a closed curve, an oval, around the origin. There is a family of such closed curves in the x_1x_2 plane for different values of k_i. The value $V(x_1,x_2) = 0$ is the point at the origin; it is the innermost curve of the family of curves representing different levels on the paraboloid for $V(x_1,x_2) = k_i$.

The gradient vector of $V(\mathbf{x})$ is defined by

$$\text{grad } V(\mathbf{x}) = \nabla V(\mathbf{x}) = \begin{bmatrix} \dfrac{\partial V(\mathbf{x})}{\partial x_1} \\[2mm] \dfrac{\partial V(\mathbf{x})}{\partial x_2} \\[2mm] \vdots \\[2mm] \dfrac{\partial V(\mathbf{x})}{\partial x_n} \end{bmatrix} \qquad (13\text{-}57)$$

The time derivative $\dot{V}(x)$ along any trajectory is

$$\frac{dV(\mathbf{x})}{dt} = \dot{V}(\mathbf{x})$$

$$= \sum_{i=1}^{n} \frac{\partial V}{\partial x_i} \frac{dx_i}{dt} = \frac{\partial V(\mathbf{x})}{\partial x_1} \frac{dx_1}{dt} + \frac{\partial V(\mathbf{x})}{\partial x_2} \frac{dx_2}{dt} + \cdots + \frac{\partial V(\mathbf{x})}{\partial x_n} \frac{dx_n}{dt}$$

$$= [\text{grad } V(\mathbf{x})]^T \dot{\mathbf{x}} = [\nabla V(\mathbf{x})]^T \dot{\mathbf{x}} = \langle \nabla V, \dot{\mathbf{x}} \rangle \qquad (13\text{-}58)$$

It is important to note that $\dot{V}(\mathbf{x})$ can be evaluated without knowing the solution to the system state equation $\dot{\mathbf{x}} = f(\mathbf{x})$.

The gradient $\nabla V(x_1,x_2)$ describes the steepness between adjacent levels of $V(x_1,x_2)$. The function $\dot{V}(x_1,x_2)$ is negative definite in Fig. 13-13, except at the origin where it is equal to zero. The state-plane trajectory shown crosses the ovals for successively smaller values of $V(\mathbf{x})$. Therefore, the system is asymptotically stable, and $V(\mathbf{x})$ is a proper Liapunov function.

The concepts described above may be summarized in the following theorem, which provides sufficient, but not necessary, conditions for stability.

Theorem 1 Liapunov asymptotic stability A system is asymptotically stable in the vicinity of the equilibrium point at the origin if there exists a scalar function $V(\mathbf{x})$ such that:

   1 $V(\mathbf{x})$ is continuous and has continuous first partial derivatives in a region S around the origin.
   2 $V(\mathbf{x}) > 0$ for $\mathbf{x} \neq 0$.
   3 $V(0) = 0$.
   4 $\dot{V}(\mathbf{x}) < 0$ for $\mathbf{x} \neq 0$.

Conditions 1 to 3 ensure that $V(\mathbf{x})$ is positive definite. Therefore, $V(\mathbf{x}) = k$ is a closed surface within the region S. Condition 4 means that $\dot{V}(\mathbf{x})$ is negative definite, and thus any trajectory in S crosses through the surface $V(\mathbf{x}) = k$ from the outside to the inside for all values of k. Therefore, the trajectory converges on the origin where $V(0) = 0$.

The condition that $\dot{V}(\mathbf{x}) < 0$ for $\mathbf{x} \neq 0$ can be relaxed in Theorem 1 under the proper conditions. Condition 4 can be changed to $\dot{V}(\mathbf{x}) \leq 0$; that is, $\dot{V}(\mathbf{x})$ is negative semidefinite. This relaxed condition is sufficient provided that $\dot{V}(\mathbf{x})$ is not equal to zero at any solution of the original differential equation except at the equilibrium point at the origin. A test for this condition is to insert the solution of $\dot{V}(\mathbf{x}) = 0$ into the state equation $\dot{\mathbf{x}} = f(\mathbf{x})$ to verify that it is satisfied only at the equilibrium point. Also, if it can be shown that no trajectory can stay forever at the points or on the line, other than the origin, at which $\dot{V} = 0$, then the origin is asymptotically stable. This is the case for the system described in Sec. 13-3. For linear systems there is only one equilibrium point which is at the origin; therefore it is sufficient for $\dot{V}(\mathbf{x})$ to be negative semidefinite. Theorem 1 may be extended so that it is applicable to the entire state space. In that case the system is said to have *global stability* or stability *in the large*. Including these conditions results in the following theorem.

Theorem 2 Liapunov global asymptotic stability A system is globally asymptotically stable if there exists a scalar function $V(\mathbf{x})$ such that:

   1 $V(\mathbf{x})$ is continuous and has continuous first partial derivatives in the entire state space.
   2 $V(\mathbf{x}) > 0$ for $\mathbf{x} \neq 0$.
   3 $V(0) = 0$.
   4 $V(\mathbf{x}) \to \infty$ as $\|\mathbf{x}\| \to \infty$.
   5 $\dot{V}(\mathbf{x}) \leq 0$.
   6 Either $\dot{V}(\mathbf{x}) \neq 0$ except at $\mathbf{x} = 0$ or any locus in the state space where $\dot{V}(\mathbf{x}) = 0$ is not a trajectory of the system.

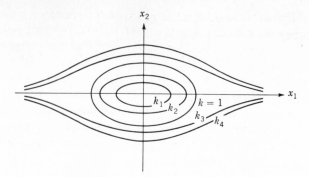

FIGURE 13-14
Contours for

$$V(\mathbf{x}) = \frac{x_1^2}{1 + x_1^2} + x_2^2 = k_i,$$

with $k_1 < k_2 < 1$ and $k_4 > k_3 > 1$.

Conditions 1 to 3 ensure that $V(\mathbf{x})$ is positive definite. Condition 4 is satisfied when $V(\mathbf{x})$ is positive definite, i.e., it is closed, in the *entire* state space. When $V(\mathbf{x})$ goes to infinity as any $x_i \rightarrow \infty$, then $V(\mathbf{x}) = k_i$ is a closed curve for any k_i. Conditions 5 and 6 mean that $V(\mathbf{x})$ is continuously decreasing along any trajectory in the entire plane and ensures that the system is asymptotically stable.

In order to check for global stability it is necessary to select $V(\mathbf{x})$ so that conditions 1 to 4 are all satisfied. For example, the following function, for a second-order system, does not satisfy condition 4:

$$V(\mathbf{x}) = \frac{x_1{}^2}{1 + x_1{}^2} + x_2{}^2 \qquad (13\text{-}59)$$

With x_2 finite and $x_1 \rightarrow \infty$, then $V(\mathbf{x}) = 1 + x_2{}^2$, which is not infinite. Therefore, this $V(\mathbf{x})$ does not satisfy condition 4 and cannot be used to determine global stability. The curves for $V(\mathbf{x}) = k$ are closed for $k \leq 1$, and they are open if $k > 1$, as shown in Fig. 13-14. This limits the region for which this function can be used to determine stability.

Finding a proper Liapunov function $V(\mathbf{x})$ means that the system is stable, but this is just a sufficient and not a necessary condition for stability. Because a function $V(\mathbf{x})$ cannot be found does not mean that it does not exist. The Liapunov stability theorems are stated above as requiring $V(\mathbf{x})$ to be positive definite and $\dot{V}(\mathbf{x})$ to be negative definite or negative semidefinite. The stability conditions can also be specified in terms of a negative definite $V(\mathbf{x})$ and a positive definite or positive semidefinite $\dot{V}(\mathbf{x})$. The only requirement is that $V(\mathbf{x})$ and $\dot{V}(\mathbf{x})$ must be of opposite sign. It is more common to start with a $V(\mathbf{x})$ which is positive definite.

The instability theorem can be used as a *sufficient* condition to positively identify an unstable system.

Theorem 3   Liapunov instability theorem   A system $\dot{\mathbf{x}} = f(\mathbf{x})$ is unstable in a region Ω about the origin if there exists a scalar function $V(\mathbf{x})$ such that:

1 $V(\mathbf{x}) \geq 0$, $V(\mathbf{0}) = 0$, and $V(\mathbf{x})$ is continuous and has continuous partial derivatives in the region Ω.
2 $\dot{V}(\mathbf{x})$ is positive definite in the region Ω.

The response of such a system is unbounded as $t \to \infty$ unless $\dot{V}(\mathbf{x})$ is globally negative semidefinite. This theorem may be rephrased in terms of negative functions. Thus, if $\dot{V}$ is negative definite, the system is unstable in any region Ω in which V is not positive definite or positive semidefinite.

This instability theorem is potentially more powerful than the stability theorem because if the conditions for instability are not satisfied, the conditions for stability are automatically satisfied. This is in contrast to the stability theorems (Theorems 1 and 2), where failure to satisfy the stability criteria does not necessarily mean that the system is unstable. It is often possible to select a function V for which $\dot{V}$ is positive (or negative) definite. Then, if V is positive (or negative) definite, semidefinite, or indefinite, the response is unstable. On the other hand, if V is negative (or positive) semidefinite, the instability conditions are not satisfied. This means that the system is stable.

The Liapunov instability theorem has the disadvantage of requiring both V and $\dot{V}$ to be positive definite over the entire region Ω. This theorem can be modified so that instability can be determined in just a region, however small, of the state space. The modifications are presented in the Cetaev theorem.

Theorem 4   Cetaev instability theorem   The origin is unstable when there is a region Ω_1, however small, inside a region Ω and when the Liapunov function $V(\mathbf{x})$ for the differential equation $\dot{\mathbf{x}} = f(\mathbf{x})$ has the following properties:

1 $V(\mathbf{x})$ has continuous first partial derivatives in Ω.
2 $V(\mathbf{x})$ and $\dot{V}(\mathbf{x})$ are positive in Ω_1.
3   The value of the function is $V(\mathbf{x}) = 0$ at the boundary of Ω_1 inside of Ω.
4   The origin is a boundary point of Ω_1.

13-7   APPLICATION OF THE LIAPUNOV METHOD TO LINEAR SYSTEMS

The second method of Liapunov is applicable to time-varying and nonlinear systems. However, there is no simple general method of developing the Liapunov function $V(\mathbf{x})$. Methods have been developed for many such systems, and the literature in this area[8] is extensive. The remaining applications in this chapter are restricted to linear systems. Since linear systems may have only one equilibrium point, the stability or instability is necessarily global in nature. The Routh-Hurwitz stability criterion is available for determining the stability of linear systems. However, the necessity of obtaining the characteristic polynomial can be a disadvantage for higher-order systems. Thus, the following material is presented to develop familiarity with the more

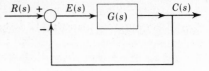

FIGURE 13-15
Unity-feedback system.

general second method of viewing stability. Confidence in the second method of Liapunov can be developed by showing that the results are identical to those obtained with the Routh-Hurwitz method.

The first approach presented is to select a $V(\mathbf{x})$ which is positive definite and to evaluate $\dot{V}(\mathbf{x})$. The coefficients of $V(\mathbf{x})$ and those restraints on the system parameters are then determined which make $\dot{V}(\mathbf{x})$ negative definite or negative semidefinite. Consider the linear feedback system presented in Fig. 13-15 with $r(t) = 0$.

EXAMPLE 1

$$G(s) = \frac{K}{s(s + a)} \qquad a > 0$$

The differential equation for the actuating signal is

$$\ddot{e} + a\dot{e} + Ke = 0 \qquad (13\text{-}60)$$

When phase variables with $x_1 = e$ are used, the state equations are

$$\dot{x}_1 = x_2 \qquad \dot{x}_2 = -Kx_1 - ax_2 \qquad (13\text{-}61)$$

A simple Liapunov function which is positive definite is

$$V(\mathbf{x}) = \tfrac{1}{2}p_1 x_1{}^2 + \tfrac{1}{2}p_2 x_2{}^2 = \tfrac{1}{2}\mathbf{x}^T \mathbf{P} \mathbf{x} \qquad (13\text{-}62)$$

where $p_1 > 0$ and $p_2 > 0$. Its derivative is

$$\dot{V}(\mathbf{x}) = p_1 x_1 \dot{x}_1 + p_2 x_2 \dot{x}_2 = p_1 x_1 x_2 - p_2 K x_1 x_2 - a p_2 x_2{}^2$$

$$= \mathbf{x}^T \begin{bmatrix} 0 & \dfrac{p_1 - p_2 K}{2} \\ \dfrac{p_1 - p_2 K}{2} & -a p_2 \end{bmatrix} \mathbf{x} = -\mathbf{x}^T \mathbf{N} \mathbf{x} \qquad (13\text{-}63)$$

The function $\dot{V}(\mathbf{x})$ is always negative semidefinite if $-\mathbf{N}$ is negative semidefinite. The matrix $-\mathbf{N}$ is negative semidefinite if all the principal minors of $\mathbf{N}$ are positive or zero.

$$a p_2 \geq 0 \qquad (13\text{-}64)$$

$$\frac{-(p_1 - p_2 K)^2}{4} \geq 0 \qquad (13\text{-}65)$$

Equation (13-64) is satisfied since it is required that both a and p_2 be greater than zero. Since the left side of Eq. (13-65) cannot be positive, only the equality condition can be satisfied, which yields $p_1 = p_2 K$. Since $p_1 > 0$ and $p_2 > 0$, this requires that

$K > 0$. Then $\dot{V}(\mathbf{x}) = -ap_2x_2^{\,2}$ which is negative semidefinite. The Liapunov global asymptotic stability theorem (Theorem 2) is satisfied since, by condition 4, $V(\mathbf{x}) \to \infty$ as $\|\mathbf{x}\| \to \infty$. The condition $\dot{V}(\mathbf{x}) = 0$ exists along the x_1 axis where $x_2 = 0$ and x_1 has any value. A way of showing that $\dot{V}(\mathbf{x})$ being negative semidefinite is sufficient for global asymptotic stability is to show that the x_1 axis is not a trajectory of the system differential equations (13-61). The first equation yields $\dot{x}_1 = 0$ or $x_1 = c$. The x_1 axis can be a trajectory only if $x_2 = 0$ and $\dot{x}_2 = 0$. But the second equation yields $-Kx_1 = -Kc = 0$. This is a contradiction since it requires $x_1 = c = 0$. Therefore, the x_1 axis is not a trajectory, and the system is asymptotically stable.

Using Liapunov's second method, the necessary condition for stability is shown above to be $K > 0$. This result is readily recognized as being correct, either from the Routh stability criterion or from the root locus.

EXAMPLE 2

$$G(s) = \frac{K}{s(s - a)} \qquad a > 0$$

The phase-variable equations, using $x_1 = e$, are

$$\dot{x}_1 = x_2 \qquad \dot{x}_2 = -Kx_1 + ax_2 \qquad (13\text{-}66)$$

A more general Liapunov function is chosen in this example

$$V(\mathbf{x}) = \tfrac{1}{2}p_{11}x_1^{\,2} + p_{12}x_1x_2 + \tfrac{1}{2}p_{22}x_2^{\,2}$$

$$= \tfrac{1}{2}\mathbf{x}^T \begin{bmatrix} p_{11} & p_{12} \\ p_{12} & p_{22} \end{bmatrix} \mathbf{x} = \tfrac{1}{2}\mathbf{x}^T\mathbf{P}\mathbf{x} \qquad (13\text{-}67)$$

The derivative $\dot{V}(\mathbf{x})$ is obtained from Eq. (13-67) with values of $\dot{x}_1$ and $\dot{x}_2$ inserted from Eq. (13-66).

$$\dot{V}(\mathbf{x}) = -p_{12}Kx_1^{\,2} + (p_{11} + p_{12}a - p_{22}K)x_1x_2 + (p_{12} + p_{22}a)x_2^{\,2} \qquad (13\text{-}68)$$

A positive definite form is selected for $\dot{V}(\mathbf{x})$. Then the coefficients $p_{11}, p_{12},$ and p_{22} are evaluated, and the definiteness of $V(\mathbf{x})$ is determined by use of Sylvester's theorem. Accordingly, select

$$\dot{V}(\mathbf{x}) = x_1^{\,2} + x_2^{\,2} \qquad (13\text{-}69)$$

Equating the coefficients of Eqs. (13-68) and (13-69) yields

$$p_{12} = -\frac{1}{K} \qquad p_{22} = \frac{1 + K}{Ka} \qquad p_{11} = \frac{a^2 + K(1 + K)}{Ka}$$

The necessary conditions that $V(\mathbf{x})$ be positive definite, using Sylvester's theorem, are

$$p_{11} = \frac{a^2 + K(1 + K)}{Ka} > 0$$

$$p_{11}p_{22} - p_{12}^{\,2} = \frac{K[a^2 + (1 + K)^2]}{K^2a^2} > 0$$

The second equation is satisfied if $K > 0$. Since $a > 0$, this requirement also satisfies the first equation; therefore, $V(\mathbf{x})$ is positive definite. Since $\dot{V}(\mathbf{x})$ is also positive definite [see Eq. (13-69)], Theorem 3 is satisfied and the system is globally unstable. This result is in agreement with the results obtained using the Routh criterion. The reader may wish to investigate (1) the results of permitting $K < 0$ and (2) the use of the negative definite function $\dot{V}(\mathbf{x})$ and $-x_1{}^2 - x_2{}^2$.

Two approaches for applying the second method of Liapunov are demonstrated in these examples. In Example 1 the Liapunov function $V(\mathbf{x})$ is selected, and the definiteness of $\dot{V}(\mathbf{x})$ is determined. In Example 2 the definiteness of $\dot{V}(\mathbf{x})$ is selected, and then the definiteness of $V(\mathbf{x})$ is determined.

Another approach presented in this section is the use of a procedure developed by Lur'e. Consider the unexcited system represented by the state equation in which $\mathbf{A}$ is of order n:

$$\dot{\mathbf{x}} = \mathbf{A}\mathbf{x} \qquad (13\text{-}70)$$

The quadratic Liapunov function $V(\mathbf{x})$ is expressed in terms of the symmetric matrix $\mathbf{P}$ by

$$V(\mathbf{x}) = \mathbf{x}^T\mathbf{P}\mathbf{x} \qquad (13\text{-}71)$$

The time derivative of $V(\mathbf{x})$ is

$$\dot{V}(\mathbf{x}) = \dot{\mathbf{x}}^T\mathbf{P}\mathbf{x} + \mathbf{x}^T\mathbf{P}\dot{\mathbf{x}} = \mathbf{x}^T(\mathbf{A}^T\mathbf{P} + \mathbf{P}\mathbf{A})\mathbf{x} = -\mathbf{x}^T\mathbf{N}\mathbf{x} \qquad (13\text{-}72)$$

where the symmetric matrix $\mathbf{N}$ is given by

$$\mathbf{N} = -(\mathbf{A}^T\mathbf{P} + \mathbf{P}\mathbf{A}) = 2(\mathbf{P}\mathbf{A}) \text{ sym.} \qquad (13\text{-}73)$$

The following procedure is used.

Step 1   Select an *arbitrary* $\mathbf{N}$ of order n which is either positive definite or positive semidefinite. A simple diagonal matrix such as $\mathbf{N} = \mathbf{I}$ or $\mathbf{N} = 2\mathbf{I}$ is positive definite. A positive semidefinite matrix $\mathbf{N}$ can be chosen which contains all zero elements except one positive element in any position along the principal diagonal.

Step 2   Determine the elements of $\mathbf{P}$ by equating terms in Eq. (13-73). Since $\mathbf{P}$ is symmetric, this requires the solution of $n(n + 1)/2$ equations.

Step 3   Use the Sylvester theorem to determine the definiteness of $\mathbf{P}$.

Step 4   Since $\mathbf{N}$ is selected as positive definite (or positive semidefinite), the necessary and sufficient condition for asymptotic stability (instability) is that $\mathbf{P}$ be positive definite (negative definite).

The Sylvester conditions for the positive definiteness of $\mathbf{P}$ are the same as the Routh-Hurwitz stability conditions. The equivalence is derived in Ref. 11. Linear systems which are asymptotically stable are globally asymptotically stable.

EXAMPLE 3   The reader may verify that Example 2 of this section is solved by using the Lur'e method with

$$\mathbf{N} = \begin{bmatrix} 1 & 0 \\ 0 & 1 \end{bmatrix}$$

EXAMPLE 4 Examine the conditions of asymptotic stability for the second-order dynamic system

$$\ddot{x} + a_1\dot{x} + a_0x = 0$$

With phase variables the coefficient matrix is

$$\mathbf{A} = \begin{bmatrix} 0 & 1 \\ -a_0 & -a_1 \end{bmatrix}$$

The symmetric matrix $\mathbf{P}$ is

$$\mathbf{P} = \begin{bmatrix} p_{11} & p_{12} \\ p_{12} & p_{22} \end{bmatrix}$$

The matrix $\mathbf{N}$ is chosen as positive semidefinite:

$$\mathbf{N} = \begin{bmatrix} 2 & 0 \\ 0 & 0 \end{bmatrix}$$

The identity in Eq. (13-73) yields three simultaneous equations

$$n_{11} = 2 = 2a_0p_{12}$$
$$n_{12} = 0 = -p_{11} + a_1p_{12} + a_0p_{22}$$
$$n_{22} = 0 = -2p_{12} + 2a_1p_{22}$$

The solution of these equations yields

$$p_{12} = \frac{1}{a_0} \qquad p_{22} = \frac{1}{a_0a_1} \qquad p_{11} = \frac{a_0 + a_1^2}{a_0a_1}$$

The necessary conditions for $\mathbf{P}$ to be positive definite are obtained by applying the Sylvester theorem:

$$p_{11} = \frac{a_0 + a_1^2}{a_0a_1} > 0 \qquad p_{11}p_{22} - p_{12}^2 = \frac{a_0}{a_0^2a_1^2} > 0$$

The second equation requires that $a_0 > 0$. Using this condition in the first equation produces the necessary condition $a_1 > 0$. These are obviously the same conditions as are obtained from the Routh-Hurwitz conditions for stability.

Krasovskii has shown[8] that a similar approach may be used with a nonlinear system. However, only sufficient conditions for local asymptotic stability in the vicinity of an equilibrium point may be determined. For the Liapunov function $V(\mathbf{x}) = \mathbf{x}^T\mathbf{P}\mathbf{x}$, the time derivative is $\dot{V}(\mathbf{x}) = -\mathbf{x}^T\mathbf{N}\mathbf{x}$, where

$$-\mathbf{N} = \mathbf{J}_x^T\mathbf{P} + \mathbf{P}\mathbf{J}_x \qquad (13\text{-}74)$$

The matrix $\mathbf{J}_x$ is the Jacobian evaluated at the equilibrium point. Selecting $\mathbf{P} = \mathbf{I}$ may often lead to a successful determination of the conditions for asymptotic stability in the vicinity of the equilibrium. This extension to nonlinear systems is presented to show the generality of the method.

13-8 ESTIMATION OF TRANSIENT SETTLING TIME[8]

When the Liapunov function $V(\mathbf{x}) = \mathbf{x}^T\mathbf{P}\mathbf{x}$ is positive definite and its derivative $\dot{V}(\mathbf{x}) = -\mathbf{x}^T\mathbf{N}\mathbf{x}$ is negative definite, it is known that the system is asymptotically stable. They may also be used to determine a measure of the settling time of the unforced transient response. The development is demonstrated by starting with the identity

$$\dot{V}(\mathbf{x}) = \frac{\dot{V}(\mathbf{x})}{V(\mathbf{x})} V(\mathbf{x}) \le -aV(\mathbf{x}) \qquad (13\text{-}75)$$

where $-a$ is defined as the maximum value of the ratio $\dot{V}(\mathbf{x})/V(\mathbf{x})$ throughout the state space $\mathbf{x}$. This definition is represented symbolically by

$$-a \equiv \max_{\mathbf{x}} \frac{\dot{V}(\mathbf{x})}{V(\mathbf{x})} \qquad (13\text{-}76)$$

where $a > 0$. Equation (13-76) can be rewritten as

$$\frac{\dot{V}(\mathbf{x})}{V(\mathbf{x})} \le -a \qquad (13\text{-}77)$$

The integration of Eq. (13-77) from $t = 0$ to the settling time $t = T_s$ yields

$$V[\mathbf{x}(T_s)] \le V[\mathbf{x}(0)]e^{-aT_s} \qquad (13\text{-}78)$$

Equation (13-78) gives the value of the upper bound of $V[\mathbf{x}(T_s)]$ at the settling time T_s. Thus, $-a$ may be considered proportional to the largest eigenvalue of the system, and $1/a$ is the upper bound on the largest time constant of the Liapunov function.

Equation (13-76) is now rewritten as

$$a = \min_{\mathbf{x}} \frac{\mathbf{x}^T \mathbf{N} \mathbf{x}}{\mathbf{x}^T \mathbf{P} \mathbf{x}} \qquad (13\text{-}79)$$

Since $\mathbf{N}$ and $\mathbf{P}$ are positive definite, the denominator is not zero except when the numerator is also zero, at $\mathbf{x} = 0$. Therefore the division is allowable. Also, the relative values of $\dot{V}(\mathbf{x})$ and $V(\mathbf{x})$ are fixed throughout the whole state space. Therefore the evaluation of a in Eq. (13-79) may be constrained to the surface in the state space where

$$\mathbf{x}^T \mathbf{P} \mathbf{x} = 1 \qquad (13\text{-}80)$$

Then Eq. (13-79) becomes

$$a = \min_{\mathbf{x}} \mathbf{x}^T \mathbf{N} \mathbf{x} \qquad (13\text{-}81)$$

The Lagrange multiplier technique is used to evaluate the function of Eq. (13-81) with the constraint of Eq. (13-80). This is done by introducing the constant λ and forming the function

$$\mathbf{x}^T \mathbf{N} \mathbf{x} - \lambda \mathbf{I} = \mathbf{x}^T \mathbf{N} \mathbf{x} - \lambda \mathbf{x}^T \mathbf{P} \mathbf{x} = \mathbf{x}^T [\mathbf{N} - \lambda \mathbf{P}] \mathbf{x} \qquad (13\text{-}82)$$

The minimization is performed by setting the derivative to zero

$$\frac{d}{d\mathbf{x}} \{\mathbf{x}^T [\mathbf{N} - \lambda \mathbf{P}] \mathbf{x}\} = 0 \qquad (13\text{-}83)$$

The result (see Prob. 13-14) is

$$[\mathbf{N} - \lambda \mathbf{P}] \mathbf{x} = 0 \qquad (13\text{-}84)$$

This confirms that $\mathbf{N}$ and $\mathbf{P}$ differ only by the proportionality constant λ. Premultiplying by $\mathbf{x}^T$ and using Eq. (13-80) gives

$$\mathbf{x}^T \mathbf{N} \mathbf{x} = \lambda > 0 \qquad (13\text{-}85)$$

From Eq. (13-84) it is seen that $[\mathbf{NP}^{-1} - \lambda\mathbf{I}] = 0$. Thus, λ is the eigenvalue of $\mathbf{NP}^{-1}$. Therefore, the right side of Eq. (13-81) is a minimum when λ is a minimum, giving

$$a = \lambda_{\min} = \text{minimum eigenvalue of } \mathbf{NP}^{-1} \qquad (13\text{-}86)$$

When the eigenvalues of $\mathbf{NP}^{-1}$ are complex, the real component is intended in Eq. (13-86).

EXAMPLE Apply the method of this section to the system (in normal form):

$$\dot{\mathbf{x}} = \begin{bmatrix} -1 & 0 & 0 \\ 0 & -2 & 0 \\ 0 & 0 & -3 \end{bmatrix} \mathbf{x} \qquad (13\text{-}87)$$

Let $\mathbf{P} = \mathbf{I}$, which is positive definite; then the matrix $\mathbf{N}$ is [see Eq. (13-73)]

$$\mathbf{N} = -[\mathbf{\Lambda}^T\mathbf{P} + \mathbf{P}\mathbf{\Lambda}] = -2\mathbf{\Lambda}$$

Since $\mathbf{NP}^{-1} = -2\mathbf{\Lambda}$, its eigenvalues are 2, 4, 6, and $\lambda_{\min} = 2 = a$. When 2 percent of the initial value is used as the condition for settling of the transient, the upper bound on the settling time of the Liapunov function is

$$T_s = \frac{4}{a} = 2 \text{ s}$$

It is obvious from Eq. (13-87) that the largest root is -1 and the settling time of the system is approximately 4 s. The response time of the Liapunov function is therefore twice as fast. These times are proportional but not equal. The Liapunov function provides a measure of the actual system performance.

13-9   SUMMARY

This chapter provides the foundation for the design of optimal feedback control systems in Chap. 14. The properties of state-space trajectories and singular points are introduced. Most of the examples used are second-order systems because they clearly and easily demonstrate the important characteristics, such as stability and the kinds of singular points. Both linear and nonlinear differential equations are considered. The Jacobian matrix is used to represent nonlinear systems by approximate linear equations in the vicinity of their singular points. The important property is demonstrated that a linear equation has only one singular point whereas a nonlinear equation may have more than one singular point. The mathematical properties of quadratic forms and the definitions of definiteness are presented. The definitions of stability then lead to the Liapunov second method. Stability is evaluated in a linear system without determining the eigenvalues. The chapter concludes with a demonstration that the Liapunov function $V(\mathbf{x})$ provides a measure of system performance. This is the justification for its use in defining a performance index upon which the conditions of optimal performance are presented in Chap. 14.

REFERENCES

1 Andronow, A. A., et al.: "Theory of Oscillations," Addison-Wesley, Reading, Mass., 1966.
2 Thaler, G. J., and M. P. Pastel: "Analysis and Design of Nonlinear Feedback Control Systems," McGraw-Hill, New York, 1962.
3 Takahashi, Y., et al.: "Control and Dynamic Systems," Addison-Wesley, Reading, Mass., 1970.
4 DeRusso, P. M., et al.: "State Variables for Engineers," Wiley, New York, 1967.
5 Gantmacher, F. R.: "Applications of the Theory of Matrices," Wiley-Interscience, New York, 1959.
6 Swamy, K. N.: On Sylvester's Criterion for Positive-Semidefinite Matrices, *IEEE Trans. Autom. Control*, vol. AC-18, p. 306, June 1973.
7 Minorsky, N.: "Theory of Nonlinear Systems," McGraw-Hill, New York, 1969.
8 Csaki, F.: "Modern Control Theories," Akademiai, Kiado, Budapest, 1972.
9 Gibson, J. E.: "Nonlinear Automatic Control," McGraw-Hill, New York, 1963.
10 Parzen, E.: "Modern Probability and Its Applications," Wiley, New York, 1960.
11 Kalman, R. E., and J. E. Bertram: Control System Analysis and Design via the Second Method of Liapunov, *J. Basic Eng.*, vol. 80, pp. 371–400, 1960.

14

INTRODUCTION TO MODERN CONTROL

14-1 INTRODUCTION

When a performance criterion or a set of performance specifications is stipulated for a system and these conditions are not met, a control problem exists. Generally, in order to obtain the desired system performance additional equipment must be used in conjunction with the basic system. Either modern control or conventional design techniques are utilized to determine the configuration and parameters of the required additional equipment. By the nature of the modern control technique a specified performance criterion (for example, PI $= \int_0^\infty e^2 \, dt$ to be minimum) is met exactly, and a unique design is obtained. The performance of the system is therefore said to be *optimal* in terms of the defined performance criterion. In contrast, the conventional technique satisfies a required set of performance specifications. These requirements may be met by a number of different designs. In general, these designs do *not* simultaneously meet a defined optimal performance criterion. Therefore, these designs may be called *suboptimal*.

The synthesis and design of conventional feedback control systems are treated extensively in the literature, and the principal methods are described in the earlier chapters of this book. The design process includes the use of compensation or stabilization to improve the transient and steady-state performance. Each of the methods described has its particular advantages and disadvantages. The frequency-response method suffers from difficulty in determining the exact transient response from the frequency-response characteristics. While a correlation is made between the frequency and time responses in Chaps. 7, 9, and 10, this is based on representing the system by an approximately equivalent second-order system. Effective damping ratio and effective undamped natural frequency are related to the maximum value of the frequency response M_m. This approximation may become less valid as the order of the system becomes higher and is of limited value for nonlinear systems. Precise methods for obtaining the time response from the frequency response generally require computers. The root-locus method has the advantage that both the transient response and the frequency response can be obtained from the location of the closed-loop poles and zeros. Therefore, the improvements obtained by additional poles and zeros can be evaluated.

The principal difficulty remaining in feedback system design is the establishment of an optimum criterion for performance. In other words, what shape of the frequency response or what location of the poles and zeros gives the "best" system performance?

Evaluation of performance, given in Sec. 3-10, is based on the time response to a step-function input. The characteristics used to judge performance are as follows:

1   Maximum overshoot M_p
2   Time for the error to reach its first zero (duplicating time) t_0
3   Time to reach the maximum overshoot t_p
4   Settling time (also called solution time), which is the time for the response to settle within a given percentage of the final value, t_s
5   Frequency of oscillation of the transient ω_d
6   Steady-state error for a given input e_{ss}

The conventional design methods of Chaps. 10 to 12 are devoted to achieving a desired system performance based upon the figures of merit such as M_p, t_p, t_s, and K_m or M_m, ω_m, and K_m. These design specifications are selected because of the convenience in graphical interpretation with respect to the root-locus or frequency plots. Thus, these design methods rely heavily on graphical analysis in achieving the desired system performance. Compensators or feedback gains are selected that yield, as closely as possible, the desired system performance. In general, it may not be possible to satisfy all the desired figures of merit. Then, through a trial-and-error procedure an acceptable system performance is achieved, in which a trade-off in the values of the figures of merit is utilized. Generally, there are many system designs that can yield this acceptable system performance. In other words, there is, in general, no unique solution. The conventional design methods work very well for single-input single-output linear systems and result in good control systems.

The ready availability of the digital computer has led to the development and exploitation of modern control theory. This permits the achievement of an optimal system performance which meets some specified performance criterion. It involves minimizing (or maximizing) a *performance index* (PI). This method, in contrast to the conventional design technique, relies on the extensive use of mathematical analysis. The selection of a performance index is often based on mathematical convenience, i.e., the selected performance index permits the mathematical design of the system. While the method yields a unique mathematical solution, the actual system performance may not have all the desired performance characteristics. In other words, the resulting optimal control satisfies the mathematical performance index but may not yield desired values of M_p, t_p, t_s, etc. A compromise must be made between specifying a performance index which includes all the desired system characteristics and a performance index which can be achieved with a reasonable amount of computation. It should further be noted that it is most difficult to analyze multiple-input multiple-output systems by conventional control theory, whereas modern control theory is quite adaptable to the analysis and design, with the use of the digital computer, of such systems.

It is desired to develop a single PI to judge the "goodness" of the time response of the system. Such a PI should have three basic properties: reliability, ease of application, and selectivity. It should be reliable for a given class of systems so that it can be applied with confidence. It should be easy to apply, and it should also be selective so that the resulting system is clearly optimal. A desirable property is for the minimization of the PI to result in the achievement of some or all of the conventional figures of merit. This chapter is devoted to presenting several PIs used for optimizing system response.

Currently most control-system designs are based upon conventional design techniques, but an increasing number of systems are being designed using modern control techniques.[1] These have played a large role in the success of many aerospace control systems, which require high precision. It is expected that as the state of the art of modern control theory is developed, it will play a larger role in the design of control systems.

14-2   DEVELOPMENT OF THE SOLUTION-TIME CRITERION

To indicate the basis for establishing a performance criterion, the characteristics of a unity-feedback system are studied. Consider first a simple linear second-order system described by the control ratio

$$\frac{Y(s)}{R(s)} = \frac{\omega_n^2}{s^2 + 2\zeta\omega_n s + \omega_n^2} \qquad (14\text{-}1)$$

This system has a zero steady-state error with a step input. Only such systems are considered. This means that the open-loop transfer function must be Type 1 for a unity-feedback system.

Figure 14-1 shows the following three quantities as a function of the damping ratio ζ for the system represented by Eq. (14-1):

1 The time for the error to reach its first zero
2 The size of the first and largest overshoot, expressed as a percentage of the final value
3 The solution time, which is the time to reach and thereafter remain within 5 percent of the final value

A commonly desired response is one in which the output has a form identical to that of the input with no error at any time. Such a response is impractical, of course. Therefore the characteristics of Fig. 14-1 are studied to determine the best perform-

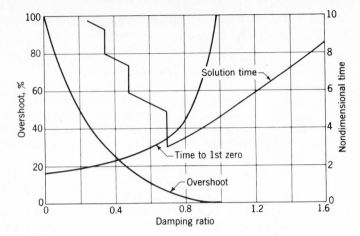

FIGURE 14-1
Transient performance of a simple second-order system with a step input.

ance. It is seen that overshoot and time to first zero are conflicting characteristics; i.e., their minimum values occur at different damping ratios. Therefore, if these two characteristics are to be used as criteria, there must be a compromise between the overshoot and rapid rise time. The solution time appears to combine both properties and can be used as a criterion of performance. The optimum solution time for this second-order system specifies a damping ratio of 0.7 as optimum, which is a commonly accepted value. Therefore the solution time may be the figure of merit to use to optimize a system. Of course, it is necessary to extend the study of this characteristic to higher-order and to nonlinear systems to make sure that it is universally applicable. Some results for linear systems are presented in Sec. 14-6. There seems to be one possible weakness in the use of this figure of merit. The sharp minimum gives an exaggerated picture of the optimum damping ratio since slightly larger or slightly smaller damping ratios will result in solution times not far removed from the optimum.

14-3   CONTROL-AREA CRITERION†

Another measure of the quality of the transient response of a control system to a step input is the *control area*,[3] which is shown as the shaded section of Fig. 14-2. The "best" system is one that has the minimum control area since $c(t)$ is then very close to $r(t)$. The control area is given by the integral of the error for some specified time. A typical performance index is

$$PI_1 = \int_0^\infty e \, dt \qquad (14\text{-}2)$$

† The material in Secs. 14-3 through 14-6 is based on Ref. 2.

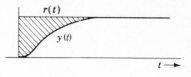

FIGURE 14-2
Control area for an overdamped system.

When this PI is a minimum for a specified input, the system performance is said to be optimal. The control ratio of the closed-loop system can be described by

$$\frac{C(s)}{R(s)} = \frac{(1 + A_1 s)(1 + A_2 s) \cdots (1 + A_w s)}{(1 + a_1 s)(1 + a_2 s) \cdots (1 + a_n s)} \tag{14-3}$$

In this equation the values of a_i and A_j can be either real or complex. The value of the control area obtained from Eq. (14-2) is

$$\mathrm{PI}_1 = (a_1 + a_2 + \cdots + a_n) - (A_1 + A_2 + \cdots + A_w) \tag{14-4}$$

If the values of the coefficients a_i in the denominator of $Y(s)/R(s)$ are complex, the minimum value of control area PI_1 occurs when the damping ratio of the dominant poles is zero. This is also shown in Fig. 14-3, which contains a plot of PI_1 vs. ζ for the system represented by Eq. (14-1). The minimum value of PI_1 occurs at $\zeta = 0$, which is obviously not a practical control system. Since the effect of minimizing the control

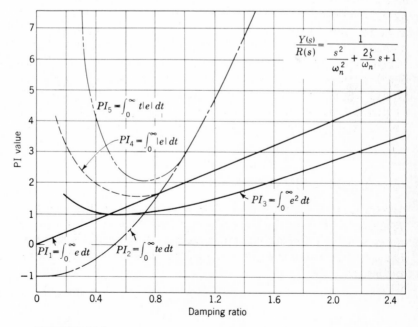

FIGURE 14-3
Criteria for second-order systems.

area without regard to the system damping may result in highly oscillatory response, this PI does not serve the intended purpose and is rejected from further consideration.

A modification of control area as a criterion, using time as a weighting factor, has also been proposed.[4] This modified criterion is defined by the integral

$$PI_2 = \int_0^\infty te\ dt \qquad (14\text{-}5)$$

The *time-weighted control area* adds a heavy penalty for errors that do not die out rapidly. It is intended that a minimum value of PI_2 should define an optimum system. Figure 14-3 contains a plot of PI_2 vs. ζ for the system represented by Eq. (14-1). Since the minimum value of PI_2 also occurs at $\zeta = 0$, this criterion fails in the same manner as the control-area criterion.

14-4   ADDITIONAL PERFORMANCE INDEXES

In this section additional PIs are investigated to demonstrate their suitability in defining optimum response. These PI are described below and are applied to the same second-order system used in Secs. 14-2 and 14-3.

$$\frac{Y(s)}{R(s)} = \frac{\omega_n^2}{s^2 + 2\zeta\omega_n s + \omega_n^2} \qquad (14\text{-}6)$$

To make a comparison between them, these additional PI are plotted vs. the damping ratio in Fig. 14-3.

One proposed figure of merit is the *integral of squared error* (ISE)

$$PI_3 = \int_0^\infty e^2\ dt \qquad (14\text{-}7)$$

Both positive and negative values of the error, which are present in underdamped response, increase the size of the integral. This eliminates the bad feature of the unmodified control area. Figure 14-3 contains a plot of PI_3 vs. ζ. The minimum value of PI_3 occurs at $\zeta = 0.5$, which may be considered a reasonable value although it results in more overshoot and a longer solution time than $\zeta = 0.7$. Also, this PI is not very selective since the plot of PI_3 has a broad minimum region.

Another figure of merit is defined by the *integral of absolute value of error* (IAE)

$$PI_4 = \int_0^\infty |e|\ dt \qquad (14\text{-}8)$$

By using the magnitude of the error, the integral increases for either positive or negative error. This should result in a good underdamped system. Figure 14-3 contains the plot of this function vs. ζ. The minimum value of PI_4 occurs at about $\zeta = 0.7$, which is known to give a good response (see Sec. 14-2). It also gives a slightly better selectivity than PI_3.

Time weighting the magnitude of error gives the PI defined by the integral

$$PI_5 = \int_0^\infty t|e|\ dt \qquad (14\text{-}9)$$

This function is known as the *integral of time multiplied by the absolute value of error* (ITAE) criterion and is plotted in Fig. 14-3. This curve has a minimum value at $\zeta = 0.7$ and is more selective than the others shown. This PI shows promise; it is used later in this chapter for more complex systems to check its suitability for determining their optimum response. This PI selects good Type I systems, but Type 2 systems have excessive overshoot.

Still other figures of merit can be formed with more complex combinations of error and time weighting. Three such PI are:

Integral of time multiplied by squared error (ITSE):

$$PI_6 = \int_0^\infty te^2 \, dt \qquad (14\text{-}10)$$

Results indicate that this PI yields good performance.

Integral of squared time multiplied by squared error (ISTSE):

$$PI_7 = \int_0^\infty t^2 e^2 \, dt \qquad (14\text{-}11)$$

Integral of squared time multiplied by absolute value of error (ISTAE):

$$PI_8 = \int_0^\infty t^2 |e| \, dt \qquad (14\text{-}12)$$

The generalization of optimizing criteria has been expanded by Schultz and Rideout.[5] PI_5 to PI_8 are not mathematically attractive, but they are used because of the availability of tables which have been derived from extensive simulation studies.

14-5 DEFINITION OF ZERO-ERROR SYSTEMS[6]

In general, the control ratio for a feedback control system has the form

$$\frac{Y(s)}{R(s)} = \frac{A(s^w + c_{w-1}s^{w-1} + \cdots + c_2 s^2 + c_1 s + c_0)}{s^n + q_{n-1}s^{n-1} + \cdots + q_2 s^2 + q_1 s + q_0} \qquad (14\text{-}13)$$

The steady-state error for this system can be shown to be

$$e_{ss} = e_0 r + e_1 \, Dr + e_2 \, D^2 r + \cdots \qquad (14\text{-}14)$$

The form of the input $r(t)$ determines the size of the steady-state error.

Since e_0 is a function of $q_0 - Ac_0$, the requirement for zero steady-state error with a step-function input is that $Ac_0 = q_0$. This also means that in a unity-feedback system the forward transfer function is Type 1 or higher. Since the degree of the numerator of $Y(s)/R(s)$ can be equal to or less than the degree of the denominator, there are many possible forms of $Y(s)/R(s)$ for which the steady-state error is zero with a step input. The system having the control ratio with only a constant in the numerator is referred to as the *zero steady-state step-error system* and is given by

$$\frac{Y(s)}{R(s)} = \frac{q_0}{s^n + q_{n-1}s^{n-1} + \cdots + q_2 s^2 + q_1 s + q_0} \qquad (14\text{-}15)$$

Since e_1 is a function of $q_0 - Ac_0$ and $q_1 - Ac_1$, zero steady-state error with a ramp-function input results when $Ac_0 = q_0$ and $Ac_1 = q_1$. Therefore the system is Type 2 or higher. *Zero steady-state ramp-error systems* are described by

$$\frac{Y(s)}{R(s)} = \frac{q_1 s + q_0}{s^n + q_{n-1}s^{n-1} + \cdots + q_2 s^2 + q_1 s + q_0} \tag{14-16}$$

Since e_2 is a function of $q_0 - Ac_0$, $q_1 - Ac_1$, and $q_2 - Ac_2$, the steady-state error is zero with a parabolic-function input if $Ac_0 = q_0$, $Ac_1 = q_1$, and $Ac_2 = q_2$. This means that the system is Type 3 or higher. Systems that have a quadratic numerator are referred to as *zero steady-state parabolic-error systems* and are described by

$$\frac{Y(s)}{R(s)} = \frac{q_2 s^2 + q_1 s + q_0}{s^n + q_{n-1}s^{n-1} + \cdots + q_2 s^2 + q_1 s + q_0} \tag{14-17}$$

These requirements for zero steady-state error, for all three standard type inputs, are also derived in Sec. 12-5.

14-6   ZERO STEADY-STATE STEP-ERROR SYSTEMS

The zero steady-state step-error system is now studied in terms of the various optimizing methods. The control ratio of the closed-loop system is given by

$$\frac{Y(s)}{R(s)} = \frac{q_0}{s^n + q_{n-1}s^{n-1} + \cdots + q_2 s^2 + q_1 s + q_0} \tag{14-18}$$

One criterion suggested as a standard for selecting the denominator coefficients is to use a pole-placement technique of having the characteristic equation composed of equal critically damped modes.[7,8] Such a system has a stable output response. This means that the coefficients are obtained from the binomial expansion for the control ratio given by

$$\frac{Y(s)}{R(s)} = \frac{\omega_0^n}{(s + \omega_0)^n} \tag{14-19}$$

Table 14-1 gives the standard binomial forms for the denominator of $Y(s)/R(s)$. This response is slow and becomes slower as the order of the system is increased. It therefore does not represent an optimum, but it can be used for comparison purposes.

The Butterworth[9] method locates the poles uniformly in the left-hand s plane on a circle of radius ω_0 with its center at the origin. Table 14-2 presents the standard Butterworth forms of the denominator of $Y(s)/R(s)$. The response to a step function using these standard forms is shown in Fig. 14-4 for systems of order 2 to 8.

With modern control, the objective is to apply an optimizing figure of merit to produce a table of standard forms in which the best value is specified for each coefficient of the denominator. Such a table of standard forms would assist the designer since he would try to duplicate these coefficients in his system. He would then be sure of having the "best" system possible.

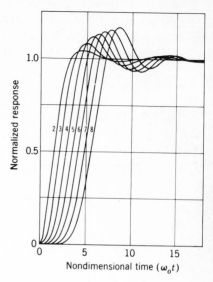

FIGURE 14-4
Response using Butterworth standard forms for zero steady-state step-error systems.

Table 14-1 BINOMIAL STANDARD FORMS

$$s + \omega_0$$
$$s^2 + 2\omega_0 s + \omega_0^2$$
$$s^3 + 3\omega_0 s^2 + 3\omega_0^2 s + \omega_0^3$$
$$s^4 + 4\omega_0 s^3 + 6\omega_0^2 s^2 + 4\omega_0^3 s + \omega_0^4$$
$$s^5 + 5\omega_0 s^4 + 10\omega_0^2 s^3 + 10\omega_0^3 s^2 + 5\omega_0^4 s + \omega_0^5$$
$$s^6 + 6\omega_0 s^5 + 15\omega_0^2 s^4 + 20\omega_0^3 s^3 + 15\omega_0^4 s^2 + 6\omega_0^5 s + \omega_0^6$$
$$s^7 + 7\omega_0 s^6 + 21\omega_0^2 s^5 + 35\omega_0^3 s^4 + 35\omega_0^4 s^3 + 21\omega_0^5 s^2 + 7\omega_0^6 s + \omega_0^7$$
$$s^8 + 8\omega_0 s^7 + 28\omega_0^2 s^6 + 56\omega_0^3 s^5 + 70\omega_0^4 s^4 + 56\omega_0^5 s^3 + 28\omega_0^6 s^2 + 8\omega_0^7 s + \omega_0^8$$

Table 14-2 BUTTERWORTH STANDARD FORMS

$$s + \omega_0$$
$$s^2 + 1.4\omega_0 s + \omega_0^2$$
$$s^3 + 2.0\omega_0 s^2 + 2.0\omega_0^2 s + \omega_0^3$$
$$s^4 + 2.6\omega_0 s^3 + 3.4\omega_0^2 s^2 + 2.6\omega_0^3 s + \omega_0^4$$
$$s^5 + 3.24\omega_0 s^4 + 5.24\omega_0^2 s^3 + 5.24\omega_0^3 s^2 + 3.24\omega_0^4 s + \omega_0^5$$
$$s^6 + 3.86\omega_0 s^5 + 7.46\omega_0^2 s^4 + 9.14\omega_0^3 s^3 + 7.46\omega_0^4 s^2 + 3.86\omega_0^5 s + \omega_0^6$$
$$s^7 + 4.49\omega_0 s^6 + 10.1\omega_0^2 s^5 + 14.6\omega_0^3 s^4 + 14.6\omega_0^4 s^3 + 10.1\omega_0^5 s^2 + 4.49\omega_0^6 s + \omega_0^7$$
$$s^8 + 5.13\omega_0 s^7 + 13.14\omega_0^2 s^6 + 21.85\omega_0^3 s^5 + 25.69\omega_0^4 s^4 + 21.85\omega_0^5 s^3 + 13.14\omega_0^6 s^2 + 5.13\omega_0^7 s + \omega_0^8$$

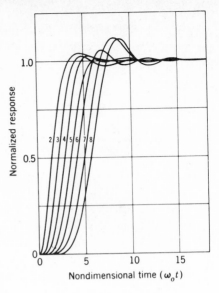

FIGURE 14-5
Response using ITAE standard forms
for zero steady-state step-error systems.

The minimum ITAE criterion was applied to this system, resulting in the standard forms given in Table 14-3. The procedure used was to vary each coefficient separately until the ITAE value became a minimum. Then the successive coefficients were varied in sequence to minimize the ITAE value. The response to a step function using these standard forms is shown in Fig. 14-5 for systems of order 2 to 8.

The next criterion used was to minimize the solution time, which is the time to reach and remain within 5 percent of the final value. Table 14-4 gives the standard forms using this criterion. The response to a step function using these standard forms is shown in Fig. 14-6 for systems of order 3 to 6. Results for higher orders can be obtained by further computation.

A comparison of the response curves of Figs. 14-4 to 14-6 shows that all three criteria give good results. It must be realized that for a system of particular order, one

Table 14-3   ITAE STANDARD FORMS FOR ZERO STEADY-STATE STEP-ERROR SYSTEMS

$$s + \omega_0$$
$$s^2 + 1.4\omega_0 s + \omega_0^2$$
$$s^3 + 1.75\omega_0 s^2 + 2.15\omega_0^2 s + \omega_0^3$$
$$s^4 + 2.1\omega_0 s^3 + 3.4\omega_0^2 s^2 + 2.7\omega_0^3 s + \omega_0^4$$
$$s^5 + 2.8\omega_0 s^4 + 5.0\omega_0^2 s^3 + 5.5\omega_0^3 s^2 + 3.4\omega_0^4 s + \omega_0^5$$
$$s^6 + 3.25\omega_0 s^5 + 6.60\omega_0^2 s^4 + 8.60\omega_0^3 s^3 + 7.45\omega_0^4 s^2 + 3.95\omega_0^5 s + \omega_0^6$$
$$s^7 + 4.475\omega_0 s^6 + 10.42\omega_0^2 s^5 + 15.08\omega_0^3 s^4 + 15.54\omega_0^4 s^3 + 10.64\omega_0^5 s^2 + 4.58\omega_0^6 s + \omega_0^7$$
$$s^8 + 5.20\omega_0 s^7 + 12.80\omega_0^2 s^6 + 21.60\omega_0^3 s^5 + 25.75\omega_0^4 s^4 + 22.20\omega_0^5 s^3 + 13.30\omega_0^6 s^2 + 5.15\omega_0^7 s + \omega_0^8$$

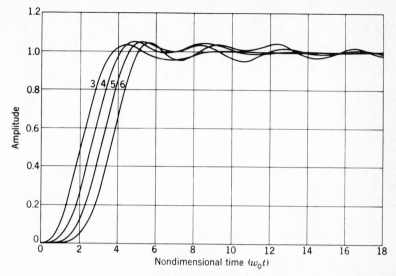

FIGURE 14-6
Response using the solution-time criterion for zero steady-state step-error systems.

of the criteria may be better in some respect. For example, for the fifth-order system the time to first zero is 4.25 s when the solution time is used and is 5.5 s for the ITAE criterion. But it must also be taken into account that the maximum overshoot is 5 percent for the solution time and only 2 percent for the ITAE criterion. Further, the solution-time response is more oscillatory than the ITAE response. The most suitable response must be determined in terms of which is most important—rise time, peak overshoot, or frequency of oscillation. Once the standard form has been chosen to yield the desired optimal performance, either the Guillemin-Truxal or the state-variable feedback method described in Chap. 12 can be used to determine the required parameters of the cascade compensator $G_c(s)$ or the state feedback vector $\mathbf{k}$, respectively. A similar analysis can be made for zero steady-state ramp-error systems.

Table 14-4 SOLUTION-TIME STANDARD FORMS FOR ZERO
STEADY-STATE STEP-ERROR SYSTEMS

$$s + \omega_0$$
$$s^2 + 1.4\omega_0 s + \omega_0{}^2$$
$$s^3 + 1.55\omega_0 s^2 + 2.10\omega_0{}^2 s + \omega_0{}^3$$
$$s^4 + 1.60\omega_0 s^3 + 3.15\omega_0{}^2 s^2 + 2.45\omega_0{}^3 s + \omega_0{}^4$$
$$s^5 + 1.575\omega_0 s^4 + 4.05\omega_0{}^2 s^3 + 4.10\omega_0{}^3 s^2 + 3.025\omega_0{}^4 s + \omega_0{}^5$$
$$s^6 + 1.45\omega_0 s^5 + 5.10\omega_0{}^2 s^4 + 5.30\omega_0{}^3 s^3 + 6.25\omega_0{}^4 s^2 + 3.425\omega_0{}^5 s + \omega_0{}^6$$

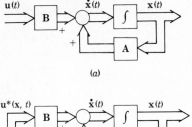

(a)

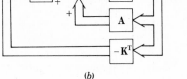

(b)

FIGURE 14-7
Various forms of control systems:
(a) open-loop optimal control system,
(b) closed-loop optimal feedback control
system $[\mathbf{r}(t) = \mathbf{0}]$, (c) conventional
unity-feedback control system.

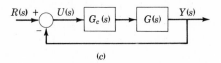

(c)

14-7 MODERN CONTROL PERFORMANCE INDEX[10]

The PIs of the previous sections represent the first applications of a PI to achieve an optimal performance based upon parameter optimization. That is, the plant parameters are fixed at values that satisfy a desired PI without constraining any variable within the system. These variables may have limits (constraints) on their maximum and/or minimum values, such as those on **u**. Because of the difficulty in constraining these variables in the parameter-optimization method another approach is necessary. Thus, to achieve an optimal performance when constraints are imposed upon the variables, modern control theory is used to design the system.

In modern control theory the PI is defined in terms of the state and control vectors, $\mathbf{x}(t)$ and $\mathbf{u}(t)$, respectively. The constraints are manifested by weighting the variables (signals) in the modern control PI. Whereas for the parameter-optimization approach the structure of the system is fixed, the modern control-theory approach requires that the structure of the system be selected. The modern optimal feedback control problem to be considered in this text depends on the following conditions.

Condition 1 A system represented by $\dot{\mathbf{x}} = f(\mathbf{x},\mathbf{u},t)$ is to be optimally controlled. (Fig. 14-7a represents the open-loop optimal control problem.)

Condition 2 The initial (starting) time t_i and the initial state $\mathbf{x}(t_i)$ are specified. Design specifications are defined for either or both the final time t_f and the final state $\mathbf{x}(t_f)$. These quantities may be described implicitly, such as when minimum time is specified or when the final state must reach within a tolerance of an ideal end state.

Condition 3 An *integral performance index* is specified to yield the optimal control

$$\text{PI} = \int_{t_i}^{t_f} L(\mathbf{x},\mathbf{u},t)\ dt \qquad (14\text{-}20)$$

This results in selecting the control **u** to minimize the PI over the entire time interval of concern. The functional $L(\mathbf{x},\mathbf{u},t)$ must be selected so that it is positive. With $L(\mathbf{x},\mathbf{u},t) > 0$, the value of the PI is a monotonically increasing function of t within $t_i \leq t \leq t_f$. If the magnitude of Eq. (14-20) is a minimum over the time interval t_i to t_f, the system performance is said to be optimal when the state is transferred from $\mathbf{x}(t_i)$ to $\mathbf{x}(t_f)$.

Condition 4 Amplitude constraints are imposed upon the control vector **u**. In physical systems there is a limit to the available energy to control the system; e.g., the amplitude constraints on $\mathbf{u}(t)$ are

$$\mathbf{u}_{\min} \leq \mathbf{u}(t) \leq \mathbf{u}_{\max} \qquad (14\text{-}21)$$

Condition 5 When Eq. (14-20) is changed to the linear quadratic performance form, the system must be controllable and observable.[11]

The object of the modern optimal feedback control problem, under the above conditions, is to determine the *optimal control* law $\mathbf{u}^*(\mathbf{x},t)$ which can transfer the system from its initial state to the final state while minimizing the PI of Eq. (14-20) (closed-loop operation). This optimal control design is based upon obtaining a system response to a set of initial conditions, $\mathbf{x}(t_i)$, with the input $\mathbf{r}(t)$ assumed to be zero. This is defined as the *optimal regulator problem*. When $\mathbf{r}(t) \neq \mathbf{0}$, it becomes the *optimal servo problem*. Thus, the *optimal control* $\mathbf{u}^*(\mathbf{x},t)$ is a function of the states $\mathbf{x}(t)$ of the system. For example, for the linear-regulator problem, to be described later, utilizing the quadratic PI yields the *optimal control law* $\mathbf{u}^*(\mathbf{x},t) = -\mathbf{K}^T(t)\mathbf{x}(t)$, where $\mathbf{K}(t)$ is the feedback matrix of order $n \times r$. For the linear system $\dot{\mathbf{x}} = \mathbf{Ax} + \mathbf{Bu}$, this optimal control law $\mathbf{u}^*(\mathbf{x},t)$ may be generated from the states $\mathbf{x}(t)$ by the use of feedback as illustrated in Fig. 14-7*b*. This is in contrast to conventional control design, which is based upon obtaining a system response to a step or sinusoidal input function with all initial conditions assumed to be zero, as illustrated in Fig. 14-7*c*. As defined by Eq. (14-20) and in the previous sections, PIs are formulated in the time domain. Although there are other forms of modern control PIs, this text is restricted to the one expressed by Eq. (14-20). The other PIs are discussed in advanced texts on optimal control.

The formulation of the particular PI to be used is based not only on mathematical convenience but also upon the particular control problem under consideration. For example, consider that the state of a system is to go from some initial value $\mathbf{x}(t_i)$ to some specified final value $\mathbf{x}(t_f)$ in minimum time but t_f is unspecified. *This is called the minimum-time optimal control problem.* The quantity to be minimized is $t_f - t_i$, which is equal to $\int_{t_i}^{t_f} dt$. Therefore $L(\mathbf{x},\mathbf{u},t) = 1$ is used to evaluate the PI. Note that L is not a function of **x** or **u**, and they are unconstrained.

$$\text{PI} = \int_{t_i}^{t_f} 1\ dt = t_f - t_i \qquad (14\text{-}22)$$

The minimum-time condition can be applied to the problem of a missile intercepting an aerospace vehicle. The time for the missile to travel from its launch point to the target-interception point depends principally on the thrust forces available. Clearly, this time can be made as small as possible if an unlimited, i.e., infinite, thrust force **u** is available—obviously an impractical situation. Usually the control **u** must be constrained, as specified by Eq. (14-21), and this in turn affects the minimum time that is possible. The minimum value of the PI of Eq. (14-22) is zero. This requires the minimum time $t_f - t_i$ to also be zero. The dynamic equations would therefore be required to yield the change of state from $\mathbf{x}(t_i)$ to $\mathbf{x}(t_f)$ in zero time. This implies the use of infinite thrust, a clearly inadmissible solution.

In contrast to the minimum-time requirement of the previous problem, consider the requirement of minimum control effort **u** for a linear system. A suitable PI for this problem with $\mathbf{x}(t_i)$, $\mathbf{x}(t_f)$, t_i, and t_f all specified is

$$PI = \int_{t_i}^{t_f} \left(\sum_{i=1}^{r} u_i^2 \right) dt = \int_{t_i}^{t_f} (\mathbf{u}^T \mathbf{u}) \, dt$$

This equation is usually generalized by introducing a proportionality factor for each control input. A more flexible performance index is achieved by introducing a weighting matrix **Z**; thus,

$$PI = \int_{t_i}^{t_f} (\mathbf{u}^T \mathbf{Z} \mathbf{u}) \, dt \qquad (14\text{-}23)$$

Thus, $L(\mathbf{x},\mathbf{u},t) = \mathbf{u}^T \mathbf{Z} \mathbf{u}$, where **Z** is a real time-varying or time-invariant weighting matrix. For analytical convenience, **Z** is often made symmetric and positive definite. Only the time-invariant case is considered in this text. If one considers the scalar case, where u may represent a variable such as a current or a voltage, this PI represents energy. Therefore, by association, problems using PIs of this form are termed *minimum-energy optimal control problems*. The PI of Eq. (14-23) is a minimum when $\mathbf{u}^*(\mathbf{x},t) = 0$, that is, PI = 0. The value of $\mathbf{u}^*(\mathbf{x},t) = 0$ leads to the analytical result for a linear system that the time to reach the desired final state $\mathbf{x}(t_f)$ must be $t_f = \infty$, which is inadmissible. This value of $\mathbf{u}(\mathbf{x},t)$ violates condition 3, which requires $L(\mathbf{x},\mathbf{u},t) > 0$.

The ISE criterion is also applicable to minimizing the error states or state deviations. When the final state $\mathbf{x}(t_f)$ is specified to be the origin and the initial state $\mathbf{x}(t_i)$ describes the initial conditions, the state is then the error and the PI is

$$PI = \int_{t_i}^{t_f} \sum_{i=1}^{n} x_i^2 \, dt = \int_{t_i}^{t_f} \mathbf{x}^T \mathbf{x} \, dt \qquad (14\text{-}24)$$

where $L(\mathbf{x},\mathbf{u},t) = \mathbf{x}^T \mathbf{x}$ and the limits of the integral are specified. A more general performance index is obtained as in Eq. (14-23) by introducing a weighting matrix **Q**. The modified minimum ISE criterion yields

$$PI = \int_{t_i}^{t_f} \mathbf{x}^T \mathbf{Q} \mathbf{x} \, dt \qquad (14\text{-}25)$$

where **Q** is a weighting matrix of order n and $L(\mathbf{x},\mathbf{u},t) = \mathbf{x}^T \mathbf{Q} \mathbf{x} > 0$. When this is applied to a specific problem, it is often necessary (see Sec. 14-8) or convenient to select $\mathbf{Q}$ as real, symmetric, and positive definite or positive semidefinite. The object of the design problem is to determine the optimal control $\mathbf{u}^*(\mathbf{x},t)$ that transfers the system from its initial state $\mathbf{x}(t_i)$ to its final state $\mathbf{x}(t_f)$ while minimizing the PI.

As noted, for the minimum-time and the minimum-energy optimal control problems, impractical results are obtained. Also, the PIs of Eqs. (14-22) to (14-25) can result in nonunique solutions. To overcome these difficulties, a more complex PI can be formulated by the combination of more than one of these simple indexes. These complex PIs lend flexibility to the optimal control design. The composite PI that is widely used in optimal control design is formed by the linear combination of Eqs. (14-23) and (14-25), i.e.,

$$\text{PI} = \int_{t_i}^{t_f} (\mathbf{x}^T\mathbf{Q}\mathbf{x} + \mathbf{u}^T\mathbf{Z}\mathbf{u})\, dt \qquad (14\text{-}26)$$

This is referred to as the *quadratic performance index.*[11] Its application to a control system achieves an optimal system which represents a compromise between, or a blending of, the minimum-error and the minimum-energy criteria. For this combined PI the matrices $\mathbf{Q}$ and $\mathbf{Z}$ are restricted to being constant and real symmetric. As stated in condition 3, the functional $L(\mathbf{x},\mathbf{u},t) = \mathbf{x}^T\mathbf{Q}\mathbf{x} + \mathbf{u}^T\mathbf{Z}\mathbf{u}$ must be selected to be positive definite. As shown later, $\mathbf{Z}$ must be positive definite in order to ensure that it has an inverse. In the solution of the optimal control problem, Sec. 14-8, the requirement appears that $\mathbf{Q}$ be made positive definite. A relaxation of the positive definite restriction on $\mathbf{Q}$ may be permitted in some cases, as described in Sec. 14-8. For a *linear* system represented by $\dot{\mathbf{x}} = \mathbf{A}\mathbf{x} + \mathbf{B}\mathbf{u}$, the determination of $\mathbf{u}$ which minimizes the quadratic performance index

$$\text{PI} = \int_0^\infty L(\mathbf{x},\mathbf{u},t)\, dt = \int_0^\infty [\mathbf{x}^T(t)\mathbf{Q}\mathbf{x}(t) + \mathbf{u}^T(t)\mathbf{Z}\mathbf{u}(t)]\, dt \qquad (14\text{-}27)$$

is often called the *linear quadratic control problem.* Note that the limits on the integral have been changed to 0 and ∞. The resulting optimal control law $\mathbf{u}^*(\mathbf{x})$ is now an explicit function of the state vector $\mathbf{x}(t)$ only. This permits the implementation of the optimal control by means of a closed-loop control. How the elements of the weighting matrices $\mathbf{Q}$ and $\mathbf{Z}$ are selected is discussed in Chap. 15. As a result of choosing the quadratic PI and an infinite range on time, the amplitudes u_i need not be constrained. Hence, in this special case, the restraints of condition 4 may be relaxed. The latter portion of this text demonstrates that the linear quadratic control problem is amenable to the utilization of conventional control-theory techniques and figures of merit.

14-8   ALGEBRAIC RICCATI EQUATION[12]

In this section the results of the second method of Liapunov are utilized to derive a method for readily determining the optimal control $\mathbf{u}^*(\mathbf{x})$ for the linear control problem. This method is applicable to multiple-input multiple-output systems. Assume[13,14] a Liapunov function of the quadratic form (see Sec. 13-6):

$$V[\mathbf{x}(t)] = \mathbf{x}(t)^T\mathbf{P}\mathbf{x}(t) \qquad (14\text{-}28)$$

where $\mathbf{P}$ is a positive-definite, symmetric, constant matrix and the control vector has the form

$$\mathbf{u}^*[\mathbf{x}(t)] = -\mathbf{K}^T\mathbf{x}(t) \qquad (14\text{-}29)$$

The feedback matrix $\mathbf{K}$ is assumed to be a constant matrix. Note that V and $\mathbf{u}^*$ are not explicit functions of time. Substitute Eq. (14-29) into Eq. (14-27) to obtain

$$\mathrm{PI} = \int_{t_i=0}^{t_f=\infty} \{\mathbf{x}(\tau)^T[\mathbf{Q} + \mathbf{K}\mathbf{Z}\mathbf{K}^T]\mathbf{x}(\tau)\}\, d\tau \qquad (14\text{-}30)$$

It is desired to determine an optimal control law that is independent of the value of the initial state $\mathbf{x}(t_i)$. The initial state can take on any value of $\mathbf{x}(t)$ for any value of t up to infinity. A generalization of Eq. (14-30) results by replacing the lower limit of the integral by t. Thus

$$\mathrm{PI} = \int_t^\infty \{\mathbf{x}(\tau)^T[\mathbf{Q} + \mathbf{K}\mathbf{Z}\mathbf{K}^T]\mathbf{x}(\tau)\}\, d\tau \qquad (14\text{-}31)$$

where τ is a dummy variable. Note that Eq. (14-31) is a quadratic performance index. The quadratic Liapunov function of Eq. (14-28) may be selected as the PI to be used in Eq. (14-31). Thus

$$V[\mathbf{x}(t)] = \mathbf{x}(t)^T\mathbf{P}\mathbf{x}(t) = \int_t^\infty \{\mathbf{x}(\tau)^T[\mathbf{Q} + \mathbf{K}\mathbf{Z}\mathbf{K}^T]\mathbf{x}(\tau)\}\, d\tau \qquad (14\text{-}32)$$

If the matrix $\mathbf{Q}$ is positive definite, a unique solution exists for a positive definite matrix $\mathbf{P}$. In the linear regulator problem, the final time is taken to be infinite. The digital-computer programs available for solving the algebraic Riccati equation involve a finite time interval, that is, $t_f < \infty$. Thus these programs yield a solution if $\mathbf{Q}$ is at least positive semidefinite. If $\mathbf{Q}$ is positive semidefinite, then a minimum to Eq. (14-27) exists iff the following condition established by Kalman,[13] is satisfied:

$$\mathrm{Rank}\ [\mathbf{H} \mid \mathbf{A}^T\mathbf{H} \mid \cdots \mid (\mathbf{A}^T)^{n-1}\mathbf{H}] = n \qquad (14\text{-}33)$$

where $\mathbf{H}\mathbf{H}^T = \mathbf{Q}$. It means that the system described by $\mathbf{A}$ and $\mathbf{H}$ is observable. This condition should be checked before proceeding with the solution of Eq. (14-32).

The following two conditions ensure that $\mathbf{P}$ is positive definite: (1) the plant is controllable, and (2) Eq. (14-33) is satisfied. These conditions also guarantee that the optimal control law is stable; i.e., all eigenvalues of $\mathbf{A} - \mathbf{B}\mathbf{K}^T$ have negative real parts. If $\mathbf{r}$ is now assumed to be a constant value other than zero and the feedback $-\mathbf{K}^T\mathbf{x}$ is used, the state equation for the closed-loop system becomes

$$\dot{\mathbf{x}} = \mathbf{A}\mathbf{x} + \mathbf{B}\mathbf{u} = [\mathbf{A} - \mathbf{B}\mathbf{K}^T]\mathbf{x} + \mathbf{B}\mathbf{r} \qquad (14\text{-}34)$$

where $\mathbf{u} = \mathbf{r} - \mathbf{K}^T\mathbf{x}$. Therefore, the poles of the control ratio, where the output is given by $\mathbf{y} = \mathbf{C}\mathbf{x}$, are the eigenvalues of $\mathbf{A} - \mathbf{B}\mathbf{K}^T$ iff the optimal closed-loop system is observable. Restricting $\mathbf{Q}$ to being positive definite ensures that $\mathbf{P}$ is positive definite.[12]

Performing the integration in Eq. (14-32) yields a function of t since the upper limit of the integral is infinity. $V[\mathbf{x}(\infty)] = 0$ since it is positive definite. Therefore differentiating Eq. (14-32) with respect to its lower limit t yields†

$$\dot{V}[\mathbf{x}(t)] = \dot{\mathbf{x}}(t)^T\mathbf{P}\mathbf{x}(t) + \mathbf{x}(t)^T\mathbf{P}\dot{\mathbf{x}}(t) = -\mathbf{x}(t)^T[\mathbf{Q} + \mathbf{K}\mathbf{Z}\mathbf{K}^T]\mathbf{x}(t) \qquad (14\text{-}35)$$

The minus sign is the result of the differentiation with respect to time, which is the lower limit of the integral. Since $\mathbf{Q}$ and $\mathbf{Z}$ are positive definite $\dot{V}(\mathbf{x})$ is negative definite and the system is asymptotically stable. Substituting from Eq. (14-34), with $\mathbf{r} = \mathbf{0}$, into Eq. (14-35) yields

$$\dot{V}(\mathbf{x}) = \mathbf{x}^T[(\mathbf{A} - \mathbf{B}\mathbf{K}^T)^T\mathbf{P} + \mathbf{P}(\mathbf{A} - \mathbf{B}\mathbf{K}^T)]\mathbf{x} = -\mathbf{x}^T[\mathbf{Q} + \mathbf{K}\mathbf{Z}\mathbf{K}^T]\mathbf{x} \qquad (14\text{-}36)$$

This equation is satisfied when the bracketed terms on each side of the equation are equal, i.e.,

$$(\mathbf{A} - \mathbf{B}\mathbf{K}^T)^T\mathbf{P} + \mathbf{P}(\mathbf{A} - \mathbf{B}\mathbf{K}^T) = -\mathbf{Q} - \mathbf{K}\mathbf{Z}\mathbf{K}^T \qquad (14\text{-}37)$$

Since $\mathbf{A}$, $\mathbf{B}$, and $\mathbf{Q}$ are known constant matrices, Eq. (14-37) represents an equation of the unknown matrix $\mathbf{K}$ as a function of the variable $\mathbf{P}$. Substituting $\mathbf{Q} + \mathbf{K}\mathbf{Z}\mathbf{K}^T$ from Eq. (14-37) into Eq. (14-32) yields

$$V[\mathbf{x}(t)] = \int_t^\infty \{-\mathbf{x}^T[(\mathbf{A} - \mathbf{B}\mathbf{K}^T)^T\mathbf{P} + \mathbf{P}(\mathbf{A} - \mathbf{B}\mathbf{K}^T)]\mathbf{x}\}\, dt \qquad (14\text{-}38)$$

Note that the function within the braces is equal to the functional $L(\mathbf{x},\mathbf{u},t)$ of Eq. (14-27), where $L(\mathbf{x},\mathbf{u},t) > 0$. The minimum value of this PI occurs when the integrand is a minimum. Therefore the problem now is to determine the values of the elements of $\mathbf{K}$ that make the integrand a minimum. This is accomplished by taking the derivative of Eq. (14-37) with respect to the matrix $\mathbf{K}$ and setting $\partial p_{vu}/\partial k_{wx} = 0$, where p_{vu} is the vuth element of $\mathbf{P}$ and k_{wx} is the wxth element of $\mathbf{K}$. In doing this, a derivative of a matrix with respect to a matrix is required. The necessary additional matrix operations are now presented.

Consider an $n \times m$ $\mathbf{G}$ matrix which is a function of the elements of an $n \times r$ $\mathbf{H}$ matrix. These matrices are partitioned into column vectors

$$\mathbf{G} = [\mathbf{g}_1 \quad \mathbf{g}_2 \quad \cdots \quad \mathbf{g}_m] \qquad \mathbf{H} = [\mathbf{h}_1 \quad \mathbf{h}_2 \quad \cdots \quad \mathbf{h}_r]$$

The derivative of a vector with respect to a vector yields the $n \times n$ *Jacobian matrix*

$$\frac{d\mathbf{g}_i}{d\mathbf{h}_j^T} = \begin{bmatrix} \dfrac{\partial g_{1i}}{\partial h_{1j}} & \dfrac{\partial g_{1i}}{\partial h_{2j}} & \cdots & \dfrac{\partial g_{1i}}{\partial h_{nj}} \\[2ex] \dfrac{\partial g_{2i}}{\partial h_{1j}} & \dfrac{\partial g_{2i}}{\partial h_{2j}} & \cdots & \dfrac{\partial g_{2i}}{\partial h_{nj}} \\[1ex] \cdots\cdots\cdots\cdots\cdots\cdots \\[1ex] \dfrac{\partial g_{ni}}{\partial h_{1j}} & \dfrac{\partial g_{ni}}{\partial h_{2j}} & \cdots & \dfrac{\partial g_{ni}}{\partial h_{nj}} \end{bmatrix} \qquad (14\text{-}39)$$

where

$$\mathbf{g}_i = \begin{bmatrix} g_{1i} \\ g_{2i} \\ \vdots \\ g_{ni} \end{bmatrix} \qquad \text{and} \qquad \mathbf{h}_j = \begin{bmatrix} h_{1j} \\ h_{2j} \\ \vdots \\ h_{nj} \end{bmatrix}$$

† The Leibnitz rule for the differentiation of an integral may also be used to obtain Eq. (14-35).

The derivative of a matrix with respect to a matrix yields the $m \times r$ *matrix of Jacobian matrices (or a matrix of Jacobians)*

$$\frac{d\mathbf{G}^T}{d\mathbf{H}} = \begin{bmatrix} \dfrac{\partial \mathbf{g}_1}{\partial \mathbf{h}_1} & \dfrac{\partial \mathbf{g}_1}{\partial \mathbf{h}_2} & \cdots & \dfrac{\partial \mathbf{g}_1}{\partial \mathbf{h}_r} \\[2mm] \dfrac{\partial \mathbf{g}_2}{\partial \mathbf{h}_1} & \dfrac{\partial \mathbf{g}_2}{\partial \mathbf{h}_2} & \cdots & \dfrac{\partial \mathbf{g}_2}{\partial \mathbf{h}_r} \\[2mm] \cdots\cdots\cdots\cdots\cdots\cdots \\[1mm] \dfrac{\partial \mathbf{g}_m}{\partial \mathbf{h}_1} & \dfrac{\partial \mathbf{g}_m}{\partial \mathbf{h}_2} & \cdots & \dfrac{\partial \mathbf{g}_m}{\partial \mathbf{h}_r} \end{bmatrix} \qquad (14\text{-}40)$$

The expanded matrix (with the Jacobian matrices inserted) is of the order $mn \times rn$; that is, it contains mrn^2 elements.

Applying Eqs. (14-39) and (14-40) to the derivative with respect to $\mathbf{K}$ of Eq. (14-37) permits equating the elements of the resulting matrices on both sides of the equation, yielding mrn^2 equations. It is found that only nr equations are independent. The nr independent equations yield

$$\mathbf{PB} = \mathbf{KZ} \qquad (14\text{-}41)$$

$\mathbf{PB}$ and $\mathbf{KZ}$ are matrices of order $n \times r$. Since $\mathbf{Z}$ is restricted to being positive definite, the existence of $\mathbf{Z}^{-1}$ is assured. Thus

$$\mathbf{K} = \mathbf{PBZ}^{-1} \qquad (14\text{-}42)$$

Since $\mathbf{P}$ and $\mathbf{Z}$ are symmetric, $\mathbf{K}^T = \mathbf{Z}^{-1}\mathbf{B}^T\mathbf{P}$. Substituting these functions into Eq. (14-37) yields

$$[\mathbf{A} - \mathbf{BZ}^{-1}\mathbf{B}^T\mathbf{P}]^T\mathbf{P} + \mathbf{P}[\mathbf{A} - \mathbf{BZ}^{-1}\mathbf{B}^T\mathbf{P}] = -\mathbf{Q} - \mathbf{PBZ}^{-1}\mathbf{ZZ}^{-1}\mathbf{B}^T\mathbf{P} \qquad (14\text{-}43)$$

To rearrange this equation it is necessary to use the matrix-transpose property, namely, $[\mathbf{G} + \mathbf{H}]^T = \mathbf{G}^T + \mathbf{H}^T$. When this matrix operation is used and the term $\mathbf{PBZ}^{-1}\mathbf{B}^T\mathbf{P}$ is canceled, Eq. (14-43) reduces to

$$\mathbf{A}^T\mathbf{P} - \mathbf{PBZ}^{-1}\mathbf{B}^T\mathbf{P} + \mathbf{PA} + \mathbf{Q} = 0 \qquad (14\text{-}44)$$

This is the *algebraic Riccati equation*, in which $\mathbf{P}$ is called the *Riccati matrix*. It is difficult to obtain the analytic solution of this equation for $\mathbf{P}$, except for low-order systems ($n \leq 3$). Fortunately, a numerical solution for $\mathbf{P}$ can be obtained on a digital computer.[15] Once $\mathbf{P}$ is known, the feedback matrix $\mathbf{K}$ is obtained from Eq. (14-42), and the optimal control law $\mathbf{u}^*[\mathbf{x}(t)] = -\mathbf{K}^T\mathbf{x}(t)$ is determined.

The solution of $\mathbf{P}$ from Eq. (14-44) minimizes the PI given by Eq. (14-27). The development of this result is based on Eq. (14-32), in which the lower limit of the integral is the initial time t so that $\mathbf{P}$ will be independent of the initial conditions $\mathbf{x}(t_i)$. Thus, the minimum value of the PI given in Eq. (14-27) is evaluated from Eq. (14-32) at $t = t_i = 0$ as

$$\min \text{PI} = \mathbf{x}^T(0)\mathbf{Px}(0)$$

The optimal control law indicates that for an optimum system to exist when the quadratic PI is used, all state variables must be fed back as shown in Fig. 14-7b.

In practice all states may not be available for measurement. Thus, in order to synthesize this optimal control law, it is necessary to construct the unavailable states. A method of doing this is the Luenberger observer theory,[11,16] not covered in this text. For linear time-invariant systems another possibility is the technique of block-diagram manipulation, discussed in Chaps. 12 and 15.

The elements of the matrix $\mathbf{K}$ are the *feedback coefficients* and are the same as those described in Chap. 12. Phase-variable representation can be used in the study of optimal control systems because of the relative ease of the mathematical operations and analysis. To convert the feedback coefficients $\mathbf{K}$, determined from a phase-variable representation, into the feedback coefficients $\mathbf{K}_p$ for a physical-variable representation, the transformation matrix $\mathbf{T}$ (see Chap. 5) is used. Inserting $\mathbf{x} = \mathbf{T}\mathbf{x}_p$ into the optimal control law yields

$$\mathbf{u}^* = -\mathbf{K}^T\mathbf{x} = -\mathbf{K}^T\mathbf{T}\mathbf{x}_p = -\mathbf{K}_p{}^T\mathbf{x}_p \qquad (14\text{-}45)$$

where
$$\mathbf{K}_p{}^T = \mathbf{K}^T\mathbf{T} \qquad (14\text{-}46)$$

When single-input time-invariant linear systems are considered, Eqs. (14-27), (14-29), (14-42), and (14-44) can be rewritten. Since $\mathbf{u}$ is now a scalar, the control weighting $\mathbf{Z}$ is also a scalar z. As a consequence $\mathbf{K}$ and $\mathbf{B}$ are both column matrices. Hence,

$$\text{PI} = \int_0^\infty \left[\mathbf{x}^T\mathbf{Q}\mathbf{x} + zu^2 \right] dt \qquad (14\text{-}47)$$

$$u^*[\mathbf{x}(t)] = -\mathbf{k}^T\mathbf{x}(t) \qquad (14\text{-}48)$$

$$\mathbf{k} = \mathbf{P}\mathbf{b}z^{-1} \qquad (14\text{-}49)$$

$$\mathbf{A}^T\mathbf{P} - \mathbf{P}\mathbf{b}z^{-1}\mathbf{b}^T\mathbf{P} + \mathbf{P}\mathbf{A} + \mathbf{Q} = 0 \qquad (14\text{-}50)$$

The direct control problem starts with the specification that either (1) $\mathbf{Q}$ is positive definite or (2) $\mathbf{Q}$ is at least positive semidefinite but the rank condition given in Eq. (14-33) must be satisfied. This restriction on $\mathbf{Q}$ is necessary to guarantee an asymptotically stable optimal system. In addition, $\mathbf{Z}$ must be positive definite in order to ensure finite control. The design leads to a feedback matrix $\mathbf{K}$ that yields roots of the system in the left-half s plane.

The inverse control problem starts with the placement of the roots of the characteristic equation so that the system has the desired stable response characteristics. The methods presented in Chap. 12 can be used to obtain the feedback gains $\mathbf{K}$ which yield the desired root placement. It is now possible to evaluate, for a given value of $\mathbf{Z}$, the matrix $\mathbf{Q}$ corresponding to the feedback matrix $\mathbf{K}$. This may result in a $\mathbf{Q}$ for which the restrictions on sign definiteness are relaxed from those required in the direct control method.

Molinari[22,23] claims that if $\mathbf{KB}$ (or $\mathbf{k}^T\mathbf{b}$) is symmetric, a stable closed-loop system is optimal. Thus, a stable single-input state-variable feedback system, with all states fed back, is an optimal system. Therefore, the state-variable-feedback design method of Chap. 12 represents a method for achieving an optimal design, satisfying the PI of Eq. (14-47). It results in the control u in the same form as given by Eq. (14-48) and does not explicitly utilize the Riccati matrix $\mathbf{P}$.

In the literature, there are design methods which simultaneously result in a desired control-ratio transfer function and an optimal quadratic performance.[17,18] The next section presents two examples which illustrate the direct control method for achieving an optimal control system.

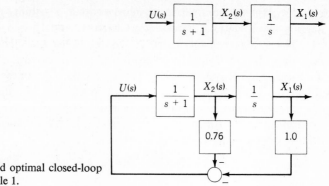

FIGURE 14-8
The open-loop and optimal closed-loop
systems of Example 1.

14-9 EXAMPLES

EXAMPLE 1   It is desired to optimize the closed-loop response of the single-input
single-output system of Fig. 14-8a utilizing the Riccati equation of Eq. (14-50), which
is based on the quadratic PI of Eq. (14-47). Because of the simplicity of the system
the phase- and physical-variable representations of this system are identical; thus

$$\mathbf{A} = \begin{bmatrix} 0 & 1 \\ 0 & -1 \end{bmatrix} \quad \mathbf{b} = \begin{bmatrix} 0 \\ 1 \end{bmatrix} \quad (14\text{-}51)$$

The weighting matrices to be used in the quadratic PI are selected as

$$\mathbf{Q} = \begin{bmatrix} 1 & 0 \\ 0 & 0.1 \end{bmatrix} \quad z = 1 \quad (14\text{-}52)$$

A procedure for selecting these matrices is presented in Chap. 15. Substituting Eqs.
(14-51) and (14-52) into Eq. (14-50) yields

$$\begin{bmatrix} 0 & 0 \\ p_{11} - p_{12} & p_{12} - p_{22} \end{bmatrix} - \begin{bmatrix} p_{12}^2 & p_{12}p_{22} \\ p_{12}p_{22} & p_{22}^2 \end{bmatrix}$$
$$+ \begin{bmatrix} 0 & p_{11} - p_{12} \\ 0 & p_{12} - p_{22} \end{bmatrix} + \begin{bmatrix} 1 & 0 \\ 0 & 0.1 \end{bmatrix} = \mathbf{0}$$

From this equation the following relationships are obtained by equating each element
of the composite matrix to zero:

$$p_{12}^2 = 1 \qquad p_{11} - p_{12} - p_{12}p_{22} = 0 \qquad 2p_{12} - 2p_{22} - p_{22}^2 + 0.1 = 0$$

A choice must be made between using $p_{12} = -1$ and $p_{12} = 1$. Utilizing the value
$p_{12} = -1$ yields values of p_{11} and p_{22} that result in a $\mathbf{P}$ matrix that is not positive
definite. Using the value $p_{12} = 1$ gives

$$p_{11} = p_{22} + 1 \qquad p_{22}^2 + 2p_{22} - 2.1 = 0$$

Only the value $p_{22} = 0.76$ yields a positive definite $\mathbf{P}$ matrix. Thus

$$\mathbf{P} = \begin{bmatrix} 1.76 & 1 \\ 1 & 0.76 \end{bmatrix}$$

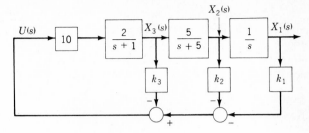

FIGURE 14-9
An optimal feedback control system, Example 2.

Therefore, from Eqs. (14-49) and (14-48), respectively,

$$\mathbf{k} = \mathbf{Pb}z^{-1} = \begin{bmatrix} 1 \\ 0.76 \end{bmatrix}$$

$$u^*(\mathbf{x}) = -\mathbf{k}^T\mathbf{x} = -(x_1 + 0.76x_2)$$

If $y(t) = x_1(t)$, the damping ratio of the resulting optimal system is 0.88. Note that this is higher than for the ITAE and Butterworth forms but lower than the binomial form. The resulting optimal feedback control system is shown in Fig. 14-8b. From the analysis in Chap. 12, the value $k_1 = 1$ is required to produce $e(t)_{ss} = 0$ for a step input. The same value is obtained in this problem because of the value assigned to q_{11}. (The criterion for this selection is described in Chap. 15.)

EXAMPLE 2   The quadratic PI is used with the phase-variable representation of the open-loop transfer function

$$G(s) = \frac{100}{s(s + 1)(s + 5)}$$

The required feedback gains† to yield an optimal closed-loop performance are $\mathbf{k}^T = \begin{bmatrix} 1 & 1.377 & 0.954 \end{bmatrix}$. Determine $\mathbf{k}_p$ required for the physical-variable representation of this system as shown in Fig. 14-9. The plant matrices and control vectors for the phase- and physical-variable representations are, respectively,

Phase Variable
Representation:
$$\mathbf{A} = \begin{bmatrix} 0 & 1 & 0 \\ 0 & 0 & 1 \\ 0 & -5 & -6 \end{bmatrix} \quad \mathbf{b} = \begin{bmatrix} 0 \\ 0 \\ 100 \end{bmatrix}$$

Physical Variable
Representation:
$$\mathbf{A}_p = \begin{bmatrix} 0 & 1 & 0 \\ 0 & -5 & 5 \\ 0 & 0 & -1 \end{bmatrix} \quad \mathbf{b}_p = \begin{bmatrix} 0 \\ 0 \\ 20 \end{bmatrix}$$

The subscript p indicates the system representation in terms of the physical variables.

† The PI uses the values

$$\mathbf{Q} = \begin{bmatrix} 1 & 0 \\ & 0.008 \\ 0 & 1.0 \end{bmatrix} \quad z = 1$$

The change from phase variables to physical variables is accomplished by a similarity transformation. The procedure is the same as that in Sec. 5-9 and requires the identification of a transformation matrix $\mathbf{T}$ so that $\mathbf{x} = \mathbf{Tx}_p$. The state equation becomes

$$\dot{\mathbf{x}}_p = \mathbf{T}^{-1}\mathbf{ATx}_p + \mathbf{T}^{-1}\mathbf{bu} = \mathbf{A}_p\mathbf{x}_p + \mathbf{b}_p u$$

Equating the elements of the equalities $\mathbf{AT} = \mathbf{TA}_p$ and $\mathbf{b} = \mathbf{Tb}_p$ gives the transformation matrix

$$\mathbf{T} = \begin{bmatrix} 1 & 0 & 0 \\ 0 & 1 & 0 \\ 0 & -5 & 5 \end{bmatrix}$$

Thus, from Eq. (14-46),

$$\mathbf{k}_p{}^T = \mathbf{k}^T\mathbf{T} = \begin{bmatrix} 1 & -3.393 & 4.770 \end{bmatrix}$$

14-10   BODE DIAGRAM ANALYSIS OF OPTIMAL STATE-VARIABLE FEEDBACK SYSTEMS[14,19]

Optimal state-variable feedback systems can be designed by a frequency-domain method. The basis for this design method is the condition of optimality represented by[13] the Kalman equation. This equation represents the Riccati equation in which $\mathbf{k}$ appears explicitly and the Riccati matrix $\mathbf{P}$ is eliminated. The form of the equation which is used in this approach for a single-input single-output system is

$$|1 + \mathbf{k}^T\mathbf{\Phi}(s)\mathbf{b}|^2 = 1 + \frac{1}{z}|G_K(s)|^2 \qquad (14\text{-}53)$$

where $s = j\omega$, $\mathbf{\Phi}(s) = [s\mathbf{I} - \mathbf{A}]^{-1}$, and $G_K(s) = \mathbf{h}^T\mathbf{\Phi}(s)\mathbf{b}$. As indicated in Sec. 14-8 the $\mathbf{Q}$ matrix may be decomposed into the matrix product $\mathbf{Q} = \mathbf{HH}^T$. For this system without loss of generality[13], let $\mathbf{Q} = \mathbf{hh}^T$. Once $\mathbf{Q}$ and z are specified, the only unknown quantity in Eq. (14-53) is $\mathbf{k}$. This equation is applicable to the linear system described by

$$\dot{\mathbf{x}} = \mathbf{Ax} + \mathbf{b}u \qquad \text{and} \qquad y = \mathbf{c}^T\mathbf{x} \qquad (14\text{-}54)$$

The quadratic PI to be satisfied is represented by

$$\text{PI} = \int_0^\infty [\mathbf{x}^T\mathbf{Qx} + zu^2]\, dt \qquad (14\text{-}55)$$

Note that $G_K(s)$ represents a transfer function which is utilized in the graphical solution for the k_i. The special case $\mathbf{h} = \mathbf{c}$ restricts the design problem; then $\mathbf{Q} = \mathbf{hh}^T$ and $G_K(s) = \mathbf{c}^T\mathbf{\Phi}(s)\mathbf{b} = G(s)$. This is not a necessary condition for use of the Bode method.

The method presented in this section is an indirect graphical solution of the Riccati equation. When the $H_{eq}(s)$ representation of the state-variable feedback as developed in Chap. 12 is used, the system is represented in Fig. 14-10. The transfer functions are

$$G(s) = \frac{Y(s)}{U(s)} = \mathbf{c}^T\mathbf{\Phi}(s)\mathbf{b} \qquad (14\text{-}56)$$

$$H_{eq}(s) = \frac{\mathbf{k}^T\mathbf{X}(s)}{Y(s)} = \frac{\mathbf{k}^T\mathbf{X}(s)}{\mathbf{c}^T\mathbf{X}(s)} = \frac{\mathbf{k}^T\mathbf{\Phi}(s)\mathbf{b}}{\mathbf{c}^T\mathbf{\Phi}(s)\mathbf{b}} \qquad (14\text{-}57)$$

$$G(s)H_{eq}(s) = \mathbf{k}^T\mathbf{\Phi}(s)\mathbf{b} \qquad (14\text{-}58)$$

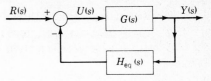

FIGURE 14-10
Equivalent feedback system.

When these relationships are used, Eq. (14-53) becomes

$$|1 + G(s)H_{eq}(s)|^2 = 1 + \frac{1}{z}|G_K(s)|^2 \qquad (14\text{-}59)$$

For $s = j\omega$ and $0 \le \omega < \infty$, the function $|G_K(j\omega)|^2/z > 0$. Thus, from Eq. (14-59),

$$|1 + G(j\omega)H_{eq}(j\omega)| > 1 \qquad (14\text{-}60)$$

A typical polar plot of $G(j\omega)H_{eq}(j\omega)$ is shown in Fig. 14-11.

Based on the condition of Eq. (14-60), the polar plot of $G(j\omega)H_{eq}(j\omega)$ must remain outside of a unit circle centered on the $-1 + j0$ point, for all values of frequency. This leads to the interesting observation that many of the control systems designed by the conventional methods of Chap. 10 are not optimum according to the quadratic PI. In other words, the methods of Chap. 10 are satisfactory design methods which yield desired conventional figures of merit. When an optimal performance (quadratic PI) is desired, the methods presented in this and succeeding sections may be used. Although the design is optimal, the time responses may not satisfy the system specifications. The method of combining these properties is presented in the following sections.

The Bode design method utilizes the following properties of Eq. (14-59):

1 The poles of $1 + G(s)H_{eq}(s)$ are known.
2 The zeros of $1 + G(s)H_{eq}(s)$ are the poles of the closed-loop system, which must be stable, and therefore they all lie in the left-half s plane.

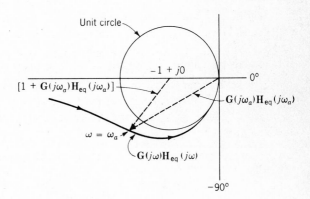

FIGURE 14-11
Polar plot of an optimal system.

3 For small ω, when $|\mathbf{G}_K(j\omega)/\sqrt{z}| \gg 1$, $|1 + \mathbf{G}(j\omega)\mathbf{H}_{eq}(j\omega)| \approx |\mathbf{G}_K(j\omega)/\sqrt{z}|$.
4 For large ω, when $|\mathbf{G}_K(j\omega)/\sqrt{z}| \ll 1$, $|1 + \mathbf{G}(j\omega)\mathbf{H}_{eq}(j\omega)| \approx 1$.
5 When $|\mathbf{G}_K(j\omega)/\sqrt{z}| = 1$, at $\omega = \omega_\phi$, $|1 + \mathbf{G}(j\omega_\phi)\mathbf{H}_{eq}(j\omega_\phi)| = \sqrt{2}$.

Properties 3 to 5 *are based upon the plant having at least one pure integration and a pole-zero excess of 1 or more.* The design method consists of matching the Bode plot of $|1 + \mathbf{G}(j\omega)\mathbf{H}_{eq}(j\omega)|$:

1 With the Bode plot of $|\mathbf{G}_K(j\omega)/\sqrt{z}|$ for $|\mathbf{G}_K(j\omega)/\sqrt{z}| > 1$
2 With unity gain (the 0-dB line) for $|\mathbf{G}_K(j\omega)/\sqrt{z}| < 1$

This matching can be accomplished by utilizing a Butterworth polynomial (see Table 14-2) with the corner frequency ω_ϕ at unity crossover for $|\mathbf{G}_K(j\omega)/\sqrt{z}|$ (see property 5).[20] A possible area of further research investigation would be to use another polynomial, such as one of the family of the Chebyshev[21] polynomials. The design procedure is as follows.

Step 1 Draw the Bode diagram of $|\mathbf{G}_K(j\omega)/\sqrt{z}|$ utilizing the straight-line approximation technique. The procedure for plotting $\mathbf{G}_K(j\omega)$ is described in Chap. 8. If necessary, *increase the gain term so that all break points occur above the 0-dB axis* (see step 6).

Step 2 For $\omega < \omega_\phi$, the Bode diagram of $|1 + \mathbf{G}(j\omega)\mathbf{H}_{eq}(j\omega)|$ is identical to that of $|\mathbf{G}_K(j\omega)/\sqrt{z}|$ (in accordance with property 3).

Step 3 For $\omega > \omega_\phi$, the Bode diagram of $|1 + \mathbf{G}(j\omega)\mathbf{H}_{eq}(j\omega)|$, by use of property 4, coincides with the 0-dB line.

Step 4 Synthesize the approximate transfer function $\mathbf{T}_s(j\omega)$ from the straight-line approximation of $|1 + \mathbf{G}(j\omega)\mathbf{H}_{eq}(j\omega)|$ which is formed in steps 2 and 3. This is accomplished by multiplying $|\mathbf{G}_K(j\omega)/\sqrt{z}|$ by a Butterworth polynomial having the order α, equal to the pole-zero excess of $\mathbf{G}_K(j\omega)$, and with $\omega_0 = \omega_\phi$. Some Butterworth polynomials are listed in Table 14-2. The term ω_0^α must be factored from these polynomials to obtain the standard form with the constant term equal to unity. The following generalized equation can be used to extend this table for values of $\alpha > 8$:

$$B_b(s) = \left(\frac{s}{\omega_0}\right)^\alpha + V_{\alpha-1}\left(\frac{s}{\omega_0}\right)^{\alpha-1} + \cdots + V_1\frac{s}{\omega_0} + 1$$

where

$$V_{\alpha-1} = \frac{1}{\sin(\pi/2\alpha)} \qquad V_{\alpha-2} = \frac{V_{\alpha-1}\cos(\pi/2\alpha)}{\sin(\pi/\alpha)}$$

$$V_{\alpha-\beta} = \frac{V_{\alpha-\beta+1}\cos[(\beta-1)(\pi/2\alpha)]}{\sin(\beta\pi/2\alpha)}$$

Choosing a Butterworth polynomial having $\omega_0 = \omega_\phi$ results in the polynomial having a value of $\sqrt{2}$ at unity crossover of $|\mathbf{G}_K(j\omega)/\sqrt{z}|$, thus satisfying property 5.

Step 5 Determine the exact expression of $1 + G(s)H_{eq}(s)$ using Eq. (14-58). All the parameters of this expression except the feedback coefficients **k** are known.

Step 6 By equating the synthesized transfer function T_s of step 4 to the exact transfer function of step 5 the unknown values of k_i are determined. In order to accomplish this the degree of the numerator and denominator polynomials of the synthesized transfer function must be equal to the degree of the numerator and denominator polynomials, respectively, of the exact transfer function. For this to be true Lm $\mathbf{G}_K(j\omega)/\sqrt{z}$ at all corner frequencies (break points) must be above the 0-dB line. The accuracy of this method is improved with larger values of Lm $\mathbf{G}_K(j\omega)/\sqrt{z}$ at the corner frequencies.

EXAMPLE   The plant of the example in Sec. 12-7, with the value of A specified, is utilized to illustrate the above procedure to solve for $\mathbf{k}_p$ of Fig. 12-6b. Assume that $z = 1$ and $\mathbf{h}_p{}^T = [1 \quad 1 \quad 1]$. Use the matrix $\mathbf{A}_p$ in Example 2, Sec. 14-9, to obtain $\mathbf{\Phi}(s)$. Then,

$$G_K(s) = \mathbf{h}_p{}^T\mathbf{\Phi}(s)\mathbf{b} = \begin{bmatrix} 1 & 1 & 1 \end{bmatrix} \begin{bmatrix} \dfrac{1}{s} & \dfrac{1}{s(s+5)} & \dfrac{5}{s(s+1)(s+5)} \\[3mm] 0 & \dfrac{1}{s+5} & \dfrac{5}{(s+1)(s+5)} \\[3mm] 0 & 0 & \dfrac{1}{s+1} \end{bmatrix} \begin{bmatrix} 0 \\[3mm] 0 \\[3mm] 2A \end{bmatrix} \quad (14\text{-}61)$$

Step 1 Selection of the value $A = 1$ for the gain is unsatisfactory because the break points of Lm $\mathbf{G}_K(j\omega)$, for the specified $\mathbf{h}_p$, do not all occur above the 0-dB line. Therefore, a value $A = 50$ is used for this example, which yields

$$G_K(s) = \frac{100(s^2 + 10s + 5)}{s(s+1)(s+5)} = \frac{100(s/9.47 + 1)(s/0.53 + 1)}{s(s+1)(0.2s + 1)} \quad (14\text{-}62)$$

The Bode plot of $\mathbf{G}_K(j\omega)$ is drawn as the solid line in Fig. 14-12. The 0-dB crossover point is identified as ω_ϕ.

Step 2 A dashed line is drawn in Fig. 14-12 to match the approximate value of Lm $[1 + \mathbf{G}(j\omega)\mathbf{H}_{eq}(j\omega)]$ with that of Lm $\mathbf{G}_K(j\omega)$ for $\omega < \omega_\phi$. Since the break points are above the 0-dB line, two roots of the system characteristic equation are already known and are equal to the zeros of $G_K(s)$, namely, $s_1 = -9.47$, $s_2 = -0.53$. The third root is given by the Butterworth polynomial as seen below.

Step 3 The dashed line is drawn along the 0-dB line, for $\omega > \omega_\phi$, to represent the approximate value of Lm $[1 + \mathbf{G}(j\omega)\mathbf{H}_{eq}(j\omega)]_{\omega > \omega_\phi} = 0$ dB.

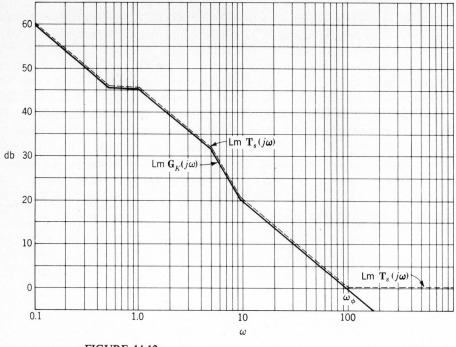

FIGURE 14-12
Plot of Lm $G_K(j\omega)$ and the synthesized Lm $T_s(j\omega)$.

Step 4   Since the pole-zero excess of $G_K(s)$ is 1, the synthesized T_s representing the approximate transfer function of $1 + G(s)H_{eq}(s)$ is obtained by multiplying $G_K(s)$ by a first-order polynomial $B_b(s)$ obtained from Table 14-2 after dividing by ω_0; thus

$$1 + G(s)H_{eq}(s) \simeq T_s = G_K(s)\left(\frac{s}{\omega_\phi} + 1\right)$$

$$= \frac{(100/\omega_\phi)[s^3 + (10 + \omega_\phi)s^2 + (10\omega_\phi + 5)s + 5\omega_\phi]}{s(s + 1)(s + 5)} \qquad (14\text{-}63)$$

Step 5   When Eq. (14-58) is used, the following exact expression of the characteristic polynomial with $A = 50$ is formed:

$$1 + G(s)H_{eq}(s) = \frac{s^3 + (6 + 100k_3)s^2 + [5 + 500(k_3 + k_2)]s + 500k_1}{s(s + 1)(s + 5)} \qquad (14\text{-}64)$$

Step 6   Equating the corresponding coefficients of Eq. (14-63) and (14-64) yields

$$\frac{100}{\omega_\phi} = 1 \qquad \omega_\phi = 100$$

$$6 + 100k_3 = 10 + \omega_\phi = 110 \qquad k_3 = 1.04$$

$$5 + 500(k_3 + k_2) = 10\omega_\phi + 5 = 1{,}005 \qquad k_2 = 0.96$$

$$500k_1 = 500 \qquad k_1 = 1$$

For comparison, the $\mathbf{k}_p$ obtained by utilizing the Riccati computer program has the values $k_1 = 1$, $k_2 = 0.96263$, and $k_3 = 1.0371$. These values compare very favorably with those obtained by the graphical method for this example. These favorable results are assured by choosing a large value of A so that all the break points of $\mathbf{G}_K(j\omega)$ occur far enough above the 0-dB line. The closer the break points are to the 0-dB line, the less accurate the graphical solution for the k_i. An explanation of this requirement and the selection of ω_0 without drawing the Bode diagram is contained in Sec. 14-12. The remaining root of the characteristic equation is $s_3 = -100$ (see Prob. 14-6).

14-11 ROOT-SQUARE-LOCUS[20] ANALYSIS OF OPTIMAL STATE-VARIABLE FEEDBACK SYSTEMS[19]

In Secs. 12-2 and 12-3 the design procedures are based upon specifying the desired poles of the closed-loop transfer function (control ratio). This section presents a procedure whereby the poles of the state-variable feedback system transfer function are determined from the synthetic transfer function $G_K(s)$ which comes from the Kalman equation (14-53). This analysis is based upon the root-square-locus (RSL) method.

When $s = j\omega$, Eq. (14-59) can be rewritten in the form

$$[1 + G(s)H_{eq}(s)][1 + G(-s)H_{eq}(-s)] = 1 + \frac{G_K(s)G_K(-s)}{z} \qquad (14\text{-}65)$$

In Sec. 12-3 it is shown that the poles and zeros of $1 + G(s)H_{eq}(s)$ for a state feedback minimum-phase plant all lie in the left-half s plane (LHP) for positive gain. Also, its poles are the poles of $G(s)$. In a similar manner, the poles and zeros of $1 + G(-s)H_{eq}(-s)$ all lie in the right-half s plane (RHP). Further, the denominator polynomials on both sides of Eq. (14-65) must be identical, as seen in the example of the last section. As the function $1 + G_K(s)G_K(-s)/z$ is also a ratio of polynomials, the technique of *spectral factorization*[20] permits factoring it so that its poles and zeros can be separated into two groups, those lying in the LHP and those lying in the RHP. These two groups are denoted by

$$\left[1 + \frac{G_K(s)G_K(-s)}{z}\right]^L \qquad \text{and} \qquad \left[1 + \frac{G_K(s)G_K(-s)}{z}\right]^R$$

respectively. Correspondingly, utilizing the spectral factorization technique leads to

$$1 + G(s)H_{eq}(s) = \left[1 + \frac{G_K(s)G_K(-s)}{z}\right]^L \qquad (14\text{-}66)$$

and

$$1 + G(-s)H_{eq}(-s) = \left[1 + \frac{G_K(s)G_K(-s)}{z}\right]^R \qquad (14\text{-}67)$$

These equations are now valid in the entire s plane. The zeros of the term on the right side of Eq. (14-66) are the poles of the control ratio. The feedback coefficients k_i are determined by equating the coefficients of like powers of s in Eq. (14-66).

The zeros of $1 + G_K(s)G_K(-s)/z$ are determined by setting Eq. (14-65) equal to zero, which yields

$$\frac{G_K(s)G_K(-s)}{z} = -1 \qquad (14\text{-}68)$$

This equation has the identical mathematical form as $G(s)H(s) = -1 = e^{-j(1+2h)\pi}$ *used for the root-locus method of Chap. 7. Therefore, the root locus can be drawn for Eq. (14-68) to determine the zeros of the numerator polynomial of* $1 + G_K(s)G_K(-s)/z$. *The function* $G_K(s)G_K(-s)/z$ *is referred to as the associated or squared transfer function.* When the root locus is drawn for Eq. (14-68), it is referred to as the root-square-locus (RSL) method. Since $G_K(s)$ has n poles and w zeros, the magnitude and angle conditions for the RSL are, respectively,

$$\frac{|G_K(s)G_K(-s)|}{z} = 1 \tag{14-69}$$

$$\underline{/G_K(s)G_K(-s)} = \begin{cases} (1 + 2h)180° & \text{for } n - w \text{ even} \\ (h)360° & \text{for } n - w \text{ odd} \end{cases} \tag{14-70}$$

where $h = 0, \pm 1, \pm 2, \ldots$. Equation (14-69) is rewritten as

$$K^2 = \frac{|s^m|^2 \cdot |s - p_1| \cdot |s + p_1| \cdot |s - p_2| \cdot |s + p_2| \cdots |s - p_u| \cdot |s + p_u|}{|s - z_1| \cdot |s + z_1| \cdots |s - z_w| \cdot |s + z_w|} \tag{14-71}$$

where $K > 0$ and K is the loop sensitivity. Note that the value of z is incorporated in K. The geometrical shortcuts for obtaining the RSL are summarized in Table 14-5. Only the branches in the second quadrant are drawn since the plot is symmetrical about both the real and imaginary axes. Sufficient points are obtained for large enough values of K so that the branches of the root locus become straight lines approaching their asymptotes. *The value of K for which the dominant branch becomes essentially a straight line is defined as* K_x.

Table 14-5 SUMMARY OF RSL GEOMETRICAL PROPERTIES

Rule 1	Number of branches of the locus $= 2(m + u) = 2n$
Rule 2	Real-axis locus: if the total number of poles and zeros to the right of the s point on the real axis is odd, this point lies on the locus for $n - w =$ even number; if the total number of poles and zeros to the right of s the point on the real axis is even, this point lies on the locus for $n - w =$ odd number
Rule 3	Locus end points: locus starting points $(K = 0)$ are the poles of $G_K(s)$; the locus ending points $(K = \infty)$ are at the zeros of $G_K(s)$ or at infinity
Rule 4	Asymptotes of locus as s approaches infinity: $\gamma = \dfrac{(1 + 2h)180°}{2(n - w)}$ for $n - w$ even $\quad$ $\gamma = \dfrac{(h)360°}{2(n - w)}$ for $n - w$ odd $\quad$ where $h = 0, \pm 1, \pm 2$
Rule 5	Real-axis intercept: $\sigma_o = 0$
Rule 6	Breakaway point on the real axis: obtained from $dW/ds = 0$, where $W(s) = -K^2$ (or K^2)
Rule 7	Complex pole (or zero): angle of departure: see Rule 7 for root locus in Chap. 7; be sure to include the poles and zeros in the RHP when applying this rule
Rule 8	Imaginary-axis crossing point: none

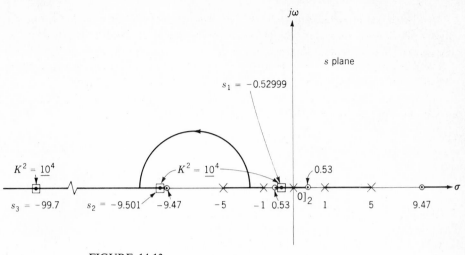

FIGURE 14-13
RSL for Eq. (14-72).

The RSL for an all-pole $G_K(s)$ with high forward gain (hfg) always has a pair of dominant complex-conjugate roots. The two asymptotes for a pole-zero $G_K(s)$ having $w = n - 1$ zeros, as determined from Eq. (14-68), lie on the real axis. Thus, under hfg operation the RSL of the system may have all real roots. A characteristic of the RSL plots is that as K becomes large, the poles and zeros of Eq. (14-68) or (14-69) can be represented by $2n - 2w$ poles at the origin. Since the asymptotes emanate from the origin, for $K > K_x$ the complex portions of the root locus, for n-$w > 1$, approach straight lines and are essentially the same for any nth-order plant. The difference between the RSL of each nth-order plant is the loop sensitivity calibration. Also, for $K > K_x$, the $2(n - w)$ branches approach the Butterworth configuration for $2(n - w)$ poles.

EXAMPLE The example of Sec. 14-10 is reworked by using the RSL method. Utilizing the transfer function from this example, [Eq. (14-62)], for $z = 1$, yields

$$G_K(s)G_K(-s) = \frac{-10^4(s \pm 9.47)(s \pm 0.53)}{s^2(s \pm 1)(s \pm 5)} \quad (14\text{-}72)$$

Since $n - w$ is odd the constant in Eq. (14-72) is negative. The RSL satisfying the $h360°$ angle condition is obtained as shown in Fig. 14-13. For $K^2 = 10^4$ the roots of the characteristic equation are shown in the figure. Using these roots to form the characteristic polynomial yields $s^3 + 109.731s^2 + 1,005.125s + 502.033$. The coefficients of this equation are equated to the corresponding coefficients of Eq. (14-64) to yield $k_1 = 1.0041$, $k_2 = 0.96294$, and $k_3 = 1.03731$. These values compare favorably with those obtained from the Bode and the Riccati computer solution methods. The advantages of the RSL are the same as for the root-locus method, i.e., the direct relationship between the s plane and the time domain. For those ranges of

K^2 for which one dominant pair of roots exists, the system loop sensitivity may be chosen to yield not only the optimal performance according to the quadratic PI but also satisfactory conventional figures of merit.

14-12 DETERMINATION OF HFG CONDITIONS

The solution for the feedback coefficients by the use of Eq. (14-53) is facilitated by using the phase-variable representation. Thus $G_K(s) = \mathbf{h}^T \mathbf{\Phi}(s) \mathbf{b}$ becomes

$$\frac{G_K(s)}{\sqrt{z}} = \frac{K_G}{\sqrt{z}} \frac{h_n s^{n-1} + \cdots + h_2 s + h_1}{s^n + a_{n-1} s^{n-1} + \cdots + a_0} \qquad (14\text{-}73)$$

The pole-zero excess of $\mathbf{G}_K(j\omega)$ is exactly 1 for $h_n \neq 0$. Property 5 in Sec. 14-10 can be achieved by multiplying $\mathbf{G}_K(j\omega)$ by a first-order Butterworth polynomial, that is, $T_s(s) = G_K(s)B_b(s)/\sqrt{z}$. Thus, using step 6 of Sec. 14-10

$$1 + G(s)H_{eq}(s) = \frac{G_K(s)B_b(s)}{\sqrt{z}} = T_s(s) \qquad (14\text{-}74)$$

Using the form of $G(s)H_{eq}(s)$ given in Eq. (12-26a), this becomes

$$1 + K_G \frac{k_n s^{n-1} + \cdots + k_2 s + k_1}{s^n + a_{n-1} s^{n-1} + \cdots + a_2 s + a_0}$$

$$= \frac{K_G}{\sqrt{z}} \left[\frac{h_n s^{n-1} + \cdots + h_1}{s^n + \cdots + a_0} \left(\frac{s}{\omega_\phi} + 1 \right) \right] \qquad (14\text{-}75)$$

The optimal closed-loop poles, represented by the zeros of the left-hand side of Eq. (14-75), asymptotically approach the w zeros of $G_K(s)$, while the remaining $n - w$ poles ($n - w = 1$ for $h_n \neq 0$) approach a Butterworth pattern. This is an important result due to Kalman.[13] Therefore Eq. (14-75) is a valid approximation for the case of hfg but is precisely true only in the limit. In other words,

$$|1 + G(s)H_{eq}(s)| \neq \frac{|G_K(s)(s/\omega_\phi + 1)|}{\sqrt{z}} \qquad (14\text{-}76)$$

For large gain

$$\lim_{K_G \to \infty} |1 + G(s)H_{eq}(s)| = \frac{\lim_{K_G \to \infty} |G_K(s)(s/\omega_\phi + 1)|}{\sqrt{z}} \qquad (14\text{-}77)$$

Equating the numerators of Eq. (14-75) gives

$$s^n + (a_{n-1} + K_G k_n) s^{n-1} + \cdots + (a_0 + K_G k_1)$$

$$= \frac{K_G}{\omega_\phi \sqrt{z}} [h_n s^n + (\omega_\phi h_n + h_{n-1}) s^{n-1} + \cdots + (\omega_\phi h_2 + h_1) s + \omega_\phi h_1] \qquad (14\text{-}78)$$

Equating coefficients yields

$$\omega_\phi = \frac{K_G h_n}{\sqrt{z}} \qquad (14\text{-}79a)$$

$$k_1 = \frac{h_1}{\sqrt{z}} - \frac{a_0}{K_G} \qquad (14\text{-}79b)$$

$$k_i = \frac{h_i}{\sqrt{z}} + \frac{h_{i-1}}{K_G h_n} - \frac{a_{i-1}}{K_G} \qquad i = 2, \ldots, n \qquad (14\text{-}79c)$$

The first expression provides an opportunity to define a minimum gain for *reproducibility*. This is the gain above which the closed-loop poles become independent of the open-loop system parameters. This value of gain can be determined by the RSL method discussed in Sec. 14-11 and in Chap. 15. For zero steady-state error with a step input, k_1 must satisfy Eq. (12-31), $k_1 = c_0 - a_0/K_G$.

Consider the accuracy of the straight-line approximation to the Bode plot. Near the corner frequencies, significant corrections to the asymptotes are required. Thus a rule of thumb for accuracy of the straight-line approximation to the Bode plot at crossover is

$$\omega_\phi \geq \xi \omega^* \qquad (14\text{-}80)$$

where ω^* represents the highest corner frequency of $\mathbf{G}_K(j\omega)$ and ξ represents the value that yields the desired separation between ω^* and ω_ϕ. For this method to apply, $\xi > 1$. For example, setting $\xi = 10$ restricts all corner frequencies of $\mathbf{G}_K(j\omega)$ to be at least 1 decade below the crossover frequency. This gives a minimum-gain value for accurate pole placement and, equivalently, a minimum gain for use of Eqs. (14-79) in computing the feedback coefficients.

Substituting Eq. (14-79a) into Eq. (14-80) gives

$$K_G h_n \geq \xi \omega^* \sqrt{z} \qquad (14\text{-}81)$$

First, consider a fixed value of K_G. Then Eq. (14-81) can be used to find the minimum value of h_n for which an exact pole placement is possible:

$$(h_n)_{\min} = \frac{\xi \omega^* \sqrt{z}}{K_G} \qquad (14\text{-}82)$$

Alternatively, consider a given value of h_n. Then the minimum value of forward gain for exact pole placement is

$$K_{G\min} = \frac{\xi \omega^* \sqrt{z}}{h_n} \qquad (14\text{-}83)$$

It should be noted from Eq. (14-79) that under hfg operation

$$\mathbf{k} \approx \mathbf{h}/\sqrt{z} \qquad (14\text{-}84)$$

EXAMPLE   The example of Sec. 14-10 is utilized to illustrate this simpler approach of obtaining $\mathbf{k}_p$. This approach solves for $\mathbf{k}$, the phase-variable feedback coefficients.

In order to compare the results of the two methods, it is necessary to determine the $\mathbf{h}$ which is equivalent to $\mathbf{h}_p$. The transformation from physical to phase variable is performed by using $\mathbf{x}_p = \mathbf{T}\mathbf{x}$ and $\mathbf{b}_p = \mathbf{T}\mathbf{b}$. Using

$$\mathbf{b}_p = \begin{bmatrix} 0 \\ 0 \\ 100 \end{bmatrix} \qquad \mathbf{b} = \begin{bmatrix} 0 \\ 0 \\ 500 \end{bmatrix}$$

$$\mathbf{A}_p = \begin{bmatrix} 0 & 1 & 0 \\ 0 & -5 & 5 \\ 0 & 0 & -1 \end{bmatrix} \qquad \mathbf{A} = \begin{bmatrix} 0 & 1 & 0 \\ 0 & 0 & 1 \\ 0 & -5 & -6 \end{bmatrix}$$

yields

$$\mathbf{T} = \begin{bmatrix} 1 & 0 & 0 \\ 0 & 1 & 0 \\ 0 & 1 & 0.2 \end{bmatrix} \qquad (14\text{-}85)$$

Substituting $\mathbf{x}_p = \mathbf{T}\mathbf{x}$ into $\mathbf{x}_p{}^T\mathbf{Q}_p\mathbf{x}_p = \mathbf{x}_p{}^T\mathbf{h}_p\mathbf{h}_p{}^T\mathbf{x}_p$ yields

$$\mathbf{x}^T\mathbf{T}^T\mathbf{h}_p\mathbf{h}_p{}^T\mathbf{T}\mathbf{x} = \mathbf{x}^T\mathbf{h}\mathbf{h}^T\mathbf{x}$$

Thus

$$\mathbf{h}^T = \mathbf{h}_p{}^T\mathbf{T} = \begin{bmatrix} 1 & 2 & 0.2 \end{bmatrix} \qquad (14\text{-}86)$$

For the transfer function

$$G(s) = \frac{10A}{s(s+1)(s+5)} = \frac{K_G}{s^3 + 6s^2 + 5s} \qquad (14\text{-}87)$$

where $K_G = 500$, utilizing Eqs. (14-79a) to (14-79c), yields

$$\omega_\phi = 500 \times 0.2 = 100 \qquad k_1 = 1$$

$$k_2 = 2 + \frac{1}{(500)(0.2)} - \frac{5}{500} = 2$$

$$k_3 = 0.2 + \frac{2}{(500)(0.2)} - \frac{6}{500} = 0.208$$

The optimal control is $\mathbf{u}^* = -\mathbf{k}^T\mathbf{x} = -\mathbf{k}^T\mathbf{T}^{-1}\mathbf{x}_p = -\mathbf{k}_p{}^T\mathbf{x}_p$ obtain

$$\mathbf{k}_p{}^T = \mathbf{k}^T\mathbf{T}^{-1} = \begin{bmatrix} 1 & 2 & 0.208 \end{bmatrix} \begin{bmatrix} 1 & 0 & 0 \\ 0 & 1 & 0 \\ 0 & -5 & 5 \end{bmatrix} = \begin{bmatrix} 1 & 0.96 & 1.04 \end{bmatrix}$$

These agree with the values of the example in Sec. 14-10. Note that for $K_G = 500$, $\mathbf{k} \approx \mathbf{h}$, and $\mathbf{k}_p{}^T = \mathbf{k}^T\mathbf{T}^{-1} \approx \mathbf{h}^T\mathbf{T}^{-1} = \mathbf{h}_p{}^T\mathbf{T}\mathbf{T}^{-1} = \mathbf{h}_p{}^T$. For this example $\mathbf{k}_p{}^T \approx \mathbf{h}_p{}^T = \begin{bmatrix} 1 & 1 & 1 \end{bmatrix}$.

14-13 SUMMARY

This chapter has established the concept of a performance index which yields an optimal system response. Typical PIs, in particular the quadratic PI, are discussed. The Riccati equation is developed, based on setting the quadratic PI equal to a Liapunov function $\mathbf{x}^T\mathbf{P}\mathbf{x}$. This equation is solved for the Riccati matrix $\mathbf{P}$, which in turn permits the determination of the optimal control law $\mathbf{u}^*(\mathbf{x}) = -\mathbf{K}^T\mathbf{x}$. Design methods are presented that satisfy the conditions for optimality and simultaneously yield satisfactory time-response characteristics. Chapter 15 presents a design method for selecting the matrix $\mathbf{Q}$ of the quadratic PI so that the quadratic conventional figures of merit can be achieved directly.

REFERENCES

1 Flügge-Lotz, I.: The Importance of Optimal Control for the Practical Engineer, *Automatica*, pp. 749–753, November 1970.

2 Graham, D., and R. C. Lathrop: The Synthesis of Optimum Response: Criteria and Standard Forms, *Trans. AIEE*, vol. 72, pt. II, pp. 273–288, November 1953.

3 Stout, Thomas M.: A Note on Control Area, *J. Appl. Phys.*, vol. 21, pp. 1129–1131, November 1950.

4 Nims, P. T.: Some Design Criteria for Automatic Controls, *Trans. AIEE*, vol. 70, pt. II, pp. 606–611, 1951.

5 Schultz, W. C., and V. C. Rideout: The Selection and Use of Servo Performance Criteria, *Trans. AIEE*, vol. 76, pt. II, pp. 383–388, January 1958.

6 King, Leonard H.: Reduction of Forced Error in Closed-Loop Systems, *Proc. IRE*, vol. 41, pp. 1037–1042, August 1953.

7 Oldenbourg, R. C., and H. Sartorius: "The Dynamics of Automatic Controls," trans. and ed. by H. L. Mason, American Society of Mechanical Engineers, New York, 1948.

8 Whitely, A. L.: The Theory of Servo Systems, with Particular Reference to Stabilization, *J. Inst. Elec. Eng. (Lond.)*, vol. 93, pp. 353–377, 1946.

9 Butterworth, S.: On the Theory of Filter Amplifiers, *Wireless Eng.*, vol. 7, pp. 536–541, 1930.

10 Kirk, D. E.: "Optimum Control Theory," Prentice-Hall, Englewood Cliffs, N.J., 1970.

11 *IEEE Trans. Autom. Control*, vol. AC-16, Special Issue on Linear-Quadratic-Gaussian Problem, December 1971.

12 Athans, M., and P. L. Falb: "Optimal Control," McGraw-Hill, New York, 1966.

13 Kalman, R. E.: "When Is a Linear Control System Optimal?," *ASME J. Basic Eng.*, pp. 51–60, March 1964.

14 Schultz, D. G., and J. L. Melsa: "State Functions and Linear Control Systems," McGraw-Hill, New York, 1967.

15 Solution of Linear-Quadratic Optimal Control Problem (OPTCON), Air Force Institute of Technology Computer Program Library, 1974.

16 Anderson, B. D. O., and J. B. Moore: "Linear Optimal Control," Prentice-Hall, Englewood Cliffs, N.J., 1971.

17 Solheim, O. A.: Design of Optimal Control Systems with Prescribed Eigenvalues, *Int. J. Control*, vol. 15, no. 1, pp. 143–160, 1972.

18 Woodhead, M. A., and B. Porter: Optimal Modal Control, *Measurement Control*, vol. 6, pp. 301–303, July 1973.

19 Leake, R. J.: "Analytical Controller Design and the Servomechanism Problem," *Syracuse Univ. Dept. Elec. Eng. Tech. Rep.* 64-4, July 1964; Return Difference Bode Diagram for Optimal System Design, *IEEE Trans. Autom. Control*, vol. AC-10, no. 3, pp. 342–344, July 1965.

20 Chang, S. S. L.: "Synthesis of Optimal Control Systems," McGraw-Hill, New York, 1961.

21 Van Valkenburg, M. E.: "Introduction to Modern Network Synthesis," Wiley, New York, 1960.

22 Willems, J. C.: Least Squares Optimal Control and the Algebraic Riccati Equation, *IEEE Trans. Autom. Control*, vol. AC-16, December 1971.

23 Molinari, B. P.: The Stable Regulator Problem and Its Inverse, *IEEE Trans. Autom. Control*, vol. AC-18, pp. 454–459, October 1973.

15

OPTIMAL DESIGN BY USE OF QUADRATIC PERFORMANCE INDEX

15-1 INTRODUCTION[1,2]

The optimal control theory associated with the quadratic PI and the Riccati equation is developed in Chap. 14. In order to determine the feedback vector $\mathbf{k}$, a weighting matrix $\mathbf{Q}$ and a weighting value z of the PI must be chosen. In that chapter these weighting quantities were assumed to be known in order to proceed with the presentation of the material. This chapter presents a method for choosing these weighting quantities for a linear time-invariant system. The development is in terms of phase variables. Then the feedback coefficients for the physical-variable representation can be determined by means of a transformation matrix. Other approaches for determining these weighting quantities are available in the literature.

15-2 BASIC SYSTEM

The state representation of a plant, with transfer function

$$G(s) = \frac{K_G(s^w + c_{w-1}s^{w-1} + \cdots + c_0)}{s^n + a_{n-1}s^{n-1} + \cdots + a_0} \qquad (15\text{-}1)$$

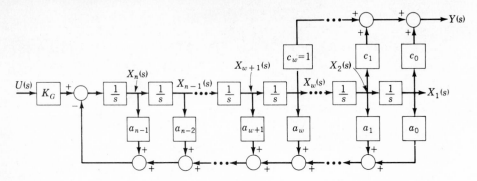

FIGURE 15-1
Block diagram representing Eq. (15-1) for phase-variable representation.

is represented in phase variables by

$$\dot{\mathbf{x}} = \mathbf{A}\mathbf{x} + \mathbf{b}u \qquad (15\text{-}2)$$

$$y = \mathbf{c}^T\mathbf{x} \qquad (15\text{-}3)$$

where

$$\mathbf{A} = \begin{bmatrix} 0 & 1 & 0 & \cdots & 0 \\ 0 & 0 & 1 & \cdots & 0 \\ \multicolumn{5}{c}{\dotfill} \\ -a_0 & -a_1 & -a_2 & \cdots & -a_{n-1} \end{bmatrix} \qquad (15\text{-}4)$$

$$\mathbf{b}^T = \begin{bmatrix} 0 & 0 & 0 & \cdots & 0 & K_G \end{bmatrix} \qquad \text{for } K_G > 0 \qquad (15\text{-}5)$$

$$\mathbf{c}^T = \begin{bmatrix} c_0 & c_1 & c_2 & \cdots & c_{w-1} & c_w & 0 & \cdots & 0 \end{bmatrix} \qquad (15\text{-}6)$$

where $c_w = 1$. For most practical systems $n > w$, thus the number of poles n of the plant is greater than the number of zeros w. The zeros of Eq. (15-1) affect only the output expression, Eq. (15-6). The block diagram of the phase-variable representation is shown in Fig. 15-1.

Note that for *an all-pole plant* $c_0 = 1$ and $c_1 = \cdots = c_w = 0$. In addition, for a Type 1 plant $a_0 = 0$, and for a Type 2 plant $a_1 = a_0 = 0$.

15-3   QUADRATIC PERFORMANCE INDEX

A suitable quadratic PI for a single-input single-output system is

$$\text{PI} = \int_0^\infty (\mathbf{x}^T\mathbf{Q}\mathbf{x} + zu^2)\, dt \qquad (15\text{-}7)$$

where z is a scalar and $\mathbf{Q}$ is at least a positive semidefinite symmetric matrix. The algebraic Riccati equation for this problem, from Eq. (14-44), is

$$\mathbf{P}\mathbf{A} + \mathbf{A}^T\mathbf{P} - \frac{1}{z}[\mathbf{P}\mathbf{b}\,\mathbf{b}^T\mathbf{P}] = -\mathbf{Q} \qquad (15\text{-}8)$$

where $\mathbf{P}$ is the Riccati matrix which is symmetric positive definite. The system is both observable and controllable. If $\mathbf{Q}$ is positive semidefinite, then Eq. (14-33) must

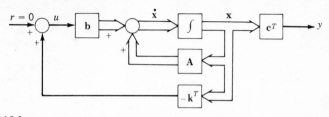

FIGURE 15-2
Optimal linear regulator system.

be satisfied for $\mathbf{P}$ to be positive definite. If $\mathbf{Q}$ is positive definite, then $\mathbf{P}$ is positive definite. The resulting optimal feedback control law is

$$u^*(\mathbf{x}) = -\mathbf{b}^T\mathbf{P}\mathbf{x} = -\mathbf{k}^T\mathbf{x} \qquad (15\text{-}9)$$

The elements of the vector $\mathbf{k}^T$ are the feedback coefficients (see Chaps. 12 and 14) and Eq. (15-9) indicates that all the state variables are to be fed back. This optimal system is represented by Fig. 15-2, with $r(t) = 0$, and is referred to as the optimal linear regulator system. In Eq. (15-8) let $\hat{\mathbf{G}} = \mathbf{PA}$; that is,

$$\hat{\mathbf{G}} \equiv \mathbf{PA} = \begin{bmatrix} p_{11} & p_{12} & \cdots & p_{1n} \\ p_{12} & p_{22} & \cdots & p_{2n} \\ \cdots\cdots\cdots\cdots\cdots \\ p_{1n} & p_{2n} & \cdots & p_{nn} \end{bmatrix} \begin{bmatrix} 0 & 1 & 0 & \cdots & 0 \\ 0 & 0 & 1 & \cdots & 0 \\ \cdots\cdots\cdots\cdots\cdots\cdots\cdots \\ -a_0 & -a_1 & -a_2 & \cdots & -a_{n-1} \end{bmatrix}$$

$$= \begin{bmatrix} -a_0 p_{1n} & \vrule & \\ -a_0 p_{2n} & \vrule & \hat{g}_{ij} \\ \cdots\cdots & \vrule & \\ -a_0 p_{nn} & \vrule & \end{bmatrix} \qquad (15\text{-}10)$$

$$\mathbf{bb}^T = \begin{bmatrix} 0 & \cdots & 0 \\ \cdots\cdots\cdots\cdots \\ 0 & \cdots & 0 \\ 0 & \cdots & K_G{}^2 \end{bmatrix} \qquad (15\text{-}11)$$

This is a matrix of order n, with all zero elements except for the nn element which is $K_G{}^2$.

$$\mathbf{Pbb}^T = \begin{bmatrix} 0 & \cdots & 0 & K_G{}^2 p_{1n} \\ 0 & \cdots & 0 & K_G{}^2 p_{2n} \\ \cdots\cdots\cdots\cdots\cdots \\ 0 & \cdots & 0 & K_G{}^2 p_{nn} \end{bmatrix} \qquad (15\text{-}12)$$

$$\mathbf{M} = \mathbf{Pbb}^T\mathbf{P} = \begin{bmatrix} K_G{}^2 p_{1n}{}^2 & \vrule & \\ K_G{}^2 p_{2n} p_{1n} & \vrule & m_{ij} \\ \vdots & \vrule & \\ K_G{}^2 p_{nn} p_{1n} & \vrule & \end{bmatrix} \qquad (15\text{-}13)$$

$$\mathbf{Q} = \begin{bmatrix} q_{11} & & & \mathbf{0} \\ & q_{22} & & \\ & & \ddots & \\ \mathbf{0} & & & q_{nn} \end{bmatrix} \qquad (15\text{-}14)$$

The matrix $\mathbf{Q}$ in Eq. (15-14) is shown in diagonal form without loss of generality because any $\mathbf{Q}$ matrix may be diagonalized by a transformation[3] Equation (15-8) is rewritten as

$$\mathbf{PA} + (\mathbf{PA})^T - \frac{\mathbf{M}}{z} = -\mathbf{Q} \qquad (15\text{-}15)$$

Adding the corresponding elements in the 1,1 position on the left-hand side of Eq. (15-15) and equating their sum to $-q_{11}$ yields the quadratic equation

$$p_{1n}{}^2 + \frac{2a_0 z}{K_G{}^2} p_{1n} - \frac{q_{11} z}{K_G{}^2} = 0 \qquad (15\text{-}16)$$

Thus

$$p_{1n} = \frac{1}{K_G} \left[-\frac{a_0 z}{K_G} \pm \sqrt{\left(\frac{a_0 z}{K_G}\right)^2 + q_{11} z} \right] \qquad (15\text{-}17)$$

For any plant higher than Type 0, $a_0 = 0$ and Eq. (15-17) reduces to

$$p_{1n} = \pm \frac{\sqrt{q_{11} z}}{K_G} \qquad (15\text{-}18)$$

The feedback coefficients [see Eq. (14-49)] are given by

$$\mathbf{k} = \mathbf{Pb}z^{-1} = \begin{bmatrix} \dfrac{K_G p_{1n}}{z} \\[2mm] \dfrac{K_G p_{2n}}{z} \\[1mm] \vdots \\[1mm] \dfrac{K_G p_{nn}}{z} \end{bmatrix} = \begin{bmatrix} k_1 \\[2mm] k_2 \\[1mm] \vdots \\[1mm] k_n \end{bmatrix} \qquad (15\text{-}19)$$

Thus, for a Type 0 system, Eqs. (15-17) to (15-19) yield

$$k_1 = \frac{K_G p_{1n}}{z} = -\frac{a_0}{K_G} \pm \sqrt{\left(\frac{a_0}{K_G}\right)^2 + \frac{q_{11}}{z}} \qquad (15\text{-}20)$$

and, for a Type 1 or higher-order plant,

$$k_1 = \pm \sqrt{\frac{q_{11}}{z}} \qquad (15\text{-}21)$$

Equation (15-17) explicitly relates the Riccati matrix element p_{1n} to the fixed plant parameters and to the weighting matrix element q_{11}. This also applies for the feedback coefficient k_1, Eq. (15-20), for a Type 0 plant. For a Type 1 or higher plant, the coefficient k_1, Eq. (15-21), is a function of only q_{11}. Thus, Eqs. (15-17) and (15-20) serve *as checks for the degree of accuracy of the resulting Riccati matrix and feedback coefficients obtained from a digital-computer program.*

15-4 STEADY-STATE ERROR REQUIREMENTS

The steady-state error characteristics for a stable unity-feedback control system obtained from the conventional control theory for the three standard inputs (step, ramp, and parabolic) are discussed in Chap. 6. A corresponding analysis is made in this section for an optimal control system which simultaneously satisfies the quadratic PI and one or more of the steady-state error requirements of Table 12-2. Thus, the overall PI is required to have *two components* in order to meet both the steady-state requirement and the optimal PI requirement. The analysis of the steady-state requirement enhances the correlation between conventional and modern control theory.

For a system having an input $r(t)$, the control signal obtained from Fig. 15-2 is

$$u = r - \mathbf{k}^T\mathbf{x} = r + u^* \quad (15\text{-}22)$$

where u^* is the optimal control and u is the control necessary to satisfy both the steady-state and optimal requirements. The feedback control structure ($r = 0$) is used to obtain the optimal control u^* as a function of a feedback vector $\mathbf{k}$, where $\mathbf{k}$ is a function of the unknown q_{ii}'s (with z fixed). When this structure is used, the additional constraint of zero steady-state error for a step input ($r \neq 0$) is imposed to obtain the overall control u. This additional constraint allows the determination of q_{11} (which is a "free" choice in the optimal criterion). The overall relationship for Fig. 15-2, see Prob. 12-3, is

$$\frac{Y(s)}{R(s)} = \mathbf{c}^T[s\mathbf{I} - (\mathbf{A} - \mathbf{b}\mathbf{k}^T)]^{-1}\mathbf{b} \quad (15\text{-}23)$$

Substituting Eqs. (15-4) to (15-6) into Eq. (15-23) yields

$$\frac{Y(s)}{R(s)} = \frac{K_G(s + b_1)(s + b_2)\cdots(s + b_w)}{s^n + (a_{n-1} + K_G k_n)s^{n-1} + \cdots + (a_0 + K_G k_1)} \quad n > w$$

$$= \frac{K_G(s^w + c_{w-1}s^{w-1} + \cdots + c_0)}{s^n + (a_{n-1} + K_G k_n)s^{n-1} + \cdots + (a_0 + K_G k_1)} \quad (15\text{-}24)$$

where $c_0 = b_1 \cdot b_2 \cdots b_w$. Equation (15-24) yields the system characteristic equation

$$s^n + (a_{n-1} + K_G k_n)s^{n-1} + \cdots + (a_0 + K_G k_1) = 0 \quad (15\text{-}25)$$

Thus, a necessary condition for stability is that all the terms

$$a_{i-1} + K_G k_i > 0 \quad (15\text{-}26)$$

Substituting Eq. (15-20) into $a_0 + K_G k_1 > 0$ yields

$$\sqrt{(a_0)^2 + \frac{K_G{}^2 q_{11}}{z}} > 0$$

This inequality is satisfied by using the positive value of k_1 in Eqs. (15-20) and (15-21).

The system error is defined as $e = r - y$. From this definition the requirements on k_1, k_2, and k_3 for step, ramp, and parabolic inputs, respectively, are developed in Chap. 12 (see Table 12-2). Remember that in order to achieve the steady-state error characteristics of a Type m system, as discussed in Chaps. 6 and 12 and Sec. 14-5, the control ratio *must have a minimum of $w = m - 1$ zeros.*

Step Input: $R(s) = R_0/s$ (*for $w \geq 0$*)

The requirement for zero steady-state error for a step input, as established by Eq. (12-31), is

$$k_1 = c_0 - \frac{a_0}{K_G} \quad (15\text{-}27)$$

Substituting this value into Eq. (15-20) and rearranging yields

$$q_{11} = \left[c_0^2 - \left(\frac{a_0}{K_G} \right)^2 \right] z \geq 0 \qquad (15\text{-}28)$$

For $c_0 \gg a_0/K_G$, hfg operation, Eqs. (15-27), (15-19), and (15-28) yield respectively

$$k_1 \approx c_0 \qquad (15\text{-}29)$$

$$p_{1n} = \frac{k_1 z}{K_G} \approx \frac{c_0 z}{K_G} \qquad (15\text{-}30)$$

$$q_{11} \approx c_0^2 z > 0 \qquad (15\text{-}31)$$

In order to achieve the Butterworth characteristics of Table 14-2 the forward gain K_G must be sufficiently high to produce roots on the asymptotic portions of the root square locus. When the value of K_G for hfg operation is determined in this manner the assumption required for Eqs. (15-29) to (15-31) is valid. It is assumed in the remainder of this chapter that the hfg value of K_G is determined in this manner. For Type 1 or higher plants, $a_0 = 0$, the approximations in (15-29) to (15-31) become exact. Since $y(t)_{ss} = 0$ for $c_0 = 0$, zero steady-state error requires that none of the zeros of the system lie at the origin. Note that for an all-pole plant, $c_0 = 1$. Stability and optimality considerations are discussed at the end of this section.

Ramp Input: $R(s) = R_1/s^2$ $(for \ w \geq 1)$

Zero steady-state error, with a step and a ramp input, requires that Eq. (12-36) be satisfied; that is, $k_2 = c_1 - a_1/K_G$. Also, Eq. (15-19) specifies that $K_G p_{2n}/z = k_2$. For a Type 1 system under hfg operation with $a/K_G \ll c_1$, the results are

$$k_2 \approx c_1 \qquad (15\text{-}32)$$

$$p_{2n} \approx \frac{c_1 z}{K_G} \qquad (15\text{-}33)$$

For a Type 2 system in which $a_1 = 0$, the results are $k_2 = c_1$, and $p_{2n} = c_1 z/K_G$.

Parabolic Input: $R(s) = R_2/s^3$ $(for \ w \geq 2)$

Zero steady-state error with a parabolic input requires that Eq. (12-41) be satisfied; that is, $k_3 = c_2 - a_2/K_G$. Also, $K_G p_{3n}/z = k_3$. Under hfg operation

$$k_3 \approx c_2 \qquad (15\text{-}34)$$

$$p_{3n} \approx \frac{c_2 z}{K_G} \qquad (15\text{-}35)$$

Steady-State Error Coefficients

See Table 12-1.

Stability and Optimality Considerations

Optimality considerations which follow are based on the condition that zero steady-state error is required for step, ramp, and parabolic inputs.

Step input Under the conditions of Eq. (12-31), $w \geq 0$ and $c_0 \neq 0$, the system characteristic equation given by Eq. (15-25), becomes

$$s^n + (a_{n-1} + K_G k_n)s^{n-1} + \cdots + (a_1 + K_G k_2)s + K_G c_0 = 0 \qquad (15\text{-}36)$$

Thus, a necessary condition for stability is $c_0 > 0$. Further, with q_{11} satisfying Eq. (15-28) and with all remaining q_{ii}'s ≥ 0, the matrix $\mathbf{Q}$ is either positive definite or positive semidefinite. With the rank requirement of Eq. (14-33) satisfied, the condition $c_0 > 0$ ensures both the stability and the optimality of the system.

Ramp input When Eqs. (12-31) and (12-36) are satisfied, and when $w \geq 1$, the characteristic equation given by Eq. (15-25) becomes

$$s^n + (a_{n-1} + K_G k_n)s^{n-1} + \cdots + (a_2 + K_G k_3)s^2 + K_G c_1 s + K_G c_0 = 0 \qquad (15\text{-}37)$$

Thus, necessary conditions for stability are $c_1 > 0$ and $c_0 > 0$. With these requirements and conditions satisfied, stability and optimality of the closed-loop system are assured when the $\mathbf{Q}$ matrix is at least positive semidefinite (see the comments under step input for positive semidefinite $\mathbf{Q}$).

Parabolic input When Eqs. (12-31), (12-36), and (12-41) are all satisfied and with $w \geq 2$, the system characteristic equation, from Eq. (15-25), is

$$s^n + (a_{n-1} + K_G k_n)s^{n-1} + \cdots$$
$$+ (a_3 + K_G k_4)s^3 + K_G c_2 s^2 + K_G c_1 s + k_G c_0 = 0 \qquad (15\text{-}38)$$

Thus, necessary conditions for stability are $c_0 > 0$, $c_1 > 0$, and $c_2 > 0$. Optimality of the system is assured with the same restraints on $\mathbf{Q}$ as listed for a step input.

EXAMPLE 1 Given an all-pole plant with Eq. (15-28) satisfied,

$$G(s) = \frac{100}{s(s+1)(s+5)} \qquad \mathbf{Q} = \begin{bmatrix} 1 & 0 & 0 \\ 0 & 0.008 & 0 \\ 0 & 0 & 0.06 \end{bmatrix} \qquad \text{and} \qquad z = 1$$

The solution of the Riccati equation, Eq. (15-8), yields

$$\mathbf{k}^T = \begin{bmatrix} 1 & 0.693 & 0.218 \end{bmatrix}$$

The resulting control ratio, from Eq. (15-24), is

$$\frac{Y(s)}{R(s)} = \frac{100}{s^3 + 27.8s^2 + 74.3s + 100}$$

For an input $r(t) = 10tu_{-1}(t)$, the result obtained from Eq. (12-37) is that the system has a steady-state error $e(t)_{ss} = 7.43$. This is expected for an all-pole plant.

EXAMPLE 2 Given

$$G(s) = \frac{100(s+2)}{s(s+1)(s+5)} \qquad \mathbf{Q} = \begin{bmatrix} 4 & 0 & 0 \\ 0 & 0.5975 & 0 \\ 0 & 0 & 0.0074 \end{bmatrix} \qquad \text{and} \qquad z = 1$$

The solution of Eq. (15-8) yields

$$\mathbf{k}^T = \begin{bmatrix} 2 & 0.95 & 0.04 \end{bmatrix}$$

The resulting control ratio, from Eq. (15-24), is

$$\frac{Y(s)}{R(s)} = \frac{100(s + 2)}{s^3 + 10s^2 + 100s + 200}$$

This system follows a step and ramp input with no steady-state error. For an input $r(t) = 10t^2u_{-1}(t)$ the result obtained from Eq. (12-42) is that the system has a steady-state error $e(t)_{ss} = 1$. There is no steady-state error with a step or a ramp input.

Special Cases of Constant, Nonzero, Steady-State Error

In Chap. 12 it is pointed out that for a ramp input a stable all-pole system has a finite nonzero steady-state error [see Eq. (12-37)]. Defining the maximum acceptable error as E_e leads to

$$e(t)_{ss} = \left(\frac{a_1}{K_G} + k_2 \right) R_1 \le E_e \qquad (15\text{-}39)$$

Normalizing (dividing by R_1) yields

$$e_N(t)_{ss} = \frac{e(t)_{ss}}{R_1} = \frac{a_1}{K_G} + k_2 \le \frac{E_e}{R_1} = E_{eN} \qquad (15\text{-}40)$$

or

$$k_2 \le E_{eN} - \frac{a_1}{K_G} \qquad (15\text{-}41)$$

For a stable system response to exist, $E_{eN} > 0$ is required. Proof of this inequality is left to the reader. Utilizing the relationship $K_G p_{2n}/z = k_2$ in Eq. (15-41) yields

$$p_{2n} \le \frac{z E_{eN}}{K_G} - \frac{a_1 z}{K_G^2} \qquad (15\text{-}42)$$

For a Type 2 plant, since $a_1 = 0$, Eqs. (15-41) and (15-42) reduce to

$$k_2 \le E_{eN} \qquad \text{and} \qquad p_{2n} \le \frac{z E_{eN}}{K_G}$$

Also, in Chap. 12 it is pointed out that for a parabolic input a stable system containing a pole-zero plant ($w = 1$) has a finite nonzero steady-state error, given by Eq. (12-42). Let the maximum acceptable error be given by

$$e(t)_{ss} = \frac{R_2(a_2 + K_G k_3)}{K_G c_0} \le E_e \qquad (15\text{-}43)$$

Normalizing (dividing by R_2) and rearranging yields

$$k_3 \le c_0 E_{eN} - \frac{a_2}{K_G} \qquad (15\text{-}44)$$

Using the relationship $K_G p_{3n}/z = k_3$ results in

$$p_{3n} \leq \frac{zc_0 E_{eN}}{K_G} - \frac{a_2 z}{K_G^2} \qquad (15\text{-}45)$$

Summary

Steady-state error requirements for step, ramp, and parabolic inputs to linear optimal control systems, based on the quadratic performance index, result in the specification of some of the elements of the feedback matrix $\mathbf{k}$. In turn, this specifies some elements of the Riccati matrix $\mathbf{P}$. These elements are specified in terms of the parameters of the plant and/or the steady-state error. Table 15-1 summarizes the results of this section for Type 0, 1, and 2 plants with the weighting factor $z = 1$. As shown, two, four, or

Table 15-1   SUMMARY OF RESULTS OF SEC. 15-4 ($z = 1$)†

Input $r(t)$	n	$e(t)_{ss}$	Fixed elements of P and Q		
			All-pole plant		
$R_0 u_{-1}(t)$	≥ 1	0	$p_{1n} = \dfrac{1}{K_G} - \dfrac{a_0}{K_G^2}$	$q_{11} = 1 - \dfrac{a_0^2}{K_G^2}$	
$R_1 t u_{-1}(t)$	≥ 2	$\left(\dfrac{a_1}{K_G} + k_2\right) R_1 \leq E_e$	$p_{1n} = \dfrac{1}{K_G} - \dfrac{a_0}{K_G^2}$	$q_{11} = 1 - \dfrac{a_0^2}{K_G^2}$	
		or $\dfrac{a_1}{K_G} + k_2 \leq E_{eN}$	$p_{2n} = \dfrac{E_{eN}}{K_G} - \dfrac{a_1}{K_G^2}$		
$\dfrac{R_2 t^2 u_{-1}(t)}{2}$		∞			
			Pole-zero plant		
$R_0 u_{-1}(t)$	≥ 1	0	$p_{1n} = \dfrac{c_0}{K_G} - \dfrac{a_0}{K_G^2}$	$q_{11} = c_0^2 - \dfrac{a_0^2}{K_G^2}$	
$R_1 t u_{-1}(t)$	≥ 2	0	$p_{1n} = \dfrac{c_0}{K_G} - \dfrac{a_0}{K_G^2}$	$q_{11} = c_0^2 - \dfrac{a_0^2}{K_G^2}$	
			$p_{2n} = \dfrac{c_1}{K_G} - \dfrac{a_1}{K_G^2}$		
$\dfrac{R_2 t^2 u_{-1}(t)}{2}$	≥ 3	$\dfrac{a_2 + K_G k_3}{c_0 K_G} R_2 \leq E_e$	$p_{1n} = \dfrac{c_0}{K_G} - \dfrac{a_0}{K_G^2}$	$q_{11} = c_0^2 - \dfrac{a_0^2}{K_G^2}$	
One zero		or $\dfrac{a_2 + K_G k_3}{c_0 K_G} \leq E_{eN}$	$p_{2n} = \dfrac{c_1}{K_G} - \dfrac{a_1}{K_G^2}$	$p_{3n} = \dfrac{c_0 E_{eN}}{K_G} - \dfrac{a_2}{K_G^2}$	
More than one zero	≥ 3	0	$p_{1n} = \dfrac{c_0}{K_G} - \dfrac{a_0}{K_G^2}$	$q_{11} = c_0^2 - \dfrac{a_0^2}{K_G^2}$	
			$p_{2n} = \dfrac{c_1}{K_G} - \dfrac{a_1}{K_G^2}$	$p_{3n} = \dfrac{c_2}{K_G} - \dfrac{a_2}{K_G^2}$	

† The requirement on q_{11} must satisfy the condition $q_{11} \geq 0$.

six elements of the **P** matrix and one element of the **Q** matrix are specified in order to achieve the desired steady-state error requirements. For an all-pole plant the optimal control system cannot follow a parabolic input. Further, for a ramp input $r(t) = R_1 t u_{-1}(t)$ the value p_{2n} is a function of the slope R_1. For a plant with one zero and a parabolic input $r(t) = (R_2 t^2/2)u_{-1}(t)$ the value p_{3n} is a function of R_2. Table 15-2 lists, for various order systems and inputs, the number of unknown elements in the matrices **P** and **Q** which exist in the Riccati equation after satisfying the steady-state error requirements. Depending on the number of requirements and the order of the system, it may be necessary to assume values for the remaining elements. Equations (12-36) and (15-28) for the all-pole plant and Eqs. (12-36), (12-41), and (15-20) for the pole-zero plant serve as additional checks for the degree of accuracy of the resulting values (computed either by hand or computer) of the elements of the Riccati matrix and of the feedback coefficients.

15-5   EXACT CORRELATION (CONVENTIONAL vs. MODERN)

The next few sections are devoted to describing the correlation between the conventional design figures of merit and the optimal control PI, based on the quadratic cost function. A correlation is presented for the simple second-order plant,[4] and from the knowledge gained for this simple plant, a correlation is presented for the higher-order plants upon the condition that z is *assumed to be unity* and that *the off-diagonal elements of the* **Q** *matrix* are taken to be zero in Eq. (15-8). The correlation is made

Table 15-2   RELATIONSHIP OF UNKNOWN QUANTITIES TO EQUATIONS AVAILABLE FOR SOLUTION SATISFYING STEADY-STATE ERROR REQUIREMENTS

Order of system n	$r(t)$	Total number of equations available from Eq. (15-8)	Unknowns		
			P	Q	Total
1	Step	0	0	0	0
2	Step	2	2	1	3
	Ramp		1		2
3	Step	5	5	2	7
	Ramp		4		6
	Parabolic†		3		5
4	Step	9	9	3	12
	Ramp		8		11
	Parabolic†		7		10

† For pole-zero plant only.

between M_p, t_p, t_s, and ω_d and the elements q_{ii}, for $i > 1$. As shown in Sec. 15-4, the element q_{11} is specified exactly in terms of the fixed parameters of the plant when zero steady-state error is required for a step input to the system. Consider the Type 1 plant transfer function

$$G(s) = \frac{Y(s)}{U(s)} = \frac{K_G}{s(s + a_1)} \tag{15-46}$$

Phase variables are utilized, and q_{11} and q_{22} are unspecified. The solution of the Riccati equation obtained by equating diagonal elements on both sides of the equation and utilizing Eq. (15-21) yields

$$k_1 = \sqrt{q_{11}} \tag{15-47}$$

$$k_2 = -\frac{a_1}{K_G} + \frac{a_1}{K_G}\sqrt{1 + \frac{2K_G k_1}{a_1{}^2} + q_{22}\frac{K_G{}^2}{a_1{}^2}} \tag{15-48}$$

The characteristic equation of the closed-loop optimal system is

$$s^2 + (a_1 + K_G k_2)s + K_G k_1 = s^2 + 2\zeta\omega_n s + \omega_n{}^2 \tag{15-49}$$

From the RSL analysis it can be shown that a stable closed-loop performance may be achieved with an indefinite **Q** matrix. This is also shown by deriving ζ of Eq. (15-49) in terms of q_{11} and q_{22}. From Eq. (15-49)

$$\omega_n = \sqrt{K_G k_1} \qquad \text{rad/s} \tag{15-50}$$

and

$$2\zeta\omega_n = a_1 + K_G k_2 \tag{15-51}$$

Hence,

$$\zeta = \frac{a_1 + K_G k_2}{2\omega_n} = \frac{a_1 + K_G k_2}{2\sqrt{K_G k_1'}} \tag{15-52}$$

From Eqs. (15-47) to (15-52) the following are obtained:

$$\omega_n = (q_{11})^{1/4}\sqrt{K_G} \tag{15-53}$$

$$\zeta = \frac{a_1\sqrt{1 + [2K_G(q_{11})^{1/2}]/a_1{}^2 + q_{22}K_G{}^2/a_1{}^2}}{2\sqrt{K_G}(q_{11})^{1/4}} \tag{15-54}$$

Let $K_1 = a_1/\sqrt{K_G} \geq 0$; then Eq. (15-54) becomes

$$\zeta = \frac{K_1}{2}\sqrt{\frac{1}{(q_{11})^{1/2}} + \frac{2}{K_1{}^2} + \frac{K_G}{K_1{}^2}\frac{q_{22}}{(q_{11})^{1/2}}} \tag{15-55}$$

Also, let

$$K_2 = \frac{q_{22}}{\sqrt{q_{11}}} \tag{15-56}$$

so that Eq. (15-55) becomes

$$\zeta = \frac{1}{2}\sqrt{\frac{K_1{}^2}{(q_{11})^{1/2}} + 2 + K_G K_2} \tag{15-57}$$

Since Eq. (15-57) contains $\sqrt{q_{11}}$, $q_{11} \geq 0$. If the restriction that **Q** be positive semi-definite is maintained, the minimum ζ that can be achieved is when $K_2 = 0$ (or

$q_{22} = 0$) and $K_1^2/\sqrt{q_{11}} \ll 2$. This yields $\zeta_{min} \approx 0.707$. Therefore, in order to achieve $0 < \zeta < 0.707$, Eqs. (15-56) and (15-57) require that $q_{22} < 0$. This in turn requires that **Q** be indefinite.

For a simple second-order plant the peak time can be derived in terms of q_{11} and q_{22} utilizing Eq. (3-60), namely,

$$t_p = \frac{\pi}{\omega_n\sqrt{1 - \zeta^2}} = \frac{\pi}{\omega_d} \quad \text{s} \quad (15\text{-}58)$$

Substituting Eqs. (15-53) and (15-57) into Eq. (15-58) yields

$$t_p = \frac{\pi}{(q_{11})^{1/4}\sqrt{K_G}\sqrt{\frac{1}{2} - \frac{1}{4}(K_1^2/\sqrt{q_{11}} + K_G K_2)}} \quad (15\text{-}59)$$

When Eq. (15-50) is used, the damped natural frequency of the response is

$$\omega_d = \omega_n\sqrt{1 - \zeta^2} = \sqrt{K_G k_1(1 - \zeta^2)} \quad \text{rad/s} \quad (15\text{-}60)$$

The corresponding peak overshoot, [Eq. (3-61)], for a unit step input is

$$M_p = y(t_p) = 1 + \exp\left(-\frac{\zeta\pi}{\sqrt{1 - \zeta^2}}\right) \quad (15\text{-}61)$$

In Sec. 15-4 it is shown that, for an all-pole plant, $e(t)_{ss} = 0$ for a step input when $q_{11} = 1$ (Type 1 or higher system) or $q_{11} \approx 1$ (Type 0 plant under hfg operation). To achieve the Butterworth characteristic, namely, that $\zeta \approx 0.707$, it may be seen from the RSL[4] that an hfg condition must exist: thus $K_1 \approx 0$. Further, from Eq. (15-57), $K_G K_2 \approx 0$, which requires that $q_{22} \approx 0$. Thus, Eqs. (15-57) and (15-59) to (15-61) respectively yield $\zeta \approx 0.707$ and

$$t_p \approx \pi\sqrt{\frac{2}{K_G}} \quad (15\text{-}62)$$

$$\omega_d = \frac{\pi}{t_p} \approx \sqrt{\frac{K_G k_1}{2}} = \sqrt{\frac{K_G}{2}} \quad (15\text{-}63)$$

$$M_p \simeq 1 + e^{-\pi} = 1.043 \quad (15\text{-}64)$$

The settling time is $T_s = 4/\zeta\omega_n$, and using Eq. (15-50) and $\zeta = 0.707$ yields

$$T_s \approx 4\sqrt{\frac{2}{K_G}} \quad (15\text{-}65)$$

Thus, curves of t_p, ω_d, and T_s vs. K_G are readily plotted for the simple second-order case under hfg operation.

For a ramp input, $R(s) = R_1/s^2$, and hfg operation, the steady-state error from Eq. (15-39) is

$$e(t)_{ss} = \left(\frac{a_1}{K_G} + k_2\right) R_1 \approx k_2 R_1 \leq E_e \quad (15\text{-}66)$$

For the Type 1 plant of Eq. (15-46) the conventional system error coefficients are $K_p = \infty$ and $K_v = K_G/a_1$. For the optimal control system the coefficients, for hfg operation, from Table 12-1 are $K'_p = \infty$ and $K'_v \approx 1/k_2$. The step-error coefficients for the conventional and optimal control systems are identical. To compare the two systems with a ramp input, let the acceptable error for the conventional system be E_e. Thus, for the conventional case

$$e(t)_{ss} = \frac{R_1}{K_v} = \frac{a_1}{K_G} R_1 = E_e \qquad (15\text{-}67)$$

If this value of E_e is substituted into Eq. (15-41) for the optimal case, $k_2 = 0$. In the event $k_2 > 0$, Eq. (15-41) implies that $E'_e/R_1 > a_1/K_G$. The prime denotes the error obtained in the optimal case. Thus,

$$E'_e > \frac{a_1}{K_G} R_1 \qquad (15\text{-}68)$$

Comparing Eqs. (15-67) and (15-68) shows that $E'_e > E_e$. Therefore, with a Type 1 plant for the condition $k_2 > 0$, the optimal control system does not perform as well in the steady state for ramp inputs as the conventional control system.

A similar analysis can be made for Type 0 and 2 plants with respect to ζ, ω_n, t_p, and t_s. This analysis is confined to the correlation of the steady-state errors achieved with the conventional and optimal control systems. Consider the Type 0 system

$$G(s) = \frac{K_G}{s^2 + a_1 s + a_0}$$

The system error coefficients are $K_p = K_G/a_0$ and $K_v = 0$. The optimal control system can be made to produce zero steady-state error with a step input by selecting $k_1 = 1 - a_0/K_G$. Then the system error coefficients obtained from Eq. (12-47) are $K'_p = \infty$ and $K'_v = K_G/(a_1 + K_G k_2)$. Thus, an optimal Type 0 control system can follow a ramp input but with a finite error. It is well known that a conventional Type 0 control system cannot follow a ramp input.

A similar analysis can be applied to the Type 2 plant described by

$$G(s) = \frac{K_G}{s^2}$$

The system error coefficients are $K_p = K_v = \infty$ and $K'_p = \infty$ and $K'_v = 1/k_2$. Therefore, with a Type 2 plant, the optimal control system does not perform as well for ramp inputs as the conventional control system.

Under the hfg condition such that $a_1/K_G \approx 0$, with $q_{11} = 1$ ($k_1 = 1$) and with $z = 1$, Eq. (15-48) reduces to

$$k_2 \approx \frac{a_1}{K_G} \sqrt{1 + \frac{2K_G}{a_1^2} + q_{22}\left(\frac{K_G}{a_1}\right)^2} = \sqrt{\left(\frac{a_1}{K_G}\right)^2 + \frac{2}{K_G} + q_{22}} \approx \sqrt{q_{22}}$$

This implies that $q_{22} > 0$. This can be generalized to $k_n \approx \sqrt{q_{nn}}$. As is shown later in this chapter, this relationship applies to all nth-order systems under the conditions of hfg and reproducibility. These properties are defined in Sec. 15-6.

15-6 CORRELATION OF HIGHER-ORDER SYSTEMS WITH ALL-POLE PLANTS

Any linear system can be characterized in the frequency domain and has values of M_m and ω_m which are functions of the system parameters. These frequency-domain properties can be correlated directly with the corresponding time-domain relationships M_p and ω_d only for a simple second-order system. This knowledge about the correlation between the frequency- and time-response specifications for a simple second-order system is often extrapolated qualitatively for higher-order systems. Experience has borne out the validity of this correlation for many higher-order systems. A similar approach can be used for an optimal control system utilizing the quadratic PI of Eq. (15-7).

For a simple second-order plant it is shown in Sec. 15-5 that the maximum overshoot attainable, with $\zeta = 0.707$, requires that $q_{22} = 0$, under the restriction that the $\mathbf{Q}$ matrix be positive semidefinite and $z = 1$. Also, direct relationships are obtained for ω_n, ω_d, ζ, and t_p in terms of the plant parameters. It is impossible to obtain similar simple relationships for higher-order plants. Thus for higher-order plants with $z = 1$, a procedure is utilized which is similar to that used for conventional control systems. In other words, the analysis technique for a simple second-order optimal control system is extrapolated qualitatively for higher-order plants such that correlation between the diagonal elements of the $\mathbf{Q}$ matrix and the time-response specifications is effected. This correlation is confirmed from results obtained by using a digital-computer simulation.

To illustrate how this correlation is made, consider a third-order all-pole plant with the $\mathbf{Q}$ matrix given by

$$\mathbf{Q} = \begin{bmatrix} q_1 & & 0 \\ & q_2 & \\ 0 & & q_3 \end{bmatrix} \qquad (15\text{-}69)$$

and $z = 1$. For simplicity, *a single-subscript notation* for the matrix elements is utilized hereafter since the off-diagonal terms are always taken to be zero. Since the analysis of a simple second-order plant shows that $q_2 \approx 0$ yields the smallest possible ζ ($=0.707$), the assumption is made that the value of q_2 in Eq. (15-69) has the greatest effect on M_p. Also, many trials with a computer simulation have shown that the value q_3 has the greatest effect on t_p, t_s, and ω_d. The values $q_2 = 0.008$ and $q_3 > 0$ are arbitrarily chosen to ensure that $\mathbf{Q}$ is positive definite. A very small value of q_2 is required in order to have an underdamped response. The correlation of the elements of $\mathbf{Q}$ with the conventional figures of merit for a step input is developed next for $n \geq 3$.

Constant M_p Criterion

Under the condition of an approximately constant M_p, the correlation requires the determination of the hfg values of $K_G \geq K_{\min}$. The value of $K_{\min}$ for a given plant can be determined from its RSL.† The dominant branch of the RSL becomes a straight line, approaching its asymptote as K_G increases. The value of K_G at which this branch becomes essentially a straight line is denoted as K_x. *It should be noted that the RSL in*

† RSL of $G(s)G(-s) = -1$.

this chapter is used solely for the purpose of determining K_x. The value of K_{min} must be greater than K_x.

The RSL, for an all-pole plant with hfg, always has a pair of dominant complex-conjugate roots, yielding an underdamped system response which approaches the response of a system with a Butterworth characteristic.[5] The asymptotes for a pole-zero plant having $w = n - 1$ zeros lie on the real axis. Thus, under hfg operation the pole-zero system may have an overdamped response. In order to ensure, under hfg operation, the existence of an underdamped response the number of zeros is restricted to $w \leq n - 2$. In both types of plants, for any given $\mathbf{Q}$, computer simulation reveals that the time-response characteristics do not change for any value of $K_G \geq K_{min}$ and that the correlation, under a constant M_p criterion, is independent of the poles and zeros of the particular plant transfer function $G(s)$ as long as $K_G \geq K_{min.}$

A known characteristic of the RSL plots is that as K_G becomes large, the poles and zeros of the plant can be replaced by $n - w$ poles at the origin. Since the asymptotes all emanate from the origin, the locus is essentially the same for any nth-order plant for $K_G \geq K_{min}$. The difference between the loci for these nth-order plants is the gain calibration of the loci. This characteristic is the basis for the correlation, under a constant M_p criterion, between the conventional figures of merit and the $\mathbf{Q}$ matrix. This basis is independent of the plant transfer function. It should be noted that as the plant poles move further to the left in the s plane, K_{min} becomes larger.

The RSL for a third-order all-pole plant is shown in Fig. 15-3. From this figure is obtained $K_x \approx 262$. The following values, namely, $K_G = 500$, $q_1 = 1$ (required value for zero steady-state error with a step input), $q_2 = 0.008$, and $0.01 \leq q_3 \leq 10$, are utilized to solve for the values k_i and for the roots of the system characteristic equation. Once the feedback coefficients are determined for various values of q_3, the corresponding output time response $y(t)$ is obtained for a unit step input. The data for M_p, t_p, and t_s obtained from such time responses are plotted in Fig. 15-4 as a function of q_3. The curve ω_d vs. q_3 represents the frequency of oscillation of the transient response as determined by the dominant roots. For comparison, similar data are obtained for $K_G = 100$ (see Table 15-3) and also plotted in Fig. 15-4. An analysis of the data and of Fig. 15-3 leads to the following conclusions.

For values of K_G equal to or greater than the value K_x at which the real part of the complex roots $s_{1,2}$ become truly dominant, the figures of merit become reproducible; i.e., *for a given* $\mathbf{Q}$, *any value of* $K_G \geq K_{min}$ *produces values of* M_p, t_p, t_s, *and* ω_d *which remain essentially constant.* (In Sec. 14-12 another method for determining K_x was presented.) Reproducibility becomes evident at about $K_G = 500$ for $q_3 \geq 0.01$ and at about $K_G = 100$ for $q_3 \geq 0.1$. Figure 15-4 shows the variation in M_p, t_p, t_s, and ω_d with values of K_G in the range 100 to 500. For the plant of Fig. 15-3 it may be inferred that reproducibility of results occurs for $K_G \geq 500$. The variation in M_p is about 0.5 percent, and its average value is 1.0418 for $0.01 \leq q_3 \leq 10$ with $K_G = 500$. For $q_3 \geq 0.1$ and $K_G \geq 100$, M_p has a variation of about 0.16 percent. The identification of $K_G \geq 500$ for reproducibility is compatible with $K_{min} \approx 262$, which is determined from Fig. 15-3. Associated with these characteristics is the fact that as the value of q_3 gets closer to q_2, the root s_3 moves closer to the imaginary axis. This results in s_3 exerting a greater effect on the time-response characteristics and increasing the variation in the value of M_p.

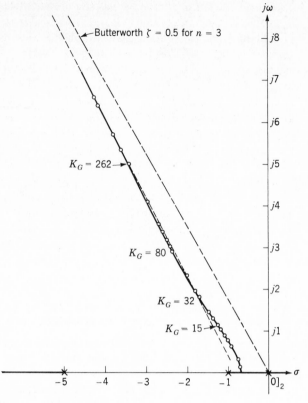

FIGURE 15-3
Third-order all-pole plant RSL for

$$G(s)G(-s) = \frac{-K_G^2}{s^2(s + 1)(s + 5)(s - 1)(s - 5)}$$

Table 15-4 presents data showing the effect of the value of q_2 on the time response of the system when K_G is not restricted to $K_G \geq K_{min}$. This table shows that increasing q_2 decreases the value of M_p and, when it is increased sufficiently, results in an overdamped system response. Note for some of the values of q_2 that $\mathbf{Q}$ is positive semidefinite. Although confirmatory data are not presented in this text, the value of q_2 affects the minimum value of K_G which yields reproducibility of the results on the lower bound of q_3. Thus, the minimum value of q_3 to be chosen is associated with reproducibility, approximately constant M_p, and the value of q_2 to be selected under hfg conditions ($K_G \geq K_{min}$). It should be noted that matrix $\mathbf{Q}_1$ in Table 15-4, for $q_3 \geq 0.1$ and $K_G \geq 100$, yields essentially the same results as those in Fig. 15-4. (For $q_3 < 0.1$ the values of M_p and ω_d are slightly higher.) This is to be expected since there is little difference between the values $q_2 = 0$ and $q_2 = 0.008$. Also, it should be noted from matrices $\mathbf{Q}_2$ and $\mathbf{Q}_3$ that for a desired value of $M_p = 1.0$ the value of q_2 should be kept as small as possible, because the value of t_s should be kept as low as possible.

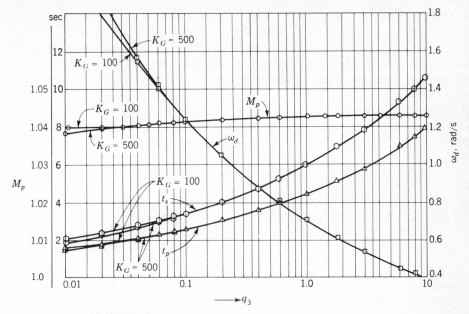

FIGURE 15-4
The parameters of the time-response characteristics for a third-order all-pole
system with constant K_G and $z = 1$.

Table 15-3   CALCULATED DATA FOR A THIRD-ORDER
ALL-POLE PLANT

$z = 1$, $q_1 = 1$, and $q_2 = 0.008$

K_G	q_3	M_p	t_p	t_s	ω_d
100	0.01	1.0395	1.65	2.05$^+$	2.099
	0.04	1.0404	2.10	2.70$^+$	1.5415
	0.08	1.0410	2.45	3.20$^+$	1.3086
	0.1	1.0413	2.55	3.40	1.2410
150	0.01	1.0391	1.55	1.95$^+$	2.150
	0.04	1.0405	2.05	2.70$^+$	1.554
	0.08	1.0412	2.40	3.20$^-$	1.315
	0.1	1.0415	2.55	3.40	1.2458
200	0.01	1.0387	1.50	1.95$^+$	2.168
	0.04	1.0405	2.05	2.65$^+$	1.5591
	0.08	1.0412	2.40	3.20$^-$	1.317
300	0.01	1.0382	1.50	1.90$^+$	2.181
	0.04	1.0405	2.05	2.65$^+$	1.563
400	0.01	1.0382	1.45	1.90$^+$	2.185
	0.04	1.0405	2.00	2.65$^+$	1.564
500	0.01	1.03813	1.45	1.85$^+$	2.187
	0.04	1.0405	2.00	2.65$^+$	1.564

Matrix Q_4 illustrates the variation of q_2 for a fixed value of q_3. The data corresponding to this matrix support the comments made about the other matrices.

The conclusions (for a Q matrix that is positive definite) drawn from the analysis
above are (1) that M_p is determined primarily by q_2 and (2) that t_p, t_s, and ω_d are
determined primarily by q_3. It should be noted that q_2 does have a secondary effect
on t_p, t_s, and ω_d.

Under the condition $q_1 = 1$ for zero steady-state error for a step input, the
following design procedure, based upon the data in Tables 15-3 and 15-4 and Fig.
15-4, may be used for a third-order all-pole system:

1 Adjust q_2 to achieve the desired approximate M_p value (from 1.0 up to about
1.043). Select $q_3 > q_2$.
2 Determine K_{min} from the RSL.
3 If an overshoot of approximately 4 percent is acceptable, go to Fig. 15-4 and
select q_3 to yield acceptable values of t_p, t_s, and ω_d. If a smaller overshoot is
desired, a set of curves similar to Fig. 15-4 must be obtained using a larger
value of q_2.

Once the values of q_2 and q_3 have been selected to satisfy the desired conventional
figures of merit, the feedback gains k_i can be determined to satisfy the optimal performance index of Eq. (15-7) with $z = 1$ by solving the Riccati equation and then
using Eq. (15-19).

Utilizing other all-pole third-order plants and their respective $K_G \geq K_{min}$ for
the corresponding $q_3|_{min} \leq q_3$ (in this example $q_3|_{min} = 0.01$) yields curves identical

Table 15-4   COMPUTED RESULTS FOR A THIRD-ORDER ALL-POLE PLANT
$K_G = 100, z = 1$

Q matrix	q_3	M_p	t_p	t_s	ω_d
$Q_1 = \begin{bmatrix} 1 & 0 \\ 0 & q_3 \end{bmatrix}$	0.1† 0.2† 0.4†	1.0428 1.0429 1.04296	2.55 3.00 3.55	3.40+ 4.00+ 4.75+	1.248 1.0528 0.8869
$Q_2 = \begin{bmatrix} 1 & 0 \\ 0.8 & q_3 \end{bmatrix}$	0.1 0.2 0.4	1.0000 1.0000 1.0013	— — 5.85	3.70+ 3.70+ 3.75+	— 0.3469 0.5383
$Q_3 = \begin{bmatrix} 1 & 0 \\ 1.274 & q_3 \end{bmatrix}$	0.1 0.2 0.4	1.00000 1.00000 1.00000	— — —	4.60+ 4.65+ 4.65+	— — —
	q_2				
$Q_4 = \begin{bmatrix} 1 & 0 \\ q_2 & 0.2 \end{bmatrix}$	0† 0.008 0.08 0.6 0.8 0.9 1.274	1.0429 1.0417 1.032 1.00086 1.0000 1.0000 1.0000	3.00 3.00 3.15 5.20 — — —	4.00+ 4.00+ 3.90+ 3.25+ 3.70+ 3.90+ 4.65+	1.053 1.0481 1.005 0.6064 0.3469 — —

† P is positive definite.

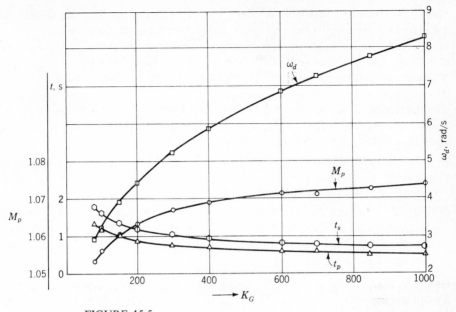

FIGURE 15-5
The parameters of the time-response characteristics for a third-order all-pole system with $z = 1$, $q_2 = q_3 = 0$.

to those in Fig. 15-4. Therefore, the concept of reproducibility of results is established; i.e., Fig. 15-4 is applicable to any third-order all-pole plant where $z = 1$, $K_G \geq K_{\min}$, and $q_3 \geq q_3|_{\min}$ is observed for each respective plant. In other words, if $M_p \approx 1.04$ is acceptable, one can select the value of q_3 to yield the acceptable values of t_p, t_s, and ω_d under the optimal control criteria of Eq. (15-7), $q_1 = 1$ (for zero steady-state error with a step input), and $q_2 = 0.008$.

Variable M_p Criteria

Under the constant M_p criterion, a desired constant M_p value for a constant value of K_G is achieved by adjusting q_2. The value of M_p can also be varied by varying K_G with either $q_3 = 0$ or $q_2 = q_3 = 0$ while maintaining the matrix **Q**. The necessary requirement is $q_1 = 1$. Table 15-5 presents the computed data for both conditions for the transfer function of $G(s)$ in Fig. 15-3. The results for the condition $q_2 = q_3 = 0$ are plotted in Fig. 15-5. For different third-order all-pole plants, under the conditions of $q_3 = 0$ and/or $q_2 = 0$, the reproducibility criterion is not satisfied. That is, Fig. 15-5, unlike Fig. 15-4 which is for the condition $q_3 > q_2 > 0$, cannot be utilized for all third-order all-pole plants. Table 15-5 indicates that

1 Values of $M_p > 1.043$ can be achieved.
2 With $q_2 = q_3 = 0$, the third-order Butterworth characteristic, $M_p \approx 1.085$ is approached as K_G is increased.
3 With variable K_G and $q_2 = q_3 = 0$, smaller values of t_p and t_s can be achieved at the expense of a larger M_p compared with the case of a positive definite matrix **Q**.

The curves of Appendix C represent the parameters of the time response for the constant M_p criterion, with $r(t) = u_{-1}(t)$, for systems with any fourth, fifth, or sixth-order all-pole plant. The conditions are $K_G \geq K_{min}$, $q_1 = 1$, and $q_n > q_i = 0.01$ $(i = 2, \ldots, n - 1)$. The plot of ω_d vs. q shows how the value of the damped natural frequency of the most dominant roots changes with changes in the value of q_n. Note the similarity between these curves and those of Fig. 15-4. An analysis of these curves and their associated data permit conclusions similar to those obtained for the all-pole third-order plant to be made for the $n > 3$ all-pole plant for both constant and variable M_p criteria. The complete results for $n \geq 3$ hfg all-pole plant system, *for the constant M_p and reproducibility criteria, reveals that there are n − 1 dominant roots and that the nondominant root is given by $s_n \approx -K_G\sqrt{q_n}$.* A general design procedure for satisfying the conventional control figures of merit and the quadratic cost is presented in Sec. 15-9.

15-7 HIGHER-ORDER SYSTEM CORRELATION (POLE-ZERO PLANT)

The control ratio, Eq. (15-24), is for a pole-zero plant whose transfer function is

$$G(s) = \frac{K_G(s + a_1)(s + a_2)\cdots(s + a_w)}{(s + b_1)(s + b_2)\cdots(s + b_n)}$$

The approach to the pole-zero correlation is based upon knowledge of the all-pole plant. Again, the hfg condition, $K_G \geq K_{min}$, is observed in the same manner as for

Table 15-5   DATA FOR A THIRD-ORDER
ALL-POLE PLANT

P is positive definite and $z = 1$†

K_G	M_p	t_p	t_s	ω_d
	$Q = \begin{bmatrix} 1 & 0 \\ & 0.008 & \\ 0 & & 0 \end{bmatrix}$			
80	1.046	1.35	1.75+	2.891
100	1.047	1.25	1.60	3.221
150	1.047	1.05	1.35+	3.883
	$Q = \begin{bmatrix} 1 & 0 \\ & 0 & \\ 0 & & 0 \end{bmatrix}$			
80	1.053	1.35	1.75+	2.919
100	1.056	1.20	1.60+	3.251
150	1.060	1.05	1.35+	3.915
200	1.063	0.90	1.20+	4.434
400	1.069	0.70	0.95−	5.875
600	1.071	0.60	0.80	6.861
850	1.073	0.55	0.70+	7.809
1,000	1.074	0.50	0.70−	8.286

† For $t_p < 1$ s the data are approximate since the calculation interval of the computer programs was 0.05 s.

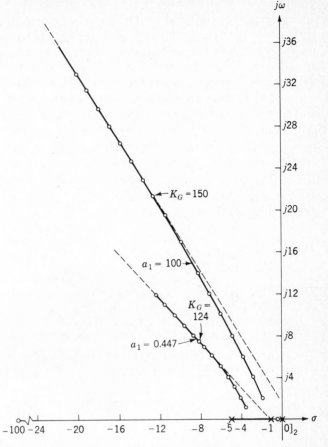

FIGURE 15-6
Third-order pole-zero plant RSL for

$$G(s)G(-s) = \frac{K_G{}^2(s \pm a_1)}{s^2(s \pm 1)(s \pm 5)}$$

the all-pole plant. As emphasized earlier, to ensure the existence of an underdamped response under hfg, the condition $w \leq n - 2$ is necessary. Also, to achieve zero steady-state error for a step input for a pole-zero plant, the value of q_1 in the **Q** matrix is specified by Eq. (15-28). The correlation analysis is presented in detail for a three-pole, one-zero plant. The RSL is shown in Fig. 15-6. From this figure $K_x \approx 124$ for $a_1 = 0.4472$ and $K_x \approx 150$ for $a_1 = 100$. Table 15-6 contains representative data $(M_p, t_p, t_s, \omega_{d_e},$ and **k**) derived by means of a digital computer, with $r(t) = u_{-1}(t)$, $z = 1$ for $a_1 = \sqrt{q_1} = 0.4472$ and also for $a_1 = \sqrt{q_1} = 10$. For each value of q_1 data are given for two values of gain, $K_G = 150$ and $K_G = 300$. The value of the damped natural frequency of the system can be estimated from the response of $y(t)$. From Eq. (15-63), the effective damped natural frequency is approximately

$$\omega_{d_e} \approx \frac{\pi}{t_p} \qquad (15\text{-}70)$$

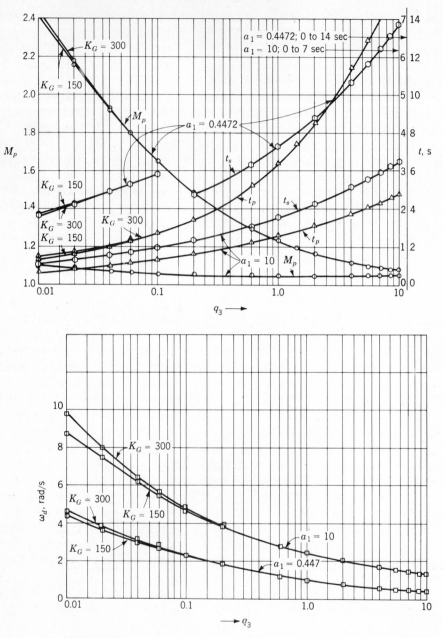

FIGURE 15-7
Time-response parameters of third-order system with one zero. (*Top*) M_p, t_p, and t_s versus q_3; (*bottom*) ω_{d_e} versus q_3.

The data for $K_G = 150$ and for $K_G = 300$ are plotted in Fig. 15-7a and b, respectively. Note that for $q_3 < 0.06$ the reproducibility characteristic for $K_G \geq K_{min}$ is not achieved. An analysis of Fig. 15-7, Table 15-6, the roots of the characteristic equation of the system, and additional time-response data for higher values of K_G (not included in this text) results in the following conclusion.

For values of K_G equal to or greater than the value for which the roots $s_{1,2}$ become definitely dominant, the figures of merit become reproducible. Reproducibility is achieved at approximately $K_G = 400$ for $q_1 = 0.2$ and $q_3 \geq 0.01$ and at $K_G = 500$ for $q_1 = 100$ and $q_3 \geq 0.01$. For $q_3 \geq 0.06$ reproducibility occurs for $K_G \geq 150$ for both values of a_1. These values of K_G for reproducibility are compatible with the values of K_x determined from the RSL of Fig. 15-6. Associated with these characteristics is the fact that as the value of q_3 becomes close to q_2, the s_3 root moves closer to the origin. The correlation data indicate that for a given value of a_1 the dominant roots remain essentially unchanged for given minimum values of K_G and q_3 or for higher values of K_G.

As in the all-pole case, the value of q_2 affects the value of K_{min} used for reproducibility of the results on the lower bound of q_3 for each transfer function, that is, $q_3 \geq q_2$.

For $q_1 = 0.2$ and $K_G = 300$, Table 15-6 indicates that M_p varies from 2.4814 to 1.0816. For $q_1 = 100$, M_p varies from a value of 1.1007 to 1.0440. This shows that as a_1 is increased, a constant M_p condition can be achieved: the value of M_p approaches the average value, 1.0418, of the third-order all-pole case. Comparing Figs. 15-4 and 15-7 ($q_1 = 100$ and $q_3 \geq 0.1$) shows that the pole-zero case has improved the values of t_s and t_p for approximately the same value of M_p. This effect is identical to that achieved in conventional control theory design when a zero is added to the original plant. For the entire range of $0.447 \leq a_1 \leq 10$ and $0.01 \leq q_3 \leq 10$, the value of t_p has been decreased by the addition of a zero to the plant. Although the data are not shown, for values of $a_1 \geq \sqrt{2}$ (approximately) the values of both t_p and t_s are less for the pole-zero plant than for the all-pole plant.

The choice of a minimum value of q_3 is associated with reproducibility ($K_G \geq K_{min}$), the value of a_1, and the value of q_2 to be selected. Table 15-7 presents additional data showing the effect of the value of q_2 on the time-response parameters of the system. Using a value of q_2 larger than 0.008 decreases the value of M_p that is obtainable. Thus, for the desired value of q_2, under reproducibility conditions, the minimum value of q_3 to be selected is based upon the maximum acceptable forward gain of the system.

The design procedures for the third-order plant with one zero can be essentially the same as those for the all-pole plant. Once the values of a_1, q_2, and q_3 have been selected to satisfy the desired conventional figures of merit, the feedback gains k_i can be determined to satisfy, for $z = 1$, the optimal performance index of Eq. (15-7). For a desired overdamped response the value of q_2 should be kept as small as possible.

For a pole-zero plant larger values of M_p are obtained than with an all-pole plant. Because the value of M_p varies for a given value of a_1 and small values of q_i, the plots of t_s and/or t_p vs. q_n can experience a discontinuity over the range of q_n. The effect is shown in Fig. 15-7a for $a_1 = 0.447$ and $q_2 = 0.008$.

Table 15-6 EXPERIMENTAL DATA: THIRD-ORDER PLANT, ONE-ZERO CASE

$K_G = 150$ and 300, $z = 1$

$$G(s) = \frac{K_G(s + \sqrt{q_1})}{s(s + 1)(s + 5)} \qquad A = \begin{bmatrix} 0 & 1 & 0 \\ 0 & 0 & 1 \\ 0 & -5 & -6 \end{bmatrix} \qquad Q = \begin{bmatrix} q_1 & & 0 \\ & 0.008 & \\ 0 & & q_3 \end{bmatrix}$$

K_G	q_3	M_p	t_p	t_s	ω_{d_e}	k_1	k_2	k_3
				$q_1 = 0.2$				
150	0.01	2.4322	0.72	3.76–3.78	4.36	0.4472	0.3153	0.0857
	0.02	2.1617	0.86	4.30–4.32	3.65	0.4472	0.3596	0.1225
	0.04	1.9182	1.05	4.90–4.95	2.99	0.4472	0.4176	0.1772
	0.06	1.792	1.16	5.32–5.34	2.71	0.4472	0.4585	0.2202
	0.1	1.6498	1.38	5.84–5.86	2.275	0.4472	0.5178	0.2894
	0.2	1.4868	1.72	4.76–4.78	1.827	0.4472	0.6138	0.4180
	0.6	1.2939	2.62–2.64	6.36–6.38	1.192	0.4472	0.8087	0.7426
	1	1.2292	3.20	7.30	0.982	0.4472	0.9268	0.9774
	2	1.1625	4.25–4.30	8.80–8.85	0.735	0.4472	1.0977	1.3799
	5.5	1.1023	6.45	11.65$^+$	0.487	0.4472	1.4195	2.3096
	10	1.0817	8.15	13.75$^+$	0.386	0.4472	1.6522	3.1260
300	0.01	2.4814	0.68	3.68–3.70	4.62	0.4472	0.3213	0.0917
	0.02	2.1839	0.86	4.24–4.26	3.83	0.4472	0.3620	0.1310
	0.04	1.9286	1.00	4.90–4.95	3.14	0.4472	0.4241	0.1879
	0.06	1.7984	1.15	5.30–5.35	2.73	0.4472	0.4668	0.2320
	0.1	1.6535	1.36	5.84–5.86	2.31	0.4472	0.5280	0.3024
	0.2	1.4886	1.74	4.74–4.76	1.805	0.4472	0.6259	0.4323
	0.6	1.2945	2.62	6.34–6.36	1.198	0.4472	0.8227	0.7584
	1	1.2292	3.20	7.25–7.30	0.982	0.4472	0.9350	0.983
	2	1.1626	4.25	8.80–8.85	0.735	0.4472	1.1128	1.3970
	10	1.0816	8.15	13.75$^+$	0.386	0.4472	1.6680	3.1441
				$q_1 = 100$				
150	0.01	1.1064	0.36	0.60–0.62	8.72	10.0000	1.9358	0.1534
	0.02	1.0865	0.42	0.66–0.68	7.48	10.0000	2.0759	0.1820
	0.04	1.0688	0.50	0.78–0.80	6.28	10.0000	2.2858	0.2285
	0.06	1.0621	0.58	0.86–0.88	5.42	10.0000	2.4462	0.2670
	0.1	1.0563	0.67	0.98–0.99	4.68	10.0000	2.6915	0.3308
	0.2	1.0513	0.82	1.16–1.17	3.83	10.0000	3.1082	0.4530
	0.6	1.0472	1.12	1.56–1.57	2.8	10.0000	3.9907	0.7692
	1	1.0461	1.30	1.75–1.80	2.415	10.0000	4.5072	0.99039
	2	1.0452	1.55	2.10–2.15	2.03	10.0000	5.3335	1.3997
	5.5	1.0443	2.05	2.75–2.80	1.53	10.0000	6.8447	2.3249
	10	1.0440	2.40	3.25$^+$	1.31	10.0000	7.9412	3.1392
300	0.01	1.1007	0.32	0.54–0.56	9.81	10.0000	1.7033	0.1275
	0.02	1.0834	0.39	0.64–0.65	8.05	10.0000	1.8918	0.1617
	0.04	1.0673	0.49	0.76–0.77	6.4	10.0000	2.1482	0.2140
	0.06	1.0612	0.56	0.84–0.85	5.62	10.0000	2.3328	0.2556
	0.1	1.0558	0.65	0.97–0.98	4.83	10.0000	2.6047	0.3232
	0.2	1.0512	0.81	1.16–1.17	3.88	10.0000	3.0500	0.4498
	0.6	1.0472	1.11	1.56–1.57	2.83	10.0000	3.9636	0.7717
	1	1.0461	1.30	1.75–1.80	2.42	10.0000	4.4899	0.9951
	2	1.0452	1.55	2.10–2.15	2.028	10.0000	5.3261	1.4069
	10	1.0440	2.40	3.25$^+$	1.31	10.0000	7.9471	3.1507

Utilizing other one-zero third-order plants and their respective $K_G \geq K_{min}$ for the corresponding values of a_1 and $q_3|_{min} \leq q_3$ (in this example $q_3|_{min} = 0.01$) yields curves identical to Fig. 15-7. Therefore, the concept of reproducibility of results is established; i.e., Fig. 15-7 is applicable to any one-zero third-order plant where $z = 1$, $K_G \geq K_{min}$, and $q_3 \geq q_3|_{min}$ is observed for each respective plant. In other words, from a reproducibility curve having an acceptable range of M_p values, one can select that value of q_3 which yields acceptable values of t_p, t_s, and ω_{d_e} under the optimal control criterion of Eq. (15-7), with $z = 1$ and $q_1 \approx c_0^2$ (for zero steady-state error for a step input).

The curves of Appendix D represent the parameters of the time response with $r(t) = u_{-1}(t)$ for systems with any fourth-, fifth-, or sixth-order plants with a single zero under the conditions of $K_G \geq K_{min}$, Eq. (15-29), $z = 1$, and $q_n > q_i$, and for the values of q_i specified in the figures. Note the similarity between these curves and those of Fig. 15-7. An analysis of these curves and their associated data for systems, of order greater than 3 and with a single zero, yield conclusions similar to those made for the one-zero third-order plant. *The entire data for higher-order systems with hfg operation reveal that, under the reproducibility condition, there are $n - 1$ dominant roots and the nondominant root is given by the expression $s_n \approx -K_G\sqrt{q_n}$, just as for the all-pole plant.* (See Sec. 15-9 for a general design procedure.)

Extending the correlation to a plant having two or more zeros in order to derive straightforward conclusions becomes more difficult. Consider the two-zero case for which the numerator of the plant transfer function is $s^2 + c_1 s + c_0$. Under the reproducibility condition (hfg operation, where $K_G \geq K_{min}$) the following general conclusions can be drawn:

1 A large value of c_0 together with a low value of c_1 provides the best overall results; i.e., acceptable values of M_p are achieved with a decrease in values of t_p and t_s.
2 Increasing the value of q_i for $i > 1$ decreases the overshoot.
3 A two-zero plant may yield better performance than an all-pole or a one-zero plant.

Table 15-7   TIME RESPONSE DATA FOR A THIRD-ORDER PLANT WITH ONE ZERO

$$K_G = 150 \qquad z = 1 \qquad Q = \begin{bmatrix} q_1 & & 0 \\ & 2 & \\ 0 & & q_3 \end{bmatrix}$$

q_1	q_3	M_p	t_p	t_s	ω_{d_e}
0.2	0.01	1.00000	—	8.60+	—
	1.0	1.00000	—	8.85	—
	10	1.00203	15.15	9.05+	0.207
100	0.01	1.00000	—	0.40+	—
	1.0	1.03322	1.35	1.70+	2.328
	10	1.03977	2.45	3.20+	1.282

15-8 ANALYSIS OF CORRELATION RESULTS

The $a_{i-1} + K_G k_i$ terms of the system characteristic equation, Eq. (15-26), under the high-gain condition ($K_G \geq K_{min}$) can be approximated by $K_G k_i$, that is, $a_{i-1} \ll K_G k_i$. Thus, Eq. (15-25) reduces to

$$s^n + K_G k_n s^{n-1} + K_G k_{n-1} s^{n-2} + \cdots + K_G k_1 = 0 \qquad (15\text{-}71)$$

An analysis of the correlation data of Secs. 15-6 and 15-7, for $z = 1$, reveals[1] that when the reproducibility condition is achieved, the actual value of the nondominant root s_n is

$$s_n \approx -K_G \sqrt{q_n} \qquad (15\text{-}72)$$

At a value of K_G sufficiently large to achieve reproducibility the actual root is very close (within approximately 0.5 percent) to the value given by Eq. (15-72). When reproducibility has been achieved, it is noted from the data that

$$k_n \approx \sqrt{q_n} \qquad (15\text{-}73)$$

As the values of q_n and n become larger, this approximation improves. *Equation (15-72) serves as a necessary and sufficient condition for satisfying the reproducibility criterion.*

The analysis of the correlation data also shows that for a given nth-order system as long as Eq. (15-72) is satisfied, the dominant roots remain unchanged, independent of $G(s)$. The only difference is the value of the nth root, given by Eq. (15-72), since each nth-order $G(s)$ function has a different value of $K_G \geq K_{min}$. This is to be expected since, if reproducibility is to occur, the dominant roots of an nth-order system utilizing any nth-order transfer function must not change. This is compatible with the development of Eq. (15-71). The $n - 1$ dominant roots, which do not change for $K_G \geq K_{min}$ and are independent of the plant, yield the polynomial

$$s^{n-1} + \varepsilon_{n-1} s^{n-2} + \cdots + \varepsilon_1 \qquad (15\text{-}74)$$

Multiplying Eq. (15-74) by $s + K_G \sqrt{q_n}$ yields the system characteristic equation

$$s^n + K_G k_n s^{n-1} + \cdots + K_G k_1$$
$$= s^n + (\varepsilon_{n-1} + K_G \sqrt{q_n}) s^{n-1} + \cdots + \varepsilon_1 K_G \sqrt{q_n} = 0 \qquad (15\text{-}75)$$

Thus

$$k_n = \frac{\varepsilon_{n-1} + K_G \sqrt{q_n}}{K_G}, \ldots, k_1 = \varepsilon_1 \sqrt{q_n} \qquad (15\text{-}76)$$

For zero steady-state error with a step input and $z = 1$, Eq. (15-76) yields

$$\varepsilon_1 = \frac{1}{\sqrt{q_n}} \qquad \text{\textit{all-pole case:} } k_1 = 1 \qquad (15\text{-}77)$$

$$\varepsilon_1 = \frac{c_0}{\sqrt{q_n}} \qquad \text{\textit{pole-zero case:} } k_1 = c_0 \qquad (15\text{-}78)$$

Consider that a reproducibility curve having the desirable M_p value (or range of values) is available for a given nth-order system and also that Eq. (15-74) is known.

For any other nth-order system determine $K_G \geq K_{\min}$, where the value $K_{\min}$ must be estimated as suitably larger than K_x, which is obtained from the RSL. From the reproducibility curve determine the desired value of q_n. The root s_n is determined from Eq. (15-72), and thence Eq. (15-75) can be obtained. The feedback coefficients $k_{n-1}, \ldots, k_2$ can be determined readily by equating coefficients, as shown in Eq. (15-76) (k_n and k_1 have been specified previously).

As a check on the value of K_G used, insert the calculated values of k_i into Eq. (15-25). The roots of this equation must yield approximately the same dominant roots as Eq. (15-74). In obtaining Eq. (15-74) about six- or seven-place accuracy must be used for the dominant roots.

15-9 A GENERAL DESIGN PROCEDURE

In the preceding sections a correlation between the conventional figures of merit with respect to the q_i elements of the $\mathbf{Q}$ matrix has been presented. For zero steady-state error with a step input to the system the q_1 element is specified in terms of the plant parameters and the weighting factor z. The correlation shows that the elements q_i ($i = 2, \ldots, n - 1$) have a primary effect on the value of M_p and that the element q_n has the primary effect on the values of t_p, t_s, and ω_d (or ω_{d_e}). The values of q_i chosen to yield a desired value (or range of values) of M_p have the secondary effect of determining the lower bound of the values of t_p and t_s and the upper bound of ω_d (or ω_{d_e}) that can be achieved by adjusting the value of q_n. From this correlation, a suitable design procedure under hfg operation is developed as follows.

Step 1 Set $q_1 = c_0^2 z$ for zero steady-state error for a step input.

Step 2 For the all-pole or the pole-zero system adjust q_i ($i = 2, \ldots, n - 1$) to achieve the desired M_p value.

Step 3 Make certain that the value of K_G is high enough at the lower bound of q_n to ensure reproducibility. This is determined by use of the reproducibility criterion, which states that the actual value of the nondominant root must be within, approximately, 0.5 percent of the value given by $-K_G\sqrt{q_n}$.

Step 4 Once steps 1 and 3 have been achieved, obtain the reproducibility curves.

Step 5 Select the value of q_n from the reproducibility curves to yield acceptable values of M_p, t_p, t_s, and ω_d (or ω_{d_e}) for the all-pole or pole-zero case.

Step 6 When the desired values of the elements of the $\mathbf{Q}$ matrix have been established, a computer program can be used to determine the feedback gains k_i to satisfy the optimal performance index of Eq. (15-7). Where applicable, the method outlined in Sec. 15-8 can be used to determine the values of k_i.

Step 7 For the overdamped case the values of q_i should be kept as low as possible to maintain $M_p = 1.0$ and to achieve the lowest value of t_s.

The all-pole system with the fastest response is achieved with the elements q_i and q_n equal to zero and with the conditions for observability and controllability satisfied.[1,5] The condition for reproducibility is not achieved with these values of the elements of $\mathbf{Q}$. The conventional figures of merit vary with the value of K_G. A one-zero plant can respond faster than an all-pole system for a suitable value of c_0. The same situation is possible for the two-zero case for reasonable values of M_p.

15-10   EFFECT OF CONTROL WEIGHTING

The preceding sections have presented a method for choosing the matrix $\mathbf{Q}$ for a control weighting $z = 1$ to achieve desired conventional figures of merit. The effect of varying the value of z is now illustrated for the third-order all-pole and one-zero plants of Secs. 15-6 and 15-7. In both cases q_1 is selected to satisfy Eq. (15-28), and the $\mathbf{Q}$ matrix utilized is

$$\mathbf{Q} = \begin{bmatrix} q_1 & & 0 \\ & 0.008 & \\ 0 & & q_3 \end{bmatrix} \qquad (15\text{-}79)$$

For the all-pole plant a value of $K_G = 500$ is used. For the one-zero plant, $c_0 = a_1 = 10$, a value of $K_G = 300$ is used.

In order to illustrate the effect of z on system performance, three values of z are selected for each plant, and the corresponding values of k_i are obtained by solving the algebraic Riccati equation. The resulting data of the associated time response are plotted in Figs. 15-8 to 15-10. Analyzing Fig. 15-8 for the all-pole third-order plant reveals that increasing the value of z:

1 Improves the constant M_p characteristic under hfg operation.
2 Improves the time response, i.e., decreases t_p and t_s.
3 Requires smaller values of feedback gains except for $k_1 = 1$, which remains fixed.

For the one-zero plant, Figs. 15-9 and 15-10 reveal that increasing the value of z:

1 Increases the value of M_p significantly for low values of q_3.
2 Decreases the values of t_p and t_s.
3 Requires smaller values of feedback gains except for $k_1 = c_0$, which remains fixed.

Therefore, as noted in Sec. 15-7, one-zero plants with a small value for a_1 have large overshoots. These large overshoots, for a given value of a_1, can be decreased by decreasing the value of z. In other words, decreasing the value of z, for a pole-zero plant, has the same effect on the value of M_p as increasing the value of q_i ($i = 2, 3, \ldots, n-1$). Similar results occur for higher-order plants. To maintain reproducibility, Eq. (14-83) requires that $K_G \geq \sqrt{z} K_{\min}$. See Sec. 15-6 for determining $K_{\min}$.

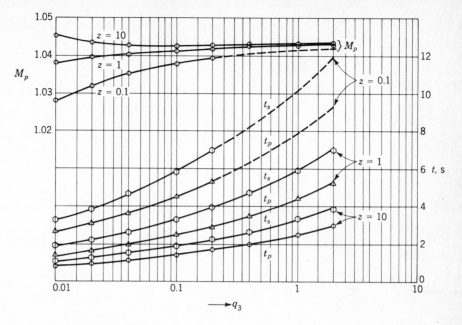

FIGURE 15-8
Time-response parameters for third-order all-pole system for $K_G = 500$ and several values of z.

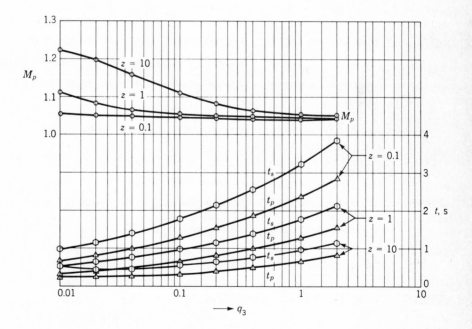

FIGURE 15-9
Time-response parameters for third-order one-zero system for $K_G = 300$, $c_0 = 10$, and several values of z.

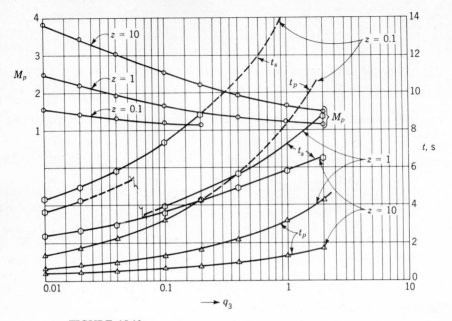

FIGURE 15-10
Time-response parameters for a third-order one-zero system for $K_G = 300$, $c_0 = \sqrt{0.2}$, and several values of z.

15-11   ALTERNATE DESIGN METHOD

The previous sections have presented a method of selecting the diagonal elements of the matrix $\mathbf{Q}$ with zero off-diagonal terms to minimize the performance index

$$\text{PI} = \int_0^\infty (\mathbf{x}^T \mathbf{Q}\mathbf{x} + zu^2)\, dt \qquad (15\text{-}80)$$

This section presents another method of obtaining the matrix $\mathbf{Q}$ by determining the elements of the vector $\mathbf{h}$, so that $\mathbf{Q} = \mathbf{h}\mathbf{h}^T$. This matrix is not diagonal; however, an equivalent diagonal matrix can be obtained.[6,10] Using the phase-variable representation, this method is based upon selecting a *defined dynamic equation* (DDE) for the desired optimal performance. This approach is used to simplify the pole-placement method of Chap. 12. Also, it is used to design a feedback system to meet desired performance specifications even though some of the parameters of the plant may vary.

If it were permissible to set $z = 0$ in Eq. (15-80), the PI would attain an absolute minimum of zero for $\mathbf{h}^T\mathbf{x}(t) = 0$. When phase-variable representation of the plant dynamics is used, this minimum can be expressed as

$$h_1 X_1(s) + h_2 X_2(s) + \cdots + h_n X_n(s) = (h_1 + h_2 s + \cdots + h_n s^{n-1})X_1(s) = 0 \qquad (15\text{-}81)$$

because

$$s X_{i-1}(s) = X_i(s) \qquad (15\text{-}82)$$

For a solution to the optimal control problem,

$$h_1 + h_2 s + \cdots + h_n s^{n-1} = 0 \qquad (15\text{-}83)$$

may be taken as the DDE of the system. The limiting form of the closed-loop system as $z \rightarrow 0$ is the model described by Eq. (15-83). For other values of z there is a mismatch. For the system design presented in this section, $z = 1$ is assumed. The necessary constraint on h_1, for zero steady-state error with a step input, using Eq. (15-28), is

$$q_{11} = h_1{}^2 = c_0{}^2 - \left(\frac{a_0}{K_G}\right)^2 \approx c_0{}^2 \qquad (15\text{-}84)$$

This constraint, under hfg operation, is observed in this section.

Equation (15-83) is one degree less than the characteristic equation of the system. The coefficients of this equation are obtained in the following manner: (1) Take all denominator factors of the desired $Y(s)/R(s)$ except the factor containing one non-dominant pole. (2) Multiply these factors and form a polynomial equation. (3) Multiply this polynomial equation by a constant so that the coefficient of the s^0 term has the required value of h_1 obtained from Eq. (15-84). This is now the DDE whose coefficients are the h_i given in Eq. (15-83).

Consider as a first example a third-order all-pole plant which is Type 1 ($a_0 = 0$). The quadratic equation formed from steps 1 and 2 above is

$$\omega_N{}^2 + 2\zeta_N \omega_N s + s^2 = 0 \qquad (15\text{-}85)$$

where the values of ζ_N and ω_N are specified. For an all-pole plant $h_1 = c_0 = 1$. When Eq. (15-85) is compared with Eq. (15-83), the value $h_1 = 1$ can be obtained by dividing Eq. (15-85) by $\omega_N{}^2$. This yields

$$1 + \frac{2\zeta_N}{\omega_N} s + \left(\frac{1}{\omega_N}\right)^2 s^2 = 1 + d_1 s + d_2 s^2 \qquad (15\text{-}86)$$

Thus
$$\mathbf{h}^T = \begin{bmatrix} 1 & d_1 & d_2 \end{bmatrix} \qquad (15\text{-}87)$$

The second example is the three-pole one-zero case. As shown in Sec. 15-7, large overshoots may exist due to the presence of the zero. Thus, it is necessary to choose $\mathbf{h}$ to minimize the effect of that zero. The obvious way to accomplish this is to place a root of the DDE near the plant zero. Thus, with the zero denoted by $-\alpha$, the quadratic equation formed from steps 1 and 2 above is

$$(s + \alpha)(s + \beta) = \alpha\beta + (\alpha + \beta)s + s^2 = 0 \qquad (15\text{-}88)$$

where the value of β is selected to yield the desired time response. In order for $h_1 = \alpha = c_0$, Eq. (15-88) is divided by β to obtain

$$\alpha + \frac{\alpha + \beta}{\beta} s + \frac{1}{\beta} s^2 = \alpha + d_1 s + d_2 s^2 \qquad (15\text{-}89)$$

Comparing Eq. (15-89) and Eq. (15-83) yields

$$\mathbf{h}^T = \begin{bmatrix} \alpha & d_1 & d_2 \end{bmatrix} \qquad (15\text{-}90)$$

The vector **h** for higher-order systems can be determined in a similar manner. For typical desired DDEs, Table 15-8 lists the vector **h** for $n = 3, 4, 5$ all-pole plants. Table 15-9 applies for the one-zero plants. These tables can readily be extended for higher-order systems. The desired DDE is obtained using the approach of Sec. 12-2. Once **h** is determined, the feedback coefficients **k** may be determined by use of a computer program.[11]

Once h_n is determined, the method of Sec. 14-12 may be applicable for obtaining a vector **k** which ensures an optimal performance. The necessary condition for this method is that the minimum value of the forward gain satisfy Eq. (14-83); i.e.,

$$K_x = \frac{\xi \omega^*}{h_n} \quad \text{for } \xi > 1 \quad (15\text{-}91)$$

Thus the value of gain used in the plant for which the DDE is to be determined must be large enough to satisfy the condition $\xi > 1$ in Eq. (15-91). An alternate approach is to obtain the Bode plot of the $G_K(s)$ and to select the gain so that $\omega_\phi \geq \xi \omega^*$, where $\xi > 1$. Once **h** and K_x are determined, the value of **k** can be calculated without a computer by utilizing Eq. (14-79).

Table 15-8 $\mathbf{h}^T = [1 \quad d_1 \quad d_2 \quad \cdots \quad d_n]$ FOR ALL-POLE PLANTS

n	Defined dynamic equation
3	$1 + \dfrac{2\zeta_N}{\omega_N} s + \left(\dfrac{1}{\omega_N}\right)^2 s^2 = 1 + d_1 s + d_2 s^2 = 0$
4	$(s + \alpha)(s^2 + 2\zeta_N \omega_N s + \omega_N^2) = 0 \quad$ or $1 + \dfrac{\omega_N^2 + 2\zeta_N \omega_N \alpha}{\alpha \omega_N^2} s + \dfrac{2\zeta_N \omega_N + \alpha}{\alpha \omega_N^2} s^2$ $\quad + \dfrac{1}{\alpha \omega_N^2} s^3 = 1 + d_1 s + d_2 s^2 + d_3 s^3 = 0$
5	$(s + \alpha)(s + \beta)(s^2 + 2\zeta_N \omega_N s + \omega_N^2) = 0 \quad$ or $1 + d_1 s + d_2 s^2 + d_3 s^3 + d_4 s^4 = 0$

Table 15-9 $\mathbf{h}^T = [\alpha \quad d_1 \quad d_2 \quad \cdots \quad d_n]$ FOR ONE-ZERO PLANTS

n	Defined dynamic equation
3	$(s + \alpha)(s + \beta) = 0 \quad$ or $\quad \alpha + d_1 s + d_2 s^2 = 0$
4	$(s + \alpha)(s^2 + 2\zeta_N \omega_N s + \omega_N^2) = 0 \quad$ or $\alpha + d_1 s + d_2 s^2 + d_3 s^3 = 0$
5	$(s + \alpha)(s + \beta)(s^2 + 2\zeta_N \omega_N s + \omega_N^2) = 0 \quad$ or $\alpha + d_1 s + d_2 s^2 + d_3 s^3 + d_4 s^4 = 0$

The following two examples illustrate the method, based upon the selection of a DDE, for achieving an optimal performance.

EXAMPLE 1   In the example in Sec. 12-7, the plant and desired control ratio are

$$G(s) = \frac{100}{s(s + 1)(s + 5)} = \frac{100}{s^3 + 6s^2 + 5s} \qquad (15\text{-}92)$$

$$\frac{Y(s)}{R(s)} = \frac{100}{(s + 100)(s^2 + 1.417s + 1)} \qquad (15\text{-}93)$$

For this third-order system the dominant poles of Eq. (15-93) are used as the roots of the desired second-order DDE. Thus, from Table 15-8

$$\mathbf{h}^T = [1 \quad 1.417 \quad 1] \qquad (15\text{-}94)$$

Use the **A** and **b** in Example 2, Sec. 14-9, to obtain $G_K(s) = \mathbf{h}^T \mathbf{\Phi}(s)\mathbf{b}$. An analysis of $G_K(j\omega)$ reveals that $\omega^* = 5$, and Eq. (15-91) yields the value $\xi = K_x h_3/\omega^* = 20$. Optimal performance is achieved for this system with the value of **k** determined from Eq. (14-79). Therefore,

$$\mathbf{k}^T = [1 \quad 1.377 \quad 0.95417] \qquad (15\text{-}95)$$

The feedback vector **k** could also be obtained by using the method of Chap. 12. This is accomplished by equating the coefficients in the denominator polynomial of Eq. (15-93) to the numerator coefficients of $1 + G(s)H_{eq}(s)$. However, the resulting system is not guaranteed to be optimal. When the transformation matrix **T** in Example 2 of Sec. 14-9 is used, the feedback coefficients for the physical-variable representation of Fig. 12-6b are

$$\mathbf{k}_p{}^T = \mathbf{k}^T \mathbf{T} = [1 \quad -3.39385 \quad 4.77085] \qquad (15\text{-}96)$$

These values agree very closely with those obtained by the method of Chap. 12.

EXAMPLE 2   From the Bode plot of $\mathbf{G}_K(j\omega)$ for the plant

$$G(s) = \frac{100(s + 2)}{s(s + 1)(s + 5)} \qquad (15\text{-}97)$$

it is evident that since $\xi = \omega_\phi/\omega^* = 16/5 = 3.2 > 1$ the condition required to achieve an optimal performance is satisfied. From Table 15-9, for $n = 3$, the DDE is

$$\alpha + d_1 s + d_2 s^2 = 0 \qquad (15\text{-}98)$$

The value $-\alpha$ is chosen to coincide with the value of the zero, $z = -2$, in Eq. (15-97). Assuming that an overdamped system response is acceptable, the particular DDE is chosen as

$$(2 + s)(4 + s) = 8 + 6s + s^2 = 0$$

Dividing this equation by 4 yields

$$2 + 1.5s + 0.25s^2 = 0 \qquad (15\text{-}99)$$

and

$$\mathbf{h}^T = [2 \quad 1.5 \quad 0.25] \qquad (15\text{-}100)$$

Equation (14-79) yields

$$\mathbf{k}^T = \begin{bmatrix} 2 & 1.53 & 0.25 \end{bmatrix} \qquad (15\text{-}101)$$

With this value of $\mathbf{k}$ the actual characteristic equation of the system, in phase-variable notation, is

$$s^3 + (a_2 + Kk_3)s^2 + (a_1 + Kk_2)s + Kk_1 = s^3 + 31s^2 + 158s^2 + 200$$
$$= (s + 2)(s + 4)(s + 25) \qquad (15\text{-}102)$$

Thus, with this design concept, a plant zero is canceled by a pole of the control ratio.

15-12 EXTENSION OF SEC. 15-11

The approach of Sec. 15-11 is a pole-placement technique which guarantees optimality. The state-variable method of Chap. 12 is also a pole-placement technique, but an optimal performance is not inherently assured. The state-variable feedback method can be applied with the DDE concept. This is illustrated for the phase-variable representation as follows. Since Eq. (15-83) is of order $n - 1$, it is multiplied by a first-order term $s + \gamma$. This yields the characteristic equation of the *actual* system, i.e.,

$$(h_n s^{n-1} + h_{n-1}s^{n-2} + \cdots + h_2 s + h_1)(s + \gamma) = 0 \qquad (15\text{-}103)$$

The values of K_G, γ, and h_i are chosen to satisfy the desired performance characteristics of the system. For example, $\mathbf{h}$ may be chosen to cancel an unwanted zero(s), and $-\gamma$ must be a nondominant pole of the system control ratio. The remaining h_i and K_G are chosen to yield the desired performance characteristics, such as $e(t)_{ss} = 0$ with a step input.

Equation (15-103) is of nth order, and it is equated to the numerator polynomial of $1 + G(s)H_{eq}(s)$; that is, with $z = 1$,

$$s^n + (a_{n-1} + K_G k_n)s^{n-1} + \cdots + (a_1 + K_G k_2)s + (a_0 + K_G k_1)$$
$$= h_n s^n + (\gamma h_n + h_{n-1})s^{n-1} + \cdots + (\gamma h_2 + h_1)s + \gamma h_1 \qquad (15\text{-}104)$$

Equating the coefficients of corresponding powers of s yields $h_n = 1$ and

$$k_1 = \frac{\gamma h_1 - a_0}{K_G} \qquad (15\text{-}105)$$

$$k_2 = \frac{\gamma h_2 + h_1 - a_1}{K_G} \qquad (15\text{-}106)$$

$$k_i = \frac{\gamma h_i + h_{i-1} - a_{i-1}}{K_G} \qquad i = 2, \ldots, n \qquad (15\text{-}107)$$

An optimal performance is achieved when the equation

$$\text{Rank } [\mathbf{h} \mid \mathbf{A}^T\mathbf{h} \mid \cdots \mid (\mathbf{A}^T)^{n-1}\mathbf{h}] = n \qquad (15\text{-}108)$$

is satisfied.

EXAMPLE   Consider the plant

$$G(s) = \frac{50(s+2)}{s^2(s+1)(s+5)} \qquad (15\text{-}109)$$

This transfer function is the same as that used in Example 2 of Sec. 12-10 but without the cascade compensation. The desired control ratio for the system is chosen to be

$$\frac{Y(s)}{R(s)} = \frac{50(s+2)}{(s+2)(s^2+2s+2)(s+25)} \qquad (15\text{-}110)$$

For this fourth order system the three dominant poles of Eq. (15-110) are used as the roots of the third-order DDE, that is, $4 + 6s + 4s^2 + s^3 = 0$. From Table 15-9

$$\mathbf{h}^T = [4 \quad 6 \quad 4 \quad 1] \qquad (15\text{-}111)$$

Since Eq. (15-91) yields $\xi = 10 > 1$ for $G_K(s)$, the hfg condition is satisfied. Note that this $\mathbf{h}^T$ is not normalized, i.e., $h_1 \neq \alpha = 2$ as specified in Table 15-9. This zero is to be canceled by a pole of Eq. (15-110). Equations (15-105) to (15-107), with $\gamma = 25$, yield

$$\mathbf{k}^T = [2 \quad 3.08 \quad 2.02 \quad 0.46] \qquad (15\text{-}112)$$

With this $\mathbf{k}$ the actual characteristic equation of the system is

$$
\begin{aligned}
s^4 + (q_3 + K_G k_4)s^3 &+ (a_2 + K_G k_3)s^2 + (a_1 + K_G k_2)s + K_G k_1 \\
&= s^4 + 29s^3 + 106s^2 + 154s + 100 \\
&= (s+2)(s^2+2s+2)(s+25)
\end{aligned} \qquad (15\text{-}113)
$$

This example illustrates the application of the DDE concept to a state-variable feedback system in which a pole of the control ratio cancels the plant zero. Equation (15-108) yields

$$\text{Rank } [\mathbf{h} | \mathbf{A}^T \mathbf{h} | (\mathbf{A}^T)^2 \mathbf{h} | (\mathbf{A}^T)^3 \mathbf{h}] = \text{rank} \begin{bmatrix} 4 & 0 & 0 & 0 \\ 6 & 4 & 0 & 0 \\ 4 & 1 & 14 & -65 \\ 1 & -2 & 13 & -64 \end{bmatrix} = 4 = n \qquad (15\text{-}114)$$

Therefore this system has an optimal performance, satisfying the quadratic PI. By means of a linear transformation the physical-variable feedback coefficients can be determined. If $\mathbf{h}$ is inserted into the Riccati equation to obtain an exact solution for $\mathbf{k}$, the normalized vector $\mathbf{h}^T = [2 \quad 3 \quad 2 \quad 0.5]$ which satisfies Table 15-9 must be used. It yields essentially the same $\mathbf{k}$ as above.

The method of Sec. 15-11 has been applied to an aircraft control problem[6,7] in which the plant parameters varied widely over the full flight range. The design included the cancellation of a plant zero with a pole of the system characteristic equation. The value of the zero varied from -0.0079 to -2.07 over the range of flight conditions. This design produced a real pole of the characteristic equation which varied along with and having essentially the same value as the plant zero. Therefore, the zero was always essentially canceled, regardless of its value. Other plant parameters also varied widely. The state-variable feedback, along with the effective zero cancellation, resulted in a design which met all the desired performance specifications of the system.

15-13 SUMMARY

This chapter presents methods for selecting the elements of the matrix $\mathbf{Q}$ or the vector $\mathbf{h}$ under hfg operation in order to achieve desired conventional figures of merit. Once the $\mathbf{Q}$ or $\mathbf{h}$ are specified, the corresponding $\mathbf{k}$ is determined in terms of the phase-variable representation. The feedback gains k_{p_i} for the physical-variable system are then solved by use of a linear transformation. When some states are not accessible, they can be generated from states that are accessible (see Sec. 12-11). The state-estimation techniques of Kalman and Luenburger may also be used. For a linear time-invariant system the block-manipulation method for generating states still results in an optimal system. An alternate design method is to correlate with the conventional figures of merit[6,7] the elements h_i of the vector $\mathbf{h}$, where $\mathbf{Q} = \mathbf{hh}^T$. The hfg design method presented in this chapter can be used for systems which have plants with varying parameters.[8] The objective is to determine a feedback vector $\mathbf{k}$ that maintains satisfactory characteristics for the time response of the system, even though the plant parameters may vary. By this method the feedback vector is determined for an appropriate choice of nominal values of the plant parameters. Optimal performance exists when the system is operating with its nominal plant parameters; otherwise the performance is suboptimal.

Chapters 12 to 15 present methods for designing state-variable feedback systems. These methods are not all-inclusive but are intended to illustrate the concepts of state-variable feedback and optimal control. Other complementary design methods, such as modal control theory,[9] are available in the literature.

REFERENCES

1 Houpis, C. H.: The Relationship between the Conventional Control Theory Figures of Merit and the Performance Indices in Optimal Control Theory, Ph.D. dissertation, Department of Electrical Engineering, University of Wyoming, 1971.

2 Houpis, C. H., and C. T. Constantinides: Functional Relationship between the Conventional Steady-State Characteristics and the Weighting Matrices in the Quadratic Performance Index, *Int. J. Control*, vol. 15, no. 6, pp. 1147–1156, 1972; Correlation between Conventional Figures of Merit and the Q Matrix of the Quadratic Cost Function: Third-Order Plant, *Int. J. Control*, vol. 16, no. 4, pp. 695–704, 1972; Relationship between Conventional Control Theory Figures of Merit and Quadratic Performance Index in Optimal Control Theory for a Single-Input/Single-Output System, *Proc. IEE*, vol. 120, no. 1, pp. 138–142, 1973.

3 Bullock, T. E., and J. M. Elder: Quadratic Performance Index Generation for Optimal Regulator Design, *1971 IEEE Conf. Decision Control*, pp. 123–124.

4 Rynaski, E. G., and R. F. Whitbeck: The Theory and Application of Linear Optimal Control, *Air Force Flight Dynam. Lab. Tech. Rep.* AFFDL-TR-65-28, Wright-Patterson Air Force Base, Ohio, 1966.

5 Kalman, R. E.: When Is a Linear Control System Optimal?, *ASME J. Basic Eng.*, pp. 51–60, March 1964.

6 Mirmak, E. V.: Some Techniques for Optimal Linear Regulator Design to Satisfy Conventional Figures of Merit, M.Sc. thesis, GA/EE/73A-4, Air Force Institute of Technology, Wright-Patterson Air Force Base, Ohio, 1974.

7 Wilt, D. C.: Design of an Optimal Linear Regulator to Satisfy Conventional Figures of Merit by Utilizing $\mathbf{Q} = \mathbf{hh}^T$ and the Defined Dynamic Equation Approach, GE/EE/74-66, Air Force Institute of Technology, Wright-Patterson Air Force Base, Ohio, 1974.

8 Ray, R. A.: A State-Variable Design Approach for a High-Performance Aerospace Vehicle Pitch Orientation System with Variable Coefficients, M.Sc. thesis, GGC/EE/73-15, Air Force Institute of Technology, Wright-Patterson Air Force Base, Ohio, 1973.

9 Porter, B., and T. R. Crossley: "Modal Control Theory and Applications," Barnes and Noble, New York, 1972.

10 Kriendler, E.: Synthesis of Flight Control Systems Subject to Vehicle Parameter Variations, *Air Force Flight Dynam. Lab. Tech. Rep.* TR-66-209, Wright-Patterson Air Force Base, Ohio, April 1967.

11 Solution of Linear-Quadratic Optimal Control Problem (OPTCON), Air Force Institute of Technology Computer Program Library, 1974.

TABLE OF LAPLACE-TRANSFORM PAIRS

$F(s)$	$f(t) \quad 0 \le t$
1. 1	$u_1(t)$ unit impulse at $t = 0$
2. $\dfrac{1}{s}$	1 or $u(t)$ unit step at $t = 0$
3. $\dfrac{1}{s^2}$	$tu(t)$ ramp function
4. $\dfrac{1}{s^n}$	$\dfrac{1}{(n-1)!}\, t^{n-1}$ n is a positive integer
5. $\dfrac{1}{s} e^{-as}$	$u(t-a)$ unit step starting at $t = a$
6. $\dfrac{1}{s}(1 - e^{-as})$	$u(t) - u(t-a)$ rectangular pulse
7. $\dfrac{1}{s+a}$	e^{-at} exponential decay
8. $\dfrac{1}{(s+a)^n}$	$\dfrac{1}{(n-1)!}\, t^{n-1}e^{-at}$ n is a positive integer
9. $\dfrac{1}{s(s+a)}$	$\dfrac{1}{a}(1 - e^{-at})$
10. $\dfrac{1}{s(s+a)(s+b)}$	$\dfrac{1}{ab}\left(1 - \dfrac{b}{b-a}e^{-at} + \dfrac{a}{b-a}e^{-bt}\right)$
11. $\dfrac{s+\alpha}{s(s+a)(s+b)}$	$\dfrac{1}{ab}\left[\alpha - \dfrac{b(\alpha-a)}{b-a}e^{-at} + \dfrac{a(\alpha-b)}{b-a}e^{-bt}\right]$
12. $\dfrac{1}{(s+a)(s+b)}$	$\dfrac{1}{b-a}(e^{-at} - e^{-bt})$
13. $\dfrac{s}{(s+a)(s+b)}$	$\dfrac{1}{a-b}(ae^{-at} - be^{-bt})$
14. $\dfrac{s+\alpha}{(s+a)(s+b)}$	$\dfrac{1}{b-a}[(\alpha-a)e^{-at} - (\alpha-b)e^{-bt}]$

$F(s)$	$f(t)$ $0 \leq t$
15. $\dfrac{1}{(s+a)(s+b)(s+c)}$	$\dfrac{e^{-at}}{(b-a)(c-a)} + \dfrac{e^{-bt}}{(c-b)(a-b)} + \dfrac{e^{-ct}}{(a-c)(b-c)}$
16. $\dfrac{s+\alpha}{(s+a)(s+b)(s+c)}$	$\dfrac{(\alpha-a)e^{-at}}{(b-a)(c-a)} + \dfrac{(\alpha-b)e^{-bt}}{(c-b)(a-b)} + \dfrac{(\alpha-c)e^{-ct}}{(a-c)(b-c)}$
17. $\dfrac{\omega}{s^2+\omega^2}$	$\sin \omega t$
18. $\dfrac{s}{s^2+\omega^2}$	$\cos \omega t$
19. $\dfrac{s+\alpha}{s^2+\omega^2}$	$\dfrac{\sqrt{\alpha^2+\omega^2}}{\omega} \sin(\omega t + \phi)$ $\phi = \tan^{-1}\dfrac{\omega}{\alpha}$
20. $\dfrac{s\sin\theta + \omega\cos\theta}{s^2+\omega^2}$	$\sin(\omega t + \theta)$
21. $\dfrac{1}{s(s^2+\omega^2)}$	$\dfrac{1}{\omega^2}(1 - \cos \omega t)$
22. $\dfrac{s+\alpha}{s(s^2+\omega^2)}$	$\dfrac{\alpha}{\omega^2} - \dfrac{\sqrt{\alpha^2+\omega^2}}{\omega^2}\cos(\omega t + \phi)$ $\phi = \tan^{-1}\dfrac{\omega}{\alpha}$
23. $\dfrac{1}{(s+a)(s^2+\omega^2)}$	$\dfrac{e^{-at}}{a^2+\omega^2} + \dfrac{1}{\omega\sqrt{a^2+\omega^2}}\sin(\omega t - \phi)$ $\phi = \tan^{-1}\dfrac{\omega}{a}$
24. $\dfrac{1}{(s+a)^2+b^2}$	$\dfrac{1}{b}e^{-at}\sin bt$
24a. $\dfrac{1}{s^2+2\zeta\omega_n s + \omega_n^2}$	$\dfrac{1}{\omega_n\sqrt{1-\zeta^2}}e^{-\zeta\omega_n t}\sin \omega_n\sqrt{1-\zeta^2}\,t$
25. $\dfrac{s+a}{(s+a)^2+b^2}$	$e^{-at}\cos bt$
26. $\dfrac{s+\alpha}{(s+a)^2+b^2}$	$\dfrac{\sqrt{(\alpha-a)^2+b^2}}{b}e^{-at}\sin(bt+\phi)$ $\phi = \tan^{-1}\dfrac{b}{\alpha-a}$
27. $\dfrac{1}{s[(s+a)^2+b^2]}$	$\dfrac{1}{a^2+b^2} + \dfrac{1}{b\sqrt{a^2+b^2}}e^{-at}\sin(bt-\phi)$ $\phi = \tan^{-1}\dfrac{b}{-a}$
27a. $\dfrac{1}{s(s^2+2\zeta\omega_n s + \omega_n^2)}$	$\dfrac{1}{\omega_n^2} - \dfrac{1}{\omega_n^2\sqrt{1-\zeta^2}}e^{-\zeta\omega_n t}\sin(\omega_n\sqrt{1-\zeta^2}\,t + \phi)$ $\phi = \cos^{-1}\zeta$
28. $\dfrac{s+\alpha}{s[(s+a)^2+b^2]}$	$\dfrac{\alpha}{a^2+b^2} + \dfrac{1}{b}\sqrt{\dfrac{(\alpha-a)^2+b^2}{a^2+b^2}}e^{-at}\sin(bt+\phi)$ $\phi = \tan^{-1}\dfrac{b}{\alpha-a} - \tan^{-1}\dfrac{b}{-a}$
29. $\dfrac{1}{(s+c)[(s+a)^2+b^2]}$	$\dfrac{e^{-ct}}{(c-a)^2+b^2} + \dfrac{e^{-at}\sin(bt-\phi)}{b\sqrt{(c-a)^2+b^2}}$ $\phi = \tan^{-1}\dfrac{b}{c-a}$
30. $\dfrac{1}{s(s+c)[(s+a)^2+b^2]}$	$\dfrac{1}{c(a^2+b^2)} - \dfrac{e^{-ct}}{c[(c-a)^2+b^2]}$ $+ \dfrac{e^{-at}\sin(bt-\phi)}{b\sqrt{a^2+b^2}\sqrt{(c-a)^2+b^2}}$ $\phi = \tan^{-1}\dfrac{b}{-a} + \tan^{-1}\dfrac{b}{c-a}$

$F(s)$	$f(t)$ $0 \leq t$

31. $\dfrac{s + \alpha}{s(s + c)[(s + a)^2 + b^2]}$

$$\dfrac{\alpha}{c(a^2 + b^2)} + \dfrac{(c - \alpha)e^{-ct}}{c[(c - a)^2 + b^2]}$$

$$+ \dfrac{\sqrt{(\alpha - a)^2 + b^2}}{b\sqrt{a^2 + b^2}\sqrt{(c - a)^2 + b^2}}\, e^{-at}\sin(bt + \phi)$$

$$\phi = \tan^{-1}\dfrac{b}{\alpha - a} - \tan^{-1}\dfrac{b}{-a} - \tan^{-1}\dfrac{b}{c - a}$$

32. $\dfrac{1}{s^2(s + a)}$

$$\dfrac{1}{a^2}(at - 1 + e^{-at})$$

33. $\dfrac{1}{s(s + a)^2}$

$$\dfrac{1}{a^2}(1 - e^{-at} - ate^{-at})$$

34. $\dfrac{s + \alpha}{s(s + a)^2}$

$$\dfrac{1}{a^2}[\alpha - \alpha e^{-at} + a(a - \alpha)te^{-at}]$$

35. $\dfrac{s^2 + \alpha_1 s + \alpha_0}{s(s + a)(s + b)}$

$$\dfrac{\alpha_0}{ab} + \dfrac{a^2 - \alpha_1 a + \alpha_0}{a(a - b)}e^{-at} - \dfrac{b^2 - \alpha_1 b + \alpha_0}{b(a - b)}e^{-bt}$$

36. $\dfrac{s^2 + \alpha_1 s + \alpha_0}{s[(s + a)^2 + b^2]}$

$$\dfrac{\alpha_0}{c^2} + \dfrac{1}{bc}[(a^2 - b^2 - \alpha_1 a + \alpha_0)^2$$

$$+ b^2(\alpha_1 - 2a)^2]^{1/2}e^{-at}\sin(bt + \phi)$$

$$\phi = \tan^{-1}\dfrac{b(\alpha_1 - 2a)}{a^2 - b^2 - \alpha_1 a + \alpha_0} - \tan^{-1}\dfrac{b}{-a}$$

$$c^2 = a^2 + b^2$$

37. $\dfrac{1}{(s^2 + \omega^2)[(s + a)^2 + b^2]}$

$$\dfrac{(1/\omega)\sin(\omega t + \phi_1) + (1/b)e^{-at}\sin(bt + \phi_2)}{[4a^2\omega^2 + (a^2 + b^2 - \omega^2)^2]^{1/2}}$$

$$\phi_1 = \tan^{-1}\dfrac{-2a\omega}{a^2 + b^2 - \omega^2} \qquad \phi_2 = \tan^{-1}\dfrac{2ab}{a^2 - b^2 + \omega^2}$$

38. $\dfrac{s + \alpha}{(s^2 + \omega^2)[(s + a)^2 + b^2]}$

$$\dfrac{1}{\omega}\left[\dfrac{\alpha^2 + \omega^2}{c}\right]^{1/2}\sin(\omega t + \phi_1)$$

$$+ \dfrac{1}{b}\left[\dfrac{(\alpha - a)^2 + b^2}{c}\right]^{1/2}e^{-at}\sin(bt + \phi_2)$$

$$c = (2a\omega)^2 + (a^2 + b^2 - \omega^2)^2$$

$$\phi_1 = \tan^{-1}\dfrac{\omega}{\alpha} - \tan^{-1}\dfrac{2a\omega}{a^2 + b^2 + \omega^2}$$

$$\phi_2 = \tan^{-1}\dfrac{b}{\alpha - a} + \tan^{-1}\dfrac{2ab}{a^2 - b^2 + \omega^2}$$

39. $\dfrac{s + \alpha}{s^2[(s + a)^2 + b^2]}$

$$\dfrac{1}{c}\left(\alpha t + 1 - \dfrac{2\alpha a}{c}\right) + \dfrac{[b^2 + (\alpha - a)^2]^{1/2}}{bc}e^{-at}\sin(bt + \phi)$$

$$c = a^2 + b^2$$

$$\phi = 2\tan^{-1}\left(\dfrac{b}{a}\right) + \tan^{-1}\dfrac{b}{\alpha - a}$$

40. $\dfrac{s^2 + \alpha_1 s + \alpha_0}{s^2(s + a)(s - b)}$

$$\dfrac{\alpha_1 + \alpha_0 t}{ab} - \dfrac{\alpha_0(a + b)}{(ab)^2} - \dfrac{1}{a - b}\left(1 - \dfrac{\alpha_1}{a} + \dfrac{\alpha_0}{a^2}\right)e^{-at}$$

$$- \dfrac{1}{a - b}\left(1 - \dfrac{\alpha_1}{b} + \dfrac{\alpha_0}{b^2}\right)e^{-bt}$$

APPENDIX B

THE SPIRULE

B-1 DESCRIPTION

The Spirule is a device used chiefly by feedback control system engineers to multiply and divide complex quantities in the application of the root-locus method of analysis. Complex quantities are multiplied by multiplying their magnitudes and adding the angles between each directed line segment and a reference axis. Division of complex quantities is the inverse operation. The word *Spirule* is the trade name given to this device by W. R. Evans and is a combination of the words *spiral* and *rule*. The device consists of a protractor and an arm which are made of transparent plastic and have a common pivot point. A commercial model made by The Spirule Company of Whittier, California, is shown in Fig. B-1.

The protractor is 4.5 in. in diameter and is graduated in 1° increments. Cross-hair lines are drawn on the 0 to 180° axis and the 90 to 270° axis. The arm extends 9.25 in. from the pivot point. One side of the arm is aligned with the diameter of the protractor. The scale printed along the edge of the arm is marked in tenths (which are $\frac{1}{2}$ in. long) as a scale factor. Each half inch is further subdivided into 10 parts. A logarithmic spiral S is drawn on the arm and is used for multiplication and division of numbers. The arm has a solid reference line marked R which starts at the pivot and is approximately diagonal on the arm. The angle between the reference line on the arm and a radial line from the pivot to any point on the spiral curve is proportional to the logarithm of the distance from the pivot to the point. The intersection of the spiral with the reference line on the arm corresponds to the unit distance 5 in. A factor of 10 for the numbers being multiplied or divided corresponds to a rotation of 90°. Therefore three of the index arrows are labeled with the factors X—0.1, X—1, and X—10. The spiral beyond 7 in. from the pivot point is reflected about the

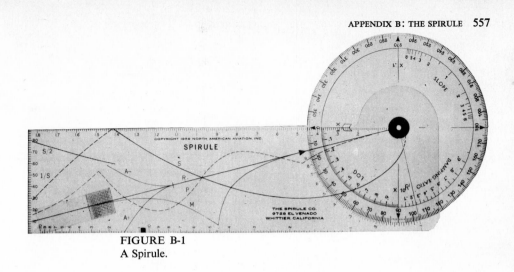

FIGURE B-1
A Spirule.

edge of the arm and is drawn as a dashed curve. The pivot permits the protractor and the arm to be moved separately. There is also enough friction between the arm and the protractor to make them rotate together.

Section 4-10 describes the procedures for using the Spirule to obtain the magnitudes and angles of the coefficients of the partial-fraction expansion of a function $F(s)$. The same basic procedures are applied in this appendix to the root locus.

B-2 ANGLE MEASUREMENT

The angle condition for points on the root locus states that the sum of the angles from the poles of $G(s)$ minus the sum of the angles from the zeros of $G(s)$ must be an odd multiple of $180°$ for positive values of gain:

$$\Sigma\underline{/\text{poles of } G(s)} - \Sigma\underline{/\text{zeros of } G(s)} = (1 + 2h)180° \qquad \text{(B-1)}$$

Note that in the root locus it is customary to take the sum of the angles of the poles minus the angles of the zeros of $G(s)$, whereas for the partial-fraction coefficients (see Sec. 4-10) the sign of the angles is reversed. The points on the complex portion of the s plane are obtained by trial and error by applying the angle condition of Eq. (B-1). For the transfer function

$$G(s) = \frac{K(s - z)}{s(s - p)} \qquad \text{(B-2)}$$

the pole-zero diagram is shown in Fig. B-2. The angle condition for this problem requires that

$$\phi_0 + \phi_1 - \psi = (1 + 2h)180° \qquad \text{(B-3)}$$

FIGURE B-2
Trial point for root locus of

$$G(s) = \frac{K(s - z)}{s(s - p)}$$

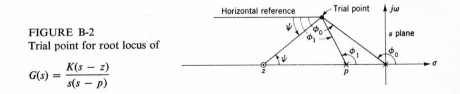

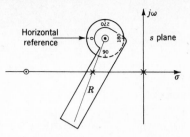

FIGURE B-3
Operation of Spirule.

Figure B-2 shows that these angles can also be measured from a horizontal reference line drawn to the left from the trial point. The use of the Spirule to measure these angles is now described.

The first step is to set the reference line R on the arm in line with the $0°$ index on the protractor. The center of the protractor is placed on a trial point with the $0°$ index aligned with a horizontal reference line of the graph paper and pointed to the left. A finger of the right hand is placed on the pivot point and the protractor is restrained from rotating by pressing on it with another finger of the same hand. The arm is rotated counterclockwise, by using the left hand, until a pole is aligned with the reference line on the arm (see Fig. B-3). The reading on the protractor is now equal to ϕ_1. The protractor is now released, but a finger is kept on the pivot point. Next, the arm and protractor are rotated together in a clockwise direction until the reference line R is again horizontal. The procedure is repeated to add the angles from the other poles.

Angles from the zeros are subtracted by rotating the arm and protractor together until the line R of the arm is aligned with a zero. The protractor is then restrained, and the arm is rotated back to the horizontal. When this procedure has included all the poles and zeros, the reading on the protractor is equal to the sum of the angles from the poles minus the sum of the angles from the zeros. If the sum of these angles is $180°$, the trial point is a point on the root locus. If the angle is not $180°$, another trial point must be chosen. A systematic procedure should be used for successive trial points. For example, successive trial points may be taken along a horizontal line. Figure B-4 shows a trial point s_1 for which the sum of the angles is $165°$. The sum of the angles increases if the second trial point is taken to the left of s_1. The second trial point s_2 shows an angle of $205°$. The point on the root locus must lie between s_1 and s_2. The point s_3 is then located for which the sum of the angles is $180°$.

B-3   SHORTCUTS IN ANGLE MEASUREMENTS

With experience, shortcut methods can be derived and used. For example, with the protractor and arm aligned with the zero, as shown in Fig. B-5, the protractor is held and the

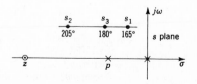

FIGURE B-4
Successive trial points for locating the root locus.

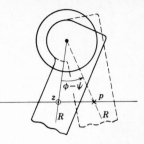

FIGURE B-5
Net angle due to a pole and a zero.

arm is rotated until the R line is over with the pole. The net angle indicated on the protractor is $\phi - \psi$. This operation measures simultaneously, in effect, the angle contributed
by both the pole and the zero. The procedure can be applied repeatedly to reduce the work
of determining the sum of the angles at a point.

The *angle of departure* from a complex pole can be determined by use of the Spirule.
To do this, the center of the Spirule is placed on the complex pole. Then the angles from
all the other poles and zeros are measured, using the procedure outlined in this appendix.
With the relative position of the arm and protractor maintained, the Spirule is rotated until
the line R is aligned with the horizontal reference and points to the left. The $0°$ index now
points along the departure angle from the complex pole. An example will illustrate the
procedure.

Consider the transfer function

$$G(s) = \frac{K(s + 3)}{s(s^2 + 2s + 2)} \qquad \text{(B-4)}$$

The center of the Spirule is placed on the complex pole $s = -1 + j1$, and the angles due
to the other two poles and the zero are measured. The Spirule reads $199°$. With the arm and
protractor in position, the line R on the arm is aligned with the horizontal reference, as shown
in Fig. B-6. The arrow on the $0°$ index points in the direction of the departure angle and is
$-19°$.

The *approach angle* to a complex zero can also be obtained by use of the Spirule. The
procedure is to place the center of the Spirule on the complex zero. The sum of the angles
from all the poles minus the angles from the other zeros is determined. Then, with the relative
position of the arm and protractor maintained, the Spirule is rotated until the $0°$ index is
directed to the left and is aligned with the horizontal reference. The reference line R on the
arm now points in the direction of the approach angle. Figure B-7 illustrates this procedure.

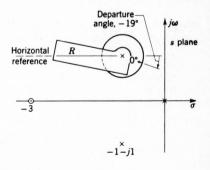

FIGURE B-6
Departure angle from complex poles
determined by use of the Spirule.

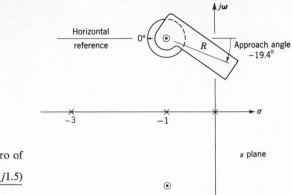

FIGURE B-7
Approach angle to the complex zero of

$$G(s) = \frac{K(s + 1 - j1.5)(s + 1 + j1.5)}{s(s + 1)(s + 3)}$$

B-4   DAMPING RATIO

The Spirule is calibrated to read damping ratio directly. To use the damping-ratio scale the protractor is placed with its center at the origin of the s plane and with the $0°$ index pointing along the horizontal line to the right. The arm is then turned until its intersection with the damping-ratio scale occurs at the desired value of ζ. The edge of the arm now describes the constant ζ line in the s plane.

B-5   MULTIPLICATION OF LENGTHS

Multiplication and division of lengths are performed by using the logarithmic spiral S drawn on the arm of the Spirule. The need to perform these operations occurs after the exact shape of the root locus has been determined by applying the angle condition. The roots of the characteristic equation are selected according to the specifications for the desired system performance. The next step is to evaluate the required magnitude of the system loop sensitivity K. This is done by use of the magnitude condition, which requires that

$$|G(s)| = 1 \qquad \text{(B-5)}$$

Solving Eq. (B-5) for the loop sensitivity gives

$$K = \frac{\prod\limits_{c=1}^{n} | s - p_c|}{\prod\limits_{h=1}^{w} |s - z_h|} \qquad \text{(B-6)}$$

For the system described by Eq. (B-2) the magnitude condition requires that

$$K = \frac{|s| \cdot |s - p|}{|s - z|} \qquad \text{(B-7)}$$

The root locus for this system is drawn in Fig. B-8, and it is assumed that the roots s_1 and s_2 have been selected. The system loop sensitivity K for these roots is determined by inserting the magnitudes of the lengths shown in Fig. B-8 into Eq. (B-7).

The value of K is determined by use of the Spirule in the following manner. First, line R on the arm is set in line with the X—1 index arrow. The pivot of the Spirule is placed

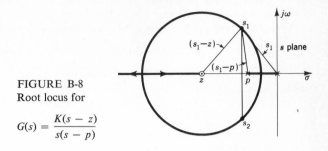

FIGURE B-8
Root locus for

$$G(s) = \frac{K(s - z)}{s(s - p)}$$

over the point s_1, to which the lengths are to be measured. To measure the length to a pole such as p, both the arm and the disk are rotated until the line R of the arm is aligned with the pole. The disk is held stationary and the arm rotated until the spiral curve S crosses over the pole. The numerical value of the magnitude $s - p$, to the given scale, can be read from the index arrow and the spiral calibration scale on the arm. The next directed line segment is multiplied by releasing the protractor so the assembly can rotate and aligning the line R with the next pole. The protractor is then restrained, and the arm is rotated until the spiral lies over the second pole. The index arrow now points to a reading on the spiral calibration scale that represents the product of the two magnitudes. Division by the length from a zero ($s - z$) is accomplished by releasing the protractor, so that the assembly can rotate, and aligning the Spirule so that the spiral curve crosses over the zero. The protractor is then held while the arm is rotated until the R line of the arm is aligned with the zero. The index arrow now points to a reading on the spiral calibration scale that is equal to the previous number divided by the length $s - z$. This procedure is continued until all the lengths have been included. When lengths greater than 1.42 units on the arm scale are involved, the reflected spiral curve which is drawn as a dashed curve may be used. However, since this is a reflected curve, the procedure for poles and zero must be modified. The curve $S/2$ can also be used but the angle associated with it must be introduced twice. The correct value of K is obtained by multiplying the reading on the spiral scale by the index factor and the scale factor:

$$K = \text{(Spirule reading)(index factor)(scale factor)}^x \qquad \text{(B-8)}$$

The index factor is read from the Spirule and is 0.1, 1, or 10, according as the X—0.1, X—1, or X—10 index arrow points to the Spirule reading on the spiral calibration scale. The scale factor is the numerical value of the scale of the plot corresponding to 5 in. (the unit length on the Spirule arm), and x is equal to the number of poles minus the number of zeros of $G(s)$.

As an example, consider the transfer function

$$G(s) = \frac{K(s + 4)}{s(s + 1)} \qquad \text{(B-9)}$$

The pole-zero diagram is drawn to the scale 1 unit equals 1 in.; the scale factor is therefore equal to 5. The point $s_1 = -1 + j1.7$ is a point on the root locus. The gain sensitivity at this point is found in the following manner: The arm of the Spirule is placed at the index X—1 and the arm is aligned with the pole at the origin. Holding the protractor and rotating the line R on the arm until the curve crosses the pole results in a reading of 0.39. The procedure is repeated for the pole at $s = -1$, and the Spirule reading on the spiral curve is now

0.133. The Spirule is then rotated until the spiral curve crosses the zero at $s = -4$. The protractor is held while the line R on the arm is aligned with the zero. The X—1 index points to the reading 0.193 on the spiral curve. The value of gain sensitivity K is

$$K = 0.193 \times 1 \times (5^{2-1}) = 0.965 \qquad \text{(B-10)}$$

Familiarity with the use of the Spirule expedites the work of plotting the root locus and the determination of loop sensitivities.

TIME-RESPONSE CHARACTERISTICS
FOR ALL-POLE SYSTEMS

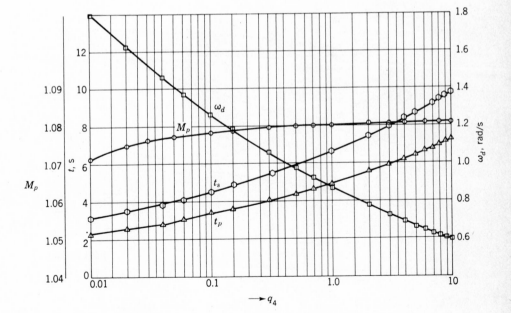

FIGURE C-1
The parameters of the time-response characteristics for a fourth-order all-pole system with constant K_G, $q_1 = 1$, $q_2 = q_3 = 0.01$, and $z = 1$.

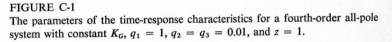

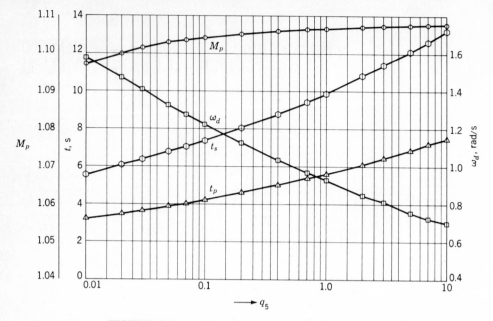

FIGURE C-2
The parameters of the time-response characteristics for a fifth-order all-pole system with constant K_G, $q_1 = 1$, $q_2 = q_3 = q_4 = 0.01$, and $z = 1$.

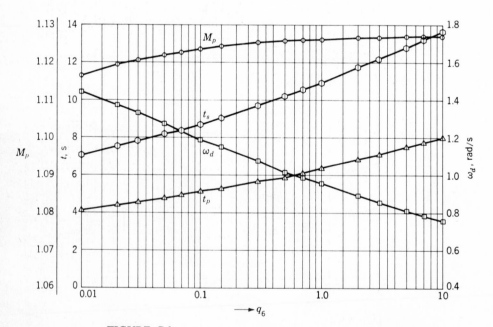

FIGURE C-3
The parameters of the time-response characteristics for a sixth-order all-pole system with constant K_G, $q_1 = 1$, $q_i = 0.01$ for $i = 2, 3, 4$ and 5, and $z = 1$.

TIME-RESPONSE CHARACTERISTICS
FOR PLANTS WITH ONE ZERO

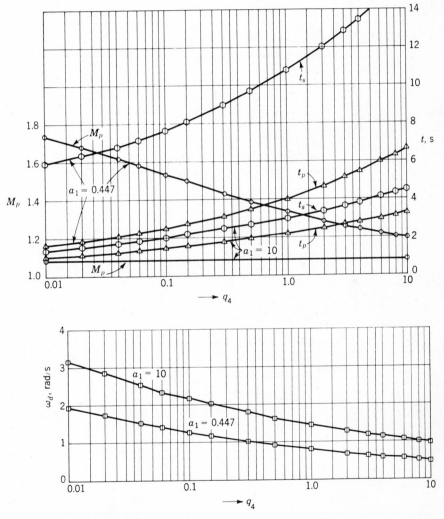

FIGURE D-1
The parameters of the time-response characteristics for a fourth-order system
with one zero, $K_G = 2,000$, $z = 1$, $q_2 = 0.01$, and $q_3 = 0.06$. (*Top*) M_p, t_p,
and t_s versus q_4; (*bottom*) ω_{d_e} versus q_4.

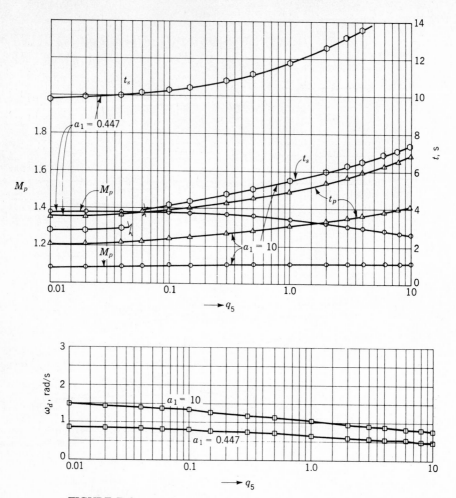

FIGURE D-2
The parameters of the time-response characteristics for a fifth-order system with one zero, $K_G = 2,000$, $z = 1$, $q_2 = 0.01$, $q_3 = 0.06$, $q_4 = 0.5$; (*top*) M_p, t_p, and t_s versus q_5; (*bottom*) ω_{d_e} versus q_5.

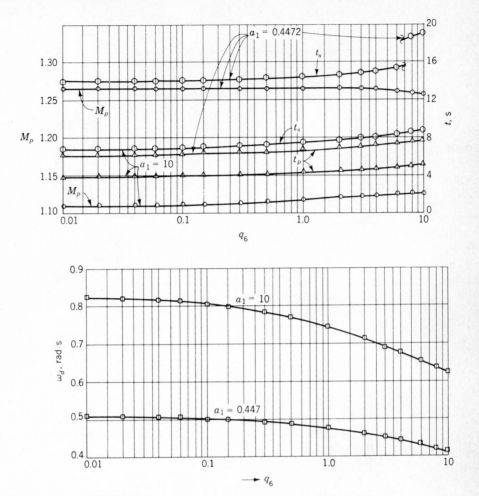

FIGURE D-3
The parameters of the time-response characteristics for a sixth-order system with one zero, $K_G = 2,400$, $z = 1$, $q_2 = 0.01$, $q_3 = 0.06$, $q_4 = 0.5$ and $q_5 = 5$. (*Top*) M_p, t_p, and t_s versus q_6; (*bottom*) ω_{de} versus q_6.

PROBLEMS

Chapter 2

2-1 Write the state equations for the circuit of Fig. 2-3.

2-2 For the spring, damper, and mass system shown: (a) draw the mechanical network; (b) from the mechanical network write the differential equations of performance; (c) write the state equations; (d) determine the transfer function x_1/f.

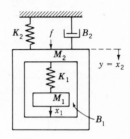

2-3 (a) Derive the differential equation relating the position $y(t)$ and the force $f(t)$. (b) Draw an analogous electric circuit. List all the analogous quantities. (c) Determine the transfer function $G(D) = y/f$. (d) Identify a suitable set of independent state variables. Write the state equation in matrix form.

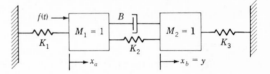

2-4 For the spring, damper, and moment-of-inertia system shown: (a) draw the mechanical network; (b) write the differential equations of performance; (c) draw the analogous electric circuit; (d) write the state equations.

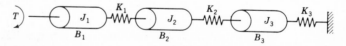

2-5 An electromagnetic actuator contains a solenoid which produces a magnetic force proportional to the current in the coil, $f = K_i i$. The coil has resistance and inductance. (a) Write the differential equations of performance. (b) Write the state equations.

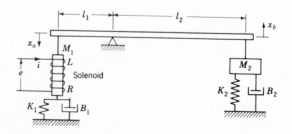

2-6 (a) Draw the mechanical network for the mechanical system shown. (b) Write the differential equations of performance. (c) Draw the analogous electric circuit in which force is analogous to current. (d) Write the state equations.

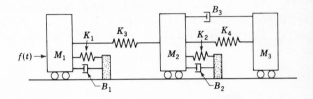

2-7 In the mechanical system shown, r_2 is the radius of the drum. (a) Write the necessary differential equations of performance for this system. (b) Obtain a differential equation expressing the relationship of the output x_a in terms of the input θ_1 and the corresponding transfer function. (c) Write the state equations with θ_1 as the input.

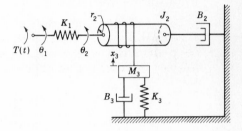

2-8 Write the state and system output equations for the rotational mechanical system of Fig. 2-19 with (a) T as the input, (b) θ_1 as the input.

2-9 For the hydraulic preamplifier write the differential equations of performance relating x_1 to y_1. (a) Neglect the load reaction. (b) Do not neglect load reaction.

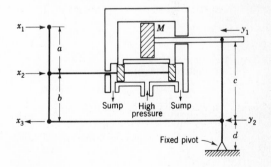

2-10 The mechanical load on the hydraulic translational actuator in Sec. 2-10 is changed to that below. Determine the new state equations and compare with Eq. (2-116).

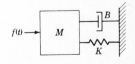

2-11 Write all the necessary equations to determine v_0. (*a*) Use nodal equations. (*b*) Use loop equations. (*c*) Write the state equations.

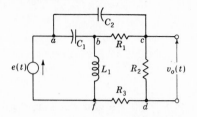

2-12 The figure represents a cylinder of inertia J_1 inside a sleeve of inertia J_2. There is viscous damping B_1 between the cylinder and the sleeve. The springs K_1 and K_2 are fastened to the inner cylinder. (*a*) Draw the mechanical network. (*b*) Write the system equations. (*c*) Draw the analogous electric circuit. (*d*) Write the state equations. (*e*) Determine the transfer function $G = \theta_2/T$.

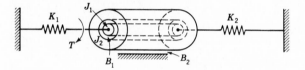

2-13 Write the (*a*) loop, (*b*) node, and (*c*) state equations for the circuit shown after the switch is closed.

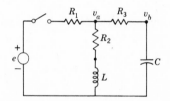

2-14 A sketch of a moving-coil microphone is shown. The diaphragm has the spring elastance K, mass M, and damping B. Fastened to the diaphragm is a coil which moves in the magnetic field produced by the permanent magnets. (*a*) Derive the differential equations of the system considering changes from the equilibrium conditions. (*b*) Draw an analogous electric circuit.

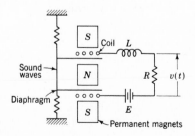

2-15 Write the (a) loop, (b) node, and (c) state equations for the circuit shown after the switch S is closed.

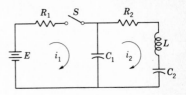

2-16 The two gear trains have an identical net reduction, have identical inertias at each stage, and are driven by the same motor. The number of teeth on each gear is indicated on the figures. At the instant of starting, the motor develops a torque T. Which system has the higher initial load acceleration?

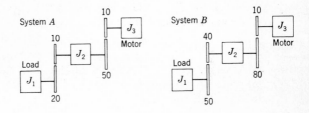

2-17 (a) Write the differential equations describing the motion of the following system, assuming small displacements. (b) Write the state equations.

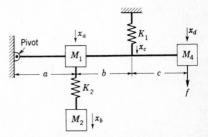

2-18 Most control systems require some type of motive power. One of the most commonly used units is the electric motor. Write the differential equation for the angular displacement of a moment of inertia, with damping, connected directly to a dc motor shaft when a voltage is suddenly applied to the armature terminals with the field separately energized.

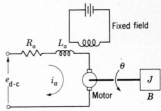

2-19 The circuit shown is in the steady state with the switch S closed. At time $t = 0$, S is opened. Write the necessary differential equations for determining $i_1(t)$.

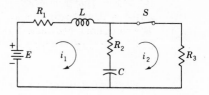

2-20 (*a*) Write the equations of motion for this system. (*b*) Using the physical energy variables, write the matrix state equation.

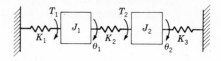

Chapter 3

3-1 (*a*) What are the initial values of current in all elements when the switch is closed? (*b*) Write the state equations. (*c*) Solve for the voltage across C as a function of time from the state equations.

$$R_1 = R_2 = 1 \text{ k}\Omega \qquad C = 50 \text{ } \mu\text{F} \qquad L = 1 \text{ H} \qquad E = 10 \text{ V}$$

3-2 In Prob. 2-4, the parameters have the following values:

$J_1 = 74{,}150 \text{ oz-in}^2 \qquad B_1 = 0.5 \text{ lb-ft/(rad/s)} \qquad K_1 = 0.5 \text{ lb-ft/rad}$

$J_2 = 1.0 \text{ lb-ft-s}^2 \qquad B_2 = 12.8 \text{ oz-ft/(rad/s)} \qquad K_2 = 8.0 \text{ oz-ft/rad}$

$J_3 = 1.0 \text{ slug-ft}^2 \qquad B_3 = 3.35 \text{ oz-in/(deg/s)} \qquad K_3 = 96.0 \text{ oz-in/rad}$

Solve for $\theta_3(t)$ if $T(t) = tu_{-1}(t)$.

3-3 In Prob. 2-5, the parameters have the following values:

$$M_1 = M_2 = 0.05 \text{ slug} \qquad L = 1 \text{ H} \qquad l_1 = 10 \text{ in}$$
$$K_1 = K_2 = 1.2 \text{ lb/in} \qquad R = 10 \text{ } \Omega \qquad l_2 = 20 \text{ in}$$
$$B_1 = B_2 = 15 \text{ oz/(in/s)} \qquad K_i = 24 \text{ oz/A}$$

Solve for $x_b(t)$ if $e(t) = u_{-1}(t)$ and the system is initially at rest.

3-4 In part (*a*) of Prob. 2-9, the parameters have the following values:

$$a = b = 6 \text{ in} \qquad C_1 = 6.0 \text{ (in/s)/in}$$
$$c = 10 \text{ in} \qquad d = 2 \text{ in}$$

Solve for $y_1(t)$ with $x_1(t) = 0.1u_{-1}(t)$ in and zero initial conditions.

3-5   (a) Given $r(t) = [(D + 1)(D^2 + 2D + 2)]c(t)$; with $r(t) = tu_{-1}(t)$ and all initial conditions zero, determine the complete solution with all constants evaluated. (b) $D^3y + 16D^2y + 85Dy + 150y = 37.5u$. Find $y(t)$ for $u(t) = u_{-1}(t)$, $D^2y(0) = Dy(0) = 0$, and $y(0) = -2$. One of the eigenvalues is $\lambda = -6$.

3-6   The rotational hydraulic transmission described in Sec. 2-11 has an inertia load coupled through a spring. The system equation is

$$4(D^2 + 1)x = (3D^2 + 3D + 1) D\theta_m$$

The system is originally stationary. With $x(t) = 5 \cos 5t$, find the motor velocity $\omega_m(t)$. Note: $\theta_m(0) = \theta_L(0) = \omega_L(0) = 0$, but $\omega_m(0) \neq 0$. Its value must be determined by using Eqs. (2-119) and (2-123).

3-7   A hydraulic motor with inertia load is driven by a variable-displacement pump. Determine (a) the undamped natural frequency ω_n; (b) the damping ratio ζ; the damped natural frequency of the system ω_d. The parameters have the following values:

$$S_p = 100 \text{ in}^3/\text{s} = d_p\omega_p \qquad L = 0.01 \ (\text{in}^3/\text{s})/(\text{lb/in}^2)$$
$$d_m = 1.0 \text{ in}^3/\text{rad} \qquad C = 1 \text{ in}^3$$
$$V = 20 \text{ in}^3 \qquad K_B = 2.5 \times 10^5 \text{ lb/in}^2$$
$$J = 200 \text{ lb-in}^2$$

3-8   The system shown is initially at rest. At time $t = 0$ the string connecting M to W is severed at X. Find $x(t)$.

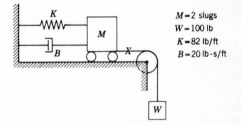

$M = 2$ slugs
$W = 100$ lb
$K = 82$ lb/ft
$B = 20$ lb-s/ft

3-9   Switch S_1 is open, and there is no energy stored in the circuit. (a) Write the state equations. (b) Draw a simulation diagram for these state equations. (c) Find $v_0(t)$ after switch S_1 is closed.

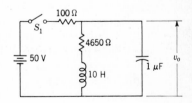

3-10 Solve the following differential equations. Assume zero initial conditions. Sketch the solutions.
(a) $D^2x + 16x = 1$ (b) $D^2x + 4Dx + 3x = 9$
(c) $D^2x + Dx + 4.25x = t + 1$ (d) $D^3x + 3D^2x + 4Dx + 2x = 10 \sin 10t$

3-11 For Prob. 2-15, solve for $i_2(t)$, where

$$E = 50 \text{ V} \qquad L = 1 \text{ H} \qquad C_1 = C_2 = 0.001 \text{ F}$$
$$R_1 = 100 \ \Omega \qquad R_2 = 100 \ \Omega$$

3-12   For the circuit of Prob. 2-15: (a) obtain the differential equation relating $i_2(t)$ to the input E; (b) write the state equations using the physical energy variables; (c) solve the state and output equations and compare with the solution of Prob. 3-11; (d) write the state and output equations using the phase variables, starting with the differential equation from part (a); (e) draw simulation diagrams corresponding to parts (b) and (d).

3-13   In Prob. 2-19, the parameters have the values $R_1 = R_2 = R_3 = 10\,\Omega$, $C = 1\,\mu F$, and $L = 100$ H. If $E = 100$ V, (a) find $i_1(t)$; (b) sketch $i_1(t)$ and label significant points; (c) determine the value of $T_s(\pm 2$ percent); (d) solve for $i_1(t)$ from the state equation.

3-14   Equation (3-101) is to be represented by a state equation using phase variables. For $w = n$, use Eqs. (3-102a) and (3-103) to develop the output equation $y = \mathbf{C}\mathbf{x} + \mathbf{B}u$.

3-15   (a) Solve the following equations. Show explicitly $\mathbf{\Phi}(t)$, $\mathbf{x}(t)$, and $\mathbf{y}(t)$.

$$\dot{\mathbf{x}} = \begin{bmatrix} -6 & 4 \\ -2 & 0 \end{bmatrix} \mathbf{x} + \begin{bmatrix} 0 \\ 1 \end{bmatrix} u$$

$$y = \begin{bmatrix} 1 & 0 \end{bmatrix}\mathbf{x}$$

$$x_1(0) = 2 \qquad x_2(0) = 0 \qquad u(t) = u_{-1}(t)$$

(b) Draw the simulation diagram.

3-16   A single-degree-of-freedom representation of the rolling dynamics of an aircraft, together with a first-order representation of the aileron servomotor, is given by

$$p(t) = \dot{\phi}(t) \qquad J_x \dot{p}(t) = L_{\delta_A}\delta_A(t) + L_p p(t)$$

$$T\dot{\delta}_A(t) = (\delta_A)_{\text{comm}}(t) - \delta_A(t)$$

(a) Derive a state-variable representation of the aircraft system.

$$(\delta_A)_{\text{comm}}(t) = KV_{\text{ref}} - \mathbf{P}\mathbf{x}(t)$$

where $\mathbf{x}(t)$ = state vector of aircraft system
      $\mathbf{P}$ = matrix of correct dimensions and with constant nonzero elements p_{ij}
      K = scalar gain

(b) Determine the aircraft state equations in response to the reference input V_{ref}.

3-17   Given:

$$D^4 y + 10D^3 y + 35D^2 y + 50Dy + 24y = 5D^2 u + 12u$$

Obtain state and output equations using (a) phase variables, (b) general programming, and (c) standard-form programming.

3-18   For the mechanical system of Fig. 2-11 the state equation for Example 2, Sec. 2-6, is given in phase-variable form. With the input as $u = x_a$, the state variables are $x_1 = x_b$ and $x_2 = \dot{x}_b$. Use $M = 1$, $K = 12$, and $B = 7$. The initial conditions are $x_b(0) = 1$ and $\dot{x}_b(0) = -2$. (a) Find the homogeneous solution for $\mathbf{x}(t)$. (b) Find the complete solution with $u(t) = u_{-1}(t)$. (c) Draw the simulation diagram.

Chapter 4

4-1   Take the inverse Laplace transform of

(a) $F(s) = \dfrac{10}{(s + 1)^2(s^2 + 2s + 2)}$ (b) $F(s) = \dfrac{13}{s(s^2 + 4s + 13)(s + 1)}$

4-2 Find $x(t)$ for

(a) $X(s) = \dfrac{1}{s^3 + 7s^2 + 20s + 24}$

(b) $X(s) = \dfrac{10}{s^5 + 4s^4 + 14s^3 + 66s^2 + 157s + 130}$

(c) $X(s) = \dfrac{0.9524(s^2 + 2s + 2.1)}{s(s + 1)(s^2 + 2s + 2)}$ (d) $X(s) = \dfrac{1.82(s + 1.1)}{s(s + 1)(s^2 + 2s + 2)}$

(e) $X(s) = \dfrac{20(s + 1)(s + 10)}{s(s + 2)^3(s^2 + 6s + 10)}$ (f) Sketch $x(t)$ for parts (a) to (e).

4-3 Given an ac servomotor with inertia load, find $\omega(t)$ by (a) the classical method; (b) the Laplace-transform method.

$$\omega(0) = 10^4 \text{ rad/s} \qquad K_\omega = -3 \times 10^{-3} \text{ oz-in/(rad/s)}$$
$$K_c = 1.2 \text{ oz-in/V} \qquad e(t) = 5u_{-1}(t) \text{ V}$$
$$J = 11.59 \text{ oz-in}^2$$

4-4 Repeat the following problems using the Laplace transform:
    (a) Prob. 3-1   (b) Prob. 3-3   (c) Prob. 3-5
    (d) Prob. 3-8   (e) Prob. 3-12   (f) Prob. 3-13(d)
    (g) Prob. 3-15   (h) Prob. 3-16

4-5 Find the partial-fraction expansions of the following:

(a) $F(s) = \dfrac{5}{(s + 1)(s + 5)}$ (b) $F(s) = \dfrac{10}{s(s + 2)(s + 5)}$

(c) $F(s) = \dfrac{10}{s(s^2 + 2s + 10)}$ (d) $F(s) = \dfrac{2(s + 2)}{s^2(s + 1)(s + 4)}$

(e) $F(s) = \dfrac{30}{s(s + 3)(s^2 + 6s + 10)}$ (f) $F(s) = \dfrac{30(s + 1)}{s(s + 3)(s^2 + 6s + 10)}$

(g) $F(s) = \dfrac{10(s + 1.01)}{s(s + 1)(s^2 + 2s + 10)}$ (h) $F(s) = \dfrac{10(s^2 + 2.2s + 10.21)}{s(s + 1)(s^2 + 2s + 10)}$

(i) $F(s) = \dfrac{20}{s(s^2 + 2s + 2)(s^2 + 6s + 10)}$ (j) $F(s) = \dfrac{52}{s(s^2 + 2s + 2)(s^2 + 10s + 26)}$

(k) $F(s) = \dfrac{10(s^2 + 2s + 2.21)}{s(s^2 + 6s + 10)(s^2 + 2s + 2)}$ (l) $F(s) = \dfrac{100}{s^2(s + 10)(s^2 + 6s + 10)}$

4-6 Solve the differential equations of Prob. 3-10 by means of the Laplace transform.

4-7 Write the Laplace transforms of the following equations and solve for $x(t)$; the initial conditions are given to the right.
    (a) $Dx + 5x = 0$ $x(0) = -1$
    (b) $D^2x + 2Dx + 5x = 10$ $x(0) = 2, Dx(0) = -4$
    (c) $D^2x + 3Dx + 2x = t$ $x(0) = 0, Dx(0) = -2$
    (d) $D^3x + 4D^2x + 8Dx + 4x = \sin 5t$ $x(0) = -4, Dx(0) = 1, D^2x(0) = 0$

4-8 Determine the final value for:
    (a) Prob. 4-1   (b) Prob. 4-2   (c) Prob. 4-5

4-9   Determine the initial value for:
  (a) Prob. 4-1   (b) Prob. 4-2   (c) Prob. 4-5
4-10   For the functions of Prob. 4-5, plot M vs. ω and α vs. ω. Use the Spirule.
4-11   Find the complete solution of $x(t)$ with zero initial conditions for

$$(D^2 + 2D + 2)(D + 4)x = (D + 2)f(t)$$

Use the Spirule to evaluate all residues. The forcing function $f(t)$ is
  (a) $u_0(t)$ (b) $10u_{-1}(t)$ (c) $tu_{-1}(t)$
4-12   A linear system is described by

$$\dot{\mathbf{x}} = \begin{bmatrix} -2 & 1 \\ 3 & -4 \end{bmatrix} \mathbf{x} + \begin{bmatrix} 0 \\ 1 \end{bmatrix} u$$

$$y = \begin{bmatrix} 1 & 0 \end{bmatrix} \mathbf{x}$$

where $u(t) = u_{-1}(t)$ and the initial conditions are $x_1(0) = 0$, $x_2(0) = 1$. (a) Using Laplace transforms, find $\mathbf{X}(s)$. Combine the elements of this matrix; i.e., put them over a common denominator. (b) Find the transfer function $G(s)$.

4-13   A linear system is represented by

$$\dot{\mathbf{x}} = \begin{bmatrix} -6 & 4 \\ -2 & 0 \end{bmatrix} \mathbf{x} + \begin{bmatrix} 1 \\ 1 \end{bmatrix} u \qquad y = \begin{bmatrix} 1 & 0 \\ 1 & 1 \end{bmatrix} \mathbf{x}$$

(a) Find the complete solution for $\mathbf{y}(t)$ when $u(t) = u_{-1}(t)$, $x_1(0) = 1$, and $x_2(0) = 0$. (b) Determine the transfer functions. (c) Draw a block diagram representing the system.

4-14   A system is described by

$$\dot{\mathbf{x}} = \begin{bmatrix} -3 & -1 \\ 2 & 0 \end{bmatrix} \mathbf{x} + \begin{bmatrix} 1 \\ 0 \end{bmatrix} u \qquad y = x_1$$

(a) Find $\mathbf{x}(t)$ with $\mathbf{x}(0) = 0$ and $u = u_{-1}(t)$. (b) Determine the transfer function $G(s) = Y(s)/U(s)$.

Chapter 5

5-1   For the temperature-control system of Fig. 5-1, some of the pertinent equations are $b = K_b\theta$, $f_s = K_s i_s$, $q = K_q x$, $\theta = K_c q/(D + a)$. The solenoid coil has resistance R and inductance L. The solenoid armature and valve have mass M, damping B, and a restraining spring K. (a) Determine the transfer function of each block in Fig. 5-1b. (b) Determine the forward transfer function $G(s)$. (c) Write the state and output equations in matrix form.
5-2   Find an example of a practical closed-loop control system not covered in this book. Briefly describe the system and show a block diagram.
5-3   A satellite-tracking system is shown in the schematic diagram. The transfer function of the tracking receiver is

$$\frac{V_1}{\Theta_{\text{comm}} - \Theta_L} = \frac{5}{1 + s/40\pi}$$

The following parameters apply:
    K_a = gain of servoamplifier = 60
    K_t = tachometer constant = 0.04 V-s
    K_T = motor torque constant = 0.5 N-m/A
    K_b = motor back-emf constant = 0.75 V-s
    J_L = antenna inertia = 2,880 kg-m^2
    J_m = motor inertia = 7×10^{-3} kg-m^2
  1:N = gearbox stepdown ratio = 1:12,000

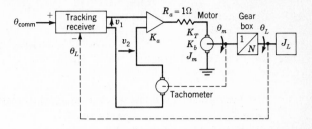

(a) Draw a detailed block diagram showing all the variables.

(b) Derive the transfer function $\Theta_L(s)/\Theta_{comm}(s)$.

5-4   A photographic control system is shown in simplified form in the diagram. The aperture slide moves to admit the light from the high-intensity lamp to the sensitive plate, the illumination of which is a linear function of the exposed area of the aperture. The maximum area of the aperture is 4 m². The plate is illuminated by 1 candela (cd) for every square meter of aperture area. This luminosity is detected by the photocell, which provides an output voltage of 1 V/cd. The motor torque constant is 1 N-m/A, the motor inertia is 0.1 kg-m², and the motor back emf constant is 1 V-s. The viscous friction is 0.2 N/(m/s), and the amplifier gain is 50 A/V. (a) Draw the block diagram of the system with the appropriate transfer function inserted in each block. (b) Derive the overall transfer function.

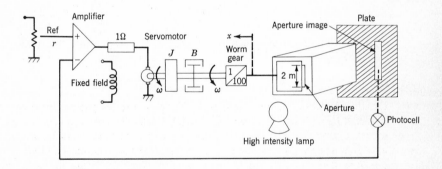

5-5   The longitudinal motion of an aircraft is represented by the vector differential equation

$$\dot{x} = \begin{bmatrix} -0.08 & 1 & -0.01 \\ -8.0 & -0.05 & -6.0 \\ 0 & 0 & -10 \end{bmatrix} x + \begin{bmatrix} 0 \\ 0 \\ 10 \end{bmatrix} \delta_E$$

where x_1 = angle of attack
        x_2 = rate of change of pitch angle
        x_3 = incremental elevator angle
        δ_E = control input into the elevator actuator

Derive the transfer function relating the system output, rate of change of pitch angle, to the control input into the elevator actuator.

5-6   (a) Use the hydraulic valve and power piston in a closed-loop position-control system. Derive the transfer function of the system. (b) Draw a diagram of a control system for the elevators on an airplane. Use a hydraulic actuator.

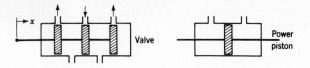

5-7 In the circuit shown, both armature and field voltages are varied to control the output θ_m. (a) Draw a block diagram that relates both inputs, v_1 and v_2, to the output θ_m. This is a nonlinear system containing a multiplication. The block diagram should be labeled in the D-operator form. (b) Linearize the system and use Laplace transforms. Draw the new block diagram. Multiplication is replaced by addition for small excursions of the variables by using

$$df(x, y) = \frac{\partial f}{\partial x}\, dx + \frac{\partial f}{\partial y}\, dy = K_x\, dx + K_y\, dy$$

(c) Write the state equation for the linearized system.

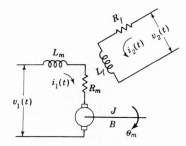

5-8 In the diagram $[q_c]$ represents compressibility flow, and the pressure p is the same in both cylinders. Assume there is no leakage flow around the pistons. Draw a block diagram that relates the output $Y(s)$ to the input $X(s)$.

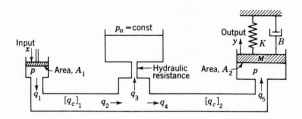

5-9 The diagram represents a gyroscope that is often used in autopilots, automatic gunsights, etc. Assume that the speed of the rotor is constant and that the total developed torque about the output axis is

$$T_0 = H \frac{d\theta_i}{dt}$$

where H is a constant. The inner gimbal has a moment of inertia J about the output axis. Draw a block diagram that relates the output $\Theta_o(s)$ to the input $\Theta_i(s)$.

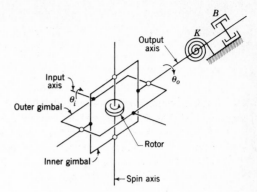

5-10 The figure shows a system for controlling the output movement $y(t)$ of a hydraulic linear actuator. The load reaction on the actuator can be considered negligible. The solenoid constant is K_c (lb/A), the output potentiometer constant is K_p (V/in), and the output of the tachometer is given by

$$e_t = K_t \frac{dy}{dt} \quad \text{V}$$

where K_t has the units of volts per (inch per second). (*a*) Draw a detailed block diagram that shows all variables of the system explicitly. (*b*) Derive the transfer functions in terms of the Laplace operator for each block in the block diagram. (*c*) Draw the SFG and use the Mason gain formula to determine the overall transfer function. *Note:* The lever at the input to the hydraulic unit has no fixed pivot. Consider that x, x_1, and y have only horizontal motion.

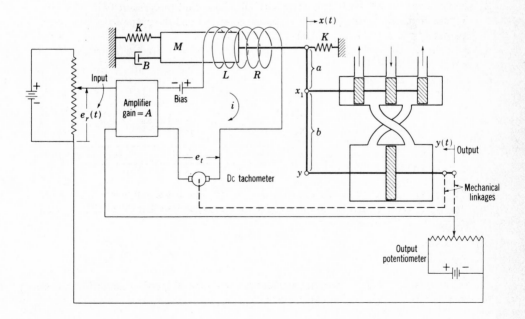

5-11 For the control system shown: (1) The force of attraction on the solenoid is given by $f_c = K_c i_c$, where K_c has the units pounds per ampere. (2) The voltage appearing across the generator field is given by $e_f = K_x x$, where K_x has the units volts per inch and x is in inches. (3) When the voltage across the solenoid coil is zero, the spring K_s is unstretched and $x = 0$. (*a*) Derive all necessary equations relating all the variables in the system. (*b*) Draw a block diagram for the control system. The diagram should include enough blocks to indicate specifically the variables $I_c(s)$, $X(s)$, $I_f(s)$, $E_g(s)$, and $T(s)$. Give the transfer function for each block in the diagram. (*c*) Draw an SFG for this system. (*d*) Determine the overall transfer function. (*e*) Write the system state and output equations in matrix form.

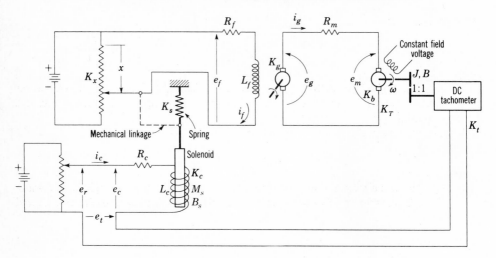

5-12 Given

$$G(s) = \frac{b_1 s + b_0}{(s - \sigma)^2 + \omega_d^{\,2}}$$

write the state equation with the **A** matrix in modified canonical form so that all elements are real. Draw the SFG.

5-13 The state and output equations of a system are

$$\dot{\mathbf{x}} = \begin{bmatrix} -6 & 1 & 0 \\ -11 & 0 & 1 \\ -6 & 0 & 0 \end{bmatrix} \mathbf{x} + \begin{bmatrix} 0 \\ 3 \\ 6 \end{bmatrix} u \qquad y = [1 \quad 0 \quad 0]\mathbf{x}$$

(*a*) Draw a state transition SFG for these equations. (*b*) Determine the eigenvalues. (*c*) Determine the modal matrix **T** which uncouples the states. (*d*) Write the new state and output equations in matrix form. (*e*) Draw a simulation diagram for the new equations.

5-14 For the system of Prob. 4-14: (*a*) draw an SFG; (*b*) from the transfer function determine a new state equation and draw the SFG using phase variables; (*c*) change the state equation to normal form and draw the SFG.

5-15 Use the state and output equations of Prob. 3-13. (*a*) Draw a simulation diagram for the given equations. (*b*) Determine the modal matrix **T**. Show that $\mathbf{T}^{-1}\mathbf{AT} = \mathbf{\Lambda}$. (*c*) Convert the equations using the similarity transformation $\mathbf{x} = \mathbf{Tz}$. (*d*) Draw the simulation diagram for (*c*).

5-16 A system has a state-variable equation $\dot{\mathbf{x}} = \mathbf{A}_p\mathbf{x} + \mathbf{B}_p u$, where $\mathbf{x}$ is the 3×1 state vector and u is the scalar control input.

$$\mathbf{A}_p = \begin{bmatrix} -1 & 10 & 0 \\ 0 & 0 & 1 \\ 0 & -20 & -10 \end{bmatrix} \quad \mathbf{B}_p = \begin{bmatrix} 0 \\ 0 \\ 5 \end{bmatrix}$$

(a) Find a nonsingular matrix $\mathbf{T}$ which transforms the state variables into phase variables. Be sure to satisfy the $\mathbf{B}_0$ matrix for the phase variable representation. (This will lead to a unique matrix $\mathbf{T}$.) (b) Write the new state equation of the system in terms of the phase variables. (c) Draw a state-variable diagram representing the system.

5-17 Given

$$\dot{\mathbf{x}} = \begin{bmatrix} -1 & 2 & -1 \\ 0 & -2 & 0 \\ 1 & 0 & -2 \end{bmatrix} \mathbf{x} + \begin{bmatrix} 0 \\ 0 \\ 1 \end{bmatrix} u$$

(a) Find a modified matrix $\mathbf{\Lambda}_m$ so that its elements are all real. (b) Determine the modified modal matrix $\mathbf{T}_m$. (c) Draw a simulation diagram for the transformed equations.

5-18 For the system described by the state equation

$$\dot{\mathbf{x}} = \begin{bmatrix} -3 & -1 \\ 2 & 0 \end{bmatrix} \mathbf{x} + \begin{bmatrix} 1 \\ 0 \end{bmatrix} u$$

(a) Derive the system transfer function if $y = x_1$. (b) Draw an appropriate state-variable diagram. (c) For zero initial state and a unit step input evaluate $x_1(t)$ and $x_2(t)$.

Chapter 6

6-1 For Prob. 5-3, (a) use the Routh criterion to check the stability of the system. (b) Compute the steady-state tracking error of the system in response to a command input which is a ramp of $0.04°/s$. (c) Repeat (b) with the gain of the tracking receiver increased from 5 to 50.

6-2 For each of the following cases, determine the range of values of K for which the response $c(t)$ is stable, where the driving function is a step function. Determine the roots that lie on the imaginary axis that yield sustained oscillations.

(a) $C(s) = \dfrac{K}{s[s(s + 2)(s^2 + s + 10) + K]}$

(b) $C(s) = \dfrac{K(s + 1)}{s[s^2(s^3 + 3s^2 + 2s + 4) + K(s + 1)]}$

(c) $C(s) = \dfrac{K(s + 5)}{s[(s + 5)(s^2 + 8s + 20) + K(s + 5)]}$

(d) $C(s) = \dfrac{K}{s[(s + 1)(s + 2)(s + 5) + K]}$

(e) $C(s) = \dfrac{K}{s[s^3 + 6s^2 + 11s + (6 + K)]}$

(f) $C(s) = \dfrac{K(s + 4)}{s[s(s + 1)(s + 2)(s + 3) + K(s + 4)]}$

6-3   Factor the following equations:
(a) $s^3 + 6s^2 + 12s + 9 = 0$ (b) $s^3 + 4s^2 + 6s + 4 = 0$
(c) $s^4 + 2s^3 + 2s^2 + 3s + 6 = 0$ (d) $s^4 + 6s^3 + 26s^2 + 56s + 80 = 0$
(e) $s^3 + 8s^2 + 17s + 10 = 0$ (f) $s^3 + 7s^2 + 16s + 10 = 0$
(g) $s^4 + 7s^3 + 9s^2 - 7s - 10 = 0$ (h) $s^4 + 7s^3 + 12.2s^2 + 11.05s = 0$
(i) $s^5 + 3s^4 + 28s^3 + 226s^2 + 600s + 400 = 0$

6-4   Use Routh's criterion to determine the number of roots in the right-half s plane for the equations of Prob. 6-3.

6-5   A unity-feedback system has the forward transfer function

$$G(s) = \frac{K(0.1s + 1)}{s(s + 2)(s + 3)}$$

To obtain the best possible system error coefficients, you might think of using the highest possible gain K. Do stability requirements limit this choice of K?

6-6   The equation relating the output $y(t)$ of a control system to its input is
(a) $[s^5 + s^4 + 2s^3 + s^2 + (K + 1)s + (K + 2)]Y(s) = 5X(s)$
(b) $[s^3 + 7s^2 + 16s + 10 + K(s + 2)]Y(s) = X(s)$
Determine the range of K for stable operation of the system. Consider both positive and negative values of K.

6-7   The output of a control system is related to its input $r(t)$ by

$$[s^4 + 2s^3 + 2s^2 + (2 + K)s + K]C(s) = K(s + 1)R(s)$$

Determine the range of K for stable operation of the system. Consider both positive and negative values of K.

6-8
$$F(s) = \frac{1}{s[(s^4 + 2s^3 + 3s^2 + s + 1) + K(s + 1)]}$$

K is real but may be positive or negative. (a) Find the range of values of K for which the time response is stable. (b) Select a value of K which will produce imaginary poles for $F(s)$. Find these poles. What is the physical significance of imaginary poles as far as the time response is concerned?

6-9   (a) For the system shown derive the transfer function $Y(s)/U(s)$. (b) If $u(t) = u_{-1}(t)$, sketch the responses of the system variables $x(t)$, $z(t)$, and $y(t)$. (c) A linear system is said to be stable if, in response to a bounded input, the system produces a bounded output. Is this system stable? Is it observable? Discuss your answer briefly.

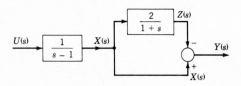

6-10   For the system shown, (a) find $C(s)/R(s)$. (b) What type of system does $C(s)/E(s)$ represent? (c) Find the step, ramp, and parabolic error coefficients. (d) Find the steady-state value of $c(t)$ if $r(t) = 10u_{-1}(t)$ rad.

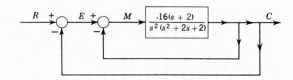

6-11 For the feedback control system shown

$$G(s) = \frac{12(s + 4)}{s(s + 1)(s + 3)(s^2 + 2s + 2)}$$

(a) Determine the step, ramp, and parabolic error coefficients (K_p, K_v, and K_a) for this system. (b) Using the appropriate error coefficients from part (a), determine the steady-state actuating signal $e(t)_{ss}$ when $r(t) = (16 + 2t)u_{-1}(t)$.

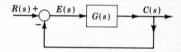

6-12 (a) Determine the open-loop transfer function $G(s)H(s)$ of Fig. (a). (b) Determine the overall transfer function. (c) Figure (b) is an equivalent block diagram of Fig. (a). What must the transfer function $H_x(s)$ be in order for Fig. (b) to be equivalent to Fig. (a)? (d) Figure (b) represents what type of system? (e) Determine the system error coefficients of Fig. (b). What is the significance of the minus sign for the ramp error coefficient? (f) If $r(t) = u_{-1}(t)$, determine the final value of $c(t)$. (g) What are the values of $e_x(t)_{ss}$ and $e(t)_{ss}$?

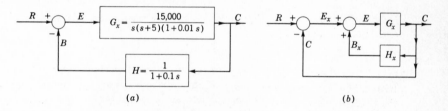

(a)   (b)

6-13 Repeat Prob. 6-12 with $H(s)$ replaced in Fig. (a) by

$$H(s) = \frac{s + 5}{s + 1}$$

6-14 Find the step, ramp, and parabolic error coefficients for unity-feedback systems which have the following forward transfer functions:

(a) $G(s) = \dfrac{10}{(0.4s + 1)(0.5s + 1)}$

(b) $G(s) = \dfrac{108}{s^2(s^2 + 4s + 4)(s^2 + 3s + 12)}$

(c) $G(s) = \dfrac{20}{s(s + 2)(0.4s + 1)}$

(d) $G(s) = \dfrac{14(s + 3)}{s(s + 6)(s^2 + 2s + 2)}$

(e) $G(s) = \dfrac{20(s + 3)}{(s + 2)(s^2 + 2s + 2)}$

(f) $G(s) = \dfrac{11(s + 30)}{s^3(s + 1)(0.2s + 1)(s^2 + 5s + 15)}$

(g) $G(s) = \dfrac{-6(1 + 0.04s)}{s(1 + 0.1s)(1 + 0.1s + 0.01s^2)}$

(h) $G(s) = \dfrac{4(s^2 + 10s + 50)}{s^2(s + 5)(s^2 + 6s + 10)}$

(i) $G(s) = \dfrac{100(s - 1)}{s(s + 2)(1 + 0.25s)}$

(j) $G(s) = \dfrac{39}{s^2(s - 1)(s^2 + 6s + 13)}$

6-15 For Prob. 6-14, find $e(t)_{ss}$ by use of the error coefficients, with the following inputs:
(a) $r(t) = 5$ (b) $r(t) = 2t$ (c) $r(t) = t^2$

6-16 A unity-feedback control system has

$$G(s) = \frac{K_1}{s(s + 1)(0.5s + 1)}$$

and $r(t) = 3t$. (a) If $K_1 = 2$ s^{-1}, determine $e(t)_{ss}$. (b) It is desired that for a ramp input $e(t)_{ss} \leq 0.1$. What minimum value must K_1 have for this condition to be satisfied? (c) For the value of K_1 determined in part (b), is the system stable?

6-17 (a) Derive the ratio $G(s) = C(s)/E(s)$. (b) Based upon $G(s)$, the figure has the characteristic of what type of system? (c) Derive the control ratio for this control system.

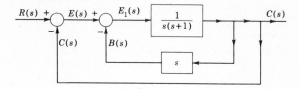

6-18 A unity-feedback system has the forward transfer function

$$G(s) = \frac{K_1(2s + 1)}{s(4s + 1)(s + 1)^2}$$

The input $r(t) = 1 + t$ is applied to the system. (a) It is desired that the steady-state value of the error be equal to or less than 0.1 for the given input function. Determine the minimum value that K_1 must have to satisfy this requirement. (b) By use of Routh's stability criterion, determine whether the system is stable for the minimum value of K_1 determined in part (a).

6-19 For the system described by

$$D^2y + 6Dy + 5y = 2u$$

(a) draw the simulation diagram. (b) Write the matrix state and output equations from the simulation diagram. (c) Rewrite the state equation and the output equation in uncoupled form. (d) From part (c) determine whether the system is completely controllable and/or observable.

6-20 Consider the following system:

$$\dot{x} = \begin{bmatrix} 0 & 1 & 0 \\ 0 & 0 & 1 \\ -6 & -11 & -6 \end{bmatrix} x + \begin{bmatrix} 1 \\ 1 \\ 0 \end{bmatrix} u \qquad y = \begin{bmatrix} 1 & 1 & 0 \end{bmatrix} x$$

Determine whether the system is (a) completely controllable, (b) completely observable, (c) stable.

6-21 A unity-feedback control system has the following forward transfer function:

$$G(s) = \frac{K(s + a)}{s(s + b)(s^2 + cs + d)}$$

The nominal values are $K = 20$, $a = 3$, $b = 6$, $c = 4$, and $d = 8$. Determine the sensitivity with respect to (a) K, (b) a, (c) b, (d) c, (e) d.

Chapter 7

7-1 Determine the pertinent geometrical properties and *sketch* the root locus for the following transfer functions for both positive and negative values of K:

(a) $G(s)H(s) = \dfrac{K}{(s + 1)(s + 5)}$

(b) $G(s)H(s) = \dfrac{K}{(s + 2)(s^2 + 8s + 20)}$

(c) $G(s)H(s) = \dfrac{K(s + 2)}{(s + 1)(s^2 + 6s + 10)}$

(d) $G(s)H(s) = \dfrac{K_0}{(1 + 0.5s)(1 + 0.2s)(1 + s)^2}$

Determine the range of values of K for which the closed-loop system is stable.

7-2 Repeat Prob. 7-1 for the following open-loop transfer functions:

(a) $G(s)H(s) = \dfrac{Ks}{(s + 1)^2}$

(b) $G(s)H(s) = \dfrac{K(s + 1)}{s^2 + 2s + 2}$

(c) $G(s)H(s) = \dfrac{K(s - 1)}{s(s^2 + 4s + 4)}$

(d) $G(s)H(s) = \dfrac{K(s^2 + 8s + 20)}{(s + 2)(s^2 + 2s + 2)}$

7-3 Draw the root locus of the following:

(a) $G(s)H(s) = \dfrac{K}{(s + 5)(s^2 + 4s + 7)}$

(b) $G(s)H(s) = \dfrac{K}{s(s^2 + 6s + 12)}$

(c) $G(s)H(s) = \dfrac{K}{(s^2 + 1)(s^2 + 4s + 5)}$

(d) $G(s)H(s) = \dfrac{K}{s(s^2 + 2s + 2)(s^2 + 6s + 10)}$

Does the root locus cross any of the asymptotes?

7-4 A system has the following transfer functions:

$$G(s) = \dfrac{K}{s(s^2 + 8s + 20)} \qquad H(s) = 1$$

(a) Plot the root locus. (b) A damping ratio of 0.58 is required for the dominant roots. Find $C(s)/R(s)$. The denominator should be in factored form. (c) With a unit step input, find $c(t)$. (d) Evaluate graphically $C(j\omega)/R(j\omega)$ vs. ω. Plot the magnitude and angle of $C(j\omega)/R(j\omega)$ vs. ω.

7-5 A feedback control system with unity feedback has a transfer function

$$G(s) = \dfrac{K_1}{s(1 + 0.02s)(1 + 0.01s)}$$

(a) Plot the locus of the roots of $1 + G(s) = 0$ as the loop sensitivity K is varied. (b) Determine the value of K_1 that just makes the system unstable. (c) From the root-locus plot determine the value of K_1 for a $\zeta = 0.5$. (d) For the value of K_1 found in part (c), determine $e(t)$ for $r(t) = u_{-1}(t)$. (e) Plot $M(j\omega)$ vs. ω for the closed-loop system. Determine the data graphically as indicated in Sec. 7-10.

7-6   (a) Sketch the root locus for the control system having the following open-loop transfer function. (b) Calculate the value of K_1 that causes instability. (c) Determine $C(s)/R(s)$ for $\zeta = 0.3$. (d) Evaluate $c(t)$ for $r(t) = u_{-1}(t)$.

$$G(s) = \frac{K_1(1 + 0.04s)}{s(1 + 0.1s)(1 + 0.1s + 0.0125s^2)}$$

7-7   (a) Sketch the root locus for a control system having the following forward and feedback transfer functions:

$$G(s) = \frac{K_2(1 + s/5)}{s^2(1 + s/12)} \qquad H(s) = 1 + \frac{s}{12}$$

(b) Choose closed-loop pole locations which produce a time constant $T = \frac{1}{3}$ s for the complex roots and indicate these locations on the root locus. Using these locations, write the factored form of the closed-loop transfer function.

7-8   For

$$G(s)H(s) = \frac{K_x}{(s + 5)(0.4s + 1.2)(0.5s^2 + 2s + 4)}$$

determine the value of K from the root locus that makes the closed-loop system a perfect oscillator.

7-9   A unity-feedback control system has the transfer function

$$G(s) = \frac{K_x}{(s + 1)(1 + 0.2s)(s^2 + 6s + 13)}$$

(a) Sketch the root locus for positive and negative values of K. (b) For what values of K does the system become unstable? (c) Determine the value of K for which all the roots are equal.

7-10   A unity-feedback control system has the transfer function

$$G(s) = \frac{K_1(1 - s)}{s(1 + s)(1 + 0.5s)(1 + 0.25s)}$$

(a) Sketch the root locus for positive and negative values of K_1. (b) What range of values of K_1 makes the system unstable?

7-11   For positive values of gain, sketch the root locus for unity-feedback control systems having the following open-loop transfer functions. For what value or values of gain does the system become unstable in each case?

(a) $G(s) = \dfrac{K(s^2 + 6s + 13)}{s(s + 1)(s + 5)^2}$ (b) $G(s) = \dfrac{K_0}{1 + 0.2s}$

(c) $G(s) = \dfrac{K_0(1 - 100s)}{(1 + 10s)(1 + 0.001s)(1 + s + s^2)}$

(d) $G(s) = \dfrac{K(s^2 + 4s + 5)}{s^2(s + 1)(s + 3)}$ (e) $G(s) = \dfrac{K(s + 2)^2}{s(s^2 - 2s + 2)}$

(f) $G(s) = \dfrac{K(s + 2)}{(s + 18)(s^2 + 2.4s + 12.33)}$

7-12   For the system of part (f) of Prob. 7-11 determine $C(s)/R(s)$ for $\zeta = 0.707$. Select the best roots.

7-13 A non-unity-feedback control system has the transfer functions

$$G(s) = \frac{K_G 1 + s/3.9)}{(1 + s/10)[1 + 2(0.7)s/23 + (s/23)^2][1 + 2(0.49)s/7.6 + (s/7.6)^2]}$$

$$H(s) = \frac{K_H(1 + s/10)}{1 + 2(0.89)s/42.7 + (s/42.7)^2}$$

(a) Sketch the root locus for the system, using as few trial points as possible. Determine the angles and locations of the asymptotes and the angles of departure of the branches from the open-loop poles. (b) Estimate the gain $K_G K_H$ at which the system becomes unstable, and determine the approximate locations of all the closed-loop poles for this value of gain. (c) Using the closed-loop configuration determined in part (b), write the expression for the closed-loop transfer function of the system.

7-14 A non-unity-feedback control system has the transfer functions

$$G(s) = \frac{10A(s^2 + 8s + 20)}{s(s + 4)} \qquad H(s) = \frac{0.2}{s + 2}$$

(a) Determine the value of the amplifier gain A that will produce complex roots having the *minimum* possible value of ζ. (b) Express $C(s)/R(s)$ in terms of its poles, zeros, and constant term.

Chapter 8

8-1 For each of the transfer functions

$$(1)\ G(s) = \frac{K_1}{s(1 + 0.2s)(1 + 0.5s)} \qquad K_1 = 10$$

$$(2)\ G(s) = \frac{K_1}{s(1 + 0.04s)(1 + s/50 + s^2/2,500)} \qquad K_1 = 1$$

(a) draw the log magnitude (exact and asymptotic) and phase diagrams; (b) from the curves of part (a) obtain the data for plotting the direct polar plots; (c) from the curves of part (a) obtain the data for plotting the inverse polar plots.

8-2 For each of the transfer functions

$$(1)\ G(s) = \frac{20}{(1 + 0.2s)(1 + 0.4s)(1 + s)} \qquad (2)\ G(s) = \frac{400}{s(s^2 + 16s + 400)}$$

$$(3)\ G(s) = \frac{2(1 + 0.4s)}{s^2(1 + 0.1s)(1 + 0.05s)} \qquad (4)\ G(s) = \frac{10}{(1 + s)(1 + 0.1s)}$$

$$(5)\ G(s) = \frac{2}{s^2(1 + 0.1s)(1 + 0.4s)} \qquad (6)\ G(s) = \frac{2(1 + 0.3s)}{s(1 + 0.1s)(1 + 0.4s)}$$

(a) draw the log magnitude (exact and asymptotic) and phase diagrams; (b) from the curves of part (a) obtain the data for plotting the direct polar plots; (c) from the curves of part (a) obtain the data for plotting the inverse polar plots.

8-3 Plot to scale the log magnitude and angle vs. log ω curves for $G(j\omega)$. Is it an integral or a derivative compensating network?

$$(a)\ G(j\omega) = \frac{10(1 + j2\omega)}{1 + j6\omega} \qquad (b)\ G(j\omega) = \frac{10(1 + j10\omega)}{(1 + j\omega)}$$

8-4   A control system with unity feedback has the forward transfer function, with $K_1 = 1$,

$$G(s) = \frac{K_1}{s(1 + 0.1s)(1 + 30s/625 + s^2/625)}$$

(a) Draw the log magnitude and angle diagrams. (b) Draw the polar plot of $G'(j\omega)$. Determine all key points of the curve. (c) Plot the polar plot of $G'(j\omega)^{-1}$. Determine all key points of the curve.

8-5   A system has

$$G(s) = \frac{2(1 + T_2s)(1 + T_3s)}{s^2(1 + T_1s)^2(1 + T_4s)}$$

where $T_1 = 4$, $T_2 = 1$, $T_3 = \frac{1}{2}$, $T_4 = \frac{1}{4}$. (a) Draw the asymptotes of $G(j\omega)$ on a decibel vs. log ω plot. Label the corner frequencies on the graph. (b) What is the *total* correction from the asymptotes at $\omega = 2$?

8-6   (a) What characteristic must the plot of magnitude in decibels vs. log ω possess if a velocity servo system (ramp input) is to have no *steady-state velocity error* for a constant velocity input $dr(t)/dt$? (b) What is true of the corresponding phase-angle characteristic?

8-7   Explain why the phase-angle curve cannot be calculated from the plot of $|G(j\omega)|$ in decibels vs. log ω if some of the factors are not minimum phase.

8-8   The asymptotic gain vs. frequency curve of the open-loop minimum-phase transfer function is shown for a unity-feedback control system. (a) Evaluate the open-loop transfer function. (b) What is the frequency at which the gain is unity? What is the phase angle at this frequency? (c) Draw the polar diagram of the open-loop control system.

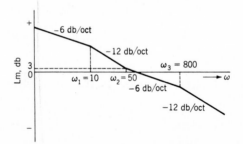

8-9   For each plot shown, (a) evaluate the transfer function; (b) find the correction that should be applied to the straight-line curve at $\omega = 8$.

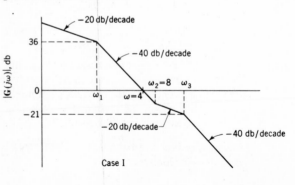

Case I

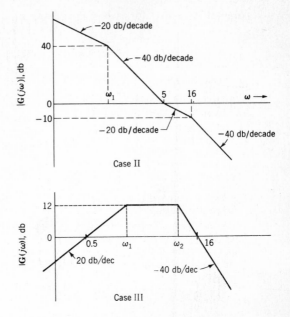

Case II

Case III

8-10 Determine the value of the error coefficient from Figs. (*a*) and (*b*).

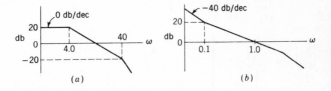

(*a*)

(*b*)

8-11 An experimental transfer function gave the following results:

ω	Lm $G(j\omega)$, dB	Angle $G(j\omega)$, deg
0.10	46.02	−179.1
0.50	18.14	−175.8
1.0	6.34	−171.8
2.0	−4.90	−166.1
4.0	−15.0	−162.9
8.0	−24.60	−173.1
10.0	−27.96	−180.0
14.0	−33.56	−192.8
20.0	−40.34	−207.9
40.0	−55.74	−234.2
80.0	−72.92	−251.2
140.0	−87.29	−259.1
200.0	−96.53	−262.4
240.0	−101.3	−263.6

(*a*) Determine the transfer function represented by the above data. (*b*) What type of system does it represent?

8-12   Repeat Prob. 8-11 for each of the following:

(a)

ω	$|G(j\omega)|$	Angle, deg
0.1	9.95	−96.9
0.3	3.19	−110.1
0.6	1.42	−127.8
0.8	0.963	−137.8
1.0	0.693	−146.3
2.0	0.21	−175.2
3.0	0.092	−192.5
5.0	0.028	−213.7
8.0	0.0082	−230.9
12.0	0.0027	−242.6
16.0	0.0012	−249.0
20.0	0.0006	−253.1

(b)

ω	Lm $G(j\omega)$, dB	Angle $G(j\omega)$, deg
1	0.08	−92.9
2	−5.71	−95.9
4	−10.77	−103.4
6	−12.55	−115.1
8	−12.68	−138.0
10	−13.98	−180.0
15	−26.80	−239.0
20	−36.00	−251.6
30	−47.75	−259.4
40	−55.64	−262.4
50	−61.63	−264.1
55	−64.17	−264.6

8-13   Determine the transfer function by using the straight-line asymptotic log plot shown and the fact that the correct angle is $-129.7°$ at $\omega = 0.8$. Assume a minimum-phase system.

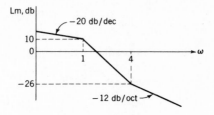

8-14   Determine whether each system shown is stable or unstable in the absolute sense by sketching the *complete* Nyquist diagrams. $H(s) = 1$.

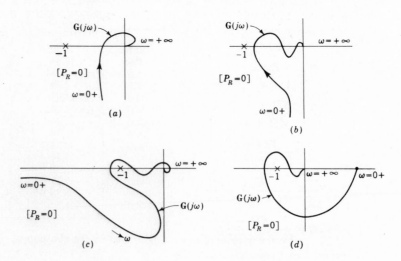

8-15 Use the Nyquist stability criterion and the polar plot to determine the range of values of K (positive or negative) for which the closed-loop system is stable.

(a) $G(s)H(s) = \dfrac{K(1 + s)^2}{s^3}$

(b) $G(s)H(s) = \dfrac{K}{s^2(-1 + 5s)(1 + s)}$

(c) $G(s)H(s) = \dfrac{K}{s^2(1 - 0.5s)}$

8-16 Shown are plots of the open-loop transfer function for a number of control systems. Only the curves for positive frequencies are given. *Using Nyquist's criterion*, determine the closed-loop system stability. $G(s)$ has no poles or zeros in the right-half plane.

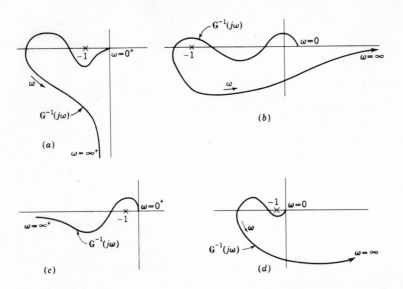

8-17 For the following transfer functions, sketch a direct and an inverse Nyquist locus to determine the closed-loop stability. Determine the values of K that correspond to stable closed-loop operation and those which correspond to unstable closed-loop operation. $H(s) = 1$.

(a) $G(s) = \dfrac{K}{1 + s}$

(b) $G(s) = K\dfrac{1 + s}{s(1 - s)}$

(c) $G(s) = \dfrac{K(1 + 5s)}{s^2(1 + s)}$

(d) $G(s) = \dfrac{K}{s(1 + s)(2 + s)}$

(e) $G(s) = \dfrac{K(s + 2)}{(s + 3)(s - 1)}$

(f) $G(s) = \dfrac{K}{(s - 1)(s + 2)(s + 4)}$

(g) $G(s) = \dfrac{K_1(1 + 2s)}{s(1 + s)(1 + s + s^2)}$

(h) $G(s) = \dfrac{K(s^2 + 2s + 2)}{s^3(1 + 0.2s)}$

(i) $G(s) = \dfrac{K(1 - s)}{(1 + 2s/3)(1 + s)}$

8-18 For the control systems having the transfer functions

(1) $G(s) = \dfrac{K_1}{s(1 + 0.01s)(1 + 0.025s)(1 + 0.10s)}$

(2) $G(s) = \dfrac{K(1 + 0.2s)}{(1 + 0.1s)(2 + 3s + s^2)}$

determine from the logarithmic curves the required value of K_m and the phase margin frequency so that each system will have (a) a positive phase margin of 45°; (b) a positive phase margin of 60°. (c) From these curves determine the maximum permissible value of K_m for stability.

8-19 A system has the transfer functions

$$G(s) = \frac{K}{s(s + 10)(s^2 + 32s + 1,600)} \qquad H(s) = 1$$

(a) Draw the log magnitude and phase diagram of $\mathbf{G}'(j\omega)$. Draw both the straight-line and the corrected log magnitude curves. (b) Draw the log magnitude–angle diagram. (c) Determine the maximum value of K_1 for stability.

8-20 What gain values would just make the systems in Prob. 8-2 unstable?

8-21 By use of the Nyquist stability criterion, determine whether the system having the following inverse polar plot is stable or unstable.

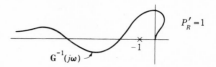

8-22 A system has the transfer function

$$G(s) = \frac{K_0}{(1 + s)^3}$$

(a) Draw the log magnitude vs. phase-angle diagram for $G'(s)$. (b) Determine the phase margin γ for $K_0 = 2$. (c) What is the maximum value of K_0 for stability?

8-23 For

$$G(s)^{-1} = \frac{s^2(0.5s + 1)(3s + 2)}{K_x(s - 1)}$$

sketch the inverse Nyquist locus and determine the closed-loop stability for (a) $K_x = 2$; (b) $K_x = -2$.

8-24 Use the Nyquist criterion to determine the maximum value of K_m for stability of the closed-loop systems having the following transfer functions:

(a) $G(s)H(s) = \dfrac{K_1 \varepsilon^{-0.5s}}{s(1 + s)(1 + 0.5s)}$ $\qquad$ (b) $G(s)H(s) = \dfrac{K \varepsilon^{-2s}}{s^2 + 2s + 2}$

(c) The transfer functions of Prob. 8-18 with the transport lag $\varepsilon^{-1.5s}$ included in the numerator.

8-25 Use the Nyquist criterion to determine the maximum value of T for stability of the closed-loop system which has the open-loop transfer function

$$G(s)H(s) = \frac{2\varepsilon^{-Ts}}{s(s^2 + 4s + 13)}$$

Hint: Determine the frequency ω_x for which $|G(j\omega)H(j\omega)| = 1$. What additional angle ωT, due to the transport lag, will make the angle of $G(j\omega_x)H(j\omega_x)$ equal to $-180°$?

$$\frac{K_1}{s(1+.2s)(1+.5s)}$$

Chapter 9

9-1 For Prob. 8-1, find K_1 and ω_m for an $M_m = 1.25$ by (a) the direct-polar-plot method; (b) the inverse-polar-plot method; (c) the log magnitude–angle diagram.

9-2 For the feedback control system of Prob. 7-4, (a) determine, by use of the polar-plot method, the value of K_1 that just makes the system unstable; (b) determine the value of K_1 that makes $M_m = 1.06$; (c) for the value of K_1 found in part (b), find $c(t)$ for $r(t) = u_{-1}(t)$; (d) determine graphically from the polar plot of $G(j\omega)$ the data for the curve of M vs. ω for the closed-loop system and plot this curve; (e) compare the results of parts (a) to (d) with the results obtained in Prob. 7-4.

9-3 Determine the value K_1 must have for an $M_m = 1.3$.

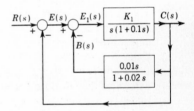

9-4 For the $1/G'(j\omega)$ curve shown it is found that two values of gain, K_a and K_b, will produce a desired M_m. The corresponding resonant frequencies are ω_a and ω_b. Which value of gain gives the better performance for the system? Give the reasons for your choice.

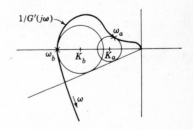

9-5 Using the plot of $G'(j\omega)$ shown, determine the number of values of gain K_m which produce the same value of M_m. Which of these values yields the best system performance? Give the reasons for your answer.

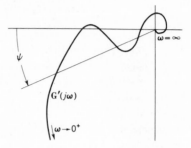

9-6 (a) Determine the values of M_m and ω_m for the transfer functions of Prob. 8-18 with $K_m = 2$. Repeat with the gain-constant values obtained in (b) part (a) of Prob. 8-18; (c) part (b) of Prob. 8-18.

9-7 (a) In Prob. 8-19, adjust the gain for $M_m = 2$ dB and determine ω_m. For this value of gain find the phase margin, and plot M versus ω and α vs. ω. (b) Repeat for Prob. 8-4.

9-8 For Prob. 8-15, part (a), with $H(s) = 1$ determine how much gain must be added to achieve an $M_m = 1.26$. What is the value of ω_m for this value of M_m? (a) Use log plots. (b) Use polar plots. (c) Use a computer.

9-9 For Prob. 8-2, determine the values of M_m and ω_m. How much gain must be added to achieve an $M_m = 1.12$? What is the value of ω_m and the phase margin for this value of M_m for each case?

9-10 Refer to Prob. 8-1. (a) Determine the values of K_1 and ω_m corresponding to the following values of M_m: 1.05, 1.1, 1.2, 1.4, 1.6, 1.8, and 2.0. For each value of M_m determine ζ_{eff} from Eq. (9-16) and plot M_m vs. ζ_{eff}. (b) For each value of K_1 determined in part (a), calculate the value of M_p. Use available computer programs. For each value of M_p determine ζ_{eff} from Fig. 3-7 and plot M_p vs. ζ_{eff}. (See the last paragraph of Sec. 7-11.) (c) What is the degree of correlation between M_m vs. ζ_{eff} and M_p vs. ζ_{eff}?

9-11 For Prob. 8-24, determine the value of K_m for an

$$M_m = 3 \text{ dB} \quad \text{with} \quad H(s) = 1$$

9-12 For the given transfer function: (a) Specify the value of K_2 that will make the peak M_m in the frequency response as small as possible. (b) At what frequency does this peak occur? (c) What value does $C(j\omega)/R(j\omega)$ have at the peak?

$$G(s) = \frac{K_2(1 + s)}{s^2(1 + 0.1s)(1 + 0.05s)}$$

9-13 The closed-loop frequency response of three simple second-order systems are sketched below. Sketch the time response to a step input for each of the three systems on the same time scale. Compare M_p, T_p, and T_s.

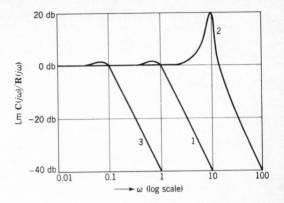

Chapter 10

10-1 It is desired that a control system have a damping ratio of 0.5 for the dominant complex roots. Using the root-locus method, (a) add a lag compensator, with $\alpha = 10$, so that this value of ζ can be obtained; (b) add a lead compensator with $\alpha = 0.1$; (c) add a lag-lead compensator with $\alpha = 10$. Indicate the time constants of the compensator in each case. Compare the results obtained by the use of each type of compensator with respect to the error coefficient, ω_d, T_s, and M_o.

$$G(s) = \frac{K}{s(s + 2)(s + 5)(s + 7)}$$

10-2 A control system has the forward transfer function

$$G_x(s) = \frac{K_x}{s^2(1 + 0.1s)}$$

The closed-loop system is to be made stable by adding a compensator $G_c(s)$ and an amplifier A in cascade with $G_x(s)$. A ζ of 0.5 is desired with a value of $\omega_n \approx 1.6$ rad/s. By use of the root-locus method, determine the following: (a) What kind of compensator is needed? (b) Select an appropriate α and T for the compensator. (c) Determine the value of the error coefficient K_2. (d) Plot the compensated locus. (e) Plot M vs. ω for the compensated system. (f) From the plot of part (e) determine the M_m and ω_m. With this value of M_m determine the effective ζ of the system by the use of $M_m = (2\zeta\sqrt{1 - \zeta^2})^{-1}$. Compare the effective ζ and ω_m with the values obtained from the dominant pair of complex roots. *Note:* The effective $\omega_m = \omega_n\sqrt{1 - 2\zeta^2}$. (g) Obtain $c(t)$ for a unit step input.

10-3 With

$$G_x(s) = \frac{K}{s(s + 10)(s^2 + 30s + 625)} \qquad H(s) = 1$$

find K_1, ω_n, M_o, T_p, T_s, N for each of the following cases: (a) original system; (b) lag compensator added, $\alpha = 10$; (c) lead compensator added, $\alpha = 0.1$; (d) lag-lead compensator added, $\alpha = 10$. Use $\zeta = 0.5$ for the dominant roots. *Note:* Except for the original system it is not necessary to obtain the complete root locus for each type of compensation. When the lead and lag-lead compensators are added to the original system, there may be other dominant roots in addition to the complex pair.

10-4 A unity-feedback system has the forward transfer function

$$G_x(s) = \frac{K_2}{s^2}$$

(a) Design a cascade compensator which will produce a stable system and which meets the following requirements without reducing the system type: (1) The dominant poles of the closed-loop control ratio are to have a damping ratio $\zeta = 0.5$. (2) The settling time is to be $T_s = 4$ s. Is it a lag or lead compensator? (b) Using the cascade compensator, determine the control ratio $C(s)/R(s)$. (c) Find $c(t)$ for a step input. What is the effect of the real pole of $C(s)/R(s)$ on the transient response?

10-5 Using the root-locus plot of Prob. 7-9, adjust the damping ratio to $\zeta = 0.5$ for the dominant roots of the system. Find K_1, ω_n, M_o, T_p, T_s, N, $C(s)/R(s)$ for (a) the original system and (b) the original system with a cascade lag compensator using $\alpha = 10$; (c) design a cascade compensator that will improve the response time, i.e., will move the dominant branch to the left in the s plane.

10-6 A control system has

$$G_x(s)H(s) = \frac{K}{(s + 1)(s + 2)(s + 5)(s + 6)} \qquad H(s) = 1$$

(a) Determine the roots of the characteristic equation of this system for $\zeta = 0.5$. (b) Determine the ratio $\sigma_{3,4}/\sigma_{1,2}$. (c) It is desired that this ratio be increased to at least 10 with no increase in T_s for $\zeta = 0.5$. A compensator that may accomplish this has the transfer function

$$G_c = A \frac{s + a}{s + b}$$

Determine appropriate positive values of a and b that can be achieved with a practical value of α.

10-7 A unity-feedback system has the transfer function $G_x(s)$. The closed-loop roots must satisfy the specifications $\zeta = 0.5$ and $T_s = 2$ s. A suggested compensator $G_c(s)$ must maintain the same degree as the characteristic equation for the basic system:

$$G_x(s) = \frac{K_x(s + 6)}{s(s + \frac{10}{3})} \qquad G_c(s) = \frac{A(s + a)}{s + b}$$

(a) Determine the values of a and b. (b) Determine the value of α. (c) Is this a lag or a lead compensator?

10-8 A unity-feedback control system contains the forward transfer function shown below. The system specifications are $\zeta = 0.6$ and $T_s \le 1.2$ s. Without compensation the value of K_x is 57.33, and the control ratio has an undesirable dominant real pole. The proposed cascade compensator has the form indicated.

$$G_x(s) = \frac{K_x(s + 2)}{s(s + 5)(s + 7)} \qquad G_c(s) = A \frac{s + a}{s + b}$$

(a) Determine the values of a and b such that the desired complex-conjugate poles of $C(s)/R(s)$ are truly dominant. (b) Determine the values of A and α for this compensator. (c) Is this $G_c(s)$ a lag or lead compensator?

10-9 A unity-feedback system has

$$G_x(s) = \frac{K}{(s + 1)(s + 3)(s + 6)}$$

(a) For $\zeta = 0.5$, determine the roots and the value of K. (b) Add a lead compensator which cancels the pole at $s = -1$. (c) Add a lead compensator which cancels the pole at $s = -3$. (d) Compare the results of parts (b) and (c). Establish a "rule" for adding a lead compensator to a Type 0 system.

10-10   For the system of Prob. 7-5, the desired roots are $s = -10 \pm j35$ and the desired gain is $K_1 = 100$. (a) Design a compensator which will achieve these characteristics. (b) Determine $c(t)$ with a unit step input. (c) Are the desired complex roots dominant? (d) Repeat this problem with the roots at $-20 \pm j35$.

10-11   A unity-feedback system has the transfer function

$$G_x(s) = \frac{K}{s^2(1 + 0.2s)}$$

The poles of the closed-loop system must be $s = -0.75 \pm j2$. (a) Design a lead compensator with the maximum possible value of α which will produce these roots. (b) Determine the control ratio for the compensated system. (c) Add a compensator to increase the gain without increasing the settling time.

10-12   A system has the transfer function

$$G_x(s) = \frac{K}{(s^2 - 2s + 2)(s + 10)}$$

The roots of the closed-loop system must have $\zeta = 0.5$. Add a lead compensator, with two zeros and two poles, to produce a stable system. Determine $C(s)/R(s)$.

10-13   In the figure the servomotor has inertia but no viscous friction. The feedback through the accelerometer is proportional to the acceleration of the output shaft. (a) When $H(s) = 1$, is the servo system stable? (b) Is the system stable if $K_1 = 10$, $K_2 = 2$, and $H(s) = 1/(2s)$? (c) Determine another $H(s)$ that causes the system to be stable. Show that the system is stable with the value of $H(s)$ selected by sketching the transfer function $C(j\omega)/E(j\omega)$ and the root locus.

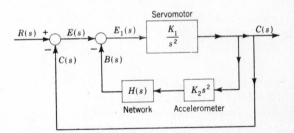

10-14   For the system shown, find (a) the transfer function $C(s)/E(s)$; (b) ω_n and ζ for the open-loop system and for the closed-loop system.

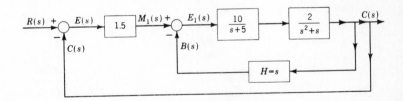

10-15 The accompanying schematic shows a method for maintaining a constant rate of discharge from a water tank by regulating the level of the water in the tank. The dynamics of the flow in and out of the tank may be taken into account by the relationships

$$Q_1 - Q_2 = 16\frac{dh}{dt} \qquad Q_2 = 4h \qquad Q_1 = 10\theta$$

where Q_1 = volumetric flow into tank
Q_2 = volumetric flow out of tank
h = pressure head in tank
θ = angular rotation of control valve

The motor damping is much smaller than the inertia. The system shown is stable only for very small values of system gain. Therefore, feedback from the control valve to the amplifier is proposed. Show conclusively which of the following feedback functions would produce the best results: (*a*) feedback signal proportional to control-valve position; (*b*) feedback signal proportional to rate of change of control-valve position; (*c*) feedback signal comprising a component proportional to valve position and a component proportional to rate of change of valve position.

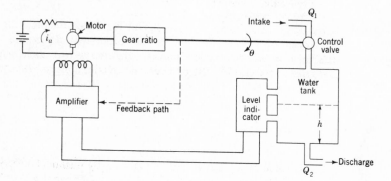

10-16–10-17 The block diagram shows a simplified form of roll control for an airplane. Overall system specifications are $t_s = 1.0$ s and $M_p = 1.3$. A_h is the gain of an amplifier in the H_2 feedback loop. G_2 is an amplifier of adjustable gain with a maximum value of 100. Restrict b between 15 and 50.

$$G_3(s) = \frac{1.7}{(1 + 0.25s)(1 + s)} \qquad H_2(s) = 0.2A_h\frac{s + a}{s + b} \qquad G_4(s) = \frac{1}{s}$$

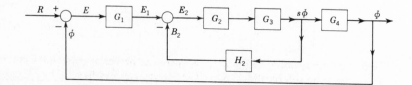

10-16   (a) By use of the root locus, method 1, design the RC network to be used in $H_2(s)$ to meet the overall system specifications with $G_1(s) = 1$. (b) Design $G_1(s)$ to improve the value of K_1 of part (a) by a factor of 5 while maintaining the desired overall system specifications. Determine all roots of the characteristic equation for Φ/R after compensation. (c) For the values obtained in part (a), determine T_s and M_p for the inner loop. Compare these values with those given in part (a) of Prob. 10-17(a).

10-17   (a) By use of the root locus, method 2, design the RC network to be used in $H_2(s)$ to meet the inner-loop specifications $M_p = 1.05$ and $T_s = 1.0$ s. Determine the necessary values of A_h and G_2. (b) Design $G_1(s)$ to improve the value of K_1, resulting from part (a), by a factor of 5 and also to meet the overall system specifications.

10-18   For the feedback-compensated system of Fig. 10-32, $A = 1$,

$$G_x(s) = \frac{K_x}{s(s + 1)}$$

$$H(s) = \frac{0.1s^2}{s + 6}$$

(a) Use method 1 to determine $C(s)/R(s)$ using $T_s = 2$ s (2 percent criterion) for the dominant roots. (b) Design a cascade compensator which yields the same dominant roots as part (a). (c) Compare the gains of the two systems.

10-19   A unity-feedback system has

$$G_x(s) = \frac{K_x}{s(s + 6)}$$

(a) For $\zeta = 0.5$, determine $C(s)/R(s)$. (b) Minor-loop and cascade compensation are to be added (see Fig. 10-32) with the amplifier replaced by $G_c(s)$:

$$G_c(s) = \frac{As}{s + b}$$

$$H(s) = \frac{K_h s}{s + a}$$

Use the value of K_x determined in part (a). By method 1 of the root-locus design procedure, determine values of a, b, K_h, and A to produce dominant roots having $\zeta = 0.5$ and $\sigma = -8$. Only positive values of K_h and A are acceptable. To minimize the effect of the third real root, it should be located near a zero of $C(s)/R(s)$.

10-20   For $G(s)$ given in Prob. 10-3, add feedback compensation containing a tachometer. Use root-locus method 1 or 2 as assigned. Design the tachometer so that an improvement in system performance is achieved while maintaining a $\zeta = 0.5$ for the dominant roots. Compare with the results of Prob. 10-3.

10-21   Refer to Fig. 10-32. It is desired that the dominant poles of $C(s)/R(s)$ have $\zeta = 0.5$ and $T_s = 2$ s. Using method 1, determine the values of a, b, and A to meet these specifications. Be careful in selecting b.

$$G_x = \frac{4}{s(s + 2)}$$

$$H(s) = \frac{s + a}{s + b}$$

10-22   Use minor-loop feedback to compensate the basic system of Prob. 10-1 to achieve the same specification. Compare the cascade and feedback performance.

10-23 Repeat Prob. 10-22 for the system of Prob. 7-9, comparing the results with those of Prob. 10-5.

10-24 Repeat Prob. 10-2 using a feedback compensator.

Chapter 11

11-1 The control system of Prob. 8-4 is to have $M_m = 1.16$. (a) Add a lag compensator, with $\alpha = 10$, so that this value of M_m can be obtained. (b) Add a lead compensator with $\alpha = 0.1$. (c) Add a lag-lead compensator with $\alpha = 10$. Indicate the time constants of the compensator in each case. Compare the results obtained by using each type of compensator. Also, compare the results of this problem with those of Prob. 10-3.

11-2 A control system has the forward transfer function

$$G_x(s) = \frac{K_2}{s^2(1 + 0.1s)} \qquad K_2 = 1$$

The closed-loop system is to be made stable by adding a compensator G_c and an amplifier A in cascade with $G_x(s)$. An $M_m = 1.5$ is desired with $\omega_m \approx 1.4$ rad/s. (a) What kind of compensator is needed? (b) Select an appropriate α and T for the compensator. (c) Select the necessary value of amplifier gain A. (d) Plot the compensated curve. (e) Compare the results of this problem with those obtained in Prob. 10-2.

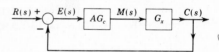

11-3 A control system with unity feedback has a forward open-loop transfer function

$$G_x(s) = \frac{K_1}{s(1 + 0.5s)(1 + 0.1s)^2}$$

(a) Find the gain K_1 for 45° phase margin, and determine the corresponding phase-margin frequency ω_ϕ. (b) For the same phase margin as in (a), it is desired to increase the phase-margin frequency to a value of $\omega_\phi = 3.0$ with the maximum possible improvement in gain. To accomplish this, a lead compensator is to be used. Determine the values of α and T that will satisfy these requirements. For these values of α and T, determine the new value of gain. (c) Repeat part (b) with the lag-lead compensator of Fig. 11-17. Select an appropriate value for T. With $G_c(s)$ inserted in cascade with $G_x(s)$, find the gain needed for 45° phase margin. (d) Show how the compensator has improved the system performance.

11-4 The mechanical system shown has been suggested for use as a compensating component in a mechanical system. (a) Determine whether it will function as a lead or a lag compensator by finding $X_2(s)/F(s)$. (b) Sketch $X_2(j\omega)/F(j\omega)$.

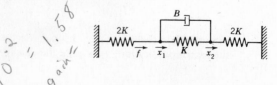

11-5 A unity-feedback control system has the transfer function

$$\frac{a}{b} = 8 \qquad G(s) = \frac{K}{(1+s)^3} G_c(s)$$

(a) What is the value of the gain of the basic system for an $M_m = 1.12$? (b) Design a cascade compensator $G_c(s)$ that will increase the step error coefficient by a factor of 8 while maintaining the same M_m. (c) What effect does the compensation have on the closed-loop response of the system? (d) Repeat the design using a minor-loop feedback compensator.

11-6 A unity-feedback control system has the transfer function

$$G(s) = G_c(s)G_1(s)$$

where $G_1(s) = 1/s^2$. It is desired to have an $M_m = 1.4$ with a damped natural frequency of about 10.0 rad/s. Design a compensator $G_c(s)$ that will help to meet these specifications.

11-7 A system has

$$G_2(s) = \frac{10(1+0.316s)}{(1+0.1s)(1+0.01s)} \qquad G_3(s) = 1$$

For the control system shown in Fig. (a), determine the cascade compensator $G_c(s)$ required to make the system meet the desired open-loop frequency-response characteristic shown in Fig. (b). What is the value of M_m for the compensated system?

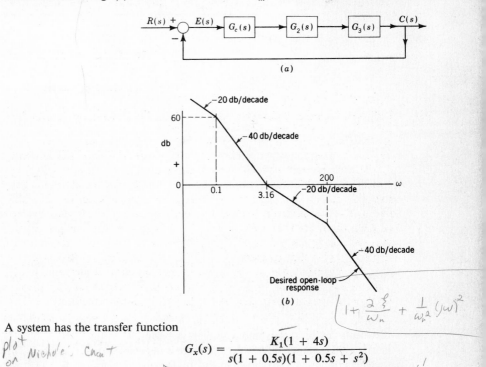

(a)

(b)

$$\left[1 + \frac{2\zeta}{w_n} + \frac{1}{w_n^2}(jw) \right]^2$$

11-8 A system has the transfer function

plot on Nichole's Chart

$$G_x(s) = \frac{K_1(1+4s)}{s(1+0.5s)(1+0.5s+s^2)}$$

(a) For $M_m = 1.26$ determine K_1 and ω_m for the original system. (b) Add a lag compensator with $\alpha = 10$ and determine T, K_1, and ω_m for the same M_m. (c) Add a lead compensator with $\alpha = 0.1$ and determine T, K_1, and ω_m for the same M_m.

$$= \frac{K_1(1+\frac{s}{.25})}{s(1+\frac{s}{2})(1+\frac{s}{2}+s^2)}$$

11-9 Repeat Prob. 10-1, using $M_m = 1.16$ as the basis of design. Compare the results.

11-10 Repeat Prob. 10-16 but solve by the use of the logarithmic plots. *Note:* Incorporate G_4 into the minor loop. Limit G_1 and G_2 to less than 100.

11-11 A system has

$$G_1(s) = 3 \qquad G_3(s) = \frac{1}{1+s} \qquad H_1(s) = \frac{s}{1+s}$$

$$G_2(s) = 10 \qquad G_4(s) = \frac{1}{s}$$

System specifications are $K_1 = 30 \text{ s}^{-1}$, $\gamma \geq +50°$, and $\omega_\phi = 3$ rad/s. Using approximate techniques, (*a*) determine whether the feedback-compensated system satisfies all the specifications; (*b*) find a cascade compensator to be added between $G_1(s)$ and $G_2(s)$, with the feedback loop omitted, to produce the same $\mathbf{C}(j\omega)/\mathbf{E}(j\omega)$ as in part (*a*); (*c*) determine whether the system of part (*b*) satisfies all the specifications.

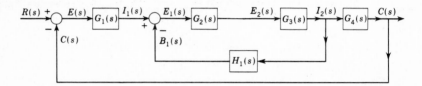

11-12 In the figure of Prob. 11-11,

$$G_1(s) = A_1 \qquad G_3(s) = \frac{1}{(1+0.1s)(1+0.01s)}$$

$$G_2(s) = A_2 \qquad G_4(s) = \frac{1}{s}$$

$H_1(s)$ is a passive network. Choose A_1, A_2, and $H_1(s)$ to meet the desired open-loop response characteristic shown in the figure of Prob. 11-7 for $\omega < 120$.

11-13 For the feedback-compensated system of Fig. 11-22

$$G_x(s) = \frac{10}{s(1+s)} \qquad H(s) = \frac{2s}{1+1.25s}$$

(*a*) For the original system, without feedback compensation, determine the values of A, K_1, and ω_ϕ for a phase margin of 45°. (*b*) Add the minor-loop compensator $H(s)$ and determine, for a phase margin of 45°, the corresponding values of A, K_1, and ω_ϕ. (*c*) Compare the results.

11-14 A servo with tachometric feedback can be represented as shown in Fig. (*a*). $G_x(s)$ represents an ideal servomotor which develops a torque $T = K_T E_1$. The total inertia on the output shaft is J_0, and there is negligible viscous friction. The torque-speed characteristics are shown in Fig. (*b*) for several values of E_1. $H_1(s)$ represents a tachometric generator which develops voltage $B = K_t Dc$ and $G_3(s) = C(s)/E(s)$. (*a*) Find, in terms of system constants, $G_x(s)$, $H_1(s)$, and $G_3(s)$. (*b*) Sketch the polar plots of $\mathbf{G}_x(j\omega)$, $\mathbf{H}_1(j\omega)$, and $\mathbf{G}_3(j\omega)$.

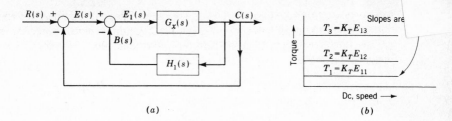

(a)   (b)

11-15 The system represented in the figure is an ac voltage regulator. (a) Calculate the response time of the compensated system, when the switch S is closed, to an output disturbance. (b) Compare this with the value obtained from the uncompensated system. Comment upon any difference in performance.

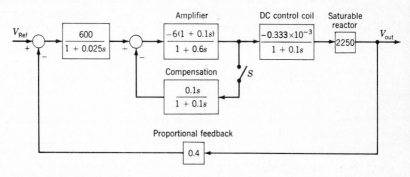

11-16 The simplified pitch-attitude-control system used in a space-launch vehicle is represented in the block diagram. (a) Determine the upper and lower gain margins. (b) Is the closed-loop system stable?

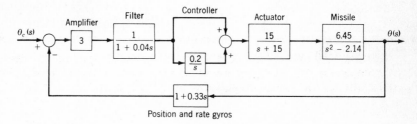

Position and rate gyros

11-17 Repeat Prob. 10-20 using the log plots for $M_m = 1.16$. Compare with the results of Probs. 10-3 and 11-1.

Chapter 12

12-1 Determine a passive cascade compensator for a unity-feedback system. The desired closed-loop time response for a step input is achieved by the system having the control ratio

$$M(s) = \frac{20(s + 3)}{(s^2 + 2s + 5)(s + 2)(s + 6)}$$

Given:

$$G_x(s) = \frac{10}{s(s^2 + 4s + 8)} \quad \text{and} \quad G_c(s) = \frac{A(s + a)}{s + b}$$

Compare M_p, t_p, and t_s for the desired and actual $M(s)$.

12-2   Determine a cascade compensator for a unity-feedback system. The desired closed-loop time response is the same as for Prob. 12-1. The open-loop transfer function is $G_x(s) = 10/s(s + 2)(s + 4)$. Determine the simplest possible approximate transfer functions that will yield the desired time response.

12-3   For a single-input single-output system (see Fig. 12-4d) the open-loop matrix state and output equations are $\dot{x} = Ax + bu$ and $y = c^T x$. Derive the overall closed-loop control ratio for a state-variable feedback system:

$$\frac{Y(s)}{R(s)} = c^T[sI - (A - bk^T)]^{-1}b$$

12-4   A control system is to use state-variable feedback. The plant has the given transfer function. The system is to meet the following requirements: (1) it must have zero steady-state error with a step input, (2) dominant poles of the closed-loop control ratio are to be $-2 \pm j2$, (3) the system is to be stable for all values $A > 0$, and (4) the control ratio is to have low sensitivity to increase in gain A. A cascade element may be added to meet this requirement.

$$G_x(s) = \frac{A}{s(s + 1)}$$

(a) Draw the block diagram showing the state-variable feedback. Use the form illustrated in Fig. 12-17. (b) Determine $H_{eq}(s)$. (c) Find $Y(s)/R(s)$ in terms of the state-variable feedback coefficients. (d) Determine the desired control ratio $Y(s)/R(s)$. (e) Determine the necessary values of the feedback coefficients. (f) Sketch the root locus for $G(s)H_{eq}(s) = -1$ and show that the system is insensitive to variations in A. (g) Determine $G_{eq}(s)$ and K_1. (h) Determine M_p, t_p, and t_s with a step input.

12-5   Design a state-variable feedback system for the given plant $G_x(s)$. The desired complex dominant roots are to have a $\zeta = 0.425$. For a unit step function the approximate specifications are $M_p = 1.2$, $t_p = 0.15$ s, and $t_s = 0.3$ s. (a) Determine a desirable $M(s)$ that will satisfy the specifications. Use steps 1 to 6 of the design procedure given in Sec. 12-4 to determine k. Draw the root locus for $G(s)H_{eq}(s) = -1$. From it show the "good" properties of state feedback. (b) Use the alternate procedure of step 7, Sec. 12-4, to evaluate k. (c) Obtain $c(t)$ for the final design. Determine the values of the figures of merit and the ramp error coefficient.

12-6   For the plant of Prob. 10-3 design a state-variable feedback system utilizing the phase-variable representation. The specifications are that dominant roots must have $\zeta = 0.5$, a zero steady-state error for a step input, and $t_s \leq \frac{1}{3}$ s. (a) Determine a desirable $M(s)$ that will satisfy the specifications. Use steps 1 to 6 of the design procedure given in Sec. 12-4 to determine k. Draw the root locus for $G(s)H_{eq}(s) = -1$. From it show the "good" properties of state feedback. (b) Use the alternate procedure of step 7, Sec. 12-4, to evaluate k. (c) Obtain $y(t)$ for the final design. Determine the values of the figures of merit and the ramp error coefficient.

12-7   Design a state-variable feedback system that satisfies the following specifications: (1) For a unit step input $M_p = 1.10$ and $T_s \leq 1$ s. (2) The system follows a ramp input with zero steady-state error. (a) Determine $y(t)$ for a step and for a ramp input. (b) For the step input determine M_p, t_p, and t_s. (c) What are the values of K_p, K_v, and K_a?

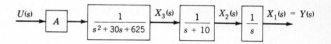

$$U(s) \rightarrow \boxed{A} \rightarrow \boxed{\dfrac{1}{s+4}} \xrightarrow{X_3(s)} \boxed{\dfrac{s+b}{s+1}} \xrightarrow{X_2(s)} \boxed{\dfrac{1}{s}} \xrightarrow{X_1(s) = Y(s)}$$

12-8   A closed-loop system with state-variable feedback is to have a control ratio of the form

$$\frac{Y(s)}{R(s)} = \frac{A(s + 2)(s + 0.5)}{(s^2 + 2s + 2)(s + 1)(s - p_1)(s - p_2)}$$

where p_1 and p_2 are not specified. (a) Find the two necessary relationships between A, p_1, and p_2 so that the system has zero steady-state error with both a step and a ramp input. (b) With $A = 100$, use the relationships developed in (a) to find p_1 and p_2. (c) Use the system shown in Fig. 12-19 with $Z_i = -0.5$. Determine the values of the feedback coefficients. (d) Draw the root locus for $G_x(s)H_{eq}(s) = -1$. Does this root locus show that the system is insensitive to changes of gain A? (e) Determine the time response with a step and with a ramp input. Discuss the response characteristics.

12-9   For the plant shown design a state-variable feedback system that satisfies the following specifications: dominant roots must have a $\zeta = 0.5$, a zero steady-state error for a step input, and $T_s = 0.8$ s. Determine M_p, t_s, t_p, and K_1.

$$U(s) \rightarrow \boxed{A} \rightarrow \boxed{\dfrac{1}{s^2 + 30s + 625}} \xrightarrow{X_3(s)} \boxed{\dfrac{1}{s+10}} \xrightarrow{X_2(s)} \boxed{\dfrac{1}{s}} \xrightarrow{X_1(s) = Y(s)}$$

12-10   For the system of Fig. 12-15 it is desired to improve t_s from that obtained with the control ratio of Eq. (12-69) while maintaining an underdamped response with a smaller overshoot and faster settling time. A desired control ratio that yields this improvement is

$$\frac{Y(s)}{R(s)} = \frac{1.428(s + 1.4)}{(s + 1)(s^2 + 2s + 2)}$$

The necessary cascade compensator can be inserted only at the output of the amplifier A. (a) Design a state-variable feedback system that yields the desired control ratio. (b) Draw the root locus for $G(s)H_{eq}(s)$ and comment on its acceptability for gain variation. (c) For a step input determine M_p, t_p, t_s, and K_1. Compare with the results of Examples 1 to 3 of Sec. 12-10.

12-11   Derive Eqs. (12-53) and (12-54).

12-12   The range of values of a is $7 \geq a \geq 3$, $M_p \leq 1.10$ for a unit step input, and $T_s = \leq 1$ s. Design a state-variable feedback system that satisfies the given specifications and is insensitive to variation of a. Assume that all states are accessible. Obtain plots of $y(t)$ for $a = 3$, $a = 5$, and $a = 7$ for the final design. Compare M_p, t_p, t_s, and K_1. Determine the sensitivity functions for variation of A and a (at $a = 3$, 5, and 7).

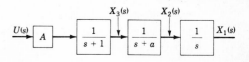

$$U(s) \rightarrow \boxed{A} \rightarrow \boxed{\dfrac{1}{s+1}} \xrightarrow{X_3(s)} \boxed{\dfrac{1}{s+a}} \xrightarrow{X_2(s)} \boxed{\dfrac{1}{s}} \xrightarrow{X_1(s)}$$

12-13 Repeat Prob. 12-12 but with state X_3 inaccessible.

12-14 Determine a cascade compensator by the Guilleman-Truxal method where

$$G_x(s) = \frac{4(s + 2)}{s(s + 1)(s + 5)}$$

$$M(s) = \frac{210(s + 1.5)}{(s + 1.75)(s + 16)(s^2 + 3s + 11.25)}$$

12-15 Determine the sensitivity factor for Example 1 in Sec. 12-10 for variations in A. [Insert the values of k_i in Eq. (12-68) and then differentiate with respect to A to obtain $S_A{}^M$.]

12-16 For Example 2, Sec. 12-10, obtain $Y(s)/R(s)$ as a function of K_G [see Eq. (12-75)]. Determine the sensitivity function $S_{K_G}{}^M$.

12-17 For Example 3, Sec. 12-10, determine $G_{eq}(s)$ and the system type. Redesign the state-variable feedback system to make it Type 2.

Chapter 13

13-1 Obtain the isocline equation for each of the following differential equations. Draw sufficient isoclines to draw trajectories starting from the point indicated. Determine the kind of singularity and the stability in each case.
  (a) $\ddot{x} + 3\dot{x} + 2x = 0$ $x_1 = 0, x_2 = 5$
  (b) $\ddot{x} + 11\dot{x} + 10x = 0$ $x_1 = 0, x_2 = 6$
  (c) $\ddot{x} + 5\dot{x} + 6x = 6$ $x_1 = 0, x_2 = 5$
  (d) $\ddot{x} + 2\dot{x} + 2x = 0$ $x_1 = 4, x_2 = 4$
  (e) $\ddot{x} + 4\dot{x} + 64x = 0$ $x_1 = 6, x_2 = 0$
  (f) $\ddot{x} - 8\dot{x} + 17x = 34$ $x_1 = 0, x_2 = 4$
  (g) $\ddot{x} - 9x = 0$ $x_1 = 2, x_2 = 1$

13-2 Construct a phase portrait using the isocline method for the following second-order differential equations. Determine the kind of singularity and the stability for each case.
  (a) $\ddot{x} + \dot{x}/|\dot{x}| + 2x = 0$ (b) $\ddot{x} + 2\dot{x} + x^2 = 0$
  (c) $\ddot{x} + 2\dot{x}^2 + x = 0$ (d) $\ddot{x} + 2\dot{x}|\dot{x}| + 2x = 0$
  (e) $\ddot{x} + \dot{x} + 4x|x| = 0$ (f) $\ddot{x} + \dot{x} - 4x|x| = 0$

13-3 Write the state equations for Prob. 13-1 and convert to normal form. Draw the trajectory in the normal plane. For Eqs. (c) and (f) it is simpler to change variables to move the equilibrium point to the origin.

13-4 The nonlinearity has a *dead-zone* characteristic, as shown. Outside the dead-zone region, $-0.1 \le E \le 0.1$, the output E of the nonlinearity is proportional to the input E. For $r(t) = 0$, determine the isocline equation for each region of the phase plane. Draw the phase portrait.

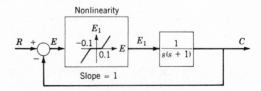

13-5 The nonlinearity in the block diagram of Prob. 13-4 is replaced by a three-position relay with the characteristic shown. (*a*) For $b = 0.1$ determine the isocline equations for each region: $E > b$, $-b \leq E \leq b$, $E < -b$. (*b*) Draw several trajectories. (*c*) Repeat for $b = 0.2$.

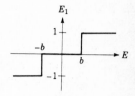

13-6 Use the Jacobian matrix to determine the stability in the vicinity of the equilibrium points for the system described by:
(*a*) $\dot{x}_1 = x_1 x_2 + 2x_2 + u_1 + u_2{}^2$
$\quad\ \dot{x}_2 = 3x_1 + x_2 + 2u_1{}^2 + x_1 u_2$ where $\mathbf{u}_0 = \begin{bmatrix} -1 & 0 \end{bmatrix}^T$
(*b*) $\dot{x}_1 = 4x_1 - x_1 x_2$
$\quad\ \dot{x}_2 = x_1{}^2 - x_2$
(*c*) $\dot{x}_1 = -4x_1 + 2x_2$
$\quad\ \dot{x}_2 = x_1{}^2 - x_2$

13-7 A system is described by the state equations

$$\dot{x}_1 = x_2$$
$$\dot{x}_2 = -x_1 - 3x_2 + 0.25x_3{}^2$$
$$\dot{x}_3 = x_1 x_3 - x_3$$

(*a*) Determine the equilibrium points for this system. (*b*) Obtain the linearized equations about the equilibrium points. (*c*) Determine whether the system is asymptotically stable or unstable in the neighborhood of its equilibrium points.

13-8 Determine the definiteness of the following quadratic forms, $V(\mathbf{x}) = \mathbf{x}^T \mathbf{A}\mathbf{x}$, where:

(*a*) $\mathbf{A} = \begin{bmatrix} 1 & 2 & -5 \\ 0 & 2 & 0 \\ 3 & 0 & 3 \end{bmatrix}$ (*b*) $\mathbf{A} = \begin{bmatrix} 0 & 1 & 0 \\ 0 & 0 & 1 \\ -10 & -5 & -6 \end{bmatrix}$

(*c*) $\mathbf{A} = \begin{bmatrix} -1 & -1 & 0 \\ 3 & -4 & -2 \\ 0 & 0 & -1 \end{bmatrix}$

13-9 (*a*) Transform the matrix $\mathbf{A}$ of Example 7, Sec. 13-4, to diagonal form. Use this matrix to verify the definiteness of $\mathbf{A}$. (*b*) Determine the Jordan matrix for the matrix $\mathbf{A}$ of Example 2, Sec. 13-4. Use this matrix to verify the definiteness of $\mathbf{A}$.

13-10 Given

$$\dot{\mathbf{x}} = \begin{bmatrix} 0 & 1 & 0 \\ 0 & 0 & 1 \\ -K & -5 & -6 \end{bmatrix} \mathbf{x}$$

and $V(\mathbf{x}) = 6Kx_1{}^2 + 2Kx_1 x_2 + 41x_2{}^2 + 12x_2 x_3 + x_3{}^2$

Show that $V(\mathbf{x})$ remains positive definite for the same range of K for which $dV(\mathbf{x})/dt$ is negative semidefinite. Also show that global asymptotic stability exists.

13-11

$$\dot{x} = \begin{bmatrix} 0 & 1 & 0 \\ 0 & -2 & 1 \\ -K & -1 & 0 \end{bmatrix} x \qquad V(x) = 2Kx_1{}^2 + 2Kx_1x_2 + x_2{}^2 + x_3{}^2$$

(a) Determine the range of K for which $\dot{V}(x)$ is negative semidefinite, while also assuring that $V(x)$ is positive definite. (b) Determine whether this system has global asymptotic stability.

13-12 Use the method of Sec. 13-7 to determine the range of values of K that yields a positive definite matrix **P** in the Liapunov function $V(x) = x^TPx$, thus assuring asymptotic stability for:

(a) $A = \begin{bmatrix} -1 & 0 & 1 \\ 2 & -2 & 0 \\ K & 0 & -2 \end{bmatrix} \qquad N = \begin{bmatrix} 0 & 0 & 0 \\ 0 & 2 & 0 \\ 0 & 0 & 0 \end{bmatrix}$

(b) $A = \begin{bmatrix} -K & 1 & 0 \\ -11 & 0 & 1 \\ -6 & 0 & 0 \end{bmatrix} \qquad N = \begin{bmatrix} 0 & 0 & 0 \\ 0 & 0 & 0 \\ 0 & 0 & 2 \end{bmatrix}$

(c) For part (a) change a_{33} to $+2$.

13-13 Determine the bound on the settling time for the system represented by

$$A = \begin{bmatrix} 0 & 1 & 0 \\ 0 & 0 & 1 \\ -8 & -14 & -7 \end{bmatrix}$$

13-14 For the quadratic form $V(x) = x^TQx$, where **Q** is a symmetric matrix, prove that

$$\frac{dV(x)}{dx} = 2Qx$$

Chapter 14

14-1 For the example of Sec. 12-7, determine the feedback coefficients and the gain A in order to yield the response corresponding to a third-order Butterworth form (Table 14-3), with $\omega_0 = 4.642$. Compare the time responses.

14-2

$$\dot{x} = \begin{bmatrix} 0 & 1 \\ -2 & -3 \end{bmatrix} x + Bu \qquad y = [1 \quad 0]x$$

$$B_1 = \begin{bmatrix} 10 \\ 0 \end{bmatrix} \qquad B_2 = \begin{bmatrix} 1 & 0 \\ 0 & 1 \end{bmatrix} \qquad Z = I$$

$$Q_a = \begin{bmatrix} 1 & 0 \\ 0 & 0 \end{bmatrix} \qquad Q_b = \begin{bmatrix} 1 & 0 \\ 0 & 1 \end{bmatrix}$$

(a) Find the feedback coefficient matrix for this system for each matrix **B**. Obtain solutions for each matrix **Q** indicated. (b) For the matrix B_1 compare the time responses with a unit step input. (See Problem 12-3 for $Y(s)/R(s)$.)

14-3 Find the feedback coefficient matrix for the system and PI indicated:

$$\mathbf{A} = \begin{bmatrix} 0 & 1 & 0 \\ 0 & 0 & 1 \\ -6 & -11 & -6 \end{bmatrix} \qquad \mathbf{B} = \begin{bmatrix} 0 \\ 0 \\ 10 \end{bmatrix} \qquad y = \begin{bmatrix} 1 & 2 & 0 \end{bmatrix} \mathbf{x}$$

$$\text{PI} = \int_0^\infty (x_1{}^2 + 0.01x_2{}^2 + 0.01x_3{}^2 + zu^2)\, dt$$

(a) $z = 1$, (b) $z = 10$, (c) $z = 0.1$. (d) Compare the time responses of the system with a step input for each value of z.

14-4 For the example in Sec. 14-10 determine $G(s)H_{eq}(s)$. Verify that the plot of $G(j\omega)H_{eq}(j\omega)$ is outside the unit circle centered at -1 (see Fig. 14-12).

14-5 Use the results of the example in Sec. 12-7 and the method of Sec. 14-12 [Eq. (14-79)] to determine the approximate value of **h**. This requires the phase-variable representation of the system. Plot the straight-line curve for Lm $G_K(j\omega)$. Show whether or not all the corner frequencies are below ω_ϕ.

14-6 For the example of Sec. 14-10, verify that the third root of the characteristic equation is $s_3 \approx -100$.

14-7 A linear regulator system is to be designed using the method of Sec. 14-10. Use $z = 1$, $\mathbf{h}^T = \begin{bmatrix} 1 & 0 & 0 \end{bmatrix} = \mathbf{c}^T$ [so that $G_K(s) = G(s)$], and $\omega_\phi = 2\omega_{cf}$, where ω_{cf} is the highest corner frequency of $G(s)$. (a) Determine the value of K_G required. (b) Using the phase-variable representation, determine the feedback coefficients.

$$G(s) = \frac{K_G}{s(s + 1)(s + 2)}$$

14-8 Repeat Prob. 14-7 using $\mathbf{h}^T = \begin{bmatrix} 1 & 1 & 1 \end{bmatrix}$.

14-9 For the system shown determine $\mathbf{k}_p$ by use of the Bode diagram. Values of K_G equal to 10 or 100 are proposed, with $z = 1$ and $\mathbf{h}_p{}^T = \begin{bmatrix} 1 & 0 & 0 \end{bmatrix}$. Give your reason for selecting one value of K_G over the other. Compare with the Riccati solution.

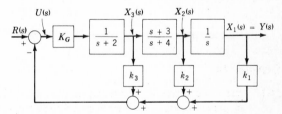

14-10 Repeat Prob. 14-9 with $\mathbf{h}_p{}^T = \begin{bmatrix} 1 & 1 & 1 \end{bmatrix}$.

14-11 In the state-variable control system of Prob. 14-9 the states x_2 and x_3 are inaccessible, and noise is present throughout. (a) Derive a minor-loop compensator that permits a physical realization of this system. (b) State reasons for the minor-loop compensator chosen.

14-12 Repeat Probs. 14-7 to 14-10 by the RSL method.†

14-13 A system is to be designed to satisfy

$$\text{PI} = \int_0^\infty (\mathbf{x}^T\mathbf{Q}\mathbf{x} + u^2)\, dt$$

where $\mathbf{Q} = \mathbf{hh}^T$. The RSL for this system for a specified **h** and phase-variable representation is shown below. The roots for the desired value of $K^2 = 400$ are shown. Determine the values of k_i.

† Use the values of K_G obtained in those problems.

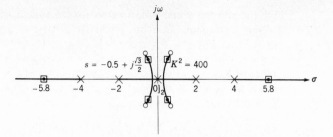

14-14 Use the system in the example of Sec. 14-10 with $A = 50$ and represent it in terms of phase variables. Use Eq. (14-79) to evaluate **k**. Transform to physical variables to obtain $\mathbf{k}_p$. Compare with the values obtained in Sec. 14-10.

14-15 Repeat Prob. 14-10 using Eq. (14-79) to evaluate **k** in terms of phase variables. Transform to physical variables to obtain $\mathbf{k}_p$. Compare with the values obtained in Prob. 14-10.

Chapter 15

15-1 Given: $G(s) = K_G/(s + 1)(s + 2)(s + 3)$. The desired characteristic equation is $(s^2 + 2s + 2)(s + \alpha)$. (a) Under the hfg condition and with $z = 1$ determine the **k** required to produce an equivalent Type 1 system. Find $G_{eq}(s)$ and verify the system type. (b) Using the **k** and K_G obtained in (a), use the method of Sec. 14-12 to determine **h**. Is the system optimal? (c) Determine $y(t)$ with a step and with a ramp input. (d) For the physical configuration shown, find $\mathbf{k}_p$.

$$U(s) \rightarrow \boxed{\dfrac{A}{s+3}} \xrightarrow{X_3(s)} \boxed{\dfrac{5}{s+1}} \xrightarrow{X_2(s)} \boxed{\dfrac{2}{s+2}} \xrightarrow{X_1(s)\,=\,Y(s)}$$

15-2 Repeat Prob. 15-1 for $G(s) = K_G/s(s + 0.1)(s + 0.5)$.

15-3 Given: $G(s) = K_G(s + 3)(s + 4)/[s(s + 1)(s + 2)]$. (a) Determine the value of **k** under hfg conditions that yields zero steady-state error for the three standard type inputs using $z = 1$. Find $G_{eq}(s)$ and determine the system type. (b) Utilizing the value of **k** and K_G determined in (a), use the design method of Sec. 14-12 to determine the value of **h**. Does this value of **h** yield an optimal performance of the system? (c) Determine $y(t)$ for a unit step input and determine M_p, t_p, and t_s. Also determine the error $e(t)$ for ramp and parabolic inputs. (d) The physical configuration of this system is shown. Determine the value of $\mathbf{k}_p$ required to implement the design values of part (a).

$$U(s) \rightarrow \boxed{\dfrac{K_G}{s+1}} \xrightarrow{X_4(s)} \boxed{\dfrac{s+4}{s+2}} \xrightarrow{X_3(s)} \boxed{\dfrac{s+3}{s+1}} \xrightarrow{X_2(s)} \boxed{\dfrac{1}{s}} \xrightarrow{X_1(s)\,=\,Y(s)}$$

15-4 Repeat parts (a) through (c) of Prob. 15-3 for $z = 0.5$ and $z = 2$.

15-5 The plant of Prob. 15-1 is to be optimized utilizing the PI of Eq. (15-7) with $z = 1$. The desired performance is $M_p = 1.04$, $t_p = 2$ s, and $e(t)_{ss} = 0$ with a unit step input. (a) Determine the minimum value of K_G required to use the nomogram of Fig. 15-4. (b) For this K_{min} determine the matrix **Q** from the nomogram, the **k** that results, and the actual $y(t)$. Compare the actual values of M_p, t_p, and t_s with those of the nomogram. Does the comparison indicate that a higher value of K_G is required? (c) For $z = 0.5$ and $z = 2$ determine the value of q_1 required to achieve $e(t)_{ss} = 0$ with a unit step input and the new value of K_G while maintaining the same value of K_{min}. (d) For each z determine **k** and $y(t)$ with a unit step input. Use the values of q_2 and q_3 obtained in part (b) and the appropriate value of q_1. (e) Compare the values of M_p, t_p, and t_s for all three values of z.

15-6 Utilize the plant of Prob. 15-1, the PI of Eq. (15-7), and the matrix $\mathbf{Q} = \begin{bmatrix} 1 & 0 \\ 0 & 0 \\ 0 & 0 \end{bmatrix}$.

Select $z = 0.5$, 1, or 2 and a set of values of K_G. Determine the corresponding **k** and $y(t)$. Each person in a class can be assigned a different value of z and K_G. For each value of z, plot the values of M_p, t_p, and t_s vs. K_G using the format of Fig. 15-5. If Probs. 15-1 and/or 15-2 have been worked, compare the values of M_p, t_p, and t_s for the different **Q** matrices used.

15-7 A plant is to be optimized by use of the quadratic cost function with $z = 1$. The desired response is $M_p = 1.1$, $t_p = 0.6$ s, and $e(t)_{ss} = 0$ with a unit step input. (a) Using Fig. 15-7 and a computer program, determine the numerator coefficient a that will satisfy these specifications. (b) Determine the feedback coefficients and $y(t)$.

$$G(s) = \frac{K_G(s + a)}{s(s + 1)(s + 5)}$$

15-8 Repeat Prob. 15-7 for $z = 0.5$ and $z = 2$ and maintain the same value of K_{min}. In each case reevaluate K_G and q_1 and use the values of q_2 and q_3 obtained in Prob. 15-7. For each z determine **k** and $y(t)$ with a unit step input. Compare the values of M_p, t_p, and t_s for all three values of z.

15-9 State-variable feedback is to be used to design a system having the plant transfer function

$$G(s) = \frac{K_G(s + 10)}{s(s^2 + 2s + 2)}$$

Assume that $K_x = 250$. For $K_G = 500$ and $r(t) =, u_{-1}(t)$ it is desired that $e(t)_{ss} = 0$, $M_p \approx 1.055$, $t_p \approx 0.67$ s, and $t_s \approx 1$. The system is to be optimal according to Eq. (15-7) with $z = 1$. (a) Determine the matrix **Q** that satisfies the desired performance specification (see Fig. 15-7). (b) Determine the value of the feedback coefficient k_1 and the approximate value of k_3. (c) Determine the approximate value of the nondominant root of the characteristic equation. (d) Determine k_2, using the method of Sec. 15-8, when the dominant roots are $-3.973 \pm j3.974$. (e) Verify the values k_i obtained and the desired time-response characteristics by using a computer solution to Eq. (15-7).

15-10 A controllable system having the plant transfer function shown is to be designed to satisfy Eq. (15-7). Find the values of q_1 and k_1 to satisfy $e(t)_{ss} = 0$ for a step input for $z = 1$.

$$G(s) = \frac{4}{(s + 1)^2(s + 2)}$$

15-11 Use the design procedure of Secs. 15-11 and 14-12. (a) Determine **h** and **k** for $z = 1$.

$$G(s) = \frac{K_G}{s^2(s + 1)^2} \qquad M(s) = \frac{200}{(s + 2)(s^2 + 1.417s + 1)(s + 100)}$$

(b) Use a computer program to determine **k** for the **h** determined in part (a). (c) Determine $y(t)$ with a unit step input for each **k** and compare with that for $M(s)$.

15-12   Repeat Prob. 15-11 with

$$G(s) = \frac{K_G(s + 0.5)}{s^2(s + 1)^2} \qquad M(s) = \frac{200}{(s^2 + 2s + 2)(s + 100)}$$

15-13   Repeat Prob. 15-11 using the method of Sec. 15-12 instead of Sec. 14-12 to evaluate **k**.

15-14   Repeat Prob. 15-12 using the method of Sec. 15-12.

ANSWERS TO SELECTED PROBLEMS

Chapter 2

2-2   (b) $(M_1D^2 + B_1D + K_1)x_1 - (B_1D + K_1)x_2 = 0$
$-(B_1D + K_1)x_1 + [M_2D^2 + (B_1 + B_2)D + (K_1 + K_2)]x_2 = f$
(c) $x_3 = \dot{x}_1, x_4 = \dot{x}_2$

$$\dot{\mathbf{x}} = \begin{bmatrix} 0 & 0 & 1 & 0 \\ 0 & 0 & 0 & 1 \\ -\dfrac{K_1}{M_1} & \dfrac{K_1}{M_1} & -\dfrac{B_1}{M_1} & \dfrac{B_1}{M_1} \\ \dfrac{K_1}{M_2} & -\dfrac{K_1 + K_2}{M_2} & \dfrac{B_1}{M_2} & -\dfrac{B_1 + B_2}{M_2} \end{bmatrix} \mathbf{x} + \begin{bmatrix} 0 \\ 0 \\ 0 \\ \dfrac{1}{M_2} \end{bmatrix} f$$

(d) $G = \dfrac{x_1}{f} = \dfrac{B_1D + K_1}{M_1M_2D^4 + (M_1B_1 + M_1B_2 + M_2B_1)D^3}$
$\qquad\qquad\qquad + (M_1K_1 + M_1K_2 + M_2K_1 + B_1B_2)D^2$
$\qquad\qquad\qquad + (B_1K_2 + B_2K_1)D + K_1K_2$

2-4   (b) (1) $T = (J_1D^2 + B_1D + K_1)\theta_1 - K_1\theta_2$
(2) $0 = -K_1\theta_1 + [J_2D^2 + B_2D + (K_1 + K_2)]\theta_2 - K_2\theta_3$
(3) $0 = -K_2\theta_2 + [J_3D^2 + B_3D + (K_2 + K_3)]\theta_3$
(d) Let $x_1 = \theta_1, x_2 = D\theta_1, x_3 = \theta_2, x_4 = D\theta_2, x_5 = \theta_3, x_6 = D\theta_3$, and $u = T$.

$$\dot{\mathbf{x}} = \begin{bmatrix} 0 & 1 & 0 & 0 & 0 & 0 \\ -\dfrac{K_1}{J_1} & -\dfrac{B_1}{J_1} & \dfrac{K_1}{J_1} & 0 & 0 & 0 \\ 0 & 0 & 0 & 1 & 0 & 0 \\ \dfrac{K_1}{J_2} & 0 & -\dfrac{K_1 + K_2}{J_2} & -\dfrac{B_2}{J_2} & \dfrac{K_2}{J_2} & 0 \\ 0 & 0 & 0 & 0 & 0 & 1 \\ 0 & 0 & \dfrac{K_2}{J_3} & 0 & -\dfrac{K_2 + K_3}{J_3} & -\dfrac{B_3}{J_3} \end{bmatrix} \mathbf{x} + \begin{bmatrix} 0 \\ \dfrac{1}{J_1} \\ 0 \\ 0 \\ 0 \\ 0 \end{bmatrix} u$$

2-5   (a) $(LD + R)i = e, K_t i = f, \dfrac{x_a}{l_1} = \dfrac{x_b}{l_2}$

$(M_2D^2 + B_2D + K_2)x_b = f_2$

$(M_1D^2 + B_1D + K_1)x_a + \dfrac{l_2}{l_1}f_2 = f = k_t i$

Let $\dfrac{l_2}{l_1} = a, \dfrac{M_1}{a} + aM_2 = b, \dfrac{B_1}{a} + aB_2 = c$, and $\dfrac{K_1}{a} + aK_2 = d$

$(bD^2 + cD + d)x_b = K_t i$

(b) There are only three independent states:

$x_1 = x_b, x_2 = \dot{x}_b, x_3 = i$, and $u = e$

$\dot{x}_1 = x_2$

$\dot{x}_2 = -\dfrac{d}{b}x_1 - \dfrac{c}{b}x_2 + \dfrac{K_i}{b}x_3$

$\dot{x}_3 = \dfrac{R}{L}x_3 + \dfrac{1}{L}u$

2-9 (a) $\left[\dfrac{C_1 ad}{(a + b)(c + d)} + D\right]y_1 = \dfrac{C_1 b}{a + b}x_1$

(b) $\dfrac{VM}{K_B C}D^3 y_1 + \left[\dfrac{VB}{K_B C} + \dfrac{M}{C}(L + C_p)\right]D^2 y_1 + \left[C_b + \dfrac{B}{C}(L + C_p)\right]Dy_1$

$$+ \dfrac{C_x ad}{(a + b)(c + d)}y_1 = \dfrac{C_x b}{a + b}x_1$$

2-15 (a) $\left(R_1 + \dfrac{1}{C_1 D}\right)i_i - \dfrac{1}{C_1 D}i_2 = e$

$-\dfrac{1}{C_1 D}i_1 + \left(LD + R_2 + \dfrac{1}{C_1 D} + \dfrac{1}{C_2 D}\right)i_2 = 0$

(b) Node 1: $\left(C_1 D + \dfrac{1}{R_1} + \dfrac{1}{R_2}\right)v_1 - \dfrac{1}{R_2}v_2 = \dfrac{1}{R_1}E$

Node 2: $-\dfrac{1}{R_2}v_1 + \left(\dfrac{1}{R_2} + \dfrac{1}{LD}\right)v_2 - \dfrac{1}{LD}v_3 = 0$

Node 3: $-\dfrac{1}{LD}v_2 + \left(C_2 D + \dfrac{1}{LD}\right)v_3 = 0$

(c) $x_1 = v_{c_1}, x_2 = v_{c_2}, x_3 = i_L$

$$\dot{\mathbf{x}} = \begin{bmatrix} -\dfrac{1}{R_1 C_1} & 0 & -\dfrac{1}{C_1} \\ 0 & 0 & \dfrac{1}{C_2} \\ \dfrac{1}{L} & -\dfrac{1}{L} & -\dfrac{R_2}{L} \end{bmatrix}\mathbf{x} + \begin{bmatrix} \dfrac{1}{R_1 C_1} \\ 0 \\ 0 \end{bmatrix}E$$

Chapter 3

3-2 $\theta_3(t) = -16.4 + 2t + 21.7e^{-0.22t} - 4.81e^{-0.48t}$
$\qquad\qquad + 0.852e^{-0.389t}\sin(0.744t - 59°) + 0.113e^{-0.411t}\sin(1.19t - 227°)$

3-6 $\omega_m(t) \approx 6.35\cos(5t + 11.5°) - 56.52e^{-0.5t}\sin 0.289t$

3-10 (b) $x(t) = 3 + 1.5e^{-3t} - 4.5e^{-t}$

3-15 (a)

$$\Phi(t) = \begin{bmatrix} -e^{-2t} + 2e^{-4t} & 2e^{-2t} - 2e^{-4t} \\ -e^{-2t} + e^{-4t} & 2e^{-2t} - e^{-4t} \end{bmatrix}$$

$$\mathbf{x}(t) = \begin{bmatrix} \frac{1}{2} - 3e^{-2t} + \frac{9}{2}e^{-4t} \\ \frac{3}{4} - 3e^{-2t} + \frac{9}{4}e^{-4t} \end{bmatrix}$$

$$y(t) = x_1 = \tfrac{1}{2} - 3e^{-2t} + \tfrac{9}{2}e^{-4t}$$

3-18 (a) $\mathbf{x}(t)_h = \begin{bmatrix} 2e^{-3t} - e^{-4t} \\ -6e^{-3t} + 4e^{-4t} \end{bmatrix}$

(b) $\mathbf{x}(t) = \begin{bmatrix} 1 - 2e^{-3t} + 2e^{-4t} \\ 6e^{-3t} - 8e^{-4t} \end{bmatrix}$

Chapter 4

4-1 (a) $f(t) = 10e^{-t}(t - \sin t)$

4-2 (b) $x(t) = [77e^{-2t} + 7.42e^{-2.035t} \sin (0.785t - 87.2°)$
$$+ 2.67e^{+1.035t} \sin (3.54t - 136.6°)]10^{-2}$$

4-4 (b) $(4.35 + 0.683e^{-10t} - 5.01e^{-0.965t} - 1.87 \times 10^{-3}e^{-239t})10^{-3}$

4-5 (d) $F(s) = \dfrac{1}{s^2} - \dfrac{3}{4s} + \dfrac{2}{3(s+1)} + \dfrac{1}{12(s+4)}$

4-6 (b) $x(t) = 3 + 3e^{-3t} - 6e^{-t}$

4-7 (b) $(s^2 + 2s + 5)X(s) = \dfrac{2(s^2 + 5)}{s}$

(d) $(s^3 + 4s^2 + 8s + 4)X(s) + 4s^2 + 15s + 28 = \dfrac{5}{s^2 + 25}$

4-8 (a) Prob. 4-1(a): 0, Prob. 4-1(b): 1
(b) Prob. 4-2(a): 0, Prob. 4-2(c): 1

4-11 (b) $x(t) = 2.5 + 0.5e^{-4t} + 3.162e^{-t} \sin (t + 251.6°)$

4-12 (a)
$$\mathbf{X}(s) = \begin{bmatrix} \dfrac{0.2}{s} - \dfrac{0.2}{s+5} \\ \dfrac{0.4}{s} + \dfrac{0.6}{s+5} \end{bmatrix}$$

(b) $G(s) = \dfrac{1}{(s+1)(s+5)}$

Chapter 5

5-5 $G(s) = \dfrac{-60s - 4}{(s+10)(s^2 + 0.13s + 9.004)}$

5-11 $(Js^2 + Bs + K)\theta_0(s) = Hs\theta_i(s)$

5-13 (b) $\lambda_1 = -1, \lambda_2 = -2, \lambda_3 = -3$

(c) $\mathbf{T} = \begin{bmatrix} 1 & 1 & 1 \\ 5 & 4 & 3 \\ 6 & 3 & 2 \end{bmatrix}$

(d) $\dot{\mathbf{z}} = \begin{bmatrix} -1 & 0 & 0 \\ 0 & -2 & 0 \\ 0 & 0 & -3 \end{bmatrix} \mathbf{z} + \begin{bmatrix} \frac{3}{2} \\ 0 \\ -\frac{3}{2} \end{bmatrix} u$

$y = [1 \ \ 1 \ \ 1]\mathbf{z}$

5-17

$$\Lambda_m = \begin{bmatrix} -1.5 & \sqrt{3}/2 & 0 \\ -\sqrt{3}/2 & -1.5 & 0 \\ 0 & 0 & -2 \end{bmatrix}$$

Chapter 6

6-2   (a) $0 < K < 35.5$ (c) $-20 < K < \infty$ (d) $-10 < K < 126$

6-3   (b) $(s + 2)(s^2 + 2s + 2)$
(c) $(s^2 + 3s + 3)(s^2 - s + 2)$
(i) $(s^2 + 6s + 10)(s^2 - 4s + 40)(s + 1)$

6-4   (b) None   (c) 2   (i) 2

6-11   (a) $K_p = \infty$, $K_v = 8$, $K_a = 0$ (b) $e(t)_{ss} = 0.25$

6-14   (g) ∞, -6, 0

6-15   (g) (a) 0, (b) $-\frac{1}{3}$, (c) ∞

6-20   (a) Yes   (b) No   (c) Yes

Chapter 7

7-3   (b) For $K > 0$, real-axis branch: 0 to $-\infty$; three branches; $\gamma = \pm 60°$, $180°$; angles of departure: $\pm 60°$, $180°$; imaginary-axis crossing: $\pm j3.464$ with $K = 73.2$. Branches are straight lines and coincide with the asymptotes.

(d) Real-axis branch: 0 to $-\infty$; five branches; $\gamma = \pm 36°$, $\pm 108°$, $180°$; $\sigma_0 = -\frac{8}{5}$; no breakaway or break-in points; angles of departure: $-90°$, $-26°$; and the imaginary-axis crossing: $\pm j0.9$ with $K = 20.6$. Locus crosses asymptotes at $\pm 36°$.

7-5   (a) For $K > 0$, real-axis branches: 0 to -50 and -100 to $-\infty$; three branches; $\gamma = \pm 60°$, $180°$; $\sigma_0 = -50$; breakaway point: -21.2 ($K_1 = 9.5$); and imaginary-axis crossing is $\pm j70.7$ ($K_1 = 150$).

(b) $K_1 = 150$ (c) $K_1 = 25.92$

(d) $e(t) = -1.29e^{-16.67t} \sin(28.9t + 223.9°) + 0.102e^{-116.7t}$

(e)

ω	M	α, deg	ω	M	α, deg
0	1	0	22.85	1.1323	-63.4
5	1.01	-11.2	25	1.126	-71.9
10	1.039	-23.2	30	1.053	-92.5
15	1.083	-36.8	35	0.908	-112.3
20	1.123	-52.9	41	0.7078	-132.0

7-9   Poles: $s_1 = -1$, $s_2 = -5$, $s_{3,4} = -3 \pm j2$

(a) For $K > 0$, real-axis branch: -1 to -5; four branches; $\gamma = \pm 45°$, $\pm 135°$; angles of departure: $-180°$, $0°$, $-90°$, $90°$; imaginary-axis crossing: $\pm j3$ with $K = 340$; break-in and breakaway points coincide at $s = -3$. For $K < 0$, real-axis branches: -5 to $-\infty$ and -1 to $+\infty$; $\gamma = 0°$, $\pm 90°$, $180°$; angles of departure: $0°$, $180°$, $90°$, $-90°$; imaginary-axis crossing: $s = 0$ with $K = -320$.

(b) $-320 > K > 340$ (c) $K = 16$

Chapter 8

8-4 (b) $G(j0+) = \infty \underline{/-90°}$, $G(j\infty) = 0\underline{/0°}$

$\omega_x = \pm 30.4$, $\omega = \pm 12.5$

8-6 (a) It must have an initial slope of -40 dB/decade.

(b) The phase-angle curve must approach an angle of 180° as $\omega \to 0$.

8-9 Case 3: $G(s) = \dfrac{2s}{(1 + 0.5s)(1 + 0.125s)^2}$

Assuming $\zeta = 1$, correction at $\omega = 8$ is -6.26 dB.

8-10 (a) $K_0 = 10$

8-11 (a) See part (3) of Prob. 8-2 for transfer function

8-14 (a) $N = 0$, $Z_R = 0$, stable; $N = -2$, $Z_R = 2$, unstable

8-17 (a) Stable for $K > -1$ (e) Stable for $0 < K < 1.5$

8-18 (c) Stable for all positive values of K

8-24 (c) (2) $K_0 \approx 7.15$ $(\omega_x \approx 1.2)$

Chapter 9

9-1 (1) $K_1 = 1.35$, $M_m = 1.247$, $\omega_m = 1.20$

(2) $K_1 \approx 13.74$, $M_m \approx 1.25$, $\omega_m \approx 14.62$

9-3 $M_m = 1.29$, $K_1 = 20$, $\omega_m = 11.50$

9-9 (1) Original system is unstable; for $M_m = 1.12$, $K_1 = 3.06$, $\omega_m = 2.38$, $\gamma = 56.8°$

9-13 (1) $M_p = 1.043$, $T_p = 4.44$ s, $T_s = 5.66$ s

(2) $M_p = 1.527$, $T_p = 0.32$ s, $T_s = 2.00$ s

(3) $M_p = 1.043$, $T_p = 44.44$ s, $T_s = 56.56$ s

Chapter 10

10-3,
10-20

System	Dominant roots	Other roots	K_1, s^{-1}	T_p, s	T_s, s	M_o
Basic	$3.57 \pm j6.19$	$-16.4 \pm j19.2$	5.23	0.558	1.19	0.167
Lag-compensated $\alpha = 10$, $T = 10$	$-3.6 \pm j6.4$	$-16.4 \pm j19.2$ -0.10175	53.2	0.54	1.55	0.195
Lead-compensated $\alpha = 0.1$, $T = 0.1$	$-9.6 \pm j16.6$	-11.5 -99.4	6.7	0.324	0.273	0.0025
Lag-lead-compensated $\alpha = 10$, $T_1 = 10$, $T_2 = 0.1$	$-9.6 \pm j16.8$	-0.101 -11.3 -99.4	67.8	0.324	0.268	0.0071
Tachometer-compensated $A = 15$, $K_t = 1$	$-8.8 \pm j15.17$	$-11.2 \pm j9.85$	6.34	0.36	0.26	0.1105

10-14 $G(s) = \dfrac{30}{s(s^2 + 6s + 45)}$

$\dfrac{C(s)}{R(s)} = \dfrac{30}{(s + 1.695)(s^2 + 4.305 + 17.703)}$

10-20 One possible solution is $A/K_t = 15$, $KK_t = 4{,}578$, dominant roots $-8.8 \pm j15.22$, other roots $-11.22 \pm j9.85$, $K_1 = 6.34$. See the solution to Prob. 10-3 for a comparison with cascade compensation. *Note:* $K_t = 1$.

Chapter 11

11-1 Basic system: $K_1 = 5.23$, $\omega_m = 5.10$
Lag-compensated: $K_1 = 47.5$, $\omega_m = 4.35$, $T_1 = 4$
Lead-compensated: $K_1 = 9.69$, $\omega_m = 14.4$, $T_2 = 0.1$
Lag-lead-compensated: $K_1 = 95.7$, $\omega_m = 14.0$, $T_1 = 4$, $T_2 = 0.1$
11-17 Tachometer-compensated with $A = 1.66$, $K_t = \frac{1}{15}$, $K_x = 5.22$: $K_1 = 6.42$, $\omega_m = 7.26$

Chapter 12

12-1 $G_c(s) = \dfrac{2(s + 3)}{s + 6}$

With this cascade compensator $M_p = 1.081$, $t_p = 1.9$ s, $t_s = 3.65$ s, and $K_1 = 1.25$. The specified $M(s)$ yields $M_p = 1.127$, $t_p = 1.95$ s, $t_s = 3.9$ s, and $K_1 = 1.36$.

12-6 (a) $M(s) = \dfrac{702{,}000}{(s + 12 \pm j21)(s + 16)(s + 75)}$

$k_1 = 1$, $k_2 = 0.108$, $k_3 = 4.34 \times 10^{-3}$, $k_4 = 1.07 \times 10^{-4}$
(c) $y(t) = 1 + 0.974e^{-12t} \sin(21t + 142.6°) - 1.627e^{-16t} + 0.036e^{-75t}$
$t_s = 0.205$ s, $M_p = 1.1018$, $t_p = 0.250$ s, $K_1 = 8.56$
12-9 The zeros of $G(s)H_{eq}(s)$ are selected as $z_{1,2} = -5 \pm j10$ and $z_3 = -30$. Then

$M(s) = \dfrac{163{,}000}{(s + 4.93 \pm j7.99)(s + 36.82 \pm j22.25)}$

$c(t) = 1 + 1.14e^{-4.93t} \sin(7.99t + 219.7°) + 0.111e^{-36.82t} \sin(22.25t - 81.3°)$

$k_1 = 1$, $k_2 = \frac{1}{30}$, $k_3 = \frac{3}{750}$, $k_4 = \frac{2}{375}$

$t_p = 0.44$ s, $M_p = 1.14$, $t_s = 0.866$ s

Chapter 13

13-1 Let $x_1 = x$ and $x_2 = \dot{x}_1$:

(a) $x_2 = \dfrac{-2x_1}{3 + N}$

(c) $x_2 = \dfrac{-6x_1 + 6}{5 + N}$

13-2 Let $x_1 = x$ and $x_2 = \dot{x}_1$:

(c) $N = \dfrac{-2x_2{}^2 - x_1}{x_2}$ vortex

(d) $N = \dfrac{2(-x_2|x_2| - x_1)}{x_2}$ stable node

13-4 Let $x_1 = e$, $x_2 = \dot{e}$:

(1) For $e < -0.1$: $N = \dfrac{-x_2 - (x_1 + 0.1)}{x_2}$

Equilibrium point is $x_1 = -0.1$, $x_2 = 0$
(2) For $-0.1 < e < 0.1$: $N = -1$

(3) For $e > 0$: $N = \dfrac{-x_2 - (x_1 - 0.1)}{x_2}$

Equilibrium point is $x_1 = 0.1$, $x_2 = 0$
13-7 (a) Three equilibrium points at

$$\mathbf{x}_a = \begin{bmatrix} 0 \\ 0 \\ 0 \end{bmatrix} \quad \mathbf{x}_b = \begin{bmatrix} 1 \\ 0 \\ 2 \end{bmatrix} \quad \mathbf{x}_c = \begin{bmatrix} 1 \\ 0 \\ -2 \end{bmatrix}$$

(b) For $\mathbf{x}_b$, $\mathbf{J}_{\mathbf{x}_b} = \begin{bmatrix} 0 & 1 & 0 \\ -1 & -3 & 1 \\ 2 & 0 & 0 \end{bmatrix}$

(c) $\mathbf{x}_b$ is unstable
13-8 (c) Negative definite
13-11 (a) $V(\mathbf{x})$ is positive definite for $0 < K < 2$.
$\dot{V}(\mathbf{x}) = (2K - 4)x_2{}^2$ is negative semidefinite for $K \leq 2$.
(b) $\dot{V}(\mathbf{x}) = 0$ only at the equilibrium point $x = 0$; therefore globally asymptotically stable.

Chapter 14

14-2 (a) Using $\mathbf{B}_1$, $\mathbf{Q}_a$, and $z = 1$ gives

$$\mathbf{P} = \begin{bmatrix} 0.098504 & 0.007423 \\ 0.007423 & 0.001556 \end{bmatrix} \quad \mathbf{k} = \begin{bmatrix} 0.98504 \\ 0.07423 \end{bmatrix}$$

(b) $\dfrac{Y(s)}{R(s)} = \dfrac{10(s + 3)}{(s + 3.076)(s + 9.774)}$

$y(t) = 0.998 + 0.0369e^{-3.076t} - 1.0347e^{-9.774t}$
$M_p \approx 1.001$, $t_p \approx 0.65$ s, $t_s = 0.354$ s
14-7 $G_K(s) = \mathbf{h}^T \mathbf{\Phi}(s)\mathbf{b} = G(s)$
Using $\omega_\phi = 2\omega_{cf} = 4$ yields $K_G = 64$. Matching coefficients of $1 + G(s)H_{eq}(s) = T_s(s)$ yields
$\mathbf{k}^T = [1 \quad 0.4688 \quad 0.07813]$

14-12   (For Prob. 14-7) the RSL with $K_G = 64$ yields roots
$s_1 = -4.2143$, $s_{2,3} = -2.0984 \pm j3.2838$.
Characteristic equation: $s^3 + 8.4111s^2 + 32.8733s + 64 = 0$
$\mathbf{k}^T = [1 \quad 0.4824 \quad 0.08455]$

Chapter 15

15-1 $G(s) = \dfrac{K_G}{(s+1)(s+2)(s+3)} = \dfrac{K_G}{s^3 + 6s^2 + 11s + 6}$

(a) From the RSL, $K_{min} = 314$. Use $K_G = 400$
From Eq. (15-28), $q_{11} = 1 - (\frac{6}{400})^2 \approx 1 \approx h_{11}$
From Eq. (15-27), $k_1 = 1 - \frac{6}{400} = 0.985$

$$G_{eq}(s) = \dfrac{400}{s[s^2 + (6 + K_G k_3)s + (11 + K_G k_2)]}$$

$\mathbf{k}^T = [0.985 \quad 0.9775 \quad 0.49]$, $\alpha = 200$
(b) $\mathbf{h}^T = [1 \quad 1.01 \quad 0.4951]$
Hint: Use Eq. (15-73) for a first estimate of h_3:
$k_3 \approx \sqrt{q_3} = h_3$
When Eq. (14-33) is used, the system is optimal.
(c) $y(t) = 1 + 1.42e^{-t} \sin(t + 224.7°) - 0.5 \times 10^{-4}e^{-200t}$
$M_p = 1.0432$, $t_p = 3.15$ s, $t_s = 4.22$ s

15-5   (a) $K_G = 400$ (see Prob. 15-1).
(b) Use $q_1 = 1$ and $q_2 = 0.008$. Then from Fig. 15-4, for $M_p = 1.04$ and $t_p \approx 1.93$ s, $t_s = 2.4^+$ s, $q_3 = 0.03$. From a computer solution of the algebraic Riccati equation,
$\mathbf{k}^T = [0.98511 \quad 0.58277 \quad 0.16704]$
$y(t) = 1 + 1.4667e^{-1.7176t} \sin(1.678t + 222.9°) - 0.001259e^{-69.381t}$
Actual values: $M_p = 1.04$, $t_p = 1.90^-$ s, $t_s \approx 2.47$ s

15-7   (a) First trial values obtained from Fig. 15-7 are $a = 9$, $q_3 = 0.06$. From the RSL, use $K_G = 300$. Then, $q_1 = c_o^2 = 81$. Also, $q_2 = 0.008$ for Fig. 15-7. Use a computer solution of the algebraic Riccati equation to evaluate $\mathbf{k}$, the roots of the characteristic equation, and the time response.
(b) $\mathbf{k}^T = [9 \quad 2.2062 \quad 0.254067]$
$s_{1,2} = 4.2778 \pm j4.2842$
$s_3 = 73.662$
$y(t) = 1 - 1.0608e^{-4.2778t} \sin(4.2842t + 263.73°) + 0.0545e^{-73.662}$
$M_p = 1.064$, $t_p = 0.58$ s, $t_s = 0.89$ s
Note: For $M_p = 1.06$ a solution using $a = 10$ is given in Table 15-6.

15-11   (a) DDE: $(s^2 + 1.417s + 1)(s + 2) = 0$
$\mathbf{h}^T = [1 \quad 1.917 \quad 1.7085 \quad 0.5]$
From Eq. (14-79), $\mathbf{k}^T = [1 \quad 1.927 \quad 1.72267 \quad 0.507085]$
(b) $\mathbf{k} = [1 \quad 1.9269 \quad 1.7226 \quad 0.50704]$
(c) $\lambda_{1,2} = 0.70489 \pm j0.70572$, $\lambda_3 = -2$, $\lambda_4 = -99.99$
$y(t) = 1 - 1.9395e^{-0.70849t} \sin(0.70572t + 195.83°)$
$$- 0.47114e^{-2t} + 0.0^+e^{-99.99t}$$
$M_p = 1.0352$, $t_p = 5.17$ s
$t_s = 6.46$ s